MILLER & FREUND'S

PROBABILITY AND STATISTICS FOR ENGINEERS

NINTH EDITION

Richard A. Johnson

University of Wisconsin–Madison

PEARSON

Boston   Columbus   Indianapolis   New York   San Francisco   Amsterdam
Cape Town   Dubai   London   Madrid   Milan   Munich   Paris   Montréal   Toronto
Delhi   Mexico City   São Paulo   Sydney   Hong Kong   Seoul   Singapore   Taipei   Tokyo

Editorial Director, Mathematics: *Christine Hoag*
Editor-in-Chief: *Deirdre Lynch*
Acquisitions Editor: *Patrick Barbera*
Project Team Lead: *Christina Lepre*
Project Manager: *Lauren Morse*
Editorial Assistant: *Justin Billing*
Program Team Lead: *Karen Wernholm*
Program Manager: *Tatiana Anacki*
Cover and Illustration Design: *Studio Montage*
Program Design Lead: *Beth Paquin*
Marketing Manager: *Tiffany Bitzel*
Marketing Coordinator: *Brooke Smith*
Field Marketing Manager: *Evan St. Cyr*
Senior Author Support/Technology Specialist: *Joe Vetere*
Senior Procurement Specialist: *Carol Melville*
Interior Design, Production Management, Answer Art, and Composition:
    iEnergizer Aptara Limited/Falls Church
Cover Image: Robots welding in a car factory: Praphan Jampala/Shutterstock

Library of Congress Cataloging-in-Publication Data
Johnson, Richard A. (Richard Arnold), 1937–
    Miller & Freund's probability and statistics for engineers / Richard A. Johnson, University of Wisconsin–Madison. — Ninth edition.
        pages cm
    Includes bibliographical references and index.
    ISBN 0-321-98624-5
    1. Engineering–Statistical methods.   2. Probabilities.   I. Miller, Irwin, 1928- II. Freund, John E.   III. Title.   IV. Title: Miller and Freund's probability and statistics for engineers.   V. Title: Probability and statistics for engineers.
    TA340.J662017
    519.202'462—dc23   2015016017

1 2 3 4 5 6 7 8 9 10—V031—19 18 17 16 15

ISBN-10: 0-321-98624-5
ISBN-13: 978-0-321-98624-5

CONTENTS

PREFACE

This book introduces probability and statistics to students of engineering and the physical sciences. It is primarily applications focused but it contains optional enrichment material. Each chapter begins with an introductory statement and concludes with a set of statistical guidelines for correctly applying statistical procedures and avoiding common pitfalls. These *Do's and Don'ts* are then followed by a checklist of key terms. Important formulas, theorems, and rules are set out from the text in boxes.

The exposition of the concepts and statistical methods is especially clear. It includes a careful introduction to probability and some basic distributions. It continues by placing emphasis on understanding the meaning of confidence intervals and the logic of testing statistical hypotheses. Confidence intervals are stressed as the major procedure for making inferences. Their properties are carefully described and their interpretation is reviewed in the examples. The steps for hypothesis testing are clearly and consistently delineated in each application. The interpretation and calculation of the *P*-value is reinforced with many examples.

In this ninth edition, we have continued to build on the strengths of the previous editions by adding several more data sets and examples showing application of statistics in scientific investigations. The new data sets, like many of those already in the text, arose in the author's consulting activities or in discussions with scientists and engineers about their statistical problems. Data from some companies have been disguised, but they still retain all of the features necessary to illustrate the statistical methods and the reasoning required to make generalizations from data collected in an experiment.

The time has arrived when software computations have replaced table lookups for percentiles and probabilities as well as performing the calculations for a statistical analysis. Today's widespread availability of statistical software packages makes it imperative that students now become acquainted with at least one of them. We suggest using software for performing some analysis with larger samples and for performing regression analysis. Besides having several existing exercises describing the use of MINITAB, we now give the R commands within many of the examples. This new material augments the basics of the freeware R that are already in Appendix C.

NEW FEATURES OF THE NINTH EDITION INCLUDE:

Large number of new examples. Many new examples are included. Most are based on important current engineering or scientific data. The many contexts further strengthen the orientation towards an applications-based introduction to statistics.

More emphasis on *P*-values. New graphs illustrating *P*-values appear in several examples along with an interpretation.

More details about using R. Throughout the book, R commands are included in a number of examples. This makes it easy for students to check the calculations, on their own laptop or tablet, while reading an example.

Stress on key formulas and downplay of calculation formulas. Generally, computation formulas now appear only at the end of sections where they can easily be skipped. This is accomplished by setting key formulas in the context of an application which only requires all, or mostly all, integer arithmetic. The student can then check their results with their choice of software.

Visual presentation of 2^2 and 2^3 designs. Two-level factorial designs have a 50-year tradition in the teaching of engineering statistics at the University of Wisconsin. It is critical that engineering students become acquainted with the key ideas of (i) systematically varying several input variables at a time and (ii) how to interpret interactions. Major revisions have produced Section 13.3 that is now self-contained. Instructors can cover this material in two or three lectures at the end of course.

New data based exercises. A large number of exercises have been changed to feature real applications. These contexts help both stimulate interest and strengthen a student's appreciation of the role of statistics in engineering applications.

Examples and now numbered. All examples are now numbered within each chapter.

This text has been tested extensively in courses for university students as well as by in-plant training of engineers. The whole book can be covered in a two-semester or three-quarter course consisting of three lectures a week. The book also makes an excellent basis for a one-semester course where the lecturer can choose topics to emphasize theory or application. The author covers most of the first seven chapters, straight-line regression, and the graphic presentation of factorial designs in one semester (see the basic applications syllabus below for the details).

To give students an early preview of statistics, descriptive statistics are covered in Chapter 2. Chapters 3 through 6 provide a brief, though rigorous, introduction to the basics of probability, popular distributions for modeling population variation, and sampling distributions. Chapters 7, 8, and 9 form the core material on the key concepts and elementary methods of statistical inference. Chapters 11, 12, and 13 comprise an introduction to some of the standard, though more advanced, topics of experimental design and regression. Chapter 14 concerns nonparametric tests and goodness-of-fit test. Chapter 15 stresses the key underlying statistical ideas for quality improvement, and Chapter 16 treats the associated ideas of reliability and the fitting of life length models.

The mathematical background expected of the reader is a year course in calculus. Calculus is required mainly for Chapter 5 dealing with basic distribution theory in the continuous case and some sections of Chapter 6.

It is important, in a one-semester course, to make sure engineers and scientists become acquainted with the least squares method, at least in fitting a straight line. A short presentation of two predictor variables is desirable, if there is time. Also, not to be missed, is the exposure to 2-level factorial designs. Section 13.3 now stands alone and can be covered in two or three lectures.

For an audience requiring more exposure to mathematical statistics, or if this is the first of a two-semester course, we suggest a careful development of the properties of expectation (5.10), representations of normal theory distributions (6.5), and then moment generating functions (5.11) and their role in distribution theory (6.6).

For each of the two cases, we suggest a syllabus that the instructor can easily modify according to their own preferences.

One-semester introduction to probability and statistics emphasizing the understanding of basic applications of statistics.		A first semester introduction that develops the tools of probability and some statistical inferences.	
Chapter 1	especially 1.6	Chapter 1	especially 1.6
Chapter 2		Chapter 2	
Chapter 3		Chapter 3	
Chapter 4	4.4–4.7	Chapter 4	4.4–4.7
			4.8 (geometric, negative binomial)
Chapter 5	5.1–5.4, 5.6, 5.12	Chapter 5	5.1–5.4, 5.6, 5.12
	5.10 Select examples of joint distribution, independence, mean and variance of linear combinations.		5.5, 5.7, 5.8 (gamma, beta) 5.10 Develop joint distributions, independence expectation and moments of linear combinations.
Chapter 6	6.1–6.4	Chapter 6	6.1–6.4
			6.5–6.7 (Representations, mgf's, transformation)
Chapter 7	7.1–7.7	Chapter 7	7.1–7.7
Chapter 8		Chapter 8	
Chapter 9	(could skip)	Chapter 9	(could skip)
Chapter 10	10.1–10.4	Chapter 10	10.1–10.4
Chapter 11	11.1–11.2		
	11.3 and 11.4 Examples		
Chapter 13	13.3 2^2 and 2^3 designs also 13.1 if possible		

Any table whose number ends in W can be downloaded from the book's section of the website

http://www.perasonhighered.com/mathstatresources/

We wish to thank MINITAB (State College, Pennsylvania) for permission to include commands and output from their *MINITAB* software package, the SAS institute (Gary, North Carolina) for permission to include output from their SAS package and the software package R (R project http://CRAN R-project.org), which we connect to many examples and discuss in Appendix C.

We wish to heartily thank all of those who contributed the data sets that appear in this edition. They have greatly enriched the presentation of statistical methods by setting each of them in the context of an important engineering problem.

The current edition benefited from the input of the reviewers.

Kamran Iqbal, University of Arakansas at Little Rock
Young Bal Moon, Syracuse University
Nabin Sapkota, University of Central Florida
Kiran Bhutani, Catholic University of America
Xianggui Qu, Oakland University
Christopher Chung, University of Houston.

All revisions in this edition were the responsibility of Richard. A. Johnson.

Richard A. Johnson

INTRODUCTION

E verything dealing with the collection, processing, analysis, and interpretation of numerical data belongs to the domain of statistics. In engineering, this includes such diversified tasks as calculating the average length of computer downtimes, collecting and presenting data on the numbers of persons attending seminars on solar energy, evaluating the effectiveness of commercial products, predicting the reliability of a launch vehicle, and studying the vibrations of airplane wings.

In Sections 1.2, 1.3, 1.4, and 1.5 we discuss the recent growth of statistics and its applications to problems of engineering. Statistics plays a major role in the improvement of quality of any product or service. An engineer using the techniques described in this book can become much more effective in all phases of work relating to research, development, or production. In Section 1.6 we begin our introduction to statistical concepts by emphasizing the distinction between a population and a sample.

1.1 Why Study Statistics?

Answers provided by statistical analysis can provide the basis for making better decisions and choices of actions. For example, city officials might want to know whether the level of lead in the water supply is within safety standards. Because not all of the water can be checked, answers must be based on the partial information from samples of water that are collected for this purpose. As another example, an engineer must determine the strength of supports for generators at a power plant. First, loading a few supports to failure, she obtains their strengths. These values provide a basis for assessing the strength of all the other supports that were not tested.

When information is sought, statistical ideas suggest a typical collection process with four crucial steps.

1. **Set clearly defined goals for the investigation.**
2. **Make a plan of what data to collect and how to collect it.**
3. **Apply appropriate statistical methods to efficiently extract information from the data.**
4. **Interpret the information and draw conclusions.**

These indispensable steps will provide a frame of reference throughout as we develop the key ideas of statistics. Statistical reasoning and methods can help you become efficient at obtaining information and making useful conclusions.

1.2   Modern Statistics

The origin of statistics can be traced to two areas of interest that, on the surface, have little in common: games of chance and what is now called political science. Mid-eighteenth-century studies in probability, motivated largely by interest in games of chance, led to the mathematical treatment of errors of measurement and the theory that now forms the foundation of statistics. In the same century, interest in the numerical description of political units (cities, provinces, countries, etc.) led to what is now called **descriptive statistics**. At first, descriptive statistics consisted merely of the presentation of data in tables and charts; nowadays, it includes the summarization of data by means of numerical descriptions and graphs.

In recent decades, the growth of statistics has made itself felt in almost every major phase of activity. The most important feature of its growth has been the shift in emphasis from descriptive statistics to **statistical inference**. Statistical inference concerns generalizations based on sample data. It applies to such problems as estimating an engine's average emission of pollutants from trial runs, testing a manufacturer's claim on the basis of measurements performed on samples of his product, and predicting the success of a launch vehicle in putting a communications satellite in orbit on the basis of sample data pertaining to the performance of the launch vehicle's components.

When making a statistical inference, namely, an inference that goes beyond the information contained in a set of data, always proceed with caution. One must decide carefully how far to go in generalizing from a given set of data. Careful consideration must be given to determining whether such generalizations are reasonable or justifiable and whether it might be wise to collect more data. Indeed, some of the most important problems of statistical inference concern the appraisal of the risks and the consequences that arise by making generalizations from sample data. This includes an appraisal of the probabilities of making wrong decisions, the chances of making incorrect predictions, and the possibility of obtaining estimates that do not adequately reflect the true situation.

We approach the subject of statistics as a science whenever possible, we develop each statistical idea from its probabilistic foundation, and immediately apply each idea to problems of physical or engineering science as soon as it has been developed. The great majority of the methods we shall use in stating and solving these problems belong to the **frequency** or **classical approach**, where statistical inferences concern fixed but unknown quantities. This approach does not formally take into account the various subjective factors mentioned above. When appropriate, we remind the reader that subjective factors do exist and also indicate what role they might play in making a final decision. This "bread-and-butter" approach to statistics presents the subject in the form in which it has successfully contributed to engineering science, as well as to the natural and social sciences, in the last half of the twentieth century, into the first part of the twenty-first century, and beyond.

1.3   Statistics and Engineering

The impact of the recent growth of statistics has been felt strongly in engineering and industrial management. Indeed, it would be difficult to overestimate the contributions statistics has made to solving production problems, to the effective use of materials and labor, to basic research, and to the development of new products. As in other sciences, statistics has become a vital tool to engineers. It enables them to understand phenomena subject to variation and to effectively predict or control them.

In this text, our attention will be directed largely toward engineering applications, but we shall not hesitate to refer also to other areas to impress upon the reader the great generality of most statistical techniques. The statistical method used to estimate the average coefficient of thermal expansion of a metal serves also to estimate the average time it takes a health care worker to perform a given task, the average thickness of a pelican eggshell, or the average IQ of first-year college students. Similarly, the statistical method used to compare the strength of two alloys serves also to compare the effectiveness of two teaching methods, or the merits of two insect sprays.

1.4   The Role of the Scientist and Engineer in Quality Improvement

During the last 3 decades, the United States has found itself in an increasingly competitive world market. This competition has fostered an international revolution in quality improvement. The teaching and ideas of W. Edwards Deming (1900–1993) were instrumental in the rejuvenation of Japanese industry. He stressed that American industry, in order to survive, must mobilize with a continuing commitment to quality improvement. From design to production, processes need to be continually improved. The engineer and scientist, with their technical knowledge and armed with basic statistical skills in data collection and graphical display, can be main participants in attaining this goal.

Quality improvement is based on the philosophy of "make it right the first time." Furthermore, one should not be content with any process or product but should continue to look for ways of improving it. We will emphasize the key statistical components of any modern quality-improvement program. In Chapter 15, we outline the basic issues of quality improvement and present some of the specialized statistical techniques for studying production processes. The experimental designs discussed in Chapter 13 are also basic to the process of quality improvement.

Closely related to quality-improvement techniques are the statistical techniques that have been developed to meet the **reliability** needs of the highly complex products of space-age technology. Chapter 16 provides an introduction to this area.

1.5   A Case Study: Visually Inspecting Data to Improve Product Quality

This study[1] dramatically illustrates the important advantages gained by appropriately plotting and then monitoring manufacturing data. It concerns a ceramic part used in popular coffee makers. This ceramic part is made by filling the cavity between two dies of a pressing machine with a mixture of clay, water, and oil. After pressing, but before the part is dried to a hardened state, critical dimensions are measured. The depth of the slot is of interest here.

Because of natural uncontrolled variation in the clay-water-oil mixture, the condition of the press, differences in operators, and so on, we cannot expect all of the slot measurements to be exactly the same. Some variation in the depth of slots is inevitable, but the depth needs to be controlled within certain limits for the part to fit when assembled.

[1]Courtesy of Don Ermer

Table 1.1 Slot depth (thousandths of an inch)							
Time	**6:30**	**7:00**	**7:30**	**8:00**	**8:30**	**9:00**	**9:30**	**10:00**
1	214	218	218	216	217	218	218	219
2	211	217	218	218	220	219	217	219
3	218	219	217	219	221	216	217	218
Sum	643	654	653	653	658	653	652	656
$\bar{x}$	214.3	218.0	217.7	217.7	219.3	217.7	217.3	218.7
Time	**10:30**	**11:00**	**11:30**	**12:30**	**1:00**	**1:30**	**2:00**	**2:30**
1	216	216	218	219	217	219	217	215
2	219	218	219	220	220	219	220	215
3	218	217	220	221	216	220	218	214
Sum	653	651	657	660	653	658	655	644
$\bar{x}$	217.7	217.0	219.0	220.0	217.7	219.3	218.3	214.7

Slot depth was measured on three ceramic parts selected from production every half hour during the first shift from 6 A.M. to 3 P.M. The data in Table 1.1 were obtained on a Friday. The sample mean, or average, for the first sample of 214, 211, and 218 (thousandths of an inch) is

$$\frac{214 + 211 + 218}{3} = \frac{643}{3} = 214.3$$

This value is the first entry in row marked $\bar{x}$.

The graphical procedure, called an **X-bar** chart, consists of plotting the sample averages versus time order. This plot will indicate when changes have occurred and actions need to be taken to correct the process.

From a prior statistical study, it was known that the process was stable and that it varied about a value of 217.5 thousandths of an inch. This value will be taken as the central line of the X-bar chart in Figure 1.1.

$$\text{central line: } \bar{\bar{x}} = 217.5$$

It was further established that the process was capable of making mostly good ceramic parts if the average slot dimension for a sample remained between certain control limits.

$$\text{Lower control limit: LCL} = 215.0$$
$$\text{Upper control limit: UCL} = 220.0$$

What does the chart tell us? The mean of 214.3 for the first sample, taken at approximately 6:30 A.M., is outside the lower control limit. Further, a measure of the variation in this sample

$$\text{range} = \text{largest} - \text{smallest} = 218 - 211 = 7$$

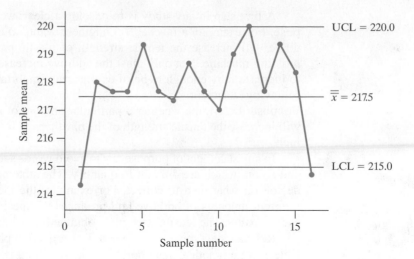

Figure 1.1
X-bar chart for depth

is large compared to the others. This evidence suggests that the pressing machine had not yet reached a steady state. The control chart suggests that it is necessary to warm up the pressing machine before the first shift begins at 6 A.M. Management and engineering implemented an early start-up and thereby improved the process. The operator and foreman did not have the authority to make this change. Deming claims that 85% or more of our quality problems are in the system and that the operator and others responsible for the day-to-day operation are responsible for 15% or less of our quality problems.

The *X*-bar chart further shows that, throughout the day, the process was stable but a little on the high side, although no points were out of control until the last sample of the day. Here an unfortunate oversight occurred. The operator did not report the out-of-control value to either the set-up person or the foreman because it was near the end of her shift and the start of her weekend. She also knew the set-up person was already cleaning up for the end of the shift and that the foreman was likely thinking about going across the street to the Legion Bar for some refreshments as soon as the shift ended. She did not want to ruin anyone's plans, so she kept quiet.

On Monday morning when the operator started up the pressing machine, one of the dies broke. The cost of the die was over a thousand dollars. But this was not the biggest cost. When a customer was called and told there would be a delay in delivering the ceramic parts, he canceled the order. Certainly the loss of a customer is an expensive item. Deming refers to this type of cost as the unknown and unknowable, but at the same time it is probably the most important cost of poor quality.

On Friday the chart had predicted a problem. Afterward it was determined that the most likely difficulty was that the clay had dried and stuck to the die, leading to the break. The chart indicated the problem, but someone had to act. For a statistical charting procedure to be truly effective, action must be taken.

1.6   Two Basic Concepts—Population and Sample

The preceding senarios which illustrate how the evaluation of actual information is essential for acquiring new knowledge, motivate the development of statistical reasoning and tools taught in this text. Most experiments and investigations conducted by engineers in the course of investigating, be it a physical phenomenon, production process, or manufactured unit, share some common characteristics.

A first step in any study is to develop a clear, well-defined **statement of purpose**. For example, a mechanical engineer wants to determine whether a new additive will increase the tensile strength of plastic parts produced on an injection molding machine. Not only must the additive increase the tensile strength, it needs to increase it by enough to be of engineering importance. He therefore created the following statement.

Purpose: Determine whether a particular amount of an additive can be found that will increase the tensile strength of the plastic parts by at least 10 pounds per square inch.

In any statement of purpose, try to avoid words such as *soft*, *hard*, *large enough*, and so on, which are difficult to quantify. The statement of purpose can help us to decide on what data to collect. For example, the mechanical engineer takes two different amounts of additive and produces 25 specimens of the plastic part with each mixture. The tensile strength is obtained for each of 50 specimens.

Relevant data must be collected. But it is often physically impossible or infeasible from a practical standpoint to obtain a complete set of data. When data are obtained from laboratory experiments, no matter how much experimentation is performed, more could always be done. To collect an exhaustive set of data related to the damage sustained by all cars of a particular model under collision at a specified speed, every car of that model coming off the production lines would have to be subjected to a collision!

In most situations, we must work with only partial information. The distinction between the data actually acquired and the vast collection of all potential observations is a key to understanding statistics.

The source of each measurement is called a **unit**. It is usually an object or a person. To emphasize the term *population* for the entire collection of units, we call the entire collection the **population of units**.

Units and population of units

> **unit:** A single entity, usually an object or person, whose characteristics are of interest.
>
> **population of units:** The complete collection of units about which information is sought.

Guided by the statement of purpose, we have a **characteristic of interest** for each unit in the population. The characteristic, which could be a qualitative trait, is called a **variable** if it can be expressed as a number.

There can be several characteristics of interest for a given population of units. Some examples are given in Table 1.2.

For any population there is the value, for each unit, of a characteristic or variable of interest. For a given variable or characteristic of interest, we call the collection of values, evaluated for every unit in the population, the **statistical population** or just the **population**. This collection of values is the population we will address in all later chapters. Here we refer to the collection of units as the **population of units** when there is a need to differentiate it from the collection of values.

Statistical population

> A **statistical population** is the set of all measurements (or record of some quality trait) corresponding to each unit in the entire population of units about which information is sought.

Generally, any statistical approach to learning about the population begins by taking a sample.

Table 1.2 Examples of populations, units, and variables		
Population	**Unit**	**Variables/Characteristics**
All students currently enrolled in school	student	GPA number of credits hours of work per week major right/left-handed
All printed circuit boards manufactured during a month	board	type of defects number of defects location of defects
All campus fast food restaurants	restaurant	number of employees seating capacity hiring/not hiring
All books in library	book	replacement cost frequency of checkout repairs needed

Samples from a population

> A **sample** from a statistical population is the subset of measurements that are actually collected in the course of an investigation.

EXAMPLE 1 **Variable of interest, statistical population, and sample**

Transceivers provide wireless communication between electronic components of consumer products, especially transceivers of Bluetooth standards. Addressing a need for a fast, low-cost test of transceivers, engineers[2] developed a test at the wafer level. In one set of trials with 60 devices selected from different wafer lots, 49 devices passed.

Identify the population unit, variable of interest, statistical population, and sample.

Solution The population unit is an individual wafer, and the population is all the wafers in lots currently on hand. There is some arbitrariness because we could use a larger population of all wafers that would arrive within some fixed period of time.

The variable of interest is pass or fail for each wafer.

The statistical population is the collection of pass/fail conditions, one for each population unit.

The sample is the collection of 60 pass/fail records, one for each unit in the sample. These can be summarized by their totals, 49 pass and 11 fail.   ∎

The sample needs both to be representative of the population and to be large enough to contain sufficient information to answer the questions about the population that are crucial to the investigation.

---

[2]G. Srinivasan, F. Taenzler, and A. Chatterjee, Loopback DFT for low-cost test of single-VCO-based wireless transceivers, *IEEE Design & Test of Computers* 25 (2008), 150–159.

EXAMPLE 2   **Self-selected samples—a bad practice**

A magazine which features the latest computer hardware and software for home-office use asks readers to go to their website and indicate whether or not they owned specific new software packages or hardware products. In past issues, this magazine used similar information to make such statements as "40% of readers have purchased software package P." Is this sample representative of the population of magazine readers?

Solution   It is clearly impossible to contact all magazine readers since not all are subscribers. One must necessarily settle for taking a sample. Unfortunately, the method used by this magazine's editors is not representative and is badly biased. Readers who regularly upgrade their systems and try most of the new software will be more likely to respond positively indicating their purchases. In contrast, those who did not purchase any of the software or hardware mentioned in the survey will very likely not bother to report their status. That is, the proportion of purchasers of software package P in the sample will likely be much higher than it is for the whole population consisting of the *purchase/not purchase* record for each reader.   ■

To avoid bias due to self-selected samples, we must take an active role in the selection process.

Using a random number table to select samples

The selection of a sample from a finite population must be done impartially and objectively. But writing the unit names on slips of paper, putting the slips in a box, and drawing them out may not only be cumbersome, but proper mixing may not be possible. However, the selection is easy to carry out using a chance mechanism called a **random number table**.

Random number table

> Suppose ten balls numbered 0, 1, ..., 9 are placed in an urn and shuffled. One is drawn and the digit recorded. It is then replaced, the balls shuffled, another one drawn, and the digit recorded. The digits in Table 7W[3] were actually generated by a computer that closely simulates this procedure. A portion of this table is shown as Table 1.3.
>
> The chance mechanism that generated the random number table ensures that each of the single digits has the same chance of occurrence, that all pairs 00, 01, ..., 99 have the same chance of occurrence, and so on. Further, any collection of digits is unrelated to any other digit in the table. Because of these properties, the digits are called *random*.

EXAMPLE 3   **Using the table of random digits**

Eighty specialty pumps were manufactured last week. Use Table 1.3 to select a sample of size $n = 5$ to carefully test and recheck for possible defects before they are sent to the purchaser. Select the sample without replacement so that the same pump does not appear twice in the sample.

Solution   The first step is to number the pumps from 1 to 80, or to arrange them in some order so they can be identified. The digits must be selected two at a time because the population size $N = 80$ is a two-digit number. We begin by arbitrarily selecting

---

[3]The W indicates that the table is on the website for this book. See Appendix B for details.

Table 1.3 Random digits (portion of Table 7W)

1306	1189	5731	3968	5606	5084	8947	3897	1636	7810
0422	2431	0649	8085	5053	4722	6598	5044	9040	5121
6597	2022	6168	5060	8656	6733	6364	7649	1871	4328
7965	6541	5645	6243	7658	6903	9911	5740	7824	8520
7695	6937	0406	8894	0441	8135	9797	7285	5905	9539
5160	7851	8464	6789	3938	4197	6511	0407	9239	2232
2961	0551	0539	8288	7478	7565	5581	5771	5442	8761
1428	4183	4312	5445	4854	9157	9158	5218	1464	3634
3666	5642	4539	1561	7849	7520	2547	0756	1206	2033
6543	6799	7454	9052	6689	1946	2574	9386	0304	7945
9975	6080	7423	3175	9377	6951	6519	8287	8994	5532
4866	0956	7545	7723	8085	4948	2228	9583	4415	7065
8239	7068	6694	5168	3117	1568	0237	6160	9585	1133
8722	9191	3386	3443	0434	4586	4150	1224	6204	0937
1330	9120	8785	8382	2929	7089	3109	6742	2468	7025

a row and column. We select row 6 and column 21. Reading the digits in columns 21 and 22, and proceeding downward, we obtain

$$41 \quad 75 \quad 91 \quad 75 \quad 19 \quad 69 \quad 49$$

We ignore the number 91 because it is greater than the population size 80. We also ignore any number when it appears a second time, as 75 does here. That is, we continue reading until five different numbers in the appropriate range are selected. Here the five pumps numbered

$$41 \quad 75 \quad 19 \quad 69 \quad 49$$

will be carefully tested and rechecked for defects.

For situations involving large samples or frequent applications, it is more convenient to use computer software to choose the random numbers. ∎

EXAMPLE 4   **Selecting a sample by random digit dialing**

Suppose there is a single three-digit exchange for the area in which you wish to conduct a phone survey. Use the random digit Table 7W to select five phone numbers.

Solution   We arbitrarily decide to start on the second page of Table 7W at row 53 and column 13. Reading the digits in columns 13 through 16, and proceeding downward, we obtain

$$5619 \quad 0812 \quad 9167 \quad 3802 \quad 4449$$

These five numbers, together with the designated exchange, become the phone numbers to be called in the survey. Every phone number, listed or unlisted, has the same chance of being selected. The same holds for every pair, every triplet, and so on. Commercial phones may have to be discarded and another number drawn from the table. If there are two exchanges in the area, separate selections could be done for each exchange. ∎

Do's and Don'ts

Do's

1. Create a clear statement of purpose before deciding upon which variables to observe.
2. Carefully define the population of interest.
3. Whenever possible, select samples using a random device or random number table.

Don'ts

1. Don't unquestioningly accept conclusions based on self-selected samples.

Review Exercises

1.1 A consumer magazine article asks "How Safe Is the Air in Airplanes?" and goes on to say that the air quality was measured on 158 different flights for U.S.-based airlines. Let the variable of interest be a numerical measure of staleness. Identify the population and the sample.

1.2 A radio-show host announced that she wanted to know which rapper was the favorite among college students in your school. Listeners were asked to call and name their favorite rapper. Identify the population, in terms of preferences, and the sample. Is the sample likely to be representative? Comment. Also describe how to obtain a sample that is likely to be more representative.

1.3 Consider the population of all laptop computers owned by students at your university. You want to know the weight of the laptop.

    (a) Specify the population unit.

    (b) Specify the variable of interest.

    (c) Specify the statistical population.

1.4 Identify the statistical population, sample, and variable of interest in each of the following situations:

    (a) Tensile strength is measured on 20 specimens of super strength thread made of the same nano-fibers. The intent is to learn about the strengths for all specimens that could conceivably be made by the same method.

    (b) Fifteen calls to the computer help desk are selected from the hundreds received one day. Only 4 of these calls ended without a satisfactory resolution of the problem.

    (c) Thirty flash memory cards are selected from the thousands manufactured one day. Tests reveal that 6 cards do not meet manufacturing specifications.

1.5 For hard drives to operate properly, the distance between the read/write head and the disk must remain between tight limits. From each hour's production, 40 drives are selected and the distance is measured.

    Identify the population unit, variable of interest, statistical population, and sample.

1.6 Ten seniors have applied to be on the team that will build a high-mileage car to compete against teams from other universities. Use Table 7 of random digits to select 5 of the 10 seniors to form the team.

1.7 Refer to the slot depth data in Table 1.1. After the machine was repaired, a sample of three new ceramic parts had slot depths 215, 216, and 213 (thousandths of an inch).

    (a) Redraw the X-bar chart and include the additional mean $\bar{x}$.

    (b) Does the new $\bar{x}$ fall within the control limits?

1.8 A Canadian manufacturer identified a critical diameter on a crank bore that needed to be maintained within a close tolerance for the product to be successful. Samples of size 4 were taken every hour. The values of the differences (measurement − specification), in ten-thousandths of an inch, are given in Table 1.4.

    (a) Calculate the central line for an X-bar chart for the 24 hourly sample means. The centerline is $\bar{\bar{x}} = (4.25 - 3.00 - \cdots - 1.50 + 3.25)/24$.

    (b) Is the average of all the numbers in the table, 4 for each hour, the same as the average of the 24 hourly averages? Should it be?

    (c) A computer calculation gives the control limits

$$\text{LCL} = -4.48$$
$$\text{UCL} = 7.88$$

Construct the X-bar chart. Identify hours where the process was out of control.

Table 1.4 The differences (measurement – specification), in ten-thousandths of an inch

Hour	1	2	3	4	5	6	7	8	9	10	11	12
	10	−6	−1	−8	−14	−6	−1	8	−1	5	2	5
	3	1	−3	−3	−5	−2	−6	−3	7	6	1	3
	6	−4	0	−7	−6	−1	−1	9	1	3	1	10
	−2	−3	−7	−2	2	−6	7	11	7	2	4	4
x	4.25	−3.00	−2.75	−5.00	−5.75	−3.75	−0.25	6.25	3.50	4.00	2.00	5.50

Hour	13	14	15	16	17	18	19	20	21	22	23	24
	5	6	−5	−8	2	7	8	5	8	−5	−2	−1
	9	6	4	−5	8	7	13	4	1	7	−4	5
	9	8	−5	1	−4	5	6	7	0	1	−7	9
	7	10	−2	0	1	3	6	10	−6	2	7	0
$\bar{x}$	7.50	7.50	−2.00	−3.00	1.75	5.50	8.25	6.50	0.75	1.25	−1.50	3.25

Key Terms

Characteristic of interest   6

Classical approach to statistics   2

Descriptive statistics   2

Population   6

Population of units   6

Quality improvement   3

Random number table   8

Reliability   3

Sample   7

Statement of purpose   6

Statistical inference   2

Statistical population   6

X-bar chart   4

Unit   6

Variable   6

CHAPTER

2

ORGANIZATION AND DESCRIPTION OF DATA

Statistical data, obtained from surveys, experiments, or any series of measurements, are often so numerous that they are virtually useless unless they are condensed, or reduced into a more suitable form. We begin with the use of simple graphics in Section 2.1. Sections 2.2 and 2.3 deal with problems relating to the grouping of data and the presentation of such groupings in graphical form. In Section 2.4 we discuss a relatively new way of presenting data.

Sometimes it may be satisfactory to present data just as they are and let them speak for themselves; on other occasions it may be necessary only to group the data and present the result in tabular or graphical form. However, most of the time data have to be summarized further, and in Sections 2.5 through 2.7 we introduce some of the most widely used kinds of statistical descriptions.

2.1 Pareto Diagrams and Dot Diagrams

Data need to be collected to provide the vital information necessary to solve engineering problems. Once gathered, these data must be described and analyzed to produce summary information. Graphical presentations can often be the most effective way to communicate this information. To illustrate the power of graphical techniques, we first describe a **Pareto diagram**. This display, which orders each type of failure or defect according to its frequency, can help engineers identify important defects and their causes.

When a company identifies a process as a candidate for improvement, the first step is to collect data on the frequency of each type of failure. For example, the performance of a computer-controlled lathe is below par so workers record the following causes of malfunctions and their frequencies:

power fluctuations	6
controller not stable	22
operator error	13
worn tool not replaced	2
other	5

These data are presented as a special case of a **bar chart** called a **Pareto diagram** in Figure 2.1. This diagram graphically depicts Pareto's empirical law that any assortment of events consists of a few major and many minor elements. Typically, two or three elements will account for more than half of the total frequency.

Concerning the lathe, 22 or $100(22/48) = 46\%$ of the cases are due to an unstable controller and $22 + 13 = 35$ or $100(35/48) = 73\%$ are due to either unstable controller or operator error. These cumulative percentages are shown in Figure 2.1 as a line graph whose scale is on the right-hand side of the Pareto diagram, as appears again in Figure 15.2.

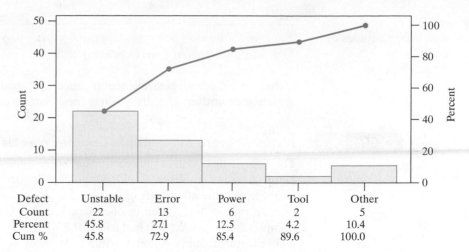

Figure 2.1
A Pareto diagram of failures

Defect	Unstable	Error	Power	Tool	Other
Count	22	13	6	2	5
Percent	45.8	27.1	12.5	4.2	10.4
Cum %	45.8	72.9	85.4	89.6	100.0

In the context of quality improvement, to make the most impact we want to select the few vital major opportunities for improvement. This graph visually emphasizes the importance of reducing the frequency of controller misbehavior. An initial goal may be to cut it in half.

As a second step toward improvement of the process, data were collected on the deviations of cutting speed from the target value set by the controller. The seven observed values of (cutting speed) − (target),

$$3 \quad 6 \quad -2 \quad 4 \quad 7 \quad 4 \quad 3$$

are plotted as a **dot diagram** in Figure 2.2. The dot diagram visually summarizes the information that the lathe is, generally, running fast. In Chapters 13 and 15 we will develop efficient experimental designs and methods for identifying primary causal factors that contribute to the variability in a response such as cutting speed.

Figure 2.2
Dot diagram of cutting speed
deviations

When the number of observations is small, it is often difficult to identify any pattern of variation. Still, it is a good idea to plot the data and look for unusual features.

EXAMPLE 1   **Dot diagrams expose outliers**

A major food processor regularly monitors bacteria along production lines that include a stuffing process for meat products. An industrial engineer records the maximum amount of bacteria present along the production line, in the units Aerobic Plate Count per square inch (APC/in^2), for $n = 7$ days. (Courtesy of David Brauch)

$$96.3 \quad 155.6 \quad 3408.0 \quad 333.3 \quad 122.2 \quad 38.9 \quad 58.0$$

Create a dot diagram and comment.

Solution   The ordered data

$$38.9 \quad 58.0 \quad 96.3 \quad 122.2 \quad 155.6 \quad 333.3 \quad 3408.0$$

are shown as the dot diagram in Figure 2.3. By using open circles, we help differentiate the crowded smaller values. The one very large bacteria count is the prominent

Figure 2.3
Maximum bacteria counts on seven days.

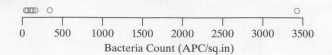

Bacteria Count (APC/sq.in)

feature. It indicates a possible health concern. Statisticians call such an unusual observation an **outlier**. Usually, outliers merit further attention. ■

EXAMPLE 2   **A dot diagram for multiple samples reveals differences**

The vessels that contain the reactions at some nuclear power plants consist of two hemispherical components welded together. Copper in the welds could cause them to become brittle after years of service. Samples of welding material from one production run or "heat" used in one plant had the copper contents 0.27, 0.35, 0.37. Samples from the next heat had values 0.23, 0.15, 0.25, 0.24, 0.30, 0.33, 0.26. Draw a dot diagram that highlights possible differences in the two production runs (heats) of welding material. If the copper contents for the two runs are different, they should not be combined to form a single estimate.

Solution   We plot the first group as solid circles and the second as open circles (see Figure 2.4). It seems unlikely that the two production runs are alike because the top two values are from the first run. (In Exercise 14.23, you are asked to confirm this fact.) The two runs should be treated separately.

The copper content of the welding material used at the power plant is directly related to the determination of safe operating life. Combining the sample would lead to an unrealistically low estimate of copper content and too long an estimate of safe life. ■

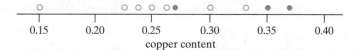

Figure 2.4
Dot diagram of copper content

copper content

When a set of data consists of a large number of observations, we take the approach described in the next section. The observations are first summarized in the form of a table.

2.2 Frequency Distributions

A **frequency distribution** is a table that divides a set of data into a suitable number of classes (categories), showing also the number of items belonging to each class. The table sacrifices some of the information contained in the data. Instead of knowing the exact value of each item, we only know that it belongs to a certain class. On the other hand, grouping often brings out important features of the data, and the gain in "legibility" usually more than compensates for the loss of information.

We shall consider mainly **numerical distributions**; that is, frequency distributions where the data are grouped according to size. If the data are grouped according to some quality, or attribute, we refer to such a distribution as a **categorical distribution**.

The first step in constructing a frequency distribution consists of deciding how many classes to use and choosing the **class limits** for each class. That is, deciding from where to where each class is to go. Generally speaking, the number of classes we use depends on the number of observations, but it is seldom profitable to use fewer than 5 or more than 15. The exception to the upper limit is when the size of the data set is several hundred or even a few thousand. It also depends on the range of the data, namely, the difference between the largest observation and the smallest.

Once the classes are set, we count the number of observations in each class, called the **class frequencies**. This task is simplified if the data are first sorted from smallest to largest.

To illustrate the construction of a frequency distribution, we consider data collected in a nanotechnology setting (see Exercise 2.44). Engineers fabricating a new transmission-type electron multiplier created an array of silicon nanopillars on a flat silicon membrane. The precise structure can influence the electrical properties, so the heights of 50 nanopillars were measured in nanometers (nm), or $10^{-9} \times$ meters. (See Figure 2.5.)[1]

Figure 2.5
Nanopillars

245	333	296	304	276	336	289	234	253	292
366	323	309	284	310	338	297	314	305	330
266	391	315	305	290	300	292	311	272	312
315	355	346	337	303	265	278	276	373	271
308	276	364	390	298	290	308	221	274	343

Since the largest observation is 391 and the smallest is 221 and the range is $391 - 221 = 170$, we might choose five classes having the limits 206–245, 246–285, 286–325, 326–365, 366–405, or the six classes 216–245, 246–275, …, 366–395. Note that, in either case, **the classes do not overlap, they accommodate all the data, and they are all of the same width**.

Initially, deciding on the first of these classifications, we count the number of observations in each class to obtain the frequency distribution:

Limits of Classes	Frequency
206–245	3
246–285	11
286–325	23
326–365	9
366–405	4
Total	50

Note that the class limits are given to as many decimal places as the original data. Had the original data been given to one decimal place, we would have used the class limits 205.9–245.0, 245.1–285.0, …, 365.1–405.0. If they had been rounded to the nearest 10 nanometers, we would have used the class limits 210–240, 250–280, 290–320, 330–360, 370–400.

In the preceding example, the data on heights of nanopillars may be thought of as values of a continuous variable which, conceivably, can be any value in an interval. But if we use classes such as 205–245, 245–285, 285–325, 325–365, 365–405, there exists the possibility of ambiguities; 245 could go into the first class or the second, 285 could go into the second class or the third, and so on. To avoid this difficulty, we take an alternative approach.

We make an **endpoint convention**. For the pillar height data, we can take (205, 245] as the first class, (245, 285] as the second, and so on through (365, 405]. That is, for this data set, we adopt the convention that the right-hand endpoint is included

[1]Data and photo from H. Qin, H. Kim, and R. Blick, Nanopillar arrays on semiconductor membranes as electron emission amplifiers, *Nanotechnology* **19** (2008), used with permission from IOP Publishing Ltd.

but the left-hand endpoint is not. For other data sets we may prefer to reverse the endpoint convention so the left-hand endpoint is included but the right-hand endpoint is not. Whichever endpoint convention is adopted, it should appear in the description of the frequency distribution.

Under the convention that the right-hand endpoint is included, the frequency distribution of the nanopillar data is

Height (nm)	Frequency
(205, 245]	3
(245, 285]	11
(285, 325]	23
(325, 365]	9
(365, 405]	4
Total	50

The **class boundaries** are the endpoints of the intervals that specify each class. As we pointed out earlier, once data have been grouped, each observation has lost its identity in the sense that its exact value is no longer known. This may lead to difficulties when we want to give further descriptions of the data, but we can avoid them by representing each observation in a class by its midpoint, called the **class mark**. In general, the class marks of a frequency distribution are obtained by averaging successive class boundaries. If the classes of a distribution are all of equal length, as in our example, we refer to the common interval between any successive class marks as the **class interval** of the distribution. Note that the class interval may also be obtained from the difference between any successive class boundaries.

EXAMPLE 3   **Class marks and class interval for grouped data**

With reference to the distribution of the heights of nanopillars, find (a) the class marks and (b) the class interval.

Solution   **(a)** The class marks are

$$\frac{205 + 245}{2} = 225 \qquad \frac{245 + 285}{2} = 265, \quad 305, \quad 345, \quad 385$$

(b) The class interval is $245 - 205 = 40$.   ∎

There are several alternative forms of distributions into which data are sometimes grouped. Foremost among these are the "less than or equal to," "less than," "or more," and "equal or more" **cumulative distributions**. A cumulative "less than or equal to" distribution shows the total number of observations that are less than or equal to the given values. These values must be class boundaries, with an appropriate endpoint convention, when the data are grouped into a frequency distribution.

EXAMPLE 4   **Cumulative distribution of the nanopillar heights**

Convert the distribution of the heights of nanopillars into a distribution according to how many observations are less than or equal to 205, less than or equal to 245, …, less than or equal to 405.

Solution   Since none of the values is less than 205, 3 are less than or equal to 245, $3 + 11 = 14$ are less than or equal to 285, $14 + 23 = 37$ are less than or equal to 325, $37 + 9 = 46$ are less than or equal to 365, and all 50 are less than or equal to 405, we have

Heights (mM)	Cumulative Frequency
(205, 245]	3
(245, 285]	14
(285, 325]	37
(325, 365]	46
(365, 405]	50

■

When the endpoint convention for a class includes the left-hand endpoint but not the right-hand endpoint, the cumulative distribution becomes a "less than" cumulative distribution.

Cumulative "more than" and "or more" distributions are constructed similarly by adding the frequencies, one by one, starting at the other end of the frequency distribution. In practice, "less than or equal to" cumulative distributions are used most widely, and it is not uncommon to refer to "less than or equal to" cumulative distributions simply as *cumulative distributions*.

2.3   Graphs of Frequency Distributions

Properties of frequency distributions relating to their shape are best exhibited through the use of graphs, and in this section we shall introduce some of the most widely used forms of graphical presentations of frequency distributions and cumulative distributions.

The most common form of graphical presentation of a frequency distribution is the **histogram**. The histogram of a frequency distribution is constructed of adjacent rectangles. Provided that the *class intervals are equal*, the heights of the rectangles represent the class frequencies and the bases of the rectangles extend between successive class boundaries. A histogram of the heights of nanopillars data is shown in Figure 2.6.

Using our endpoint convention, the interval (205, 245] that defines the first class has frequency 3, so the rectangle has height 3, the second rectangle, over the interval

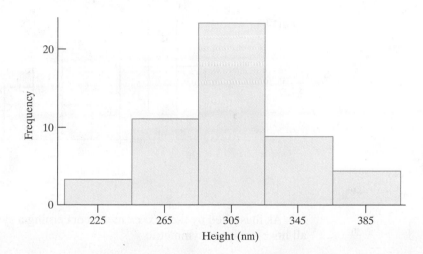

Figure 2.6
Histogram of pillar height

(245, 285], has height 9, and so on. The tallest rectangle is over the interval (285, 325] and has height 23. The histogram has a single peak and is reasonably symmetric. Almost half of the area, representing half of the observations, is over the interval 285 to 325 nanometers.

The choice of frequency, or relative frequency, for the vertical scale is only valid when all of the classes have the same width.

Inspection of the graph of a frequency distribution as a histogram often brings out features that are not immediately apparent from the data themselves. Aside from the fact that such a graph presents a good overall picture of the data, it can also emphasize irregularities and unusual features. It can reveal outlying observations which somehow do not fit the overall picture. Their distruption of the overall pattern of variation in the data may be due to errors of measurement, equipment failure, and similar causes. Also, the fact that a histogram exhibits two or more *peaks* (maxima) can provide pertinent information. The appearance of two peaks may imply, for example, a shift in the process that is being measured, or it may imply that the data come from two or more sources. With some experience one learns to spot such irregularities or anomalies, and an experienced engineer would find it just as surprising if the histogram of a distribution of integrated-circuit failure times were symmetrical as if a distribution of American men's hat sizes were bimodal.

Sometimes it can be enough to draw a histogram in order to solve an engineering problem.

EXAMPLE 5   **A histogram reveals the solution to a grinding operation problem**

A metallurgical engineer was experiencing trouble with a grinding operation. The grinding action was produced by pellets. After some thought he collected a sample of pellets used for grinding, took them home, spread them out on his kitchen table, and measured their diameters with a ruler. His histogram is displayed in Figure 2.7. What does the histogram reveal?

Solution   The histogram exhibits two distinct peaks, one for a group of pellets whose diameters are centered near 25 and the other centered near 40.

By getting his supplier to do a better sort, so all the pellets would be essentially from the first group, the engineer completely solved his problem. Taking the action to obtain the data was the big step. The analysis was simple.   ∎

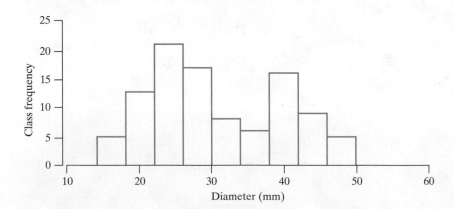

Figure 2.7
Histogram of pellet diameter

As illustrated by the next example concerning a system of supercomputers, not all histograms are symmetric.

EXAMPLE 6   **A histogram reveals the pattern of a supercomputer systems data**

A computer scientist, trying to optimize system performance, collected data on the time, in microseconds, between requests for a particular process service.

2,808	4,201	3,848	9,112	2,082	5,913	1,620	6,719	21,657
3,072	2,949	11,768	4,731	14,211	1,583	9,853	78,811	6,655
1,803	7,012	1,892	4,227	6,583	15,147	4,740	8,528	10,563
43,003	16,723	2,613	26,463	34,867	4,191	4,030	2,472	28,840
24,487	14,001	15,241	1,643	5,732	5,419	28,608	2,487	995
3,116	29,508	11,440	28,336	3,440				

Draw a histogram using the equal length classes [0, 10,000), [10,000, 20,000), ..., [70,000, 80,000) where the left-hand endpoint is included but the right-hand endpoint is not.

Solution   The histogram of this interrequest time data, shown in Figure 2.8, has a long right-hand tail. Notice that, with this choice of equal length intervals, two classes are empty. To emphasize that it is still possible to observe interrequest times in these intervals, it is preferable to regroup the data in the right-hand tail into classes of unequal lengths (see Exercise 2.62). ∎

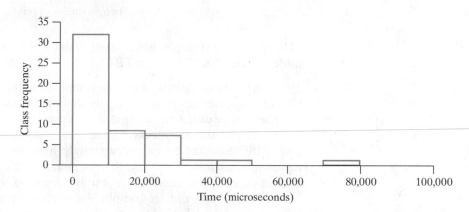

Figure 2.8
Histogram of interrequest time

When a histogram is constructed from a frequency table having classes of unequal lengths, the height of each rectangle must be changed to

$$\text{height} = \frac{\text{relative frequency}}{\text{width}}$$

The area of the rectangle then represents the relative frequency for the class and the total area of the histogram is 1. We call this a **density histogram**.

EXAMPLE 7   **A density histogram has total area 1**

Compressive strength was measured on 58 specimens of a new aluminum alloy undergoing development as a material for the next generation of aircraft.

66.4	67.7	68.0	68.0	68.3	68.4	68.6	68.8	68.9	69.0	69.1
69.2	69.3	69.3	69.5	69.5	69.6	69.7	69.8	69.8	69.9	70.0
70.0	70.1	70.2	70.3	70.3	70.4	70.5	70.6	70.6	70.8	70.9
71.0	71.1	71.2	71.3	71.3	71.5	71.6	71.6	71.7	71.8	71.8
71.9	72.1	72.2	72.3	72.4	72.6	72.7	72.9	73.1	73.3	73.5
74.2	74.5	75.3								

Draw a density histogram, that is, a histogram scaled to have a total area of 1 unit. For reasons to become apparent in Chapter 6, we call the vertical scale **density**.

Solution   We make the height of each rectangle equal to *relative frequency / width*, so that its area equals the relative frequency. The resulting histogram, constructed by computer, has a nearly symmetric shape (see Figure 2.9). We have also graphed a continuous curve that approximates the overall shape. In Chapter 5, we will introduce this bell-shaped family of curves.

∎

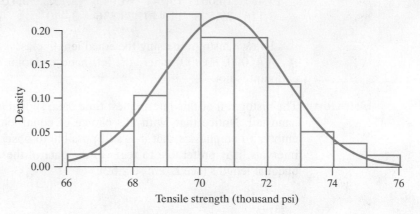

Figure 2.9
Histogram of aluminum alloy
tensile strength

[Using **R: with (sample, hist (strength,prob=TRUE,las=1))** after **sample=read. table ("C2Ex.TXT",header=TRUE)**]

This example suggests that histograms, for observations that come from a continuous scale, can be approximated by smooth curves.

Cumulative distributions are usually presented graphically in the form of **ogives**, where we plot the cumulative frequencies at the class boundaries. The resulting points are connected by means of straight lines, as shown in Figure 2.10, which represents the cumulative "less than or equal to" distribution of nanopillar height data on page 15. The curve is steepest over the class with highest frequency.

When the endpoint convention for a class includes the left-hand endpoint but not the right-hand endpoint, the ogive represents a "less than" cumulative distribution.

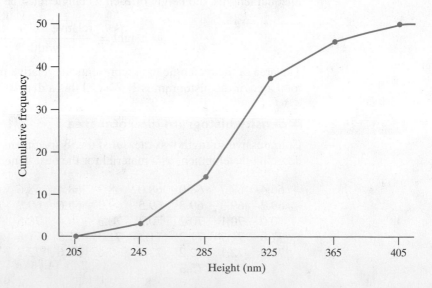

Figure 2.10
Ogive of heights of nanopillars

2.4   Stem-and-Leaf Displays

In the two preceding sections we directed our attention to the grouping of relatively large sets of data with the objective of putting such data into a manageable form. As we saw, this entailed some loss of information. Similar techniques have been proposed for the preliminary explorations of small sets of data, which yield a good overall picture of the data without any loss of information.

To illustrate, consider the following humidity readings rounded to the nearest percent:

29	44	12	53	21	34	39	25	48	23
17	24	27	32	34	15	42	21	28	37

Proceeding as in Section 2.2, we might group these data into the following distribution:

Humidity Readings	Frequency
10–19	3
20–29	8
30–39	5
40–49	3
50–59	1

If we wanted to avoid the loss of information inherent in the preceding table, we could keep track of the last digits of the readings within each class, getting

10–19	2 7 5
20–29	9 1 5 3 4 7 1 8
30–39	4 9 2 4 7
40–49	4 8 2
50–59	3

This can also be written as

1	2 7 5		1	2 5 7
2	9 1 5 3 4 7 1 8		2	1 1 3 4 5 7 8 9
3	4 9 2 4 7	or	3	2 4 4 7 9
4	4 8 2		4	2 4 8
5	3		5	3

where the left-hand column, the **stem**, gives the tens digits 10, 20, 30, 40, and 50. The numbers in a row, the leaves, have the unit 1.0. In the last step, the leaves are written in ascending order. The three numbers in the first row are 12, 15, and 17. This table is called a **stem-and-leaf display** or simply a **stem-leaf display**. The left-hand column forms the *stem*, and the numbers to the left of the vertical line are the **stem labels**, which in our example are 1, 2, . . . , 5. Each number to the right of the vertical line is a **leaf**. There should not be any gaps in the stem even if there are no leaves for that particular value.

Essentially, a stem-and-leaf display presents the same picture as the corresponding tally, yet it retains all the original information. For instance, if a stem-and-leaf display has the two-digit stem

1.2 | 0 2 3 5 8

where the leaf unit = 0.01, the corresponding data are 1.20, 1.22, 1.23, 1.25, and 1.28. If a stem-and-leaf display has the two digit leaves

$$0.3 \mid 03 \quad 17 \quad 55 \quad 89$$

where the first leaf digit unit = 0.01, the corresponding data are 0.303, 0.317, 0.355, and 0.389.

There are various ways in which stem-and-leaf displays can be modified to meet particular needs (see Exercises 2.25 and 2.26), but we shall not go into this here in any detail as it has been our objective to present only one of the relatively new techniques, which come under the general heading of **exploratory data analysis**.

Exercises

2.1 Accidents at a potato-chip plant are categorized according to the area injured.

fingers	17
eyes	5
arm	2
leg	1

Draw a Pareto chart.

2.2 Damages at a paper mill (in thousands of dollars) due to breakage can be divided according to the product manufactured.

toilet paper	132
hand towels	85
napkins	43
12 other products	50

(a) Draw a Pareto chart.

(b) What percent of the loss occurs in making:

(1) toilet paper?

(2) toilet paper or hand towels?

2.3 Probes for noninvasive measurement of blood flow must not give off too much energy (milliwatts per square centimeter). A sample of 12 probes yielded the energy values:

9, 10, 6, 7, 9, 7, 9, 6, 8, 5, 10, and 12

Construct a dot diagram.

2.4 The following are 14 measurements on the strength (psi) of paper to be used in cardboard tubes:

121, 128, 129, 132, 135, 133, 127, 115, 131, 125, 118, 114, 120, 116

Construct a dot diagram.

2.5 Civil engineers help municipal wastewater treatment plants operate more efficiently by collecting data on the quality of the effluent. On seven occasions, the amounts of suspended solids (parts per million) at one plant were

14   12   21   28   30   65   26

Display the data in a dot diagram. Comment on your findings.

2.6 Jump River Electric serves part of Northern Wisconsin, and because much of the area is forested, it is prone to outages. One August there were 11 power outages. Their durations (in hours) are

2.5 2.0 1.5 3.0 1.0 1.5 2.0 1.5 1.0 10.0 1.0

Display the data in a dot diagram.

2.7 Physicists first observed neutrinos from a supernova that occurred outside of our solar system when the detector near Kamiokande, Japan, recorded twelve arrivals. The times(seconds) between the neutrinos are

0.107   0.196   0.021   0.281   0.179   0.854   0.58
0.19   7.30   1.18   2.00

(a) Draw a dot diagram.

(b) Identify any outliers.

2.8 The breaking forces (lbf) of salt pellets, given to the nearest tenth, are grouped into a table having the classes [140.0, 160.0), [160.0, 180.0), [180.0, 200.0), [200.0, 220.0), [220.0, 240.0), and [240.0, 260.0), where the left-hand endpoint is included but the right-hand endpoint is not. Find

(a) the class marks

(b) the class interval

2.9 With reference to the preceding exercise, is it possible to determine from the grouped data how many pellets had breaking force

(a) less than 160.0?

(b) more than 160.0?

(c) at least 220.0?

(d) at most 240.0?

(e) from 220.0 to 260.0 inclusive?

2.10 To continually increase the speed of computers, electrical engineers are working on ever-decreasing scales.

The size of devices currently undergoing development is measured in nanometers (nm), or $10^{-9}\times$ meters. Engineers fabricating a new transmission-type electron multiplier[2] created an array of silicon nanopillars on a flat silicon membrane. Subsequently, they measured the diameters (nm) of 50 pillars.

62	68	69	80	68	79	83	70	74	73
74	75	80	77	80	83	73	79	100	93
92	101	87	96	99	94	102	95	90	98
86	93	91	90	95	97	87	89	100	93
92	98	101	97	102	91	87	110	106	118

Group these measurements into a frequency distribution and construct a histogram using (60,70], (70, 80], (80,90], (90,100], (100, 110], (110,120], where the right-hand endpoint is included but the left-hand endpoint is not.

2.11 Convert the distribution obtained in the preceding exercise into a cumulative "less than or equal to" distribution and graph its ogive.

2.12 The following are the ignition times of certain upholstery materials exposed to a flame (given to the nearest hundredth of a second):

2.58	2.51	4.04	6.43	1.58	4.32	2.20	4.19
4.79	6.20	1.52	1.38	3.87	4.54	5.12	5.15
5.50	5.92	4.56	2.46	6.90	1.47	2.11	2.32
6.75	5.84	8.80	7.40	4.72	3.62	2.46	8.75
2.65	7.86	4.71	6.25	9.45	12.80	1.42	1.92
7.60	8.79	5.92	9.65	5.09	4.11	6.37	5.40
11.25	3.90	5.33	8.64	7.41	7.95	10.60	3.81
3.78	3.75	3.10	6.43	1.70	6.40	3.24	1.79
4.90	3.49	6.77	5.62	9.70	5.11	4.50	2.50
5.21	1.76	9.20	1.20	6.85	2.80	7.35	11.75

Group these figures into a table with a suitable number of equal classes and construct a histogram.

2.13 Convert the distribution obtained in Exercise 2.12 into a cumulative "less than" distribution and plot its ogive.

2.14 An engineer uses a thermocouple to monitor the temperature of a stable reaction. The ordered values of 50 observations (Courtesy of Scott Sanders), in tenths of °C, are

1.11	1.21	1.21	1.21	1.23	1.24	1.25	1.25	1.27	1.27	1.28
1.29	1.31	1.31	1.31	1.32	1.34	1.34	1.35	1.36	1.36	1.36
1.36	1.36	1.36	1.36	1.37	1.39	1.40	1.41	1.42	1.42	1.42
1.42	1.43	1.43	1.43	1.44	1.44	1.44	1.47	1.48	1.48	1.50
1.50	1.56	1.56	1.60	1.60	1.68

Group these figures into a distribution having the classes 1.10–1.19, 1.20–1.29, 1.30–1.39,..., and 1.60–1.69, and plot a histogram using [1.10, 1.20),...,

[1.60, 1.70), where the left-hand endpoint is included but the right-hand endpoint is not.

2.15 Convert the distribution obtained in Exercise 2.14 into a cumulative "less than" distribution and plot its ogive.

2.16 The following are the number of automobile accidents that occurred at 60 major intersections in a certain city during a Fourth of July weekend:

0	2	5	0	1	4	1	0	2	1
5	0	1	3	0	0	2	1	3	1
1	4	0	2	4	1	2	4	0	4
3	5	0	1	3	6	4	2	0	2
0	2	3	0	4	2	5	1	1	2
2	1	6	5	0	3	3	0	0	4

Group these data into a frequency distribution showing how often each of the values occurs and draw a bar chart.

2.17 Given a set of observations $x_1, x_2, \ldots,$ and x_n, we define their **empirical cumulative distribution** as the function whose values $F(x)$ equal the proportion of the observations less than or equal to x. Graph the empirical cumulative distribution for the 12 measurements of Exercise 2.3.

2.18 Referring to Exercise 2.17, graph the Empirical cumulative distribution for the data in Exercise 2.16.

2.19 The pictogram of Figure 2.11 is intended to illustrate the fact that per capita income in the United States doubled from \$21,385 in 1993 to \$42,643 in 2012. Does this pictogram convey a fair impression of the actual change? If not, state how it might be modified.

Per capita income

Figure 2.11   Pictogram for Exercise 2.19

2.20 Categorical distributions are often presented graphically by means of **pie charts**, in which a circle is divided into sectors proportional in size to the frequencies (or percentages) with which the data are distributed among the categories. Draw a pie chart to represent the following data, obtained in a study in

[2]H. Qin, H. Kim, and R. Blick, *Nanotechnology* **19** (2008), 095504. (5pp)

which 40 drivers were asked to judge the maneuverability of a certain make of car:

Very good, good, good, fair, excellent, good, good, good, very good, poor, good, good, good, good, very good, good, fair, good, good, very good, very good, fair, good, good, excellent, very good, good, good, good, fair, fair, very good, good, very good, excellent, very good, fair, good, good, and very good.

2.21 Convert the distribution of nanopillar heights on page 16 into a distribution having the classes (205, 245], (245, 325], (325, 365], (365, 405], where the right-hand endpoint is included. Draw two histograms of this distribution, one in which the class frequencies are given by the heights of the rectangles and one in which the class frequencies are given by the area of the rectangles. Explain why the first of these histograms gives a very misleading picture.

2.22 The following are figures on an oil well's daily production in barrels: 214, 203, 226, 198, 243, 225, 207, 203, 208, 200, 217, 202, 208, 212, 205, and 220. Construct a stem-and-leaf display with the stem labels 19, 20, . . . , and 24.

2.23 The following are determinations of a river's annual maximum flow in cubic meters per second: 405, 355, 419, 267, 370, 391, 612, 383, 434, 462, 288, 317, 540, 295, and 508. Construct a stem-and-leaf display with two-digit leaves.

2.24 List the data that correspond to the following stems of stem-and-leaf displays:

(a) 1 | 1 2 3 4 5 7 8. Leaf unit = 1.0.

(b) 23 | 0 0 1 4 6. Leaf unit = 1.0.

(c) 2 | 03 18 35 57. First leaf digit unit = 10.0

(d) 3.2 | 1 3 4 4 7. Leaf unit = 0.01

2.25 To construct a stem-and-leaf display with more stems than there would be otherwise, we might repeat each

stem. The leaves 0, 1, 2, 3, and 4 would be attached to the first stem and leaves 5, 6, 7, 8, and 9 to the second. For the humidity readings on page 21, we would thus get the **double-stem display**:

1	2
1	5 7
2	1 1 3 4
2	5 7 8 9
3	2 4 4
3	7 9
4	2 4
4	8
5	3

where we doubled the number of stems by cutting the interval covered by each stem in half. Construct a double-stem display with one-digit leaves for the data in Exercise 2.14.

2.26 If the double-stem display has too few stems, we create 5 stems where the first holds leaves 0 and 1, the second holds 2 and 3, and so on. The resulting stem-and-leaf display is called a **five-stem display**.

(a) The following are the IQs of 20 applicants to an undergraduate engineering program: 109, 111, 106, 106, 125, 108, 115, 109, 107, 109, 108, 110, 112, 104, 110, 112, 128, 106, 111, and 108. Construct a five-stem display with one-digit leaves.

(b) The following is part of a five-stem display:

53	4 4 4 4 5 5	Leaf unit = 1.0
53	6 6 6 7	
53	8 9	
54	1	

List the corresponding measurements.

2.5 Descriptive Measures

Histograms, dot diagrams, and stem-and-leaf diagrams summarize a data set pictorially so we can visually discern the overall pattern of variation. Numerical measures can augment visual displays when describing a data set. To proceed, we introduce the notation

$$x_1, x_2, \ldots, x_i, \ldots, x_n$$

for a general sample consisting of n measurements. Here x_i is the ith observation in the list so x_1 represents the value of the first measurement, x_2 represents the value of the second measurement, and so on.

Given a set of n measurements or observations, $x_1, x_2, \ldots, x_n$, there are many ways in which we can describe their center (middle, or central location). Most popular among these are the **arithmetic mean** and the **median**, although other kinds of "averages" are sometimes used for special purposes. The arithmetic mean—or, more succinctly, the **mean**—is defined as the sum of the observations divided by sample size.

Sample mean

$$\bar{x} = \frac{\sum\limits_{i=1}^{n} x_i}{n}$$

The notation $\bar{x}$, read x bar, represents the mean of the x_i. To emphasize that it is based on the observations in a data set, we often refer to $\bar{x}$ as the **sample mean**.

Sometimes it is preferable to use the **sample median** as a descriptive measure of the center, or location, of a set of data. This is particularly true if it is desired to minimize the calculations or if it is desired to eliminate the effect of extreme (very large or very small) values. The median of n observations $x_1, x_2, \ldots, x_n$ can be defined loosely as the "middlemost" value once the data are arranged according to size. More precisely, if the observations are arranged according to size and n is an odd number, the median is the value of the observation numbered $\frac{n+1}{2}$; if n is an even number, the median is defined as the mean (average) of the observations numbered $\frac{n}{2}$ and $\frac{n+2}{2}$.

Sample median

Order the n observations from smallest to largest.

sample median = observation in position $\frac{n+1}{2}$,   if n odd.

= average of two observations in

positions $\frac{n}{2}$ and $\frac{n+2}{2}$,   if n even.

EXAMPLE 8   **Calculation of the sample mean and median**

A sample of five university students responded to the question "How much time, in minutes, did you spend on the social network site yesterday?"

$$100 \quad 45 \quad 60 \quad 130 \quad 30$$

Find the mean and the median.

Solution   The mean is

$$\bar{x} = \frac{100 + 45 + 60 + 130 + 30}{5} = 73 \text{ minutes}$$

and, ordering the data from smallest to largest

$$30 \quad 45 \quad \underbrace{60} \quad 100 \quad 130$$

the median is the third largest value, namely, 60 minutes.

The two very large values cause the mean to be much larger than the median. ∎

EXAMPLE 9 **Calculation of the sample median with even sample size**

An engineering group receives e-mail requests for technical information from sales and service. The daily numbers of e-mails for six days are

$$11 \quad 9 \quad 17 \quad 19 \quad 4 \quad 15$$

Find the mean and the median.

Solution   The mean is

$$\bar{x} = \frac{11 + 9 + 17 + 19 + 4 + 15}{6} = 12.5 \text{ requests}$$

and, ordering the data from the smallest to largest

$$4 \quad 9 \quad \underline{11 \quad 15} \quad 17 \quad 19$$

the median, the mean of the third and fourth largest values, is 13 requests.   ∎

The sample mean has a physical interpretation as the balance point, or center of mass, of a data set. Figure 2.12 is the dot diagram for the data on the number of e-mail requests given in the previous example. In the dot diagram, each observation is represented by a ball placed at the appropriate distance along the horizontal axis. If the balls are considered as masses having equal weights and the horizontal axis is weightless, then the mean corresponds to the center of inertia or balance point of the data. This interpretation of the sample mean, as the balance point of the observations, holds for any data set.

Figure 2.12
The interpretation of the sample mean as a balance point

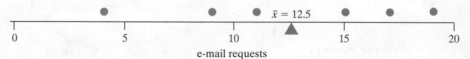

Although the mean and the median each provide a single number to represent an entire set of data, the mean is usually preferred in problems of estimation and other problems of statistical inference. An intuitive reason for preferring the mean is that the median does not utilize all the information contained in the observations.

The following is an example where the median actually gives a more useful description of a set of data than the mean.

EXAMPLE 10 **The median is unaffected by a few outliers**

A small company employs four young engineers, who each earn $80,000, and the owner (also an engineer), who gets $200,000. Comment on the claim that on the average the company pays $104,000 to its engineers and, hence, is a good place to work.

Solution   The mean of the five salaries is $104,000, but it hardly describes the situation. The median, on the other hand, is $80,000, and it is most representative of what a young engineer earns with the firm. Moneywise, the company is not such a good place for young engineers.   ∎

This example illustrates that there is always an inherent danger when summarizing a set of data in terms of a single number.

One of the most important characteristics of almost any set of data is that the values are not all alike; indeed, the extent to which they are unlike, or vary among themselves, is of basic importance in statistics. The mean and median describe one important aspect of a set of data—their "middle" or their "average"—but they tell us nothing about the extent of variation.

We observe that the dispersion of a set of data is small if the values are closely bunched about their mean, and that it is large if the values are scattered widely about their mean. It would seem reasonable, therefore, to measure the variation of a set of data in terms of the amounts by which the values deviate from their mean.

If a set of numbers $x_1, x_2, \ldots, x_n$ has mean $\bar{x}$, the differences

$$x_1 - \bar{x}, x_2 - \bar{x}, \ldots, x_n - \bar{x}$$

are called the **deviations from the mean**. We might use the average of the deviations as a measure of variation in the data set. Unfortunately, this will not do. For instance, refer to the observations 11, 9, 17, 19, 4, 15, displayed above in Figure 2.12, where $\bar{x} = 12.5$ is the balance point. The six deviations are $-1.5, -3.5, 4.5, 6.5, -8.5$, and 2.5. The sum of positive deviations

$$4.5 + 6.5 + 2.5 = 13.5$$

exactly cancels the sum of the negative deviations

$$-1.5 - 3.5 - 8.5 = -13.5$$

so the sum of all the deviations is 0.

As you will be asked to show in Exercise 2.50, the sum of the deviations is always zero. That is,

$$\sum_{i=1}^{n} (x_i - \bar{x}) = 0$$

so the mean of the deviations is always zero. Because the deviations sum to zero, we need to remove their signs. Absolute value and square are two natural choices. If we take their absolute value, so each negative deviation is treated as positive, we would obtain a measure of variation. However, to obtain the most common measure of variation, we square each deviation. The **sample variance**, s^2, is essentially the average of the squared deviations from the mean, $\bar{x}$, and is defined by the following formula.

Sample Variance

$$s^2 = \frac{\sum_{i=1}^{n} (x_i - \bar{x})^2}{n - 1}$$

Our reason for dividing by $n-1$ instead of n is that there are only $n-1$ independent deviations $x_i - \bar{x}$. Because their sum is always zero, the value of any particular one is always equal to the negative of the sum of the other $n - 1$ deviations.

If many of the deviations are large in magnitude, either positive or negative, their squares will be large and s^2 will be large. When all the deviations are small, s^2 will be small.

EXAMPLE 11   **Calculation of sample variance**

The delay times (handling, setting, and positioning the tools) for cutting 6 parts on an engine lathe are 0.6, 1.2, 0.9, 1.0, 0.6, and 0.8 minutes. Calculate s^2.

Solution   First we calculate the mean:

$$\bar{x} = \frac{0.6 + 1.2 + 0.9 + 1.0 + 0.6 + 0.8}{6} = 0.85$$

To find $\sum (x_i - \bar{x})^2$, we set up the table:

x_i	$x_i - \bar{x}$	$(x_i - \bar{x})^2$
0.6	−0.25	0.0625
1.2	0.35	0.1225
0.9	0.05	0.0025
1.0	0.15	0.0225
0.6	−0.25	0.0625
0.8	−0.05	0.0025
5.1	0.00	0.2750

where the total of the third column $0.2750 = \sum (x_i - \bar{x})^2$.
    We divide 0.2750 by $6 - 1 = 5$ to obtain

$$s^2 = \frac{0.2750}{5} = 0.055 \text{ (minute)}^2$$

By calculating the sum of deviations in the second column, we obtain a check on our work. For all data sets, this sum should be 0 up to rounding error. ∎

Notice that the units of s^2 are not those of the original observations. The data are delay times in minutes, but s^2 has the unit $(\text{minute})^2$. Consequently, we define the **standard deviation** of n observations $x_1, x_2, \ldots, x_n$ as the square root of their variance, namely

Sample standard deviation

$$s = \sqrt{\frac{\sum_{i=1}^{n} (x_i - \bar{x})^2}{n - 1}}$$

The standard deviation is by far the most generally useful measure of variation. Its advantage over the variance is that it is expressed in the same units as the observations.

EXAMPLE 12 **Calculation of sample standard deviation**

With reference to the previous example, calculate s.

Solution From the previous example, $s^2 = 0.055$. Take the square root and get

$$s = \sqrt{0.055} = 0.23 \text{ minute}$$

[Using **R**: Enter data $\mathbf{x = c(.6, 1.2, .9, 1, .6, .8)}$. Then **mean(x)**, **var(x)**, and **sd(x)**] ∎

The standard deviation s has a rough interpretation as the average distance from an observation to the sample mean.

The standard deviation and the variance are measures of **absolute variation**; that is, they measure the actual amount of variation in a set of data, and they depend on the scale of measurement. To compare the variation in several sets of data, it is generally desirable to use a measure of **relative variation**, for instance, the **coefficient of variation**, which gives the standard deviation as a percentage of the mean.

Coefficient of variation

$$V = \frac{s}{\bar{x}} \cdot 100\%$$

EXAMPLE 13   **The coefficient of variation for comparing relative preciseness**

Measurements made with one micrometer of the diameter of a ball bearing have a mean of 3.92 mm and a standard deviation of 0.0152 mm, whereas measurements made with another micrometer of the unstretched length of a spring have a mean of 1.54 inches and a standard deviation of 0.0086 inch. Which of these two measuring instruments is relatively more precise?

Solution   For the first micrometer the coefficient of variation is

$$V = \frac{0.0152}{3.92} \cdot 100 = 0.39\%$$

and for the second micrometer the coefficient of variation is

$$V = \frac{0.0086}{1.54} \cdot 100 = 0.56\%$$

Thus, the measurements made with the first micrometer are relatively more precise. ∎

In this section, we have limited the discussion to the sample mean, median, variance, and standard deviation. However, there are many other ways of describing sets of data.

2.6 Quartiles and Percentiles

In addition to the median, which divides a set of data into halves, we can consider other division points. When an ordered data set is divided into quarters, the resulting division points are called sample **quartiles**. The *first quartile*, Q_1, is a value that has one-fourth, or 25%, of the observations below its value. The first quartile is also the sample 25th **percentile** $P_{0.25}$. More generally, we define the sample 100 pth percentile as follows.

Sample percentiles

The sample 100 pth percentile is a value such that at least $100p\%$ of the observations are at or below this value, and at least $100(1 - p)\%$ are at or above this value.

As in the case of the median, which is the 50th percentile, this may not uniquely define a percentile. Our convention is to take an observed value for the sample percentile unless two adjacent values both satisfy the definition. In this latter case, take their mean. This coincides with the procedure for obtaining the median when the sample size is even. (Most computer programs linearly interpolate between the two adjacent values. For moderate or large sample sizes, the particular convention used to locate a sample percentile between the two observations is inconsequential.)

The following rule simplifies the calculation of sample percentiles.

Calculating the sample 100 pth percentile:

1. Order the n observations from smallest to largest.

2. Determine the product np.

If np is not an integer, round it up to the next integer and find the corresponding ordered value.

If np is an integer, say k, calculate the mean of the kth and $(k + 1)$st ordered observations.

The quartiles are the 25th, 50th, and 75th percentiles.

Sample quartiles

first quartile	$Q_1 = $ 25th percentile
second quartile	$Q_2 = $ 50th percentile
third quartile	$Q_3 = $ 75th percentile

EXAMPLE 14   **Calculation of percentiles for the strength of green materials**

Of all the waste materials entering landfills, a substantial proportion consists of construction and demolition materials. From the standpoint of green engineering, before incorporating these materials into the base for new or rehabilitated roadways, engineers must assess their strength. Generally, higher values imply a stiffer base which increases pavement life.

Measurements of the resiliency modulus (MPa) on $n = 18$ specimens of recycled concrete aggregate produce the ordered values (Courtesy of Tuncer Edil)

136	143	147	151	158	160
161	163	165	167	173	174
181	181	185	188	190	205

Obtain the quartiles and the 10th percentile.

Solution   According to our calculation rule, $np = 18 \left(\frac{1}{4} \right) = 4.5$, which we round up to 5. The first quartile is the 5th ordered observation

$$Q_1 = 158 \, \text{MPa}$$

Since $p = \frac{1}{2}$ for the second quartile, or median,

$$np = 18 \left(\frac{1}{2} \right) = 9$$

which is an integer. Therefore, we average the 9th and 10th ordered values

$$Q_2 = \frac{165 + 167}{2} = 166 \, \text{MPa}$$

The third quartile is the 14th observation, $Q_3 = 181$ seconds. We could also have started at the largest value and counted down to the 5th position.

To obtain the 10th percentile, we determine that $np = 18 \times 0.10 = 1.8$, which we round up to 2. Counting to the 2nd position, we obtain

$$P_{0.10} = 143 \, \text{MPa}$$

The 10th percentile provides a useful description regarding the resiliency modulus of the lowest 10% green pavement specimens.

In the context of monitoring green materials we also record that the maximum resiliency modulus measured was 205 MPa.

[Using **R**: **with(*x*, quantile(resiliency, c(.25,.5,.75,.10),type=2))** after ***x*=read. table("C2Ex14.TXT",header=TRUE)**] ∎

The **minimum** and **maximum** observations also convey information concerning the amount of variability present in a set of data. Together, they describe the interval containing all of the observed values and whose length is the

$$\textbf{range} = \text{maximum} - \text{minimum}$$

Care must be taken when interpreting the range since a single large or small observation can greatly inflate its value.

The amount of variation in the middle half of the data is described by the

$$\textbf{interquartile range} = \text{third quartile} - \text{first quartile} = Q_3 - Q_1$$

EXAMPLE 15   **The range and interquartile range for the materials data**

Obtain the range and interquartile range for the resiliency modulus data in Example 14.

Solution   The minimum $= 136$. From the previous example, the maximum $= 205$, $Q_1 = 158$, and $Q_3 = 181$.

$$\text{range} = \text{maximum} - \text{minimum} = 205 - 136 = 69\,\text{MPa}$$

$$\text{interquartile range} = Q_3 - Q_1 = 181 - 158 = 23\,\text{MPa}$$ ∎

Boxplots

The summary information contained in the quartiles is highlighted in a graphic display called a **boxplot**. The center half of the data, extending from the first to the third quartile, is represented by a rectangle. The median is identified by a bar within this box. A line extends from the third quartile to the maximum, and another line extends from the first quartile to the minimum. (For large data sets the lines may only extend to the 95th and 5th percentiles.)

Figure 2.13 gives the boxplot for the green pavement data. The median is closer to Q_1 than Q_3.

A **modified boxplot** can both identify outliers and reduce their effect on the shape of the boxplot. The outer line extends to the largest observation only if it is not too far from the third quartile. That is, for the line to extend to the largest observation, it must be within $1.5 \times$ (interquartile range) units of Q_3. The line from

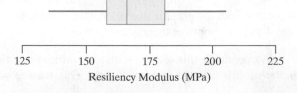

Figure 2.13
Boxplot of the resiliency
modulus of green pavement.

125   150   175   200   225
Resiliency Modulus (MPa)

Q_1 extends to the smallest observation if it is within that same limit. Otherwise the line extends to the next most extreme observations that fall within this interval.

EXAMPLE 16   **A modified boxplot—possible outliers are detached**

Physicists, trying to learn about neutrinos, detected twelve of them coming from a supernova outside of our solar system. The $n = 11$ times (seconds) between the arrivals are presented in their original order in Exercise 2.7, page 22.

The ordered interarrival times are

 0.021   0.107   0.179   0.190   0.196   0.283   0.580   0.854   1.18   2.00   7.30

Construct a modified boxplot.

Solution   Since $n/4 = 11/4 = 2.75$, the first quartile is the third ordered time 0.179 and $Q_3 = 1.18$, so the interquartile range is $1.18 - 0.179 = 1.001$. Further, $1.5 \times 1.001 = 1.502$ and the smallest observation is closer than this to $Q_1 = 0.179$, but

$$\text{maximum} - Q_3 = 7.30 - 1.18 = 6.12$$

exceeds $1.502 = 1.5 \times$ (interquartile range)

As shown in Figure 2.14, the line to the right extends to 2.00, the most extreme observation within 1.502 units, but not to the largest observation, which is shown as detached from the line.

[Using **R**: **with(x, boxplot(time,horizontal=TRUE)** after **x=read.table ("C2Exl4.TXT",header=TRUE)**]

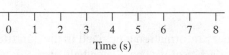

Figure 2.14
Modified boxplot for neutrino data

0   1   2   3   4   5   6   7   8
Time (s)

Boxplots are particularly effective for graphically portraying comparisons among sets of observations. They are easy to understand and have a high visual impact.

EXAMPLE 17   **Multiple boxplots can reveal differences and similarities**

Sometimes, with rather complicated components like hard-disk drives or random access memory (RAM) chips for computers, quality is quantified as an index with target value 100. Typically, a quality index will be based upon the deviations of several physical characteristics from their engineering specifications. Figure 2.15 shows the quality index at 4 manufacturing plants.

Comment on the relationships between quality at different plants.

Solution   It is clear from the graphic that plant 2 needs to reduce its variability and that plants 2 and 4 need to improve their quality level.

We conclude this section with a warning. Sometimes it is a trend over time that is the most important feature of data. This feature would be lost entirely if the set

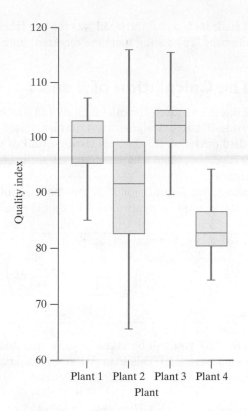

Figure 2.15
Boxplot of the quality index

of data were summarized in a dot diagram, stem-and-leaf display, or boxplot. In one instance, a major international company purchased two identical machines to rapidly measure the thickness of the material and test its strength. The machines were expensive but much faster than the current testing procedure. Before sending one across the United States and the other to Europe, engineers needed to confirm that the two machines were giving consistent results. Following one failed comparison, the problem machine was worked on for a couple of months by the engineers. In the second series of comparative trials, the average value from this machine was appropriate, but fortunately the individual values were plotted as in Figure 2.16. The

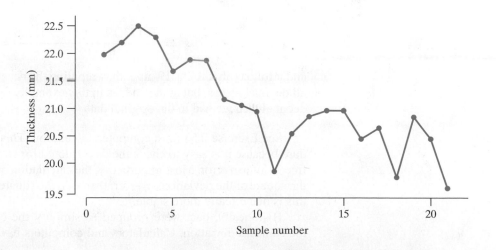

Figure 2.16
Machine measurement of
thickness shows trend

time plot made it clear that the trend was the key feature, not the average, which was a poor summary. The testing machine required more work.

2.7   The Calculation of $\bar{x}$ and s

Here, we discuss methods for calculating $\bar{x}$ and s from data that are already grouped into intervals. These calculations are, in turn, based on the formulas for the mean and standard deviation for data consisting of all of the individual observations. In this latter case, we obtain $\bar{x}$ by summing all of the observations and dividing by the sample size n.

An alternative formula for s^2 forms the basis of the grouped data formula for variance. It was originally introduced to simplify hand calculations.

Variance (handheld calculator formula)

$$s^2 = \frac{\sum_{i=1}^{n} x_i^2 - \left(\sum_{i=1}^{n} x_i \right)^2 / n}{n - 1}$$

(In Exercise 2.51 you will be asked to show that this formula is, in fact, equivalent to the one on page 27.) This expression for variance is without $\bar{x}$, which reduces roundoff error when using a handheld calculator.

EXAMPLE 18   **Calculating variance using the handheld calculator formula**

Find the mean and the standard deviation of the following miles per gallon (mpg) obtained in 20 test runs performed on urban roads with an intermediate-size car:

19.7	21.5	22.5	22.2	22.6
21.9	20.5	19.3	19.9	21.7
22.8	23.2	21.4	20.8	19.4
22.0	23.0	21.1	20.9	21.3

Solution   Using a calculator, we find that the sum of these figures is 427.7 and that the sum of their squares is 9,173.19. Consequently,

$$\bar{x} = \frac{427.7}{20} = 21.39 \text{ mpg}$$

and

$$s^2 = \frac{9,173.19 - (427.7)^2/20}{19} = 1.412$$

and it follows that $s = 1.19$ mpg. In computing the necessary sums we usually retain all decimal places, but at the end, as in this example, we usually round to one more decimal than we had in the original data. ∎

See Exercise 2.58 for a computer calculation. This is the recommended procedure because it is easy to check the data entered for accuracy, and the calculation is free of human error. Most importantly, the calculation of variance can be done using the square of the deviations $x_i - \bar{x}$ rather than the squares of the observations x_i, and this is numerically more stable.

Historically, data were grouped to simplify the calculation of the mean and the standard deviation. Calculators and computers have eliminated the calculation problem. Nevertheless, it is sometimes necessary to calculate $\bar{x}$ and s from grouped data since some data (for instance, from government publications) is available only in grouped form.

To calculate $\bar{x}$ and s from grouped data, we must assume something about the distribution of the values within each class. We represent each value within a class by the corresponding class mark. Then the sum of the x's and the sum of their squares can be written

$$\sum_{i=1}^{k} x_i f_i \quad \text{and} \quad \sum_{i=1}^{k} x_i^2 f_i$$

where x_i is the class mark of the ith class, f_i is the corresponding class frequency, and k is the number of classes in the distribution. Substituting these sums into the formula for $\bar{x}$ and the computing formula for s^2, we get

Mean and variance (grouped data)

$$\bar{x} = \frac{\sum_{i=1}^{k} x_i f_i}{n}$$

$$s^2 = \frac{\sum_{i=1}^{k} x_i^2 f_i - \left(\sum_{i=1}^{k} x_i f_i \right)^2 / n}{n-1}$$

EXAMPLE 19   **Calculating a mean and variance from grouped data**

Use the distribution obtained on page 17 to calculate the mean, variance, and standard deviation of the nanopillar heights data.

Solution   Recording the class marks and the class frequencies in the first two columns and the products $x_i f_i$ and $x_i^2 f_i$ in the third and fourth columns, we obtain

x_i	f_i	$x_i f_i$	$x_i^2 f_i$
225	3	675	151,875
265	11	2,915	772,475
305	23	7,015	2,139,575
345	9	3,105	1,071,225
385	4	1,540	592,900
Total	50	15,250	4,728,050

Then, substitution into the formula yields

$$\bar{x} = \frac{15,250}{50} = 305.0$$

and

$$s^2 = \frac{4,728,050 - 15,250^2 / 50}{49} = 1,567.35 \quad \text{so} \quad s = 39.6$$

For comparison, the original data have mean $= 305.6$ and standard deviation $= 37.0$.

■

Exercises

2.27 In each of the following situations, should your value be near the average or an outlier? If an outlier, should it be too large or too small?

    (a) Income on your starting job

    (b) Your score on the final exam in a physics class

    (c) Your weight in 10 years

2.28 In each of the following situations, should your value be near the average or an outlier? If outlier, should it be too large or too small?

    (a) The time you take to complete a lab assignment next week

    (b) Your white blood cell count

2.29 Is the influence of a single outlier greater on the mean or the median? Explain.

2.30 Is the influence of a single outlier greater on the sample range or the interquartile range? Explain.

2.31 Referring to Exercise 1.8 in Chapter 1, we see that the sample of 4 deviations (observation − specification) during the second hour for a critical crank-bore diameter is

$$-6 \quad 1 \quad -4 \quad -3$$

ten-thousandths of an inch. For these 4 deviations

    (a) calculate the sample mean $\bar{x}$

    (b) calculate the sample standard deviation s

    (c) On average, is the hole too large or too small?

2.32 At the end of 2012, nine skyscrapers in the world were over 300 meters tall. The ordered values of height are

$$366 \quad 381 \quad 442 \quad 452 \quad 484 \quad 492 \quad 508 \quad 601 \quad 828$$

The tallest is in Dubai.

    (a) Calculate the sample mean

    (b) Drop the largest value and re-calculate the mean.

    (c) Comment on effect of dropping the single very large value.

2.33 Engineers[3] are developing a miniaturized robotic capsule for exploration of a human gastrointestinal tract. One novel solution uses motor-driven legs. The engineers' best design worked for a few trials, and then debris covered the tip of the leg and performance got worse. After cleaning, the next trial resulted in

$$35 \quad 37 \quad 38 \quad 34 \quad 30 \quad 24 \quad 13$$

distances covered (mm/min).

    (a) Calculate the sample mean distance.

    (b) Does the sample mean provide a good summary of these trials? If not, write a sentence or two to summarize more accurately.

2.34 A contract for the maintenance of a national railway's high-horsepower locomotives was given to a large private company. After one year of experience with the maintenance program, those in charge of the program felt that major improvements could be made in the reliability of the locomotives. To document the current status, they collected data on the cost of materials for rebuilding traction motors. Use the data below to

    (a) calculate the sample mean $\bar{x}$,

    (b) calculate the sample standard deviation s.

Materials costs for rebuilding traction motors (thousands of dollars):

1.41	1.70	1.03	0.99	1.68	1.09	1.68	1.94
1.53	2.25	1.60	3.07	1.78	0.67	1.76	1.17
1.54	0.99	0.99	1.17	1.54	1.68	1.62	0.67
0.67	1.78	2.12	1.52	1.01			

2.35 If the mean annual compensation paid to the chief executives of three engineering firms is $175,000, can one of them receive $550,000?

2.36 Records show that in Phoenix, Arizona, the normal daily maximum temperature for each month is 65, 69, 74, 84, 93, 102, 105, 102, 98, 88, 74, and 66 degrees Fahrenheit. Verify that the mean of these figures is 85 and comment on the claim that, in Phoenix, the average daily maximum temperature is a very comfortable 85 degrees.

2.37 The output of an instrument is often a waveform. With the goal of developing a numerical measure of closeness, scientists asked 11 experts to look at two waveforms on the same graph and give a number between 0 and 1 to quantify how well the two waveforms agree.[4] The agreement numbers for one pair of waveforms are

$$0.50 \quad 0.40 \quad 0.04 \quad 0.45 \quad 0.65 \quad 0.40 \quad 0.20 \quad 0.30 \quad 0.60 \quad 0.45$$

    (a) Calculate the sample mean $\bar{x}$.

    (b) Calculate sample standard deviation s.

[3]M. Quirini and S. Scapellato, Design and fabrication of a motor legged capsule for the active exploration of the gastrointestinal tract. *IEEE/ASME Transactions on Mechatronics* (2008) **13**, 169–179.

[4]L. Schwer, Validation metrics for response histories: Perspectives and case studies. *Engineering with Computers* **23** (2007), 295–309.

2.38 With reference to the preceding exercise, find s using

(a) the formula that defines s;

(b) the handheld calculator formula for s.

2.39 Meat products are regularly monitored for freshness. A trained inspector selects a sample of the product and assigns an offensive smell score between 1 and 7 where 1 is very fresh. The resulting offensive smell scores, for each of 16 samples, are (Courtesy of David Brauch)

3.2 3.9 1.7 5.0 1.9 2.6 2.4 5.3
1.0 2.7 3.8 5.2 1.0 6.3 3.3 4.3

(a) Find the mean.

(b) Find the median.

(c) Draw a boxplot.

2.40 With reference to Exercise 2.31, find s^2 using

(a) the formula that defines s^2;

(b) the handheld calculator formula for s^2.

2.41 Material manufactured continuously before being cut and wound into large rolls must be monitored for thickness (caliper). A sample of 10 measurements on paper, in millimeters, yielded

32.2 32.0 30.4 31.0 31.2 31.2 30.3 29.6 30.5 30.7

Find the mean and quartiles for this sample.

2.42 For the four observations 9 7 15 5

(a) calculate the deviations $(x_i - \bar{x})$ and check that they add to 0;

(b) calculate the variance and the standard deviation.

2.43 With reference to Exercise 2.14 on page 24, draw a boxplot.

2.44 A company was experiencing a chronic weld-defect problem with a water-outlet-tube assembly. Each assembly manufactured is leak tested in a water tank. Data were collected on a gap between the flange and the pipe for 6 bad assemblies that leaked and 6 good assemblies that passed the leak test.

Leaker 0.290 0.104 0.207 0.145 0.104 0.124

(a) Calculate the sample mean $\bar{x}$.

(b) Calculate the sample standard deviation s.

2.45 Refer to Exercise 2.44. The measurements for 6 assemblies that did not leak were

Good 0.207 0.124 0.062 0.301 0.186 0.124

(a) Calculate the sample mean $\bar{x}$.

(b) Calculate the sample standard deviation s.

(c) Does there appear to be a major difference in gap between assemblies that leaked and those that did not? The quality improvement group turned their focus to welding process variables.

2.46 Find the mean and the standard deviation of the 20 humidity readings on page 21 by using

(a) the raw (ungrouped) data

(b) the distribution obtained in that example

2.47 Use the distribution in Exercise 2.10 on page 22 to find the mean and the variance of the nanopillar diameters.

2.48 Use the distribution obtained in Exercise 2.12 on page 23 to find the mean and the standard deviation of the ignition times. Also determine the coefficient of variation.

2.49 Use the distribution obtained in Exercise 2.14 on page 23 to find the coefficient of variation of the temperature data.

2.50 Show that

$$\sum_{i=1}^{n} (x_i - \bar{x}) = 0$$

for any set of observations $x_1, x_2, \ldots, x_n$.

2.51 Show that the computing formula for s^2 on page 34 is equivalent to the one used to define s^2 on page 27.

2.52 If data are coded so that $x_i = c \cdot u_i + a$, show that $\bar{x} = c \cdot \bar{u} + a$ and $s_x = |c| \cdot s_u$.

2.53 **Median of grouped data** To find the *median* of a distribution obtained for n observations, we first determine the class into which the median must fall. Then, if there are j values in this class and k values below it, the median is located $\frac{(n/2) - k}{j}$ of the way into this class, and to obtain the median we multiply this fraction by the class interval and add the result to the lower boundary of the class into which the median must fall. This method is based on the assumption that the observations in each class are "spread uniformly" throughout the class interval, and this is why we count $\frac{n}{2}$ of the observations instead of $\frac{n+1}{2}$ as on page 25.

To illustrate, let us refer to the nanopillar height data on page 15 and the frequency distribution on page 16. Since $n = 50$, it can be seen that the median must fall in class $(285, 325]$, which contains $j = 23$ observations. The class has width 40 and there are $k = 3 + 11 = 14$ values below it, so the median is

$$285 + \frac{25 - 14}{23} \times 40 = 264.13$$

(a) Use the distribution obtained in Exercise 2.10 on page 22 to find the median of the grouped nanopillar diameters.

(b) Use the distribution obtained in Exercise 2.12 on page 23 to find the median of the grouped ignition times.

2.54 For each of the following distributions, decide whether it is possible to find the mean and whether it is possible to find the median. Explain your answers.

(a)

Grade	Frequency
40–49	5
50–59	18
60–69	27
70–79	15
80–89	6

(b)

IQ	Frequency
less than 90	3
90–99	14
100–109	22
110–119	19
more than 119	7

(c)

Weight	Frequency
110 or less	41
101–110	13
111–120	8
121–130	3
131–140	1

2.55 To find the first and third quartiles Q_1 and Q_3 for grouped data, we proceed as in Exercise 2.53, but count $\frac{n}{4}$ and $\frac{3n}{4}$ of the observations instead of $\frac{n}{2}$.

(a) With reference to the distribution of the nanopillar height data on page 15 and the frequency distribution on page 16, find Q_1, Q_3, and the interquartile range.

(b) Find Q_1 and Q_3 for the distribution of the ignition time data obtained in Exercise 2.12.

2.56 If k sets of data consist, respectively, of $n_1, n_2, \ldots, n_k$ observations and have the means $\bar{x}_1, \bar{x}_2, \ldots, \bar{x}_k$, then the overall mean of all the data is given by the formula

$$\bar{x} = \frac{\sum_{i=1}^{k} n_i \bar{x}_i}{\sum_{i=1}^{k} n_i}$$

(a) The average annual salaries paid to top-level management in three companies are $264,000, $272,000, and $269,000, If the respective numbers of top-level executives in these companies are 4, 15, and 11, find the average salary paid to these 30 executives.

(b) In a nuclear engineering class there are 22 juniors, 18 seniors, and 10 graduate students. If the juniors averaged 71 in the midterm examination, the seniors averaged 78, and the graduate students averaged 89, what is the mean for the entire class?

2.57 The formula for the preceding exercise is a special case of the following formula for the **weighted mean**:

$$\bar{x}_w = \frac{\sum_{i=1}^{k} w_i x_i}{\sum_{i=1}^{k} w_i}$$

where w_i is a weight indicating the relative importance of the ith observation.

(a) If an instructor counts the final examination in a course four times as much as each 1-hour examination, what is the weighted average grade of a student who received grades of 69, 75, 56, and 72 in four 1-hour examinations and a final examination grade of 78?

(b) From 2010 to 2015, the cost of food in a certain city increased by 60%, the cost of housing increased by 30%, and the cost of transportation increased by 40%. If the average salaried worker spent 24% of his or her income on food, 33% on housing, and 15% on transportation, what is the combined percentage increase in the total cost of these items.

2.58 Modern computer software programs have come a long way toward removing the tedium of calculating statistics. *MINITAB* is one common and easy-to-use program. We illustrate the use of the computer using *MINITAB* commands. Other easy-to-use programs have a quite similar command structure.

The lumber used in the construction of buildings must be monitored for strength. Data for the strength of 2×4 pieces of lumber in pounds per square inch are in the file 2-58.TXT. We give the basic commands that calculate n, $\bar{x}$, and s as well as the quartiles.

The session commands require the data to be set in the first column, C1, of the *MINITAB* work sheet. The command for creating a boxplot is also included.

Data in 2-58.TXT
strength
Dialog box:
Stat> Basic Statistics > Descriptive Statistics
Type *strength* in **Variables**.
Click **OK**.

Output (partial)

Variable	N	Mean	Median	StDev
Strength	30	1908.8	1863.0	327.1

Variable	Minimum	Maximum	Q1	Q3
Strength	1325.0	2983.0	1711.5	2071.8

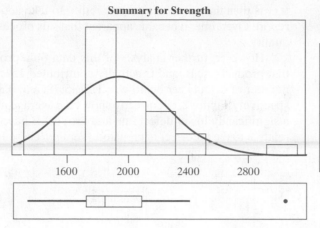

Figure 2.17
MINITAB 14 output

Use *MINITAB*, or some other statistical package, to find $\bar{x}$ and s for

(a) the decay times on page 146

(b) the interrequest times on page 19

2.59 (Further *MINITAB* calculation and graphs.) With the observations on the strength (in pounds per square inch) of 2 × 4 pieces of lumber already set in C1, the sequence of choices and clicks produces an even more complete summary (see Figure 2.17).

Stat> **Basic Statistics** > **Graphical Summary**

Type *strength* in **Variables**. Click **OK**.

The ordered strength data are

1325 1419 1490 1633 1645 1655 1710 1712 1725 1727 1745
1828 1840 1856 1859 1867 1889 1899 1943 1954 1976 2046
2061 2104 2168 2199 2276 2326 2403 2983

From the ordered data

(a) obtain the quartiles

(b) construct a histogram and locate the mean, median, Q_1, and Q_3 on the horizontal axes

(c) repeat parts (a) and (b) with the aluminum alloy data on page 19.

2.8   A Case Study: Problems with Aggregating Data

As circuit boards and other components move through a company's surface mount technology assembly line, a significant amount of data is collected for each assembly. The data (courtesy of Don Ermer) are recorded at several stages of manufacture in a serial tracking database by means of computer terminals located throughout the factory. The data include the board serial number, the type of defect, number of defects, and their location. The challenge here is to transform a large amount of data into manageable and useful information. When there is a variety of products and lots of data are collected on each, record management and the extraction of appropriate data for product improvement must be done well.

Originally, an attempt was made to understand this large database by *aggregating*, or grouping together, data from all products and performing an analysis of the data as if it were one product! This was a poor practice that decreased the resolution of the information obtained from the database. The products on the assembly line ranged in complexity, maturity, method of processing, and lot size.

To see the difficulties caused by aggregation, consider a typical week's production, where 100 printed circuit boards of Product A were produced, 40 boards of Product B, and 60 boards of Product C. Following a wave-soldering process, a total of 400 solder defects was reported. This translates to an overall average of $400/200 = 2$ defects per board. It was this company's practice to circulate the weekly aggregate average throughout the factory floor for review and comment.

It was then the operator's responsibility to take action according to the *misleading* report. Over time, it became apparent that this process was ineffective for improving quality.

However, further analysis of this data on a product-by-product basis revealed that products A, B, and C actually contributed 151, 231, and 18 defects. Thus, the number of defects per board was 1.51, 5.78, and 0.30 for products A, B, and C, respectively. Figure 2.18 correctly shows the average number of defects. Product C has a significantly lower defect rate and Product B has a significantly higher defect rate relative to the incorrect aggregated average. These latter are also the more complex boards.

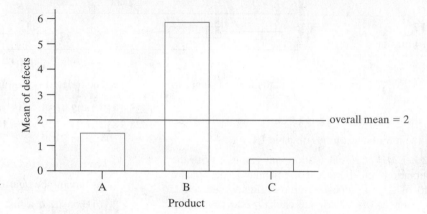

Figure 2.18
Average number of defects per product type

These data concern the number of defects that occurred when boards were wave-soldered after an assembly stage. The next step was to implement control charts for the number of defects for each of the three products. The numbers of defects for Product B were

$$
\begin{array}{cccccccccccccc}
10 & 8 & 8 & 4 & 6 & 8 & 8 & 10 & 6 & 7 & 4 & 2 & 4 & 5 & 5 \\
 & 5 & 2 & 11 & 6 & 6 & 5 & 7 & 3 & 4 & 3 & 2 & 6 & 5 & 1 & 7 \\
 & 3 & 1 & 1 & 5 & 4 & 5 & 12 & 13 & 11 & 8 & & & & &
\end{array}
$$

The appropriate control chart is a time plot where the serial numbers of the product or sample are on the horizontal axis and the corresponding number of defects on the vertical axis. In this C-chart, the central line labeled $\overline{C}$ is the average number of defects over all cases in the plot. The dashed lines are the control limits set at three standard deviations about the central line. (For reasons explained in Section 15.6, we use $\sqrt{\overline{C}}$ rather than s when the data are numbers of defects.)

$$\text{LCL} = \overline{C} - 3\sqrt{\overline{C}}$$
$$\text{UCL} = \overline{C} + 3\sqrt{\overline{C}}$$

Figure 2.19(a) gives a C-chart constructed for Product B, but where the centerline is incorrectly calculated from the aggregated data is $\overline{C} = 2.0$. This is far too low and so is the upper control limit 6.24. The lower control limit is negative so we use 0. It looks like a great many of the observations are out of control because they exceed the upper control limit.

When the C-chart is correctly constructed on the basis of data from Product B alone, the centerline is $\overline{C} = 231/40 = 5.775$ and the upper control limit is 12.98. The lower control limit is again negative so we use 0. From Figure 2.19(b), the correct C-chart, the wave soldering process for Product B appears to be in control except for time 38 when 13 defects were observed.

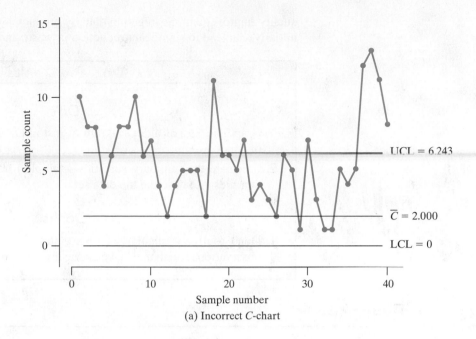

(a) Incorrect C-chart

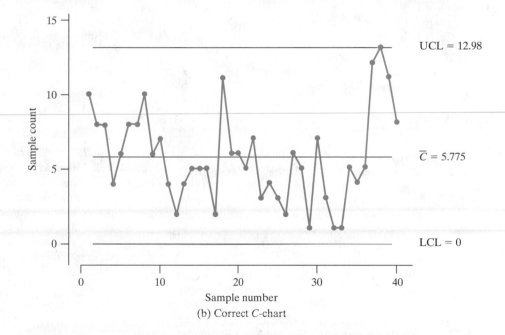

(b) Correct C-chart

Figure 2.19
C-charts for defects

With the data segregated into products, separate charts were constructed for each of the three products. With this new outlook on data interpretation, a number of improvement opportunities surfaced that were previously disguised by aggregation. For example, by reducing the dimensions of an electrical pad, a significant reduction was achieved in the number of solder bridges between pins. This same design change was added to all of the board specifications and improvements were obtained on all products.

In summary, the aggregation of data from different products, or more generally from different sources, can lead to incorrect conclusions and mask opportunities for quality improvement. Segregating data by product, although more time-consuming initially, can lead to significant reduction in waste and manufacturing costs.

Do's and Don'ts

Do's

1. Graph the data as a dot diagram or histogram to assess the overall pattern of data.
2. Calculate the summary statistics—sample mean, standard deviation, and quartiles—to describe the data set.

Don'ts

1. Don't routinely calculate summary statistics without identifying unusual observations which may have undue influence on the values of the statistics.

Review Exercises

2.60 From 2,000 computer chips inspected by the manufacturer, the following numbers of defects were recorded.

holes not open	182
holes too large	55
poor connections	31
incorrect size chip	5
other	7

Create a Pareto chart.

2.61 Create

(a) a frequency table of the aluminum alloy strength data on page 19 using the classes [66.0, 67.5), [67.5, 69.0), [69.0, 70.5), [70.5, 72.0), [72.0, 73.5), [73.5, 75.0), [75.0, 76.5), where the right-hand endpoint is excluded

(b) a histogram using the frequency table in part (a)

2.62 Create

(a) a frequency table of the interrequest time data on page 19 using the intervals [0, 2,500), [2,500, 5,000), [5,000, 10,000), [10,000, 20,000), [20,000, 40,000), [40,000, 60,000), [60,000, 80,000), where the left-hand endpoint is included but the right-hand endpoint is not

(b) a histogram using the frequency table in part (a) (Note that the intervals are unequal, so make the height of the rectangle equal to relative frequency divided by width.)

2.63 Direct evidence of Newton's universal law of gravitation was provided from a renowned experiment by Henry Cavendish (1731–1810). In the experiment, masses of objects were determined by weighing, and the measured force of attraction was used to calculate the density of the earth. The values of the earth's density, in time order by row, are

5.36	5.29	5.58	5.65	5.57	5.53	5.62	5.29
5.44	5.34	5.79	5.10	5.27	5.39	5.42	5.47
5.63	5.34	5.46	5.30	5.75	5.68	5.85	

(*Source*: *Philosophical Transactions* 17 (1798); 469.)

(a) Find the mean and standard deviation.

(b) Find the median, Q_1, and Q_3.

(c) Plot the observations versus time order. Is there any obvious trend?

2.64 J. J. Thomson (1856–1940) discovered the electron by isolating negatively charged particles for which he could measure the mass/charge ratio. This ratio appeared to be constant over a wide range of experimental conditions and, consequently, could be a characteristic of a new particle. His observations, from two different cathode-ray tubes that used air as the gas, are

Tube 1	0.57	0.34	0.43	0.32	0.48	0.40	0.40
Tube 2	0.53	0.47	0.47	0.51	0.63	0.61	0.48

(*Source*: *Philosophical Magazine* 44; 5 (1897): 293.)

(a) Draw a dot diagram with solid dots for Tube 1 observations and circles for Tube 2 observations.

(b) Calculate the mean and standard deviation for the Tube 1 observations.

(c) Calculate the mean and standard deviation for the Tube 2 observations.

2.65 With reference to Exercise 2.64,

(a) calculate the median, maximum, minimum, and range for Tube 1 observations;

(b) calculate the median, maximum, minimum, and range for the Tube 2 observations.

2.66 A. A. Michelson (1852–1931) made many series of measurements of the speed of light. Using a revolving mirror technique, he obtained

12   30   30   27   30   39   18   27   48   24   18

for the differences

$$(\text{velocity of light in air}) - (229,700) \text{ km/s}$$

(*Source: The Astrophysical Journal* 65 (1927): 11.)

(a) Create a dot diagram.

(b) Find the median and the mean. Locate both on the dot diagram.

(c) Find the variance and standard deviation.

2.67 With reference to Exercise 2.66,

(a) find the quartiles;

(b) find the minimum, maximum, range, and interquartile range;

(c) create a boxplot.

2.68 A civil engineer monitors water quality by measuring the amount of suspended solids in a sample of river water. Over 11 weekdays, she observed

14   12   21   28   30   63   29   63   55   19   20

suspended solids (parts per million).

(a) Create a dot diagram.

(b) Find the median and the mean. Locate both on the dot diagram.

(c) Find the variance and standard variation.

2.69 With reference to Exercise 2.68,

(a) find the quartiles;

(b) find the minimum, maximum, range, and interquartile range;

(c) construct a boxplot.

2.70 The weight (grams) of meat in a pizza product produced by a large meat processor is measured for a sample of $n = 20$ packages. The ordered values are (Courtesy of Dave Brauch)

16.12   16.77   16.87   16.91   16.96   16.99   17.02
17.19   17.20   17.26   17.36   17.39   17.39   17.62
17.63   17.76   17.85   17.86   17.91   19.00

(a) find the quartiles;

(b) find the minimum, maximum, range, and interquartile range;

(c) find the 10th percentile and 20th percentile.

2.71 With reference to Exercise 2.70, construct

(a) a boxplot.

(b) a modified boxplot.

2.72 With reference to the aluminum-alloy strength data in Example 7, make a stem-and-leaf display.

2.73 During the manufacture of hard disks, the height between the disk and the head, the headlift, must be controlled. One manufacturer recorded

0.239   0.246   0.245   0.243   0.239   0.241
0.248   0.246   0.243   0.242   0.251   0.246

Find the

(a) sample mean

(b) sample standard deviation

(c) coefficient of variation

(d) Measurements from a larger hard disk have sample mean 0.280 and standard deviation 0.05. Which is relatively more variable?

2.74 With reference to the lumber-strength data in Exercise 2.59, the statistical software package *SAS* produced the output in Figure 2.20. Using this output,

(a) identify the mean and standard deviation and compare these answers with the values given in Exercise 2.59.

(b) Create a boxplot.

The UNIVARIATE Procedure
Variable: Strength

Moments

N	30	Sum Weights	30
Mean	1908.76667	Sum Observations	57263
Std Deviation	327.115047	Variance	107004.254

Basic Statistical Measures

Location		Variability	
Mean	1908.767	Std Deviation	327.11505
Median	1863.000	Variance	107004
		Range	1658
		Interquartile Range	349.00000

Quantiles (Definition 5)

Level	Quantile
100% Max	2983.0
99%	2983.0
95%	2403.0
90%	2301.0
75% Q3	2061.0
50% Median	1863.0
25% Q1	1712.0
10%	1561.5
5%	1419.0
1%	1325.0
0% Min	1325.0

Figure 2.20   Selected SAS output to describe the lumber strength data from Exercise 2.59

2.75 Civil engineers must monitor flow on rivers where power is generated. The following are the daily mean flow rates in millions of gallons per day (MGD) on the Namekagon River during the month of May for 47 years.

602.0	517.5	572.5	392.4	505.8	547.5	389.1	497.2
794.8	657.6	904.7	595.5	611.9	482.9	698.6	606.7
986.4	567.7	400.1	634.9	448.4	479.1	1156.0	718.5
575.6	743.3	1146.0	461.6	644.0	480.8	429.1	626.9
833.9	889.0	752.6	516.5	817.2	895.8	572.2	563.7
679.3	738.0	618.9	390.8	550.9	425.9	760.6	

(a) Obtain the quartiles.

(b) Obtain the 90th percentile.

(c) Construct a histogram.

2.76 The National Highway Traffic Safety Administration reported the relative speed (rounded to the nearest 5 mph) of automobiles involved in accidents one year. The percentages at different speeds were

20 mph or less	2.0%
25 or 30 mph	29.7%
35 or 40 mph	30.4%
45 or 50 mph	16.5%
55 mph	19.2%
60 or 65 mph	2.2%

(a) From these data, can we conclude that it is safe to drive at high speeds? Why or why not?

(b) Why do most accidents occur in the 35 or 40 mph and in the 25 or 30 mph ranges?

(c) Construct a density histogram using the endpoints 0, 22.5, 32.5, 42.5, 52.5, 57.5, 67.5 for the intervals.

2.77 Given a five-number summary,

$$\text{minimum} \quad Q_1 \quad Q_2 \quad Q_3 \quad \text{maximum}$$

is it possible to determine whether or not an outlier is present? Explain.

2.78 Given a stem-and-leaf display, is it possible to determine whether or not an outlier is present? Explain.

2.79 Traversing the same section of interstate highway on 11 different days, a driver recorded the number of cars pulled over by the highway patrol:

$$0 \quad 1 \quad 3 \quad 0 \quad 2 \quad 0 \quad 1 \quad 0 \quad 2 \quad 1 \quad 0$$

(a) Create a dot plot.

(b) There is a long tail to the right. You might expect the sample mean to be larger than the median. Calculate the sample mean and median and compare the two measures of center. Comment.

2.80 An experimental study of the atomization characteristics of biodiesel fuel[5] was aimed at reducing the pollution produced by diesel engines. Biodiesel fuel is recyclable and has low emission characteristics. One aspect of the study is the droplet size (μm) injected into the engine, at a fixed distance from the nozzle. From data provided by the authors on droplet size, we consider a sample of size 41 that has already been ordered.

2.1	2.2	2.2	2.3	2.3	2.4	2.5	2.5	2.5
2.8	2.9	2.9	2.9	3.0	3.1	3.1	3.2	3.3
3.3	3.3	3.4	3.5	3.6	3.6	3.6	3.7	3.7
4.0	4.2	4.5	4.9	5.1	5.2	5.3	5.7	6.0
6.1	7.1	7.8	7.9	8.9				

(a) Group these droplet sizes and obtain a frequency table using [2, 3), [3, 4), [4, 5) as the first three classes, but try larger classes for the other cases. Here the left-hand endpoint is included but the right-hand endpoint is not.

(b) Construct a density histogram.

(c) Obtain $\bar{x}$ and s^2.

(d) Obtain the quartiles.

[5]H. Kim, H. Suh, S. Park, and C. Lee, An experimental and numerical investigation of atomization characteristics of biodiesel, dimethyl ether, and biodiesel-ethanol blended fuel, *Energy and Fuels*, **22** (2008), 2091–2098.

Key Terms

Absolute variation 28	Coefficient of variation 29	Frequency distribution 14
Arithmetic mean 24	Cumulative distribution 16	Histogram 17
Bar chart 12	Density histogram 19	Interquartile range 31
Boxplot 31	Deviation from the mean 27	Leaf 21
Categorical distribution 14	Dot diagram 13	Maximum 31
Class boundary 16	Double-stem display 24	Mean 25
Class frequency 15	Empirical cumulative distribution 23	Median 25
Class interval 16	Endpoint convention 15	Minimum 31
Class limit 14	Exploratory data analysis 22	Modified boxplot 31
Class mark 16	Five-stem display 24	Numerical distribution 14

3

PROBABILITY

In the study of probability there are basically three kinds of questions: (1) What do we mean when we say that the probability of an event is, say, 0.50, 0.02, or 0.81? (2) How are the numbers we call probabilities determined, or measured in actual practice? (3) What are the mathematical rules that probabilities must obey?

After some mathematical preliminaries in Sections 3.1 and 3.2, we study the first two kinds of questions in Section 3.3 and the third kind of question in Sections 3.4 through 3.7.

3.1 Sample Spaces and Events

Probability allows us to quantify the variability in the outcome of any experiment whose exact outcome cannot be predicted with certainty. However, before we can introduce probability, it is necessary to specify the space of outcomes and the events on which it will be defined.

In statistics, a set of all possible outcomes of an experiment is called a **sample space**, because it consists of all the things that can happen when one takes a sample. Sample spaces are usually denoted by a distinctive font $\mathcal{S}$. To avoid misunderstandings about the words *experiment* and *outcome* as we have used them here, it should be understood that statisticians use these terms in a very wide sense. An **experiment** may consist of the simple process of noting whether a switch is turned on or off; it may consist of determining the time it takes a car to accelerate to 30 miles per hour; or it may consist of the complicated process of finding the mass of a sub atomic particle. Thus, the **outcome** of an experiment may be a simple choice between two possibilities: it may be the result of a direct measurement or count, or it may be an answer obtained after extensive measurements and calculations.

When we study the outcomes of an experiment, we usually identify the various possibilities with numbers, points, or some other kinds of symbols. For instance, if four contractors bid on a highway construction job and we let a, b, c, and d denote that it is awarded to Mr. Adam, Mrs. Brown, Mr. Clark, or Ms. Dean, then the sample space for this experiment is the set $\mathcal{S} = \{a, b, c, d\}$.

Also, if a government agency must decide where to locate two new computer research facilities and that (for a certain purpose) it is of interest to indicate how many of them will be located in Texas and how many in California, we can write the sample space as

$$\mathcal{S} = \{(0, 0), (1, 0), (0, 1), (2, 0), (1, 1), (0, 2)\}$$

where the first coordinate is the number of research facilities that will be located in Texas and the second coordinate is the number that will be located in California. Geometrically, this sample space may be pictured as in Figure 3.1, from which it is apparent, for example, that in two of the six possibilities Texas and California will get an equal number of the new research facilities.

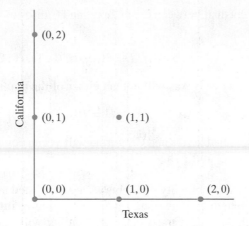

Figure 3.1

Sample space for the number of new computer research facilities to be located in Texas and in California

The use of points rather than letters or numbers has the advantage that it makes it easier to visualize the various possibilities, and perhaps discover some special features which several of the outcomes may have in common.

Generally, sample spaces are classified according to the number of elements (points) that they contain. In the two preceding examples, the sample spaces had four and six elements, and they are both referred to as **finite sample spaces**. Other examples of finite sample spaces are the one for the various ways in which a president and a vice president can be selected from among the 25 members of a union local and the one for the various ways in which a student can answer the 12 questions on a true-false test. As we see on page 51, the first of these sample spaces has 600 elements and the other has 4,096.

The following are examples of sample spaces that are not finite. If persons checking the nitrogen-oxide emission of cars are interested in the number of cars they have to inspect before they observe the first one that does not meet government regulations, it could be the first, the second, . . . , the fiftieth, . . . , and for all we know they may have to check thousands of cars before they find one that does not meet government regulations. Not knowing how far they may have to go, it is appropriate in an example like this to take as the sample space the whole set of natural numbers, of which there is a countable infinity. To go one step further, if they were interested in the nitrogen oxide emission of a given car in grams per mile, the sample space would have to consist of all the points on a continuous scale (a certain interval on the line of real numbers), of which there is a continuum.

In general, a sample space is said to be a **discrete sample space** if it has finitely many or a countable infinity of elements. If the elements (points) of a sample space constitute a continuum—for example, all the points on a line, all the points on a line segment, or all the points in a plane—the sample space is said to be a **continuous sample space**.

In the remainder of this chapter we shall consider only discrete and mainly finite sample spaces.

In statistics, any subset of a sample space is called an **event**. By subset we mean any part of a set, including the whole set and, trivially, a set called the **empty set** and denoted by ϕ, which has no elements at all. For instance, with reference to Figure 3.1,

$$C = \{(1, 0), (0, 1)\}$$

is the event that, between them, Texas and California will get one of the two research facilities,

$$D = \{(0, 0), (0, 1), (0, 2)\}$$

is the event that Texas will not get either of the two research facilities, and

$$E = \{(0, 0), (1, 1)\}$$

is the event that Texas and California will get an equal number of the facilities. Note that events C and E have no elements in common—they are **mutually exclusive events**.

In many probability problems we are interested in events which can be expressed in terms of two or more events by forming **unions, intersections**, and **complements**. Although the reader must surely be familiar with these terms, let us review briefly that if A and B are any two sets in a sample space $\mathcal{S}$, their union $A \cup B$ is the subset of $\mathcal{S}$ that contains all elements that are either in A, in B, or in both; their intersection $A \cap B$ is the subset of $\mathcal{S}$ that contains all elements that are in both A and B; and the complement $\overline{A}$ of A is the subset of $\mathcal{S}$ that contains all the elements of $\mathcal{S}$ that are not in A.

EXAMPLE 1   **Combining events by union, intersection, and complement**

With reference to the sample space of Figure 3.1 and the events C, D, and E just defined, list the outcomes comprising each of the following events and also express the events in words:

(a) $C \cup E$;
(b) $C \cap D$;
(c) $\overline{D}$.

Solution   (a) Since $C \cup E$ contains all the elements that are in C, in E, or in both,

$$C \cup E = \{(1, 0), (0, 1), (0, 0), (1, 1)\}$$

is the event that neither Texas nor California will get both of the new research facilities.

(b) Since $C \cap D$ contains all the elements that are in both C and D,

$$C \cap D = \{(0, 1)\}$$

is the event that Texas will not get either of the two new facilities and California will get one.

(c) Since $\overline{D}$ contains all the elements of the sample space that are not in D,

$$\overline{D} = \{(1, 0), (1, 1), (2, 0)\}$$

is the event that Texas will get at least one of the new computer research facilities.   ∎

Sample spaces and events, particularly relationships among events, are often depicted by means of **Venn diagrams** like those of Figures 3.2–3.4. In each case the sample space is represented by a rectangle, whereas events are represented by regions within the rectangle, usually by circles or parts of circles. The shaded regions of the four Venn diagrams of Figure 3.2 represent event A, the complement of event A, the union of events A and B, and the intersection of events A and B.

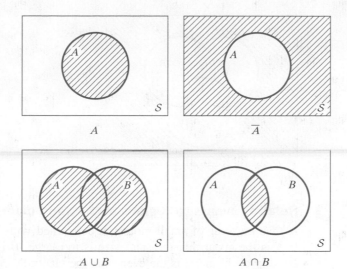

Figure 3.2
Venn diagrams showing complement, union, and intersection

A $\overline{A}$

$A \cup B$ $A \cap B$

EXAMPLE 2   **Relation of regions in Venn diagrams to events**

If A is the event that a certain student is taking a course in calculus and B is the event that the student is taking a course in applied mechanics, what events are represented by the shaded regions of the four Venn diagrams of Figure 3.2?

Solution   The shaded region of the first diagram represents the event that the student is taking a course in calculus. That of the second diagram represents the event that the student is not taking a course in calculus. The shaded area of the third diagram represents the event that the student is taking a course in calculus and/or a course in applied mechanics. Finally, that of the fourth diagram represents the event that the student is taking a course in calculus as well as a course in applied mechanics. ∎

Venn diagrams are often used to verify relationships among sets, thus making it unnecessary to give formal proofs based on the algebra of sets. To illustrate, let us show that $\overline{A \cup B} = \overline{A} \cap \overline{B}$, which expresses the fact that the complement of the union of two sets equals the intersection of their complements. To begin, note that the shaded region of the first Venn diagram of Figure 3.3 represents the set $\overline{A \cup B}$ (compare this diagram with the third diagram of Figure 3.2). The cross-hatched region of the second Venn diagram of Figure 3.3 was obtained by shading the region representing $\overline{A}$ with lines going in one direction and that representing $\overline{B}$ with lines going in another direction. Thus, the cross-hatched region represents the intersection of $\overline{A}$ and $\overline{B}$. Clearly, the cross-hatched area is identical with the shaded region of the first Venn diagram of Figure 3.3.

When we deal with three events, we draw the circles as in Figure 3.4. In this diagram, the circles divide the sample space into eight regions, numbered 1 through 8, and it is easy to determine whether the corresponding events are parts of A or $\overline{A}$, B or $\overline{B}$, and C or $\overline{C}$.

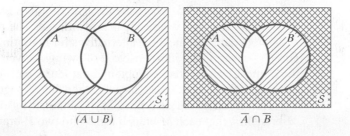

Figure 3.3
Use of Venn diagrams to show that $\overline{A \cup B} = \overline{A} \cap \overline{B}$

$\overline{(A \cup B)}$ $\overline{A} \cap \overline{B}$

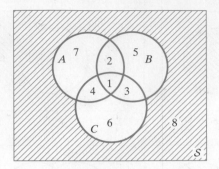

Figure 3.4
Venn diagram

EXAMPLE 3   **Relating events to regions of the Venn diagram**

A manufacturer of small motors is concerned with three major types of defects. If A is the event that the shaft size is too large, B is the event that the windings are improper, and C is the event that the electrical connections are unsatisfactory, express in words what events are represented by the following regions of the Venn diagram of Figure 3.4:

(a) region 2
(b) regions 1 and 3 together
(c) regions 3, 5, 6, and 8 together

Solution   (a) Since this region is contained in A and B but not in C, it represents the event that the shaft is too large and the windings improper, but the electrical connections are satisfactory.

(b) Since this region is common to B and C, it represents the event that the windings are improper and the electrical connections are unsatisfactory.

(c) Since this is the entire region outside A, it represents the event that the shaft size is not too large. ∎

3.2 Counting

At times it can be quite difficult, or at least tedious, to determine the number of elements in a finite sample space by direct enumeration. To illustrate, suppose all newer used cars in a large city can be classified as low, medium, or high current mileage; moderate or high priced; and be inexpensive, average, or expensive to operate. In how many ways can a used car be categorized?

Clearly, there are many possibilities; a used car can have low current mileage, be moderately priced, and be inexpensive to operate; have neither low or high mileage, be high priced, and be average cost to operate; and so on. Continuing in this way, we may be able to list all 18 possibilities, but the chances are that we will omit at least one or two.

To handle this kind of problem systematically, it helps to draw a **tree diagram** like that of Figure 3.5, where the three alternatives for current mileage are denoted by M_1, M_2, and M_3, where M_1 is low mileage. The price is either P_1 or P_2, where P_1 is moderate; and the three alternatives for operating costs are denoted by C_1, C_2, and C_3, where C_1 is inexpensive. Following a given path from left to right along the branches of the tree, we obtain a particular categorization, namely a particular element of the sample space. It can be seen that all together there are 18 possibilities.

This result could also have been obtained by observing that there are three M-branches, that each M-branch forks into two P-branches, and that each P-branch

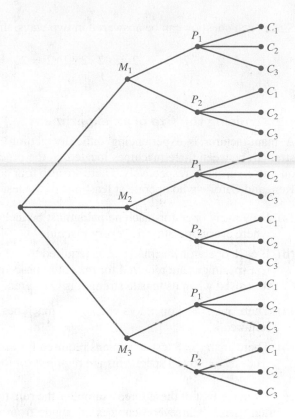

Figure 3.5
Tree diagram for used cars

forks into three C-branches. Thus, there are $3 \cdot 2 \cdot 3 = 18$ combinations of branches, or paths. This result is a special case of the following theorem often called the *fundamental theorem of counting*.

Multiplication of choices

> **Theorem 3.1**   If sets $A_1, A_2, \cdots, A_k$ contain, respectively, $n_1, n_2, \cdots, n_k$ elements, there are $n_1 \cdot n_2 \cdots n_k$ ways of choosing first an element of A_1, then an element of $A_2, \cdots$, and finally an element of A_k.

In our example we had $n_1 = 3$, $n_2 = 2$, and $n_3 = 3$, and hence, $3 \cdot 2 \cdot 3 = 18$ possibilities.

EXAMPLE 4   **The multiplication rule for $k = 2$ stages of choices**

In how many different ways can a union local with a membership of 25 choose a vice president and a president?

Solution   Since the vice president can be chosen in 25 ways and, subsequently, the president in 24 ways, there are altogether $25 \cdot 24 = 600$ ways in which the whole choice can be made.   ∎

EXAMPLE 5   **The multiplication rule with $k = 12$ stages of choices**

If a test consists of 12 true-false questions, in how many different ways can a student mark the test paper with one answer to each question?

Solution   Since each question can be answered in two ways, there are all together

$$2 \cdot 2 \cdot 2 \cdot 2 \cdot 2 \cdot 2 \cdot 2 \cdot 2 \cdot 2 \cdot 2 \cdot 2 \cdot 2 = 2^{12} = 4{,}096 \text{ possibilities}$$ ∎

EXAMPLE 6   **Determining the size of an experiment**

A manufacturer is experiencing difficulty getting consistent readings of tensile strength between three machines located on the production floor, research lab, and quality control lab, respectively. There are also four possible technicians—Tom, Joe, Ken, and Carol—who operate at least one of the test machines regularly.

(a) How many operator-machine pairs must be included in a designed experiment where every operator tries every machine?

(b) If each operator-machine pair is required to test eight specimens, how many test specimens are required for the entire procedure? Note: A specimen is destroyed when its tensile strength is measured.

Solution   (a) There are $n_1 = 4$ operators and $n_2 = 3$ machines, so $4 \cdot 3 = 12$ pairs are required.

(b) There are $n_3 = 8$ test specimens required for each operator-machine pair, so $8 \cdot 12 = 96$ test specimens are required for the designed experiment.   ∎

As in the first of these three examples, the rule for the multiplication of choices is often used when several choices are made from one set and we are concerned with the order in which they are made. In general, if r objects are chosen from a set of n distinct objects, any particular arrangement, or order, of these objects is called a **permutation**. For instance, 4 1 2 3 is a permutation of the first four positive integers, and Maine, Vermont, and Connecticut is a permutation, a particular ordered arrangement, of three of the six New England states.

To find a formula for the total number of permutations of r objects selected from a set of n distinct objects, we observe that the first selection is made from the whole set of n objects, the second selection is made from the $n - 1$ objects which remain after the first selection has been made, ..., and the rth selection is made from the $n - (r - 1) = n - r + 1$ objects which remain after the first $r - 1$ selections have been made. Therefore, by the rule for the multiplication of choices, the total number of permutations of r objects selected from a set of n distinct objects is

$$_nP_r = n(n - 1)(n - 2) \cdots (n - r + 1)$$

for $r = 1, 2, \ldots, n$.

Since products of consecutive integers arise in many problems relating to permutations or other kinds of special selections, it will be convenient to introduce here the **factorial notation**, where $1! = 1$, $2! = 2 \cdot 1 = 2$, $3! = 3 \cdot 2 \cdot 1 = 6$, $4! = 4 \cdot 3 \cdot 2 \cdot 1 = 24$.

factorial notation

For any integer n, **n factorial** is defined as

$$n! = n(n - 1)(n - 2) \cdots 2 \cdot 1$$

Also, to make various formulas more generally applicable, we let $0! = 1$ by definition.

To express the formula for $_nP_r$ in terms of factorials, we multiply and divide by $(n-r)!$, getting

$$_nP_r = \frac{n(n-1)(n-2)\cdots(n-r+1)(n-r)!}{(n-r)!} = \frac{n!}{(n-r)!}$$

To summarize:

Number of permutations of n objects taken r at a time

> **Theorem 3.2**   The number of permutations of r objects selected from a set of n distinct objects is
>
> $$_nP_r = n(n-1)(n-2)\cdots(n-r+1)$$
>
> or, in factorial notation,
>
> $$_nP_r = \frac{n!}{(n-r)!}$$

Note that the second formula also holds for $r=0$.

EXAMPLE 7   **The number of ways to assemble chips in a controller**

An electronic controlling mechanism requires 5 distinct, but interchangeable, memory chips. In how many ways can this mechanism be assembled

(a) by placing the 5 chips in the 5 positions within the controller?

(b) by placing 3 chips in the odd numbered positions within the controller?

Solution

(a) When all 5 chips must be placed, the answer is 5!. Alternatively, in the permutation notation with $n=5$ and $r=5$, the first formula yields

$$_5P_5 = 5 \cdot 4 \cdot 3 \cdot 2 \cdot 1 = 120$$

and the second formula yields

$$_5P_5 = \frac{5!}{(5-5)!} = \frac{5!}{0!} = 5! = 120$$

The first formula for $_nP_r$ is generally easier to use unless we can use a calculator which directly yields factorials and/or ratios of factorials.

(b) For $n=5$ chips placed in $r=3$ positions, the permutation is

$$_5P_3 = \frac{5!}{2!} = \frac{5 \cdot 4 \cdot 3 \cdot 2 \cdot 1}{2 \cdot 1} = 5 \cdot 4 \cdot 3 = 60$$

[Using **R**: (a) **factorial(5)** (b) **factorial(5) / factorial (2)**]   ■

There are many problems in which we must find the number of ways in which r objects can be selected from a set of n objects, but we do not care about the order in which the selection is made. For instance, we may want to know in how many ways 3 of 20 laboratory assistants can be chosen to assist with an experiment. In general, there are $r!$ permutations of any r objects we select from a set of n distinct objects. So, the $_nP_r$ permutations of r objects, selected from a set of n objects, contains each set of r objects $r!$ times. Therefore, to find the number of ways in which r objects can be selected from a set of n distinct objects, also called the number of **combinations** of n objects taken r at a time and denoted by $_nC_r$ or $\binom{n}{r}$, we divide $_nP_r$ by $r!$ and get

---

Number of combinations of n objects taken r at a time

Theorem 3.3   The number of ways in which r objects can be selected from a set of n distinct objects is

$$\binom{n}{r} = \frac{n(n-1)(n-1)\cdots(n-r+1)}{r!}$$

or, in factorial notation,

$$\binom{n}{r} = \frac{n!}{r!(n-r)!}$$

EXAMPLE 8   **Evaluating a combination**

In how many different ways can 3 of 18 automotive engineers be chosen for a team to develop a new ceramic diesel engine.

Solution   For $n = 18$ and $r = 3$, the first formula for $\binom{n}{r}$ yields

$$\binom{18}{3} = \frac{18 \cdot 17 \cdot 16}{3!} = 816$$

∎

EXAMPLE 9   **Selection of machines for an experiment**

A calibration study needs to be conducted to see if the readings on 15 test machines are giving similar results. In how many ways can 3 of the 15 be selected for the initial investigation?

Solution
$$\binom{15}{3} = \frac{15 \cdot 14 \cdot 13}{3 \cdot 2 \cdot 1} = 455 \text{ ways}$$

Note that selecting which 3 machines to use is the same as selecting which 12 not to include. That is, according to the second formula,

$$\binom{15}{12} = \frac{15!}{12! \, 3!} = \frac{15!}{3! \, 12!} = \binom{15}{3}$$

∎

EXAMPLE 10   **The number of choices of new researchers**

In how many different ways can the director of a research laboratory choose 2 chemists from among 7 applicants and 3 physicists from among 9 applicants?

Solution   The 2 chemists can be chosen in $\binom{7}{2} = 21$ ways and the 3 physicists can be chosen in $\binom{9}{3} = 84$ ways. By the multiplication rule, the whole selection can be made in $21 \cdot 84 = 1,764$ ways.

∎

Exercises

3.1 An environmental engineer suspects mercury contamination in an area which contains three lakes and two streams. He will check all five for mercury contamination.

(a) Express each outcome using two coordinates, so (2, 1), for example, represents the event that two of the lakes and one of the streams will be contaminated. Draw a diagram similar to that of Figure 3.1 showing the 12 outcomes in the sample space.

(b) If R is the event that equally many lakes and streams are contaminated, T is the event that none of the streams is contaminated, and U is the event that fewer lakes than streams are contaminated, express each of these events symbolically by listing its elements.

3.2 With reference to Exercise 3.1, which of the three pairs of events, R and T, R and U, and T and U, are mutually exclusive?

3.3 With reference to Exercise 3.1, list the outcomes comprising each of the following events, and also express the events in words.

(a) $R \cup U$

(b) $R \cap T$

(c) $\overline{T}$

3.4 With reference to the sample space of Figure 3.1, express each of the following events in words.

(a) $F = \{(1, 0), (1, 1)\}$

(b) $G = \{(0, 2), (1, 1), (2, 0)\}$

(c) $F \cap G$

3.5 To construct sample spaces for experiments in which we deal with nonnumerical data, we often code the various alternatives by assigning them numbers. For instance, if a mechanic is asked whether work on a certain model car is very easy, easy, average, difficult, or very difficult, we might assign these alternatives the codes, 1, 2, 3, 4, and 5. If $A = \{3, 4\}$, $B = \{2, 3\}$, and $C = \{4, 5\}$, express each of the following symbolically by listing its elements and also in words.

(a) $A \cup B$

(b) $A \cap B$

(c) $A \cup \overline{B}$

(d) $\overline{C}$

3.6 With reference to Exercise 3.5, which of the three pairs of events, A and B, A and C, and B and C, are mutually exclusive?

3.7 Two professors and 3 graduate assistants are responsible for the supervision of a physics lab, and at least one professor and one graduate assistant have to be present at all times.

(a) Using two coordinates so that (1, 3), for example, represents the event that one professor and 3 graduate assistants are present, draw a diagram similar to that of Figure 3.1 showing the points of the corresponding sample space.

(b) Describe in words the events which are represented by $B = \{(1, 3), (2, 3)\}$, $C = \{(1, 1), (2, 2)\}$, and $D = \{(1, 2), (2, 1)\}$.

(c) With reference to part (b), express $C \cup D$ symbolically by listing its elements, and also express this event in words.

(d) With reference to part (b), are B and D mutually exclusive?

3.8 For each of the following experiments, decide whether it would be appropriate to use a sample space which is finite, countably infinite, or continuous.

(a) A Geiger counter, located adjacent to a building containing a reactor, will record the total number of alpha particles during a one-hour period.

(b) Five of the members of a professional society with 12,600 members are chosen to serve on a nominating committee.

(c) An experiment is conducted to measure the thickness of a new synthetic silk thread in nanometers.

(d) A study is made to determine in how many of 450 airplane accidents the main cause is pilot error.

(e) Measurements are made to determine the uranium content of a certain ore.

(f) In a torture test, a watch is dropped a number of times from a tall building until it stops running.

3.9 In Figure 3.6, C is the event that an ore contains copper and U is the event that it contains uranium. Explain in

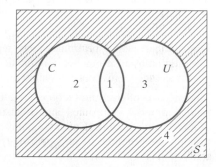

Figure 3.6   Venn diagram for Exercises 3.9 and 3.10

words what events are represented by regions 1, 2, 3, and 4.

3.10 With reference to Exercise 3.9, what events are represented by

(a) regions 1 and 3 together;

(b) regions 3 and 4 together;

(c) regions 1, 2, and 3 together?

3.11 With reference to Figure 3.4, what events are represented by

(a) region 5;

(b) regions 4 and 6 together;

(c) regions 7 and 8 together;

(d) regions 1, 2, 3 and 5 together?

3.12 With reference to Figure 3.4, what regions or combinations of regions represent the events that a motor will have

(a) none of the major defects;

(b) a shaft that is large and windings improper;

(c) a shaft that is large and/or windings improper but the electrical connections are satisfactory;

(d) a shaft that is large and the windings improper and/or the electrical connections are unsatisfactory?

3.13 Use Venn diagrams to verify that

(a) $\overline{A \cap B} = \overline{A} \cup \overline{B}$

(b) $A \cup (A \cap B) = A$

(c) $(A \cap B) \cup (A \cap \overline{B}) = A$

(d) $A \cup B = (A \cap B) \cup (A \cap \overline{B}) \cup (\overline{A} \cap B)$

(e) $A \cup (B \cap C) = (A \cup B) \cap (A \cup C)$

3.14 A building inspector has to check the wiring in a new apartment building either on Monday, Tuesday, Wednesday, or Thursday, and at 8 A.M., 1 P.M., or 2 P.M. Draw a tree diagram which shows the various ways in which the inspector can schedule the inspection of the wiring of the new apartment building.

3.15 If the five finalists in an international volleyball tournament are Spain, the United States, Uruguay, Portugal, and Japan, draw a tree diagram that shows the various possible first- and second-place finishers.

3.16 If a number cannot be immediately repeated, how many different three number combinations are possible for a combination lock with numbers 0, 1, ..., 29.

3.17 Students are offered three cooperative training programs at local companies and four training programs outside the state. Count the number of possible training opportunities if an opportunity consists of training at

(a) one local company or one company outside of the state.

(b) one local company and one company outside of the state.

3.18 You are required to choose a four digit personal identification number (PIN) for a new debit card. Each digit is selected from 0, 1, ..., 9. How many choices do you have.

3.19 One engineering group consists of 6 men and 4 women.

(a) How many different project teams can be formed consisting of 2 men and 2 women?

(b) If 2 women have the same boyfriend and refuse to be on the same team together, how many different project teams can be formed consisting of 2 men and 2 women?

3.20 If there are 9 cars in a race, in how many different ways can they place first, second, and third?

3.21 In how many ordered ways can a television director schedule 6 different commercials during the 6 time slots allocated to commercials during the telecast of the first period of a hockey game?

3.22 If among n objects k are alike and the others are all distinct, the number of permutations of these n objects taken all together is $n!/k!$.

(a) How many permutations are there of the letters of the word *class*?

(b) In how many ways can the television director of Exercise 3.21 fill the 6 time slots allocated to commercials, if there are 4 different commercials, of which a given one is to be shown 3 times while each of the others is to be shown once?

3.23 Determine the number of ways in which a manufacturer can choose 2 of 15 locations for a new warehouse.

3.24 How many ways can a company select 4 candidates to interview from a short list of 12 engineers?

3.25 A carton of 12 rechargeable batteries contains one that is defective. In how many ways can an inspector choose 3 of the batteries and

(a) get the one that is defective;

(b) not get the one that is defective?

3.26 With reference to Exercise 3.25, suppose that two of the batteries are defective. In how many ways can the inspector choose 3 of the batteries and get

(a) none of the defective batteries;

(b) one of the defective batteries;

(c) both of the defective batteries?

3.27 The supply department has 8 different electric motors and 5 different starting switches. In how many ways can 2 motors and 2 switches be selected for an experiment concerning a tracking antenna?

3.3 Probability

So far we have studied only what is possible in a given situation. Now we go one step further and judge also what is probable and what is improbable. Historically, the oldest way of measuring uncertainties is the **classical probability concept**, which was developed originally in connection with games of chance. It applies when all possible outcomes are equally likely.

The classical probability concept

> If there are m equally likely possibilities, of which one must occur and s are regarded as favorable, or as a "success," then the probability of a "success" is given by $\dfrac{s}{m}$.

In the application of this rule, the terms *favorable* and *success* are used rather loosely—favorable may mean that a television set does not work and success may mean that someone catches the flu.

EXAMPLE 11

Well-shuffled cards are equally likely to be selected

What is the probability of drawing an ace from a well-shuffled deck of 52 playing cards?

Solution   There are $s = 4$ aces among the $m = 52$ cards, so we get

$$\frac{s}{m} = \frac{4}{52} = \frac{1}{13}$$

∎

Although equally likely possibilities are found mostly in games of chance, the classical probability concept applies also to a great variety of situations where gambling devices are used to make random selections. They occur when offices are assigned to research assistants by lot, when laboratory animals are chosen for an experiment so that each one has the same chance of being selected, or when washing-machine parts are chosen for inspection so that each part produced has the same chance of being selected.

EXAMPLE 12

Random selection results in the equally likely case

The next generation of miniaturized wireless capsules with active locomotion will require two miniature electric[1] motors to maneuver each capsule. Suppose 10 motors have been fabricated but that, in spite of tests performed on the individual motors, 2 will not operate satisfactorily when placed into a capsule.

To fabricate a new capsule, 2 motors will be randomly selected (that is, each pair of motors has the same chance of being selected). Find the probability that

(a) both motors will operate satisfactorily in the capsule

(b) one motor will operate satisfactorily and the other will not

Solution   **(a)** There are $\dbinom{10}{2} = 45$ equally likely ways of choosing 2 of 10 motors, so $m = 45$.

---

[1] M. Quirini et al, Design and fabrication of a motor legged capsule for the active exploration of the gastrointestinal tract, *IEEE/ASME Transactions on Mechatronics* (2008), **13**, 169–179.

The number of favorable outcomes is the number of ways in which two good motors can be selected from eight:

$$s = \binom{8}{2} = 28$$

so the probability that both motors will operate satisfactorily in the capsule is

$$\frac{s}{m} = \frac{28}{45}$$

or approximately 0.622.

(b) The number of favorable outcomes is the number of ways in which one satisfactory motor and one unsatisfactory motor can be selected, or

$$s = \binom{8}{1}\binom{2}{1} = 8 \cdot 2 = 16$$

It follows that the probability is

$$\frac{s}{m} = \frac{16}{45} = 0.356$$ ∎

A major shortcoming of the classical probability concept is its limited applicability, for there are many situations in which the various possibilities cannot all be regarded as equally likely. This would be the case, for example, if we are concerned with the question of whether it will rain the next day, whether a missile launching will be a success, whether a newly designed engine will function for at least 1,000 hours, or whether a certain candidate will win an election.

Among the various probability concepts, most widely held is the **frequency interpretation**.

The frequency interpretation of probability

> The probability of an event (or outcome) is the proportion of times the event will occur in a long run of repeated experiments.

If we say that the probability is 0.78 that a jet from New York to Boston will arrive on time, we mean that such flights arrive on time 78% of the time. Also, if the Weather Service predicts that there is a 40% chance for rain (that the probability is 0.40), this means that under the same weather conditions it will rain 40% of the time.

We illustrate the long run behavior of relative frequency by performing an experiment where an event A occurs with probability 0.4. This experiment could be as simple of reading a random digit from Table 7W and deciding the event has occurred if 1, 2, 3, or 4 are selected. Instead, we use computer software to generate a 1, with probability 0.4, to indicate that A occurs and a 0 otherwise. We then repeat this experiment a large number of times. After each time, or trial, we calculate the relative frequency of A. Let r_N be the relative frequency of an event A after the experiment has been performed N times.

$$r_N = \frac{\text{Number of times } A \text{ occurs in } N \text{ trials}}{N}.$$

In our sequence of experiments, the event does occur on the first trial and third trial but not of the second. The first three relative frequencies are then 1, 0.5, and 0.667.

Figure 3.7 displays the typical behavior of r_N as the number of repetitions grows. This relative frequency begins to stabilize for large N. Figure 3.7 actually has two parts. Figure 3.7(a) shows the results for the first 50 trials and the fluctuations are

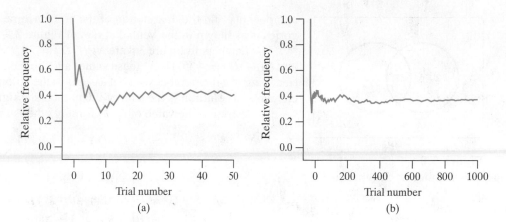

Figure 3.7
Relative frequency stabilizes
after many trials.

quite large. Figure 3.7(b) shows 1000 trials where it is clear that the fluctuations in r_N become damped with increasing N. Even with 1000 trials, the relative frequency is approaching the probability 0.4.

This behavior of relative frequency, after many repeated trials of an experiment, is a key fundamental in statistics. You are encouraged to conduct your own computer based simulation experiment (See Exercise 3.100 for the *MINITAB* and Appendix C for the R commands).

In accordance with the frequency interpretation of probability, we will this change estimate the probability of an event by observing what fraction of the time similar events have occurred in the past.

EXAMPLE 13   **Long-run relative frequency approximation to probability**

If records show that 294 of 300 ceramic insulators tested were able to withstand a certain thermal shock, what is the probability that any one untested insulator will be able to withstand the thermal shock?

Solution   Among the insulators tested, $\dfrac{294}{300} = 0.98$ were able to withstand the thermal shock, and we use this figure as an estimate of the probability. ∎

An alternative point of view is to interpret probabilities as *personal* or *subjective evaluations*. Such **subjective probabilities** express the strength of one's belief with regard to the uncertainties that are involved, and they apply especially when there is little or no direct evidence, so that there is no choice but to consider collateral (indirect) evidence, educated guesses, and perhaps intuition and other subjective factors. Subjective probabilities are best determined by referring to risk taking, or betting situations, as will be explained in Exercise 3.53.

3.4   The Axioms of Probability

In this section we define probabilities mathematically as the values of **additive set functions**. Since the reader is probably most familiar with functions for which the elements of the domain and the range are all numbers, let us first give a very simple example where the elements of the domain are sets, while the elements of the range are nonnegative integers, namely, a **set function** that assigns to each subset A of a finite sample space S the number of elements in A, written $N(A)$. Suppose that 500 machine parts are inspected before they are shipped, that I denotes that a machine part is improperly assembled, D denotes that it contains one or more defective

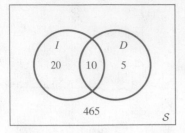

Figure 3.8
Classification of 500 machine parts

components, and the distribution of the 500 machine parts among the various categories is as shown in the Venn diagram of Figure 3.7.

The numbers in Figure 3.8 are $N(I \cap \overline{D}) = 20$, $N(I \cap D) = 10$, $N(\overline{I} \cap D) = 5$, and $N(\overline{I} \cap \overline{D}) = 465$. Using these values and the fact that the set function is *additive* (meaning that the number which it assigns to the union of two subsets which have no elements in common is the sum of the numbers assigned to the individual subsets), we can determine the value of $N(A)$ for any other subset A of S. For instance,

$$N(\overline{I}) = N(\overline{I} \cap D) + N(\overline{I} \cap \overline{D}) = 5 + 465 = 470$$
$$N(I \cup D) = N(I \cap \overline{D}) + N(I \cap D) + N(\overline{I} \cap D)$$
$$= 20 + 10 + 5 = 35$$
$$N(\overline{I} \cup D) = N(I \cap D) + N(\overline{I} \cap D) + N(\overline{I} \cap \overline{D})$$
$$= 10 + 5 + 465 = 480$$

and

$$N(D) = N(I \cap D) + N(\overline{I} \cap D) = 10 + 5 = 15$$

Using the concept of an additive set function, let us now explain what we mean by the probability of an event. Given a finite sample space S and an event A in S, we define $P(A)$, the probability of A, to be a value of an additive set function that satisfies the following three conditions.

The axioms of probability for a finite sample space

Axiom 1 $0 \leq P(A) \leq 1$ for each event A in S.
Axiom 2 $P(S) = 1$.
Axiom 3   If A and B are mutually exclusive events in S, then

$$P(A \cup B) = P(A) + P(B)$$

The first axiom states that probabilities are real numbers on the interval from 0 to 1, inclusive. The second axiom states that the sample space as a whole is assigned a probability of 1. Since S contains all possible outcomes, and one of these must always occur, S is certain to occur. The third axiom states that probability functions must be additive—the probability of the union is the sum of the two probabilities when the two events have no outcomes in common.

Axioms for a mathematical theory require no proof, but if such a theory is to be applied to the physical world, we must show somehow that the axioms are "realistic." Thus, let us show that the three postulates are consistent with the classical probability concept and the frequency interpretation.

So far as the first axiom is concerned, fractions of the form $\frac{s}{m}$, where $0 \leq s \leq m$ and m is a positive integer, cannot be negative or exceed 1, and the same is true also for the proportion of the time that an event will occur. To show that the second axiom is consistent with the classical probability concept and the frequency interpretation for a long series of repeated experiments, we have only to observe that for the whole sample space

$$P(S) = \frac{m}{m} = 1$$

and for the frequency interpretation that some outcome must happen 100% of the time.

So far as the third axiom is concerned, if

$$P(A) = \frac{s_1}{m}, \ P(B) = \frac{s_2}{m}$$

and A and B are mutually exclusive, then

$$P(A \cup B) = \frac{s_1 + s_2}{m} = P(A) + P(B)$$

Also, if one event occurs in proportion 0.36 or 36% of the time, another event occurs 41% of the time, and the two events are mutually exclusive, then one or the other will occur in proportion $0.36 + 0.41 = 0.77$ or 77%.

Before we go any further, it is important to stress the point that the axioms of probability do not tell us how to assign probabilities to the various outcomes of an experiment, they merely restrict the ways in which it can be done. In actual practice, probabilities are assigned on the basis of past experience, on the basis of a careful analysis of conditions underlying the experiment, on the basis of subjective evaluations, or on the basis of assumptions—say, the common assumption that all the outcomes are equiprobable.

EXAMPLE 14   **Checking possible assignments of probability**
If an experiment has the three possible and mutually exclusive outcomes A, B, and C, check in each case whether the assignment of probabilities is permissible:

(a) $P(A) = \frac{1}{3}$, $P(B) = \frac{1}{3}$, and $P(C) = \frac{1}{3}$
(b) $P(A) = 0.64$, $P(B) = 0.38$, and $P(C) = -0.02$
(c) $P(A) = 0.35$, $P(B) = 0.52$, and $P(C) = 0.26$
(d) $P(A) = 0.57$, $P(B) = 0.24$, and $P(C) = 0.19$

Solution   (a) The assignment of probabilities is permissible because the values are all on the interval from 0 to 1, and their sum is $\frac{1}{3} + \frac{1}{3} + \frac{1}{3} = 1$.
(b) The assignment is not permissible because $P(C)$ is negative.
(c) The assignment is not permissible because $0.35 + 0.52 + 0.26 = 1.13$, which exceeds 1.
(d) The assignment is permissible because the values are all on the interval from 0 to 1 and their sum is $0.57 + 0.24 + 0.19 = 1$.   ∎

The approach in the last example extends to any experiment where the sample space S is discrete so the outcomes can be arranged in a sequence. An amount of probability p_i is assigned to the ith outcome, where

$$0 \le p_i \quad \text{and} \quad \sum_{\text{all outcomes in } S} p_i = 1$$

and then the probability of any event A is defined as

$$P(A) = \sum_{\text{all outcomes in } A} p_i$$

When probability is assigned in this manner, the axioms of probability are always satisfied.

Intuitively, we can think of the scientist as starting with a unit amount of clay (probability) and placing a proportion p_1 on the first outcome, p_2 on the second outcome, and so on. Some outcomes can be assigned a large amount and others lesser amounts. The total unit amount of clay (probability) is assigned to the outcomes in the sample space. Then, an event A is assigned the total of all the clay (probability) assigned to each outcome in A.

3.5 Some Elementary Theorems

With the use of mathematical induction, the third axiom of probability can be extended to include any number of mutually exclusive events; in other words, the following can be shown.

Generalization of the third axiom of probability

> **Theorem 3.4**   If $A_1, A_2, \ldots, A_n$ are mutually exclusive events in a sample space $\mathcal{S}$, then
>
> $$P(A_1 \cup A_2 \cup \cdots \cup A_n) = P(A_1) + P(A_2) + \cdots + P(A_n)$$

In the next chapter we shall see how the third axiom of probability must be modified so that the axioms apply also to sample spaces which are not finite.

EXAMPLE 15   **Probabilities add for mutually exclusive events**

The probability that a consumer testing service will rate a new antipollution device for cars very poor, poor, fair, good, very good, or excellent are 0.07, 0.12, 0.17, 0.32, 0.21, and 0.11. What are the probabilities that it will rate the device

(a) very poor, poor, fair, or good;

(b) good, very good, or excellent?

Solution   Since the probabilities are all mutually exclusive, direct substitution into the formula of Theorem 3.4 yields

$$0.07 + 0.12 + 0.17 + 0.32 = 0.68$$

for part (a) and

$$0.32 + 0.21 + 0.11 = 0.64$$

for part (b).   ∎

As it can be shown that a sample space of n points (outcomes) has 2^n subsets, it would seem that the problem of specifying a probability function (namely, a probability for each subset or event) can easily become very tedious. Indeed, for $n = 20$ there are already more than 1 million possible events. Fortunately, this task can be simplified considerably by the use of the following theorem:

Rule for calculating probability of an event

> **Theorem 3.5**   If A is an event in the finite sample space $\mathcal{S}$, then $P(A)$ equals the sum of the probabilities of the individual outcomes comprising A.

To prove this theorem, let $E_1, E_2, \ldots, E_n$ be the n outcomes comprising event A, so that we can write $A = E_1 \cup E_2 \cup \cdots \cup E_n$. Since the E's are individual outcomes, they are mutually exclusive, and by Theorem 3.4 we have

$$P(A) = P(E_1 \cup E_2 \cup \cdots \cup E_n)$$
$$= P(E_1) + P(E_2) + \cdots + P(E_n)$$

which completes the proof.

EXAMPLE 16   **Using a Venn diagram to visualize probability calculations**

Refer to the used car classification example on page 50. Suppose that the probabilities of the 18 outcomes are as shown in Figure 3.9 (which, except for the the probabilities, is identical to Figure 3.5).

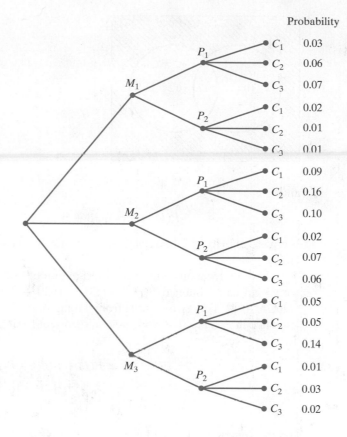

Figure 3.9
Used car classifications and
their probabilities

Find $P(M_1)$, $P(P_1)$, $P(C_3)$, $P(M_1 \cap P_1)$, and $P(M_1 \cap C_3)$.

Solution   Adding the probabilities of the outcomes comprising the respective events, we get

$$P(M_1) = 0.03 + 0.06 + 0.07 + 0.02 + 0.01 + 0.01 = 0.20$$
$$P(P_1) = 0.03 + 0.06 + 0.07 + 0.09 + 0.16 + 0.10 + 0.05$$
$$0.05 + 0.14 = 0.75$$
$$P(C_3) = 0.07 + 0.01 + 0.10 + 0.06 + 0.14 + 0.02 = 0.40$$
$$P(M_1 \cap P_1) = 0.03 + 0.06 + 0.07 = 0.16$$

and

$$P(M_1 \cap C_3) = 0.07 + 0.01 = 0.08$$ ■

In Theorem 3.4 we saw that the third axiom of probability can be extended to include more than two mutually exclusive events. Another useful and important extension of this axiom allows us to find the probability of the union of any two events in $\mathcal{S}$ regardless of whether or not they are mutually exclusive. To motivate the theorem which follows, let us consider the Venn diagram of Figure 3.10, which concerns the job offers received by recent engineering-school graduates. The letters I and G stand for a job offer from industry and a job offer from the government, respectively.

It follows from the Venn diagram that

$$P(I) = 0.18 + 0.12 = 0.30$$
$$P(G) = 0.12 + 0.24 = 0.36$$

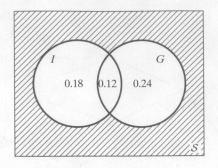

Figure 3.10
Venn diagram for job offers

and

$$P(I \cup G) = 0.18 + 0.12 + 0.24 = 0.54$$

We were able to add the various probabilities because they represent mutually exclusive events.

Had we erroneously used the third axiom of probability to calculate $P(I \cup G)$, we would have obtained $P(I) + P(G) = 0.30 + 0.36$, which exceeds the correct value by 0.12. This error results from adding in $P(I \cap G)$ twice, once in $P(I) = 0.30$ and once in $P(G) = 0.36$ and, we could correct for it by subtracting 0.12 from 0.66. Thus, we would get

$$P(I \cup G) = P(I) + P(G) - P(I \cap G)$$
$$= 0.30 + 0.36 - 0.12$$
$$= 0.54$$

and this agrees, as it should, with the result obtained before.

In line with this motivation, let us now state and prove the following theorem:

General addition rule for probability

> **Theorem 3.6**   If A and B are any events in $\mathcal{S}$, then
>
> $$P(A \cup B) = P(A) + P(B) - P(A \cap B)$$

To prove this theorem, we make use of the identities of parts (c) and (d) of Exercise 3.13, getting

$$P(A \cup B) = P(A \cap B) + P(A \cap \overline{B}) + P(\overline{A} \cap B)$$
$$= [P(A \cap B) + P(A \cap \overline{B})]$$
$$+ [P(A \cap B) + P(\overline{A} \cap B)] - P(A \cap B)$$
$$= P(A) + P(B) - P(A \cap B).$$

where, in the third line, we add and subtract $P(A \cap B)$. Note that when A and B are mutually exclusive so that $P(A \cap B) = 0$, Theorem 3.6 reduces to the third axiom of probability. For this reason, we sometimes refer to the third axiom of probability as the **special addition rule**.

EXAMPLE 17   **Using the general addition rule for probability**

With reference to the used car example of page 50, find the probability that a car will have low mileage or be expensive to operate, namely $P(M_1 \cup C_3)$.

Solution   Making use of the results obtained on page 63, $P(M_1) = 0.20$, $P(C_3) = 0.40$, and $P(M_1 \cap C_3) = 0.08$, we substitute into the general addition rule of

Theorem 3.6 to get

$$P(M_1 \cup C_3) = P(M_1) + P(C_3) - P(M_1 \cap C_3)$$
$$= 0.20 + 0.40 - 0.08$$
$$= 0.52$$

EXAMPLE 18   **The probability of requiring repair under warranty**

If the probabilities are 0.87, 0.36, and 0.29 that, while under warranty, a new car will require repairs on the engine, drive train, or both, what is the probability that a car will require one or the other or both kinds of repairs under the warranty?

Solution   Substituting these given values into the formula of Theorem 3.6, we get

$$0.87 + 0.36 - 0.29 = 0.94$$

Note that the general addition rule, Theorem 3.6, can be generalized further so that it applies to more than two events (see Exercise 3.49).

Using axioms of probability, we can derive many other theorems which play important roles in applications. For instance, let us show the following:

Probability rule of the complement

> **Theorem 3.7**   If A is any event in S, then $P(\overline{A}) = 1 - P(A)$.

To prove this theorem, we make use of the fact that A and $\overline{A}$ are mutually exclusive by definition, and that $A \cup \overline{A} = S$ (namely, that among them A and $\overline{A}$ contain all the elements in S). Hence we can write

$$P(A) + P(\overline{A}) = P(A \cup \overline{A})$$
$$= P(S)$$
$$= 1$$

so that $P(\overline{A}) = 1 - P(A)$. As a special case we find that $P(\phi) = 1 - P(S) = 0$ since the empty set ϕ is the complement of S.

EXAMPLE 19   **Using the probability rule of the complement**

Referring to the used car example of page 50 and the results on page 63, find

(a) the probability that a used car will not have low mileage

(b) the probability that a used car will either not have low mileage or not be expensive to operate

Solution   By the rule of the complement

(a) $P(\overline{M}_1) = 1 - P(M_1) = 1 - 0.20 = 0.80$

(b) Since $\overline{M}_1 \cup \overline{C}_3 = \overline{M_1 \cap C_3}$ according to the identity of part (a) of Exercise 3.13, by the rule of the complement we get

$$P(\overline{M}_1 \cup \overline{C}_3) = 1 - P(M_1 \cap C_3) = 1 - 0.08 = 0.92$$

Exercises

3.28 (a) Among 880 smart phones sold by a retailer, 72 required repairs under the warranty. Estimate the probability that a new phone, which has just been sold, will require repairs under the warranty. Explain your reasoning.

(b) Last year 8,400 students applied for the 6,000 student season tickets available for football games. Next year you will apply and would like to estimate the probability of receiving a season ticket. Give your estimate and comment on one factor that might influence the accuracy of your estimate.

3.29 When we roll a pair of balanced dice, what are the probabilities of getting

(a) 7;

(b) 11;

(c) 7 or 11;

(d) 3;

(e) 2 or 12;

(f) 2, 3, or 12?

3.30 A lottery sells tickets numbered from 00001 through 50000. What is the probability of drawing a number that is divisible by 200?

3.31 A car rental agency has 19 compact cars and 12 intermediate-size cars. If four of the cars are randomly selected for a safety check, what is the probability of getting two of each kind?

3.32 Last year; the maximum daily temperature in a plants' server room exceeded 68°F in 12 days. Estimate the probability that the maximum temperature will exceed 68°F tomorrow.

3.33 In a group of 160 graduate engineering students, 92 are enrolled in an advanced course in statistics, 63 are enrolled in a course in operations research, and 40 are enrolled in both. How many of these students are not enrolled in either course?

3.34 Among 150 persons interviewed as part of an urban mass transportation study, some live more than 3 miles from the center of the city (A), some now regularly drive their own car to work (B), and some would gladly switch to public mass transportation if it were available (C). Use the information given in Figure 3.11 to find

(a) $N(A)$;

(b) $N(B)$;

(c) $N(C)$;

(d) $N(A \cap B)$;

(e) $N(A \cap C)$;

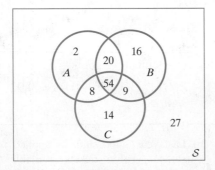

Figure 3.11   Diagram for Exercise 3.34

(f) $N(A \cap B \cap C)$;

(g) $N(A \cup B)$;

(h) $N(B \cup C)$;

(i) $N(\overline{A} \cup \overline{B} \cup C)$;

(j) $N[B \cap (A \cup C)]$.

3.35 An experiment has the four possible mutually exclusive outcomes A, B, C, and D. Check whether the following assignments of probability are permissible:

(a) $P(A) = 0.38, P(B) = 0.16, P(C) = 0.11, P(D) = 0.35$;

(b) $P(A) = 0.27, P(B) = 0.30, P(C) = 0.28, P(D) = 0.16$;

(c) $P(A) = 0.32, P(B) = 0.27, P(C) = -0.06, P(D) = 0.47$;

(d) $P(A) = \frac{1}{2}, P(B) = \frac{1}{4}, P(C) = \frac{1}{8}, P(D) = \frac{1}{16}$;

(e) $P(A) = \frac{5}{18}, P(B) = \frac{1}{6}, P(C) = \frac{1}{3}, P(D) = \frac{2}{9}$.

3.36 With reference to Exercise 3.1, suppose that the points (0, 0), (0, 1), (0, 2), (1, 0), (1, 1), (1, 2), (2, 0), (2, 1), (2, 2), (3, 0), (3, 1), and (3, 2) have the probabilities 0.060, 0.012, 0.006, 0.067, 0.014, 0.092, 0.260, 0.027, 0.080, 0.166, 0.110, and 0.106.

(a) Verify that this assignment of probabilities is permissible.

(b) Find the probabilities of events R, T, and U given in part (b) of that exercise.

(c) Calculate the probabilities that zero, one, or two streams are contaminated.

3.37 With reference to Exercise 3.7, suppose that each point (i, j) of the sample space is assigned the probability $\dfrac{15/28}{i + j}$.

(a) Verify that this assignment of probabilities is permissible.

(b) Find the probabilities of events B, C, and D described in part (b) of that exercise.

(c) Find the probabilities that one, two, or three of the graduate students will be supervising the physics lab.

3.38 Explain why there must be a mistake in each of the following statements:

(a) The probability that a mineral sample will contain silver is 0.38 and the probability that it will not contain silver is 0.52.

(b) The probability that a drilling operation will be a success is 0.34 and the probability that it will not be a success is -0.66.

(c) An air-conditioning repair person claims that the probability is 0.82 that the compressor is all right, 0.64 that the fan motor is all right, and 0.41 that they are both all right.

3.39 Refer to parts (c) and (d) of Exercise 3.13 to show that

(a) $P(A \cap B) \leq P(A)$;

(b) $P(A \cup B) \geq P(A)$.

3.40 Explain why there must be a mistake in each of the following statements:

(a) The probability that a student will get an A in a geology course is 0.3, and the probability that he or she will get either an A or a B is 0.27.

(b) A company is working on the construction of two shopping centers; the probability that the larger one will be completed on time is 0.35 and the probability that both will be completed on time is 0.42.

3.41 If A and B are mutually exclusive events, $P(A) = 0.45$, and $P(B) = 0.30$, find

(a) $P(\overline{A})$;

(b) $P(A \cup B)$;

(c) $P(A \cap \overline{B})$;

(d) $P(\overline{A} \cap \overline{B})$.

3.42 With reference to Exercise 3.34, suppose that the questionnaire filled in by one of the 150 persons is to be double-checked. If it is chosen in such a way that each questionnaire has a probability of $\frac{1}{150}$ of being selected, find the probabilities that the person

(a) lives more than 3 miles from the center of the city;

(b) regularly drives his or her car to work;

(c) does not live more than 3 miles from the center of the city and would not want to switch to public mass transportation if it were available;

(d) regularly drives his or her car to work but would gladly switch to public mass transportation if it were available.

3.43 A campus police department needs new bicycles for its patrol persons, and the probabilities are 0.17, 0.22, 0.03, 0.29, 0.21, and 0.08 that it will buy Bianshe, Cannonhill, Fishim, Giante, Trec, or HT. Find the probabilities that it will buy

(a) Cannonhill or Trec;

(b) Bianshe, Giante, or Trec;

(c) Fishim or Trec;

(d) Cannonhill, Giante, or HT.

3.44 The probabilities that a TV station will receive 0, 1, 2, 3, ..., 8 or at least 9 complaints after showing a controversial program are, respectively,

0.01, 0.03, 0.07, 0.15, 0.19, 0.18, 0.14, 0.12, 0.09, and 0.02. What are the probabilities that after showing such a program the station will receive

(a) at most 4 complaints;

(b) at least 6 complaints;

(c) from 5 to 8 complaints?

3.45 If each point of the sample space of Figure 3.12 represents an outcome having the probability $\frac{1}{32}$, find

(a) $P(A)$;

(b) $P(B)$;

(c) $P(A \cap B)$;

(d) $P(A \cup B)$;

(e) $P(\overline{A} \cap B)$;

(f) $P(\overline{A} \cap \overline{B})$.

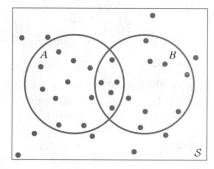

Figure 3.12 Diagram for Exercise 3.45

3.46 The probability that an integrated circuit chip will have defective etching is 0.06, the probability that it will have a crack defect is 0.03, and the probability that it has both defects is 0.02.

(a) What is the probability that a newly manufactured chip will have either an etching or a crack defect?

(b) What is the probability that a newly manufactured chip will have neither defect?

3.47 The probability that a new airport will get an award for its design is 0.16, the probability that it will get an award for the efficient use of materials is 0.24, and the probability that it will get both awards is 0.11.

(a) What is the probability that it will get at least one of the two awards?

(b) What is the probability that it will get only one of two awards?

3.48 Given $P(A) = 0.30$, $P(B) = 0.62$, and $P(A \cap B) = 0.12$, find

(a) $P(A \cup B)$;

(b) $P(\overline{A} \cap B)$;

(c) $P(A \cap \overline{B})$;

(d) $P(\overline{A} \cup \overline{B})$.

3.49 It can be shown that for any three events A, B, and C, the probability that at least one of them will occur is given by

$$P(A \cup B \cup C) = P(A) + P(B) + P(C)$$
$$- P(A \cap B) - P(A \cap C)$$
$$- P(B \cap C) + P(A \cap B \cap C)$$

Verify that this formula holds for the probabilities of Figure 3.13.

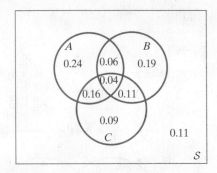

Figure 3.13 Diagram for Exercise 3.49

3.50 Suppose that in the maintenance of a large medical-records file for insurance purposes the probability of an error in processing is 0.0010, the probability of an error in filing is 0.0009, the probability of an error in retrieving is 0.0012, the probability of an error in processing as well as filing is 0.0002, the probability of an error in processing as well as retrieving is 0.0003, the probability of an error in filing as well as retrieving is 0.0003, and the probability of an error in processing and filing as well as retrieving is 0.0001. What is the probability of making at least one of these errors?

3.51 If the probability of event A is p, then the **odds** that it will occur are given by the ratio of p to $1 - p$. Odds are usually given as a ratio of two positive integers having no common factor, and if an event is more likely not to occur than to occur, it is customary to give the odds that it will not occur rather than the odds that it will occur. What are the odds for or against the occurrence of an event if its probability is

(a) $\dfrac{4}{7}$; (b) 0.05; (c) 0.80?

3.52 Use the definition of Exercise 3.51 to show that if the odds for the occurrence of event A are a to b, where a and b are positive integers, then

$$p = \frac{a}{a+b}$$

3.53 The formula of Exercise 3.52 is often used to determine subjective probabilities. For instance, if an applicant for a job "feels" that the odds are 7 to 4 of getting the job, the subjective probability the applicant assigns to getting the job is

$$p = \frac{7}{7+4} = \frac{7}{11}$$

(a) If a businessperson feels that the odds are 3 to 2 that a new venture will succeed (say, by betting $300 against $200 that it will succeed), what subjective probability is he or she assigning to its success?

(b) If a student is willing to bet $30 against $10, but not $40 against $10, that he or she will get a passing grade in a certain course, what does this tell us about the subjective probability the student assigns to getting a passing grade in the course?

3.54 Subjective probabilities may or may not satisfy the third axiom of probability. When they do, we say that they are **consistent**; when they do not, they ought not to be taken too seriously.

(a) The supplier of delicate optical equipment feels that the odds are 7 to 5 against a shipment arriving late, and 11 to 1 against it not arriving at all. Furthermore, he feels that there is a 50/50 chance (the odds are 1 to 1) that such a shipment will either arrive late or not at all. Are the corresponding probabilities consistent?

(b) There are two Ferraris in a race, and an expert feels that the odds against their winning are, respectively, 2 to 1 and 3 to 1. Furthermore, she claims that there is a less-than-even chance that either of the two Ferraris will win. Discuss the consistency of these claims.

3.6 Conditional Probability

As we have defined probability, it is meaningful to ask for the probability of an event only if we refer to a given sample space $\mathcal{S}$. To ask for the probability that an engineer earns at least $90,000 a year is meaningless unless we specify whether we are referring to all engineers in the western hemisphere, all engineers in the United States, all those in a particular industry, all those affiliated with a university, and so forth. Thus, when we use the symbol $P(A)$ for the probability of A, we really mean the probability of A given some sample space $\mathcal{S}$. Since the choice of $\mathcal{S}$ is not always evident, or we are interested in the probabilities of A with respect to more than one sample space, the notation $P(A|\mathcal{S})$ makes it clear that we are referring to a particular sample space $\mathcal{S}$. We read $P(A|\mathcal{S})$ as the conditional probability of A relative to $\mathcal{S}$, and every probability is thus a conditional probability. Of course, we use the simplified notation $P(A)$ whenever the choice of $\mathcal{S}$ is clearly understood.

To illustrate some of the ideas connected with conditional probabilities, let us consider again the 500 machine parts of which some are improperly assembled and some contain one or more defective components as shown in Figure 3.8. Assuming equal probabilities in the selection of one of the machine parts for inspection, it can be seen that the probability of getting a part with one or more defective components is

$$P(D) = \frac{10 + 5}{500} = \frac{3}{100}$$

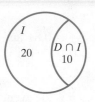

Figure 3.14
Reduced sample space

To check whether the probability is the same if the choice is restricted to the machine parts that are improperly assembled, we have only to look at the reduced sample space of Figure 3.14 and assume that each of the 30 improperly assembled parts has the same chance of being selected. We thus get

$$P(D \mid I) = \frac{N(D \cap I)}{N(I)} = \frac{10}{30} = \frac{1}{3}$$

and it can be seen that the probability of getting a machine part with one or more defective components has increased from $\frac{3}{100}$ to $\frac{1}{3}$. Note that if we divide the numerator and denominator of the preceding formula for $P(D \mid I)$ by $N(\mathcal{S})$, we get

$$P(D \mid I) = \frac{\dfrac{N(D \cap I)}{N(\mathcal{S})}}{\dfrac{N(I)}{N(\mathcal{S})}} = \frac{P(D \cap I)}{P(I)}$$

where $P(D \mid I)$ is given by the ratio of $P(D \cap I)$ to $P(I)$.

Looking at this example in another way, note that with respect to the whole sample space $\mathcal{S}$ we have

$$P(D \cap I) = \frac{10}{500} = \frac{1}{50} \quad \text{and} \quad P(\overline{D} \cap I) = \frac{20}{500} = \frac{2}{50}$$

assuming, as before, that each of the 500 machine parts has the same chance of being selected. Thus, the probabilities that the machine part selected will or will not contain one or more defective components, given that it is improperly assembled, should be in the ratio 1 to 2. Since the probabilities of D and $\overline{D}$ in the reduced sample space must add up to 1, it follows that

$$P(D \mid I) = \frac{1}{3} \quad \text{and} \quad P(\overline{D} \mid I) = \frac{2}{3}$$

which agrees with the result obtained before. This explains why, in the last step, we had to divide by $P(I)$ to

$$P(D \mid I) = \frac{P(D \cap I)}{P(I)}$$

Division by $P(I)$, or multiplication by $1/P(I)$, takes care of the proportionality factor, which makes the sum of the probabilities over the reduced sample space equal to 1.

Following these observations, let us now make the following general definition:

Conditional probability

If A and B are any events in S and $P(B) \neq 0$, the conditional probability of A given B is

$$P(A \mid B) = \frac{P(A \cap B)}{P(B)}$$

EXAMPLE 20 **Calculating a conditional probability**

If the probability that a communication system will have high fidelity is 0.81 and the probability that it will have high fidelity and high selectivity is 0.18, what is probability that a system with high fidelity will also have high selectivity?

Solution If A is the event that a communication system has high selectivity and B is the event that it has high fidelity, we have $P(B) = 0.81$ and $P(A \cap B) = 0.18$, and substitution into the formula yields

$$P(A \mid B) = \frac{0.18}{0.81} = \frac{2}{9}$$ ∎

EXAMPLE 21 **The conditional probability that a used car has low mileage given that it is expensive to operate**

Referring to the used car example, for which the probabilities of the individual outcomes are given in Figure 3.9, use the results on page 63 to find $P(M_1 \mid C_3)$.

Solution Since we had $P(M_1 \cap C_3) = 0.08$ and $P(C_3) = 0.40$, substitution into the formula for conditional probability yields

$$P(M_1 \mid C_3) = \frac{P(M_1 \cap C_3)}{P(C_3)} = \frac{0.08}{0.40} = 0.20$$

It is of interest to note that the value of the conditional probability obtained here, $P(M_1 \mid C_3) = 0.20$, equals the value for $P(M_1)$ obtained on page 63. This means that the probability a used car has low mileage is the same whether or not it is expensive to operate. We say that M_1 is **independent** of C_3. As the reader is asked to verify in Exercise 3.59, it also follows from the results on page 63 that M_1 is not independent of P_1, namely, that low mileage is related to the car's price. ∎

In general, if A and B are any two events in a sample space S, we say that **A is independent of B if and only if $P(A \mid B) = P(A)$**, but as it can be shown that B is independent of A whenever A is independent of B, it is customary to say simply that **A and B are independent events**.

General multiplication rule of probability

Theorem 3.8 If A and B are any events in S, then
$$P(A \cap B) = P(A) \cdot P(B \mid A) \qquad \text{if} \quad P(A) \neq 0$$
$$= P(B) \cdot P(A \mid B) \qquad \text{if} \quad P(B) \neq 0$$

The second of these rules is obtained directly from the definition of conditional probability by multiplying both sides by $P(B)$; the first is obtained from the second by interchanging the letters A and B.

EXAMPLE 22   **Using the general multiplication rule of probability**

The supervisor of a group of 20 construction workers wants to get the opinion of 2 of them (to be selected at random) about certain new safety regulations. If 12 workers favor the new regulations and the other 8 are against them, what is the probability that both of the workers chosen by the supervisor will be against the new safety regulations?

Solution   Assuming equal probabilities for each selection (which is what we mean by the selections being random), the probability that the first worker selected will be against the new safety regulations is $\frac{8}{20}$, and the probability that the second worker selected will be against the new safety regulations given that the first one is against them is $\frac{7}{19}$. Thus, the desired probability is

$$\frac{8}{20} \cdot \frac{7}{19} = \frac{14}{95}$$ ∎

In the special case where A and B are independent so $P(A \mid B) = P(A)$, Theorem 3.8 leads to the following result:

Special product rule of probability

> **Theorem 3.9**   Two events A and B are independent events if and only if
> $$P(A \cap B) = P(A) \cdot P(B)$$

Thus, the probability that two independent events will both occur is simply the product of their probabilities. This rule is sometimes used as the definition of independence. It applies even when $P(A)$ or $P(B)$ or both equal 0. In any case, it may be used to determine whether two given events are independent.

EXAMPLE 23   **The outcomes to unrelated parts of an experiment can be treated as independent**

What is the probability of getting two heads in two flips of a balanced coin?

Solution   Since the probability of heads is $\frac{1}{2}$ for each flip and the two flips are not physically connected, we treat them as independent. The probability is

$$\frac{1}{2} \cdot \frac{1}{2} = \frac{1}{4}$$ ∎

EXAMPLE 24   **Independence and selection with and without replacement**

Two cards are drawn at random from an ordinary deck of 52 playing cards. What is the probability of getting two aces if

(a) the first card is replaced before the second card is drawn;

(b) the first card is not replaced before the second card is drawn?

Solution   **(a)** Since there are four aces among the 52 cards, we get

$$\frac{4}{52} \cdot \frac{4}{52} = \frac{1}{169}$$

(b) Since there are only three aces among the 51 cards that remain after one ace has been removed from the deck, we get

$$\frac{4}{52} \cdot \frac{3}{51} = \frac{1}{221}$$

Note that

$$\frac{1}{221} \neq \frac{4}{52} \cdot \frac{4}{52}$$

so independence is violated when the sampling is without replacement.   ∎

EXAMPLE 25   **Checking if two events are independent under an assigned probability**

If $P(C) = 0.65$, $P(D) = 0.40$, and $P(C \cap D) = 0.24$, are the events C and D independent?

Solution   Since $P(C) \cdot P(D) = (0.65)(0.40) = 0.26$ and not 0.24, the two events are not independent.   ∎

In the preceding examples we have used the assigned probabilities to check if two events are independent. The concept of independence can be—and frequently is—employed when probabilities are assigned to events that concern unrelated parts of an experiment.

EXAMPLE 26   **Assigning probability by the special product rule**

Let A be the event that raw material is available when needed and B be the event that the machining time is less than 1 hour. If $P(A) = 0.8$ and $P(B) = 0.7$, assign probability to the event $A \cap B$.

Solution   Since the events A and B concern unrelated steps in the manufacturing process, we invoke independence and make the assignment

$$P(A \cap B) = P(A)P(B) = 0.8 \times 0.7 = 0.56$$ ∎

The special product rule can easily be extended so that it applies to more than two independent events—again, we multiply together all the individual probabilities.

EXAMPLE 27   **The extended special product rule of probability**

What is the probability of not rolling any 6's in four rolls of a balanced die?

Solution   The probability is $\dfrac{5}{6} \cdot \dfrac{5}{6} \cdot \dfrac{5}{6} \cdot \dfrac{5}{6} = \dfrac{625}{1,296}$ ∎

For three or more dependent events the multiplication rule becomes more complicated, as is illustrated in Exercise 3.70.

EXAMPLE 28   **The probability of falsely signaling a pollution problem**

Many companies must monitor the effluent that is discharged from their plants into rivers and waterways. In some states, it is the law that some substances have water-quality limits that are below the limit of detection, L, for the current method of measurement. The effluent is judged to satisfy the quality limit if every test specimen is below the limit of detection L. Otherwise it will be declared to fail compliance with the quality limit. Suppose the water does not contain the contaminant of interest but that the variability in the chemical analysis still gives a 1% chance that a measurement on a test specimen will exceed L.

(a) Find the probability that neither of two test specimens, both free of the contaminant, will fail to be in compliance.

(b) If one test specimen is taken each week for two years, and they are all free of the contaminant, find the probability that none of the test specimens will fail to be in compliance.

(c) Comment on the incorrect reasoning of having a fixed limit of detection no matter how many tests are conducted.

Solution   (a) If the two samples are not taken too closely in time or space, we treat them as independent. We use the special product rule to obtain the probability that both are in compliance:

$$0.99 \times 0.99 = 0.9801$$

(b) Treating the results for different weeks as independent,

$$(0.99)^{104} = 0.35$$

so, even with excellent water quality, there is almost a two-thirds chance that at least once the water quality will be declared to fail to be in compliance with the law.

(c) With this type of law, no company would want to collect test specimens more than maybe once a year. This is in direct opposition to the scientific idea that more information is better than less information on water quality. Some effort should be made to allow for higher limits when the testing is more frequent. ∎

EXAMPLE 29   **Using probability to compare the accuracy of two schemes for sending messages**

Electrical engineers are considering two alternative systems for sending messages. A message consists of a word that is either a 0 or a 1. However, because of random noise in the channel, a 1 that is transmitted could be received as a 0 and vice versa. That is, there is a small probability, p, that

$$P[\text{A transmitted 1 is received as 0}] = p$$
$$P[\text{A transmitted 0 is received as 1}] = p$$

One scheme is to send a single digit. The message is short but may be unreliable. A second scheme is to repeat the selected digit three times in succession. At the receiving end, the majority rule will be used to decode. That is, when any of 101, 110, 011, or 111 are received, it is interpreted to mean a 1 was sent.

(a) Evaluate the probability that a transmitted 1 will be received as a 1 under the three-digit scheme when $p = 0.01, 0.02$, or 0.05. Compare this with the scheme where a single digit is transmitted as a word. Treat the results for different digits as independent.

(b) Suppose a message, consisting of the two words, a 1 followed by 0, is to be transmitted using the three-digit scheme. What is the probability that the total message will be correctly decoded under the majority rule with $p = 0.05$? Compare with the scheme where a single digit is transmitted as a word.

Solution   (a) The three digits 111 are transmitted. By independence, the sequence 111 has probability $(1 - p)(1 - p)(1 - p)$ of being received as 111. Also the probability of receiving 011 is $p(1-p)(1-p)$, so the probability of exactly one 0 among the three received is $3p(1-p)^2$. Using the majority rule,

$$P[\text{Correct}] = P[\text{transmitted 1 received as 1}] = (1-p)^3 + 3p(1-p)^2$$

p	0.01	0.02	0.05
$P[\text{Correct}]$	0.9997	0.9988	0.9928

All three probabilities are considerably above the corresponding single-digit scheme probabilities 0.99, 0.98, and 0.95, respectively.

(b) Both of the words 1 and 0 must be received correctly. As in part (a), the probability that a 0 is received correctly is also 0.9928. Consequently, using independence, the probability that the total message is correctly received is $(0.9928)^2 = 0.986$. This improves over the scheme where single digits are sent for each word since that scheme has only the probability $(0.95)^2 = 0.903$ of correctly receiving the total message.

Redundancy helps improve accuracy, but more digits need to be transmitted, which results in a significantly lower throughput.   ∎

3.7 Bayes' Theorem

The general multiplication rules are useful in solving many problems in which the ultimate outcome of an experiment depends on the outcomes of various intermediate stages. A manufacturer of tablets receives its LED screens from three different suppliers, 60% from supplier B_1, 30% from supplier B_2, and 10% from supplier B_3. In other words, the probabilities that any one LED screens received by the plant comes from these three suppliers are 0.60, 0.30, and 0.10. Also suppose that 95% of the LED screens from B_1, 80% of those from B_2, and 65% of those from B_3 perform according to specifications. We would like to know the probability that any one LED screen received by the plant will perform according to specifications.

If A denotes the event that a LED screen received by the plant performs according to specifications, and B_1, B_2, and B_3 are the events that it comes from the respective suppliers, we can write

$$A = A \cap [B_1 \cup B_2 \cup B_3]$$
$$= (A \cap B_1) \cup (A \cap B_2) \cup (A \cap B_3)$$

where B_1, B_2, and B_3 are mutually exclusive events of which one must occur. It follows that $A \cap B_1, A \cap B_2$, and $A \cap B_3$ are also mutually exclusive. By the generalization of the third axiom of probability on page 60, we get

$$P(A) = P(A \cap B_1) + P(A \cap B_2) + P(A \cap B_3)$$

Then, if we apply the second of the general multiplication rules to $P(A \cap B_1)$, $P(A \cap B_2)$, and $P(A \cap B_3)$, we get

$$P(A) = P(B_1) \cdot P(A \mid B_1) + P(B_2) \cdot P(A \mid B_2) + P(B_3) \cdot P(A \mid B_3)$$

and substitution of the given numerical values yields

$$P(A) = (0.60)(0.95) + (0.30)(0.80) + (0.10)(0.65)$$
$$= 0.875$$

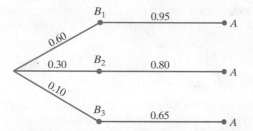

Figure 3.15
Tree diagram for example dealing with three suppliers of LED screens

for the probability that any one LED screen received by the plant will perform according to specifications.

To visualize this result, we have only to construct a tree diagram like that of Figure 3.15, where the probability of the final outcome is given by the sum of the products of the probabilities corresponding to each branch of the tree.

In the preceding example there were only 3 possibilities at the intermediate stage, but if there are n mutually exclusive possibilities $B_1, B_2, \ldots, B_n$ at the intermediate stage, a similar argument will lead to the following result, sometimes called the **rule of elimination** or the **rule of total probability**:

Rule of total probability

> **Theorem 3.10**   If $B_1, B_2, \ldots, B_n$ are mutually exclusive events of which one must occur, then
>
> $$P(A) = \sum_{i=1}^{n} P(B_i) \cdot P(A \mid B_i)$$

The tree diagram like that of Figure 3.16, where the probability of the final outcome is again given by the sum of the products of the probabilities corresponding to each branch of the tree, graphically explains the calculation.

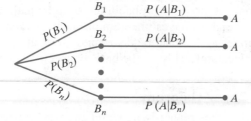

Figure 3.16
Tree diagram for rule of elimination

To consider a problem that is closely related to the one we have just discussed, suppose we want to know the probability that a particular LED screen, which is known to perform according to specifications, came from supplier B_3. Symbolically, we want to know the value of $P(B_3 \mid A)$, and to find a formula for this probability we first write

$$P(B_3 \mid A) = \frac{P(A \cap B_3)}{P(A)}$$

Then, substituting $P(B_3) \cdot P(A \mid B_3)$ for $P(A \cap B_3)$ and $\sum_{i=1}^{3} P(B_i) \cdot P(A \mid B_i)$ for $P(A)$ in accordance with Theorems 3.8 and 3.10, we get

$$P(B_3 \mid A) = \frac{P(B_3) \cdot P(A \mid B_3)}{\sum_{i=1}^{3} P(B_i) \cdot P(A \mid B_i)}$$

which expresses $P(B_3 \mid A)$ in terms of given probabilities. Substituting the values from page 74 (or from Figure 3.15), we finally obtain

$$P(B_3 \mid A) = \frac{(0.10)(0.65)}{(0.60)(0.95) + (0.30)(0.80) + (0.10)(0.65)}$$
$$= 0.074$$

Note that the probability that an LED screen is supplied by B_3 decreases from 0.10 to 0.074 once it is known that it performs according to specifications.

The method used to solve the preceding example can easily be generalized to yield the following formula, called **Bayes' theorem**:

Bayes' theorem

> **Theorem 3.11**   If $B_1, B_2, \ldots, B_n$ are mutually exclusive events of which one must occur, then
>
> $$P(B_r \mid A) = \frac{P(B_r) \cdot P(A \mid B_r)}{\sum_{i=1}^{n} P(B_i) \cdot P(A \mid B_i)}$$
>
> for $r = 1, 2, \ldots, n$.

Note that the expression in the numerator is the probability of reaching A via the rth branch of the tree and that the expression in the denominator is the sum of the probabilities of reaching A via the n branches of the tree.

Bayes' theorem provides a formula for finding the probability that the "effect" A was "caused" by the event B_r. For instance, in our example we found the probability that an acceptable LED screen was made by supplier B_3. The probabilities $P(B_i)$ are called the *prior*, or *a priori*, probabilities of the "causes" B_i, and in practice it is often difficult to assign them numerical values. For many years Bayes' theorem was looked upon with suspicion because it was used with the often erroneous assumption that the prior probabilities are all equal. A good deal of the controversy once surrounding Bayes' theorem has been cleared up with the realization that the probabilities $P(B_i)$ must be determined separately in each case from the nature of the problem, preferably on the basis of specific knowledge or past experience.

EXAMPLE 30   **Using Bayes' theorem**

Four technicians regularly make repairs when breakdowns occur on an automated production line. Janet, who services 20% of the breakdowns, makes an incomplete repair 1 time in 20; Tom, who services 60% of the breakdowns, makes an incomplete repair 1 time in 10; Georgia, who services 15% of the breakdowns, makes an incomplete repair 1 time in 10; and Peter, who services 5% of the breakdowns, makes an incomplete repair 1 time in 20. For the next problem with the production line diagnosed as being due to an initial repair that was incomplete, what is the probability that this initial repair was made by Janet?

Solution   Let A be the event that the initial repair was incomplete, B_1 that the initial repair was made by Janet, B_2 that it was made by Tom, B_3 that it was made by Georgia, and B_4 that it was made by Peter.

Substituting the various probabilities into the formula of Theorem 3.11, we get

$$P(B_1 \mid A) = \frac{(0.20)(0.05)}{(0.20)(0.05) + (0.60)(0.10) + (0.15)(0.10) + (0.05)(0.05)}$$
$$= 0.114$$

and it is of interest to note that although Janet makes an incomplete repair only 1 out of 20 times, namely, 5% of the breakdowns she services, more than 11% of the incomplete repairs are her responsibility. ∎

Bayes' Theorem plays an integral part in most schemes for filtering spam. It gives the probability that a chance message is really spam given the presence of certain words.

EXAMPLE 31   **Identifying spam using Bayes' Theorem**

A first step towards identifying spam is to create a list of words that are more likely to appear in spam than in normal messages. For instance, words like buy or the brand name of an enhancement drug are more likely to occur in spam messages than in normal messages. Suppose a specified list of words is available and that your data base of 5000 messages contains 1700 that are spam. Among the spam messages, 1343 contain words in the list. Of the 3300 normal messages, only 297 contain words in the list.

Obtain the probability that a message is spam given that the message contains words in the list.

Solution   Let A = [message contains words in list] be the event a message is identified as spam and let B_1 = [message is spam] and B_2 = [message is normal]. We use the observed relative frequencies from the data base as approximations to the probabilities.

$$P(B_1) = \frac{1700}{5000} = .34 \qquad P(B_2) = \frac{3300}{5000} = .66$$

$$P(A \mid B_1) = \frac{1343}{1700} = .79 \qquad P(A \mid B_2) = \frac{297}{3300} = .09$$

Bayes' Theorem expresses the probability of being spam, given that a message is identified as spam, as

$$P(B_1 \mid A) = \frac{P(A \mid B_1)P(B_1)}{P(A \mid B_1)P(B_1) + P(A \mid B_2)P(B_2)}$$

The updated, or posterior probability, is

$$P(B_1 \mid A) = \frac{.79 \times .34}{.79 \times .34 + .09 \times .66} = \frac{.2686}{.328} = .819$$

Because this posterior probability of being spam is quite large, we suspect that this message really is spam. Since $P(B_1) = .34$, or 34% of the incoming messages are spam, we likely would want the spam filter to remove this message. Existing spam filer programs learn and improve as you mark your incoming messages spam. ∎

Exercises

3.55 With reference to Figure 3.8, find $P(I \mid D)$ and $P(I \mid \overline{D})$, assuming that originally each of the 500 machine parts has the same chance of being chosen for inspection.

3.56 (a) Would you expect the probability that a randomly selected car will need major repairs in the next year to be smaller, remain the same, or increase if you are told it already has high mileage? Explain.

(b) Would you expect the probability that a randomly selected senior would know the second law of thermodynamics, to be smaller, remain the same, or increase if the person selected is a mechanical engineering major? Explain.

(c) In Part (a), identify the two events with symbols A and B and the conditional probability of interest.

3.57 With reference to Exercise 3.34 and Figure 3.11, assume that each of 150 persons has the same chance of being selected, and find the probabilities that he or she

  (a) lives more than 3 miles from the center of the city given that he or she would gladly switch to public mass transportation;

  (b) regularly drives his or her car to work given that he or she lives more than 3 miles from the center of the city;

  (c) would not want to switch to public mass transportation given that he or she does not regularly drive his or her car to work.

3.58 With reference to Figure 3.13, find

  (a) $P(A\,|\,B)$;

  (b) $P(B\,|\,\overline{C})$;

  (c) $P(A\cap B\,|\,C)$;

  (d) $P(B\cup C\,|\,\overline{A})$;

  (e) $P(A\,|\,B\cup C)$;

  (f) $P(A\,|\,B\cap C)$;

  (g) $P(A\cap B\cap C\,|\,B\cap C)$;

  (h) $P(A\cap B\cap C\,|\,B\cup C)$.

3.59 With reference to the used car example and the probabilities given in Figure 3.9, find

  (a) $P(M_1\,|\,P_1)$ and compare its value with that of $P(M_1)$;

  (b) $P(C_3\,|\,P_2)$ and compare its value with that of $P(C_3)$;

  (c) $P(M_1\,|\,P_1\cap C_3)$ and compare its value with that of $P(M_1)$.

3.60 With reference to Exercise 3.47, find the probabilities that the airport will get the design award given that

  (a) it got the award for the efficient use of materials;

  (b) it did not get the award for the efficient use of materials.

3.61 Prove that $P(A\,|\,B)=P(A)$ implies that $P(B\,|\,A)=P(B)$ provided that $P(A)\neq 0$ and $P(B)\neq 0$.

3.62 In one area of the state, red cars are targeted by law enforcement for special attention. Assume that the probability of a red car being stopped for speeding is .06 while the probability is only .02 for non-red cars. Taking, as an approximation for all cars in that area, the nation wide proportion 0.09 of red, find

  (a) probability that a car will be stopped for speeding.

  (b) probability that a car stopped for speeding is red.

3.63 Given that $P(A)=0.60$, $P(B)=0.40$, and $P(A\cap B)=0.24$, verify that

  (a) $P(A\,|\,B)=P(A)$;

  (b) $P(A\,|\,\overline{B})=P(A)$;

  (c) $P(B\,|\,A)=P(B)$;

  (d) $P(B\,|\,\overline{A})=P(B)$.

3.64 Among the 24 invoices prepared by a billing department, 4 contain errors while the others do not. If we randomly check 2 of these invoices, what are the probabilities that

  (a) both will contain errors;

  (b) neither will contain an error?

3.65 Among 60 automobile repair parts loaded on a truck in San Francisco, 45 are destined for Seattle and 15 for Vancouver. If two of the parts are unloaded in Portland by mistake and the "selection" is random, what are the probabilities that

  (a) both parts should have gone to Seattle;

  (b) both parts should have gone to Vancouver;

  (c) one should have gone to Seattle and one to Vancouver?

3.66 A large firm has 85% of its service calls made by a contractor, and 10% of these calls result in customer complaints. The other 15% of the service calls are made by their own employees, and these calls have a 5% complaint rate. Find the

  (a) probability of receiving a complaint.

  (b) probability that the complaint was from a customer serviced by the contractor.

3.67 If $P(A)=0.60$, $P(B)=0.45$, and $P(A\cap B)=0.27$, are A and B independent?

3.68 If the odds are 5 to 3 that an event M will not occur, 2 to 1 that event N will occur, and 4 to 1 that they will not both occur, are the two events M and N independent?

3.69 Find the probabilities of getting

  (a) eight heads in a row with a balanced coin;

  (b) three 3's and then a 4 or a 5 in four rolls of a balanced die;

  (c) five multiple-choice questions answered correctly, if for each question the probability of answering it correctly is $\frac{1}{3}$.

3.70 For three or more events which are not independent, the probability that they will all occur is obtained by multiplying the probability that one of the events will occur, times the probability that a second of the events will occur given that the first event has occurred, times the probability that a third of the events will occur given that the first two events have occurred, and so on. For instance, for three events we can write

$$P(A\cap B\cap C)=P(A)\cdot P(B\,|\,A)\cdot P(C\,|\,A\cap B)$$

and we find that the probability of drawing without replacement three aces in a row from an ordinary deck

of 52 playing cards is

$$\frac{4}{52} \cdot \frac{3}{51} \cdot \frac{2}{50} = \frac{1}{5,525}$$

(a) If six bullets, of which three are blanks, are randomly inserted into a gun, what is the probability that the first three bullets fired will all be blanks?

(b) In a certain city during the month of May, the probability that a rainy day will be followed by another rainy day is 0.80 and the probability that a sunny day will be followed by a rainy day is 0.60. Assuming that each day is classified as being either rainy or sunny and that the weather on any given day depends only on the weather the day before, find the probability that in the given city a rainy day in May is followed by two more rainy days, then a sunny day, and finally another rainy day.

(c) A department store which bills its charge-account customers once a month has found that if a customer pays promptly one month, the probability is 0.90 that he will also pay promptly the next month; however, if a customer does not pay promptly one month, the probability that he will pay promptly the next month is only 0.50. What is the probability that a customer who has paid promptly one month will not pay promptly the next three months?

(d) If 5 of a company's 12 delivery trucks do not meet emission standards and 4 of the 12 trucks are randomly picked for inspection, what is the probability that none of them meets emission standards?

3.71 Use the information on the tree diagram of Figure 3.17 to determine the value of

(a) $P(A)$;

(b) $P(B|A)$;

(c) $P(B|\bar{A})$.

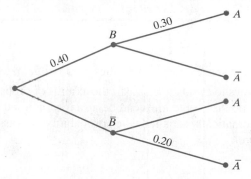

Figure 3.17 Diagram for Exercise 3.71

3.72 There are over twenty thousand objects orbiting in space. For a given object, let A be the event that the charred remains do hit the earth. Suppose experts, using their knowledge of the size and composition of the object as well as its re-entry angle, determine that $P(A) = 0.25$.

Next, let B be the event that your city is hit, given that charred remains reach the earth. The probability $P(B|A)$ will depend on both the size of your city and its location relative to the current orbit of the object. Suppose that, the experts conclude that $P(B|A) = 0.0002$.

(a) Find the probability that your city is hit with charred remains.

(b) Change $P(B|A)$ to 0.0004 and repeat the calculation.

3.73 An insurance company's records of 12,299 automobile insurance policies showed that 2073 policy holders made a claim (Courtesy J. Hickman). Among insured drivers under age 25, there were 1032 claims out of 5192 policies. For person selected at random from the policy holders, let $A =$ [Claim was filed] and $B =$ [Under age 25] .

(a) Fill in the four probabilities, and the marginal totals, in the table

(b) Use Bayes' Theorem to obtain the probability that the person is under age 25 given that a claim was filed .

(c) Check your answer using directly from your table in Part (a) and the definition of $P(B|A)$.

3.74 Identity theft is a growing problem in the United States. According to a Federal Trade Commission Report about 280,000 identity complaints were filed for 2011. Among the 43.2 million persons in the 20–29 year old age group, 56,689 complaints were filed. The 20–29 year old age group makes up proportion .139 of the total population. Use the relative frequencies to approximate the probability, that for the current year,

(a) a person in the 20–29 age group files an identity theft complaint.

(b) a person not in the 20–29 age group files an identity theft complaint. Comment on your answers to Parts (a) and (b).

(c) a random person will file an identity theft complaint.

(d) If a complaint is filed, what is the probability it was by someone in the 20–29 age group.

3.75 Refer to the example on page 74 but suppose the manufacturer has difficulty getting enough LED screens. Because of the shortage, the manufacturer had to obtain 40% of the screens from the second supplier and 15% from the third supplier. Find the

(a) probability that a LED screen will meet specifications.

(b) probability that a LED screen that meets specifications, was sent by the second supplier.

3.76 Refer to Example 31 concerning spam but now suppose that among the 5000 messages, the 1750 spam messages have 1570 that contain the words on a new list and that the 3250 normal messages have 300 that contain the words.

(a) Find the probability that a message is spam given that the message contains words on the new list.

(b) Would you prefer the new list here or the one in Example 31? Why?

3.77 With reference to the Example 30, for a problem diagnosed as being due to an incomplete initial repair, find the probability that the initial repair was made by

(a) Tom;

(b) Georgia;

(c) Peter.

3.78 Two firms V and W consider bidding on a road-building job, which may or may not be awarded depending on the amounts of the bids. Firm V submits a bid and the probability is 0.8 that it will get the job provided firm W does not bid. The probability is 0.7 that W will bid, and if it does, the probability that V will get the job is only 0.4.

(a) What is the probability that V will get the job?

(b) If V gets the job, what is the probability that W did not bid?

3.79 Engineers in charge of maintaining our nuclear fleet must continually check for corrosion inside the pipes that are part of the cooling systems. The inside condition of the pipes cannot be observed directly but a nondestructive test can give an indication of possible corrosion. This test is not infallible. The test has probability 0.7 of detecting corrosion when it is present but it also has probability 0.2 of falsely indicating internal corrosion. Suppose the probability that any section of pipe has internal corrosion is 0.1.

(a) Determine the probability that a section of pipe has internal corrosion, given that the test indicates its presence.

(b) Determine the probability that a section of pipe has internal corrosion, given that the test is negative.

3.80 An East Coast manufacturer of printed circuit boards exposes all finished boards to an online automated verification test. During one period, 900 boards were completed and 890 passed the test. The test is not infallible. Of 30 boards intentionally made to have noticeable defects, 25 were detected by the test. Use the relative frequencies to approximate the conditional probabilities needed below.

(a) Give an approximate value for $P[\text{Pass test} \mid \text{board has defects}]$.

(b) Explain why your answer in part a may be too small.

(c) Give an approximate value for the probability that a manufactured board will have defects. In order to answer the question, you need information about the conditional probability that a good board will fail the test. This is important to know but was not available at the time an answer was required. To proceed, you can assume that this probability is zero.

(d) Approximate the probability that a board has defects given that it passed the automated test.

Do's and Don'ts

Do's

1. Begin by creating a sample space $\mathcal{S}$ which specifies all possible outcomes.

2. Always assign probabilities to events that satisfy the axioms of probability. In the discrete case, the possible outcomes can be arranged in a sequence. The axioms are then automatically satisfied when probability p_i is assigned to the ith outcome, where

$$0 \leq p_i \quad \text{and} \quad \sum_{\text{all outcomes in } \mathcal{S}} p_i = 1$$

and the probability of any event A is defined as

$$P(A) = \sum_{\text{all outcomes in } A} p_i$$

3. Combine the probabilities of events according to rules of probability.

General Addition Rule: $P(A \cup B) = P(A) + P(B) - P(A \cap B)$

Rule of the Complement: $P(\overline{A}) = 1 - P(A)$

General Multiplication Rule: $P(A \cap B) = P(A)P(B \mid A)$ if $P(A) \neq 0$
$= P(B)P(A \mid B)$ if $P(B) \neq 0$

Conditional Probability: $P(A \mid B) = \dfrac{P(A \cap B)}{P(B)}$ if $P(B) \neq 0$

Don'ts

1. Don't confuse independent events with mutually exclusive events. When A and B are mutually exclusive, only one of them can occur. Their intersection is empty and so has probability 0.

2. Don't assign probability to $A \cap B$ according to special product rule

$$P(A \cap B) = P(A)P(B)$$

unless the conditions for independence hold. Independence may be plausible when the events A and B pertain to physically unrelated parts of a large system and there are no common causes that jointly affect the occurrence of both events.

Review Exercises

3.81 (a) In a recent random check of 300 books housed in the Engineering Library, 27 were checked out. For a randomly selected book today, what is the probability that it is currently checked out?

(b) Last year, 380 students applied for 28 internships administered by the university. Next year you will apply for one of the 28 internships and would like to estimate the probability of receiving one. Give your estimate and comment on one factor that might influence your estimate.

3.82 A salesperson of industrial chemicals has four customers in Sacramento, whom he may or may not be able to visit on a 2-day trip to this city. He will not visit any of these customers more than once.

(a) Using two coordinates so that (2,1), for example, represents the event that he will visit two of his customers on the first day and one on the second day, draw a diagram similar to that of Figure 3.1 showing the points of the corresponding sample space.

(b) List the points of the sample space that constitute the events A, B, and C that he will visit all four of his customers, that he will visit more of this customers on the first day than on the second day, and that he will visit at least three of his customers on the second day.

(c) Which of the three pairs of events, A and B, B and C, or A and C, are mutually exclusive?

3.83 With reference to the preceding exercise, express each of the following events symbolically by listing its elements, and also express it in words:

(a) $\overline{A}$;

(b) $A \cup B$;

(c) $A \cap C$;

(d) $\overline{A} \cap B$.

3.84 Use Venn diagrams to verify that

(c) $\overline{A \cup \overline{B}} = A \cap B$;

(b) $\overline{A \cap \overline{B}} = \overline{A} \cup B$.

3.85 The quality of surround sound from four digital movie systems is to be rated superior, average, or inferior, and we are interested only in how many of the systems get each of these ratings. Draw a tree diagram which shows the 12 different possibilities.

3.86 Alarm units are to be connected at four fixed positions along a pipeline. In how many ways can the four available alarm units be connected to the four positions along the line?

3.87 In how many ways can two out of seven chemical engineers be assigned to a new project?

3.88 Refer to Example 12 of motors for miniaturized capsules, but instead suppose that 20 motors are available and that 4 will not operate satisfactorily, when placed in a capsule. If the scientist wishes to fabricate two capsules, with two motors each, find the probability that among the four randomly selected motors

(a) all four operate satisfactorily;

(b) three operate satisfactorily and one does not.

3.89 Given $P(A) = 0.30$, $P(B) = 0.40$, and $P(A \cap B) = 0.20$, find

(a) $P(A \cup B)$;

(b) $P(\overline{A} \cap B)$;

(c) $P(A \cap \overline{B})$;

(d) $P(\overline{A} \cup \overline{B})$.

(e) Are A and B independent?

3.90 In a sample of 446 cars stopped at a roadblock, only 67 of the drivers had their seatbelts fastened. Estimate the probability that a driver stopped on that road will have his or her seatbelt fastened.

3.91 The marketing manager reported to the head engineer regarding a survey concerning the company's portable cleaning tool. He claims that, among the 200 customers surveyed, 165 said the product is reliable, 117 said it is easy to use, 88 said it is both reliable and easy to use, and 33 said it is neither reliable nor easy to use. Explain why the head engineer should question this claim.

3.92 The probabilities that a satellite launching rocket will explode during lift-off or have its guidance system fail in flight are 0.0002 and 0.0005. Assuming independence find the probabilities that such a rocket will

(a) not explode during lift-off;

(b) explode during lift-off or have its guidance system fail in flight;

(c) neither explode during lift-off nor have its guidance system fail in flight.

3.93 Given $P(A) = 0.20$, $P(B) = 0.45$, and $P(A \cap B) = 0.09$, verify that

(a) $P(A \mid B) = P(A)$;

(b) $P(A \mid \overline{B}) = P(A)$;

(c) $P(B \mid A) = P(B)$;

(d) $P(B \mid \overline{A}) = P(B)$.

3.94 If events A and B are independent and $P(A) = 0.50$ and $P(B) = 0.30$, find

(a) $P(A \cap B)$;

(b) $P(A \mid B)$;

(c) $P(A \cup B)$;

(d) $P(\overline{A} \cap \overline{B})$.

3.95 The following frequency table shows the classification of 58 landfills in a state according to their concentration of the three hazardous chemicals arsenic, barium, and mercury.

		Barium			
		High		Low	
		Mercury		Mercury	
		High	Low	High	Low
Arsenic	High	1	3	5	9
	Low	4	8	10	18

If a landfill is selected at random, find the probability that it has

(a) a high concentration of mercury;

(b) a high concentration of barium and low concentrations of arsenic and mercury;

(c) high concentrations of any two of the chemicals and low concentration of the third;

(d) a high concentration of any one of the chemicals and low concentrations of the other two.

3.96 Refer to Exercise 3.95. Given that a landfill, selected at random, is found to have a high concentration of barium, what is the probability that its concentration is

(a) high in mercury?

(b) low in both arsenic and mercury?

(c) high in either arsenic or mercury?

3.97 An explosion in an LNG storage tank in the process of being repaired could have occurred as the result of static electricity, malfunctioning electrical equipment, an open flame in contact with the liner, or purposeful action (industrial sabotage). Interviews with engineers who were analyzing the risks involved led to estimates that such a explosion would occur with probability 0.25 as a result of static electricity, 0.20 as a result of malfunctioning electric equipment, 0.40 as a result of an open flame, and 0.75 as a result of purposeful action. These interviews also yielded subjective estimates of the prior probabilities of these four causes of 0.30, 0.40, 0.15, and 0.15, respectively. What was the most likely cause of the explosion?

3.98 During the semester, students in a statistics class do their homework on a laptop, tablet, or neither. They could also use both. The probability of using a laptop is 0.9 and the probability of using a tablet is 0.3. Also, the probability of using both during the semester is 0.2.

(a) For a randomly selected student, what is the probability the student does not use a laptop but uses a tablet?

(b) What is the probability the student uses exactly one of the devices?

(c) Given that the student uses at least one of the devices, what is the probability of using a laptop.

3.99 Amy commutes to work by two different routes A and B. If she comes home by route A, then she will be home no later than 6 P.M. with probability 0.8, but if she comes home by route B, then she will be home no later than 6 P.M. with probability 0.7. In the past, the proportion of times that Amy chose route A is 0.4.

(a) What proportion of times is Amy home no later than 6 P.M.?

(b) If Amy is home after 6 P.M. today, what is the probability that she took route B?

3.100 **Long run relative frequency interpretation of probability. A simulation.** A long series of experiments can be simulated using *MINITAB* and then the relative frequencies plotted as in Figure 3.7b.

Dialog box:

To start, enter 0 and 1 in $C1$, .6 and .4 in $C2$ to represent the values and the probabilities.

Label C3 *Trial no.* and select

Calc > Make Patterned Data > Simple Set of Numbers.

Type $C3$ in **Store**, 1000 in **last value** and 1 in the other three boxes.

Next, label C4 *Outcomes* and select **Calc > Random Data > Discrete**.

Type 1000 in Number, $C4$ in **Store**, $C1$ in **Values** and $C2$ in **Probabilities**.

Label C5 *Relative frequency* and select **Calc > Calculator > partial sum**.

Type $C5$ in **Store** and $C4$ in **Expression** to read *PARS*($C4$) and then /$C3$ to read *PARS*($C4$)/$C3$.

Click **O K**.

Select **Graph > Scatterplot > With Connect Lines**.

Type $C5$ in **Y** and $C3$ in **X**. Click **OK**.

Change the probability of 1 to .7 and repeat.

Key Terms

Additive set function   59
Axioms of probability   60
Bayes' theorem   76
Classical probability concept   57
Combination   54
Complement   48
Conditional probability   70
Consistency of probabilities   68
Continuous sample space   47
Discrete sample space   47
Empty set   47
Event   47

Experiment   46
Factorial notation   52
Finite sample space   47
Frequency interpretation   58
General addition rule   64
General multiplication rule   70
Independent events   70
Intersection   48
Multiplication of choices   51
Mutually exclusive events   48
Odds   68
Outcome   46

Permutation   52
Rule of complement   65
Rule of total probability   75
Sample space   46
Set function   59
Special addition rule   64
Special product rule   71
Subjective probability   59
Tree diagram   50
Union   48
Venn diagram   48

4

PROBABILITY DISTRIBUTIONS

In most statistical problems we are concerned with one number or a few numbers that are associated with the outcomes of experiments. When inspecting a manufactured product we may be interested only in the number of defectives; in the analysis of a road test we may be interested only in the average speed and the average fuel consumption; and in the study of the performance of a miniature rechargeable battery we may be interested only in its power and lifelength. All these numbers are associated with situations involving an element of chance—in other words, they are values of random variables.

In the study of random variables we are usually interested in their probability distributions, namely, in the probabilities with which they take on the various values in their range. The introduction to random variables and probability distributions in Section 4.1 is followed by a discussion of various special probability distributions in Sections 4.2, 4.3, 4.6, 4.7, 4.8, and 4.9, and descriptions of the most important features of probability distributions in Sections 4.4 and 4.5.

4.1   Random Variables

To be more explicit about the concept of a random variable, let us refer again to the used car example of page 50 and the corresponding probabilities shown in Figure 3.9. Now let us refer to M_1 (low current mileage), P_1 (moderate price), and C_1 (inexpensive to operate) as preferred attributes. Suppose we are interested only in the number of preferred attributes a used car possesses. To find the probabilities that a used car will get 0, 1, 2, or 3 preferred attributes, let us refer to Figure 4.1, which is like Figure 3.9 in Chapter 3 except that we indicate for each outcome the number of preferred attributes. Adding the respective probabilities, we find that for 0 preferred attributes the probability is

$$0.07 + 0.06 + 0.03 + 0.02 = 0.18$$

and for one preferred attribute the probability is

$$0.01 + 0.01 + 0.16 + 0.10 + 0.02 + 0.05 + 0.14 + 0.01 = 0.50$$

For two preferred attributes, the probability is

$$0.06 + 0.07 + 0.02 + 0.09 + 0.05 = 0.29$$

and for three preferred attributes the probability is 0.03.

These results may be summarized, as in the following table, where x denotes a possible number of preferred attributes

x	0	1	2	3
Probability	0.18	0.50	0.29	0.03

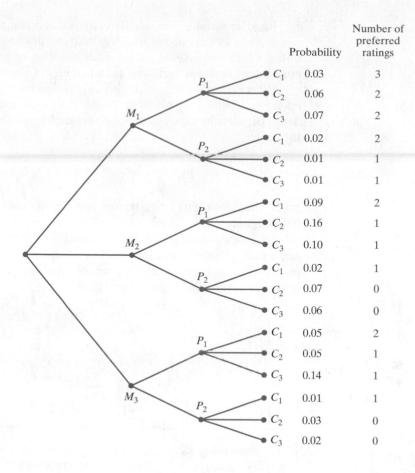

Figure 4.1
Used cars and numbers of
preferred attributes

	Probability	Number of preferred ratings
C_1	0.03	3
C_2	0.06	2
C_3	0.07	2
C_1	0.02	2
C_2	0.01	1
C_3	0.01	1
C_1	0.09	2
C_2	0.16	1
C_3	0.10	1
C_1	0.02	1
C_2	0.07	0
C_3	0.06	0
C_1	0.05	2
C_2	0.05	1
C_3	0.14	1
C_1	0.01	1
C_2	0.03	0
C_3	0.02	0

The numbers 0, 1, 2, and 3 in this table are values of a **random variable**—the number of preferred attributes. Corresponding to each elementary outcome in the sample space there is exactly one value x for this random variable. That is, the random variable may be thought of as a function defined over the elements of the sample space. This is how we define random variables in general; they are functions defined over the elements of a sample space.

Random variables | A **random variable** is any function that assigns a numerical value to each possible outcome.

The numerical value should be chosen to quantify an important characteristic of the outcome.

Random variables are denoted by capital letters X, Y, and so on, to distinguish them from their possible values given in lowercase x, y.

To find the probability that a random variable will take on any one value within its range, we proceed as in the above example. Indeed, the table which we obtained displays another function, called the **probability distribution** of the random variable. To denote the values of a probability distribution, we shall use such symbols as $f(x), g(x), \varphi(y), h(z)$, and so on. Strictly speaking, it is the function $f(x) = P(X = x)$ which assigns probability to each possible outcome x that is called the probability distribution. However, we will follow the common practice of also calling $f(x)$ the probability distribution, with the understanding that we are referring to the function and that the range of x values is part of the definition.

Random variables are usually classified according to the number of values they can assume. In this chapter we shall limit our discussion to **discrete random variables**, which can take on only a finite number, or a countable infinity of values; **continuous random variables** are taken up in Chapter 5.

Whenever possible, we try to express probability distributions by means of equations. Otherwise, we must give a table that actually exhibits the correspondence between the values of the random variable and the associated probabilities. For instance,

$$f(x) = \frac{1}{6} \qquad \text{for } x = 1, 2, 3, 4, 5, 6$$

gives the probability distribution for the number of points we roll with a balanced die.

Of course, not every function defined for the values of a random variable can serve as a probability distribution. Since the values of probability distributions are probabilities and one value of a random variable must always occur, it follows that if $f(x)$ is a probability distribution, then

$$f(x) \geq 0 \qquad \text{for all } x$$

and

$$\sum_{\text{all } x} f(x) = 1$$

Probability distributions

The **probability distribution** of a discrete random variable X is a list of the possible values of X together with their probabilities

$$f(x) = P[X = x]$$

The probability distribution always satisfies the conditions

$$f(x) \geq 0 \quad \text{and} \quad \sum_{\text{all } x} f(x) = 1$$

EXAMPLE 1   **Checking for nonnegativity and total probability equals one**

Check whether the following can serve as probability distributions:

(a) $f(x) = \dfrac{x-2}{2} \qquad$ for $x = 1, 2, 3, 4$

(b) $h(x) = \dfrac{x^2}{25} \qquad$ for $x = 0, 1, 2, 3, 4$

Solution   (a) This function cannot serve as a probability distribution because $f(1)$ is negative.

(b) The function cannot serve as a probability distribution because the sum of the five probabilities is $\dfrac{6}{5}$ and not 1. ∎

It is often helpful to visualize probability distributions by means of graphs like those of Figure 4.2. The one on the left is called a **probability histogram**; the areas

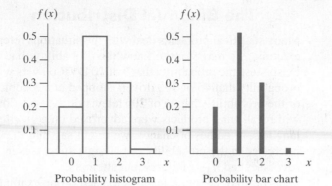

Figure 4.2

Graphs of the probability distribution of the number of preferred attributes

Probability histogram

Probability bar chart

of the rectangles are equal to the corresponding probabilities so their heights are proportional to the probabilities. The bases touch so that there are no gaps between the rectangles representing the successive values of the random variable. The one on the right is called a **bar chart**; the heights of the rectangles are also proportional to the corresponding probabilities, but they are narrow and their width is of no significance.

Besides the probability $f(x)$ that the value of a random variable is x, there is an important related function. It gives the probability $F(x)$ that the value of a random variable is less than or equal to x. Specifically,

$$F(x) = P[X \leq x] \quad \text{for all } -\infty < x < \infty$$

and we refer to the function $F(x)$ as the **cumulative distribution function** or just the **distribution function** of the random variable. For any value x, it adds up, or accumulates, all the probability assigned to that value and smaller values.

Referring to the used car example and basing our calculations on the table on page 84, we get

x	0	1	2	3
$F(x)$	0.18	0.68	0.97	1.00

for the cumulative distribution function of the number of preferred attributes.

The cumulative distribution jumps the amount $f(x)$ at $x = 0, 1, 2, 3$ and is constant between the values in the table as illustrated in Figure 4.3. The solid dots emphasize the fact that $F(x)$ takes the upper value at jumps and this makes it continuous from the right.

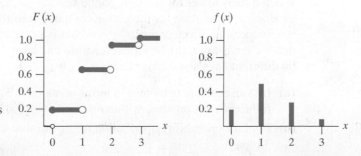

Figure 4.3

The cumulative distribution has jumps corresponding to $f(x) = P[X = x]$

4.2 The Binomial Distribution

Many statistical problems deal with the situations referred to as **repeated trials**. For example, we may want to know the probability that 1 of 5 rivets will rupture in a tensile test, the probability that 9 of 10 DVR players will run at least 1,000 hours, the probability that 45 of 300 drivers stopped at a roadblock will be wearing seatbelts, or the probability that 66 of 200 television viewers (interviewed by a rating service) will recall what products were advertised on a given program. To borrow from the language of games of chance, we might say that in each of these examples we are interested in the probability of getting x **successes** in n **trials**, or, in other words, x successes and $n - x$ failures in n attempts.

There are common features to each of the examples. They are all composed of a series of trials which we refer to as **Bernoulli trials** if the following assumptions hold.

1. **There are only two possible outcomes for each trial** (arbitrarily called "success" and "failure," without inferring that a success is necessarily desirable).
2. **The probability of success is the same for each trial**.
3. **The outcomes from different trials are independent**.

If the assumptions cannot be met, the theory we develop does not apply.

EXAMPLE 2   **Checking the adequacy of the Bernoulli trials assumptions**

Can the following be treated as Bernoulli trials? Drivers stopped at a roadblock will be checked for failure to wear a seatbelt.

Solution   There are only two outcomes, and we call not wearing a seatbelt a success. (Success in this context does not mean success in life.)

If all cars are treated alike, their drivers would all have the same probability of not wearing a seatbelt. If drivers are grouped by age, you may need different probabilities for persons under 20 than for those 30 to 40 years old. Then you would not have Bernoulli trials.

The results on seatbelt wear, for different drivers, should be independent. There is no obvious common cause. If someone caught without a seatbelt were to inform oncoming cars about the checkpoint, that would introduce dependence. ∎

In the problems we study in this section, we add the additional assumption that the number of trials is fixed in advance.

4. **There are a fixed number n of Bernoulli trials conducted**.

EXAMPLE 3   **Binomial probability distribution $n = 3$**

When a relay tower for wireless phone service breaks down, it quickly becomes an expensive proposition for the phone company, and the cost increases with the time it is inoperable. From company records, it is postulated that the probability is 0.90 that the breakdown can be repaired within one hour. For the next three breakdowns, on different days and different towers,

(a) List all possible outcomes in terms of success, S, repaired within one hour, and failure, F, not repaired within one hour.

(b) Find the probability distribution of the number of successes, X, among the 3 repairs.

Solution   (a) We write *FSS* for the outcome where the first repair is not made within one hour and the second and third are successful. The $2 \times 2 \times 2 = 8$ possible outcomes can be systematically arranged as follows:

FFF	*FFS*	*FSS*	*SSS*
	FSF	*SFS*	
	SFF	*SSF*	

$$X = 0 \quad X = 1 \quad X = 2 \quad X = 3$$

where the number of successes X is the same for each outcome in a column. This value is recorded at the bottom of the column.

(b) The results of repairs on different days and different towers should be independent. Also, the probability of success 0.90 is the same for each repair. Therefore, the probability that $X = 0$ is $0.1 \times 0.1 \times 0.1 = 0.001$. Next, the probability of *FFS* is $0.1 \times 0.1 \times 0.9 = 0.009$ and both *SFS* and *FSS* have the same probability. Consequently, the probability that $X = 1$ is $3 \times 0.009 = 0.027$. Note that the number of outcomes where $X = 1$ is just the number of ways to select 1 or 3 trials for an *S* and the others are *F*. By similar reasoning, the probability that $X = 2$ is

$$3\,(0.1 \times 0.9 \times 0.9) = 0.243 = \binom{3}{2} (0.9)^2 (0.1)^1$$

Finally, the probability that $X = 3$, no repair takes over one hour, is

$$0.9 \times 0.9 \times 0.9 = 0.729 = \binom{3}{3} (0.9)^3 (0.1)^0$$

All of these probabilities can be expressed by the formula

$$f(x) = P(X = x) = \binom{3}{x}(0.9)^x (0.1)^{3-x} \qquad \text{for } x = 0, 1, 2, 3$$

This is the probability distribution for a binomial random variable when the success probability is $p = 0.9$ and there are $n = 3$ trials. ∎

Let X be the random variable that equals the number of successes in n trials. To obtain probabilities concerning X, we proceed as follows: If p and $1 - p$ are the probabilities of success and failure on any one trial, then the probability of getting x successes and $n - x$ failures, in some specific order, is $p^x(1 - p)^{n-x}$. Clearly, in this product of p's and $(1-p)$'s there is one factor p for each success, one factor $1 - p$ for each failure. The x factors p and $n - x$ factors $1 - p$ are all multiplied together by virtue of the generalized multiplication rule for more than two independent events. Since this probability applies to any point of the sample space that represents x successes and $n - x$ failures (in any specific order), we have only to count how many points of this kind there are, and then multiply $p^x(1 - p)^{n-x}$ by this. The number of ways in which we can select the x trials on which there is to be a success is $\binom{n}{x}$, the number of combinations of x objects selected from a set of n objects. Multiplying, we arrive at the following result:

Binomial distribution

$$b(x; n, p) = \binom{n}{x} p^x(1 - p)^{n-x} \qquad x = 0, 1, 2, \ldots, n$$

This probability distribution is called the **binomial distribution** because for $x = 0, 1, 2, \ldots$, and n, the values of the probabilities are the successive terms of the binomial expansion of $[p + (1 - p)]^n$. For the same reason, the combinatorial quantities $\binom{n}{x}$ are referred to as **binomial coefficients**. Actually, the preceding equation defines a family of probability distributions, with each member characterized by a given value of the **parameter** p and the number of trials n.

Important information about the shape of binomial distributions is shown in Figures 4.4. First, if $p = 0.50$, the equation for the binomial distribution is

$$b(x; n, 0.50) = \binom{n}{x}(0.5)^n$$

and since

$$\binom{n}{n-x} = \binom{n}{x}$$

it follows that $b(x; n, 0.50) = b(n - x; n, 0.50)$. For any n, the binomial distribution with $p = 0.5$ is a **symmetrical distribution**. This means that the probability histograms of such binomial distributions are symmetrical, as is illustrated in Figure 4.4(b). Note, however, that if p is less than 0.50, it is more likely that X will be small rather than large compared to $n/2$ and that the opposite is true if p is greater than 0.50. This is illustrated in Figure 4.4(a) and (c), showing binomial distributions with $n = 5$ and $p = 0.30$ and $p = 0.70$. These two are mirror images of each other as can be verified more generally in Exercise 4.7.

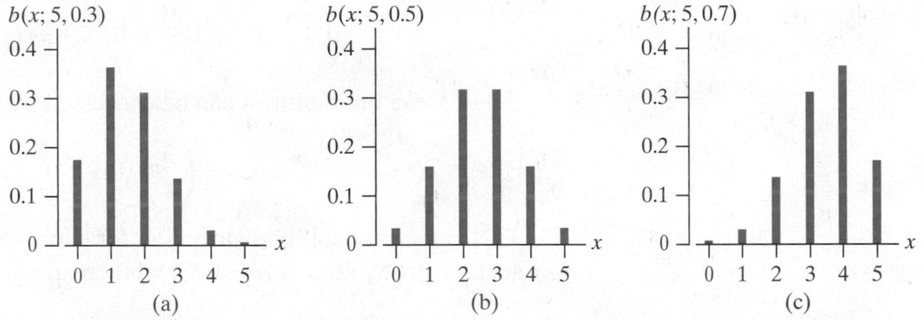

Figure 4.4
Binomial distributions for
$n = 5$ and (a) $p = 0.3$
(b) $p = 0.5$ (c) $p = 0.7$

Finally, a probability distribution that has a probability bar chart like those in Figure 4.4(a) and 4.4(c) is said to be a long-tailed or **skewed distribution**. It is said to be a **positively skewed distribution** if the tail is on the right, and it is said to be **negatively skewed** if the tail is on the left.

EXAMPLE 4

Evaluating binomial probabilities

It has been claimed that in 60% of all solar-heat installations the utility bill is reduced by at least one-third. Accordingly, what are the probabilities that the utility bill will be reduced by at least one-third in

(a) four of five installations;

(b) at least four of five installations?

Solution   **(a)** Substituting $x = 4$, $n = 5$, and $p = 0.60$ into the formula for the binomial distribution, we get

$$b(4; 5, 0.60) = \binom{5}{4}(0.60)^4(1 - 0.60)^{5-4}$$

$$= 0.259$$

(b) Substituting $x = 5$, $n = 5$, and $p = 0.60$ into the formula for the binomial distribution, we get

$$b(5; 5, 0.60) = \binom{5}{5} (0.60)^5 (1 - 0.60)^{5-5}$$

$$= 0.078$$

and the answer is $b(4; 5, 0.60) + b(5; 5, 0.60) = 0.259 + 0.078 = 0.337$. ∎

If n is large, the calculation of binomial probabilities can become quite tedious. Many statistical software programs have binomial distribution commands (see Exercises 4.30 and 4.31), as do some statistical calculators. Otherwise it is convenient to refer to special tables. Table 1 at the end of the book gives the values of

$$B(x; n, p) = \sum_{k=0}^{x} b(k; n, p) \qquad \text{for } x = 0, 1, 2, \ldots, n$$

for $n = 2$ to $n = 20$ and $p = 0.05, 0.10, 0.15, \ldots, 0.90, 0.95$. We tabulated the **cumulative probabilities** rather than the values of $b(x; n, p)$, because the values of $B(x; n, p)$ are the ones needed more often in statistical applications. Note, however, that the values of $b(x; n, p)$ can be obtained by subtracting adjacent entries in Table 1. Because the two cumulative probabilities $B(x; n, p)$ and $B(x-1; n, p)$ differ by the single term $b(x; n, p)$

$$b(x; n, p) = B(x; n, p) - B(x - 1; n, p)$$

where $B(-1) = 0$. The examples that follow illustrate the direct use of Table 1 and the use of this relationship.

EXAMPLE 5   **Evaluating cumulative binomial probabilities**

If the probability is 0.05 that a certain wide-flange column will fail under a given axial load, what are the probabilities that among 16 such columns

(a) at most two will fail;

(b) at least four will fail?

Solution   **(a)** Table 1 shows that $B(2; 16, 0.05) = 0.9571$.

(b) Since

$$\sum_{x=4}^{16} b(x; 16, 0.05) = 1 - B(3; 16, 0.05)$$

Table 1 yields $1 - 0.9930 = 0.0070$. ∎

EXAMPLE 6   **Finding a binomial probability using cumulative binomial probabilities**

Sport stories and financial reports, written by algorithms based on artificial intelligence, have become common place. One company fed its algorithm with box scores and play-by-play information and created over one million on-line reports of little league games in 2011. They now write many stories on the Big Ten Network site.[1]

---

[1]S. Levy, "The Rise of the Robot Reporter", *Wired*, May, (2012) pp.

Suppose that the algorithm, or robot reporter, typically writes proportion 0.65 of the stories on the site. If 15 new stories are scheduled to appear on a web site next weekend, find the probability that

(a) 11 will be written by the algorithm.

(b) at least 10 will be written by the algorithm

(c) between 8 and 11 inclusive will be written by the algorithm.

Solution   (a) Using the relationship to cumulative probabilities and then looking up these probabilities in Table 1, we get

$$b(11, 15, 0.65) = B(11, 15, 0.65) - B(10, 15, 0.65)$$
$$= 0.8273 - 0.6481 = 0.1792$$

(b) $1 - B(9, 18, 0.65) = 1 - 0.4357 = 0.5643$

(c) $B(11, 15, 0.65) - B(7, 15, 0.65) = 0.8273 - 0.1132 = 0.7141$
This last calculation, depicted in Figure 4.5, visually demonstrates the calculation of the probability of an interval as the cumulative probabilities up to the upper value minus the cumulative probabilities up to one less than the lower limit.

[Using **R**: (a) **dbinom(11, 15, .65)**   (b) **1 - pbinom(9, 15, .65)**
(c) **pbinom(11, 15, .65) - pbinom(7, 15, .65)**]

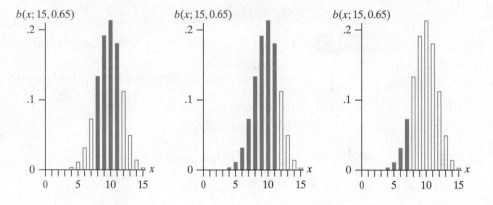

Figure 4.5
The calculation of
$P(8 \leq X \leq 11)$

$$P(8 \leq X \leq 11) = B(11; 15, 0.65) - B(7; 15, 0.65)$$

The following example illustrates the use of the binomial distribution in a problem of decision making.

EXAMPLE 7   **A binomial probability to guide decision making**

A manufacturer of external hard drives claims that only 10% of his drives require repairs within the warranty period of 12 months. If 5 of 20 of his drives required repairs within the first year, does this tend to support or refute the claim?

Solution   Let us first find the probability that 5 or more of 20 of the drives will require repairs within a year when the probability that any one will require repairs within a year is

0.10. Using Table 1, we get

$$\sum_{x=5}^{20} b(x;\ 20, 0.10) = 1 - B(4;\ 20, 0.10)$$

$$= 1 - 0.9568$$

$$= 0.0432$$

Since this probability is very small, it would seem reasonable to reject the hard drive manufacturer's claim. ∎

4.3 The Hypergeometric Distribution

Suppose that we are interested in the number of defectives in a sample of n units drawn without replacement from a lot containing N units, of which a are defective. Let the sample be drawn in such a way that at each successive drawing, whatever units are left in the lot have the same chance of being selected. The probability that the first drawing will yield a defective unit is $\dfrac{a}{N}$, but for the second drawing it is $\dfrac{a-1}{N-1}$ or $\dfrac{a}{N-1}$, depending on whether or not the first unit drawn was defective. Thus, the trials are not independent, the third assumption underlying the binomial distribution is not met, and the binomial distribution does not apply. Note that the binomial distribution would apply if we do **sampling with replacement**, namely, if each unit selected for the sample is replaced before the next one is drawn.

To solve the problem of **sampling without replacement** (that is, as we originally formulated the problem), let us proceed as follows: The x successes (defectives) can be chosen in $\dbinom{a}{x}$ ways, the $n - x$ failures (nondefectives) can be chosen in $\dbinom{N-a}{n-x}$ ways, and hence, x successes and $n - x$ failures can be chosen in $\dbinom{a}{x}\dbinom{N-a}{n-x}$ ways. Also, n objects can be chosen from a set of N objects in $\dbinom{N}{n}$ ways, and if we consider all the possibilities as equally likely, it follows that for sampling without replacement the probability of getting "x successes in n trials" is

Hypergeometric distribution

$$h(x;\ n, a, N) = \frac{\dbinom{a}{x}\dbinom{N-a}{n-x}}{\dbinom{N}{n}} \qquad \text{for } x = 0, 1, \ldots, n$$

where x cannot exceed a and $n - x$ cannot exceed $N - a$. This equation defines the **hypergeometric distribution**, whose parameters are the sample size n, the lot size (or population size) N, and the number of "successes" in the lot a.

EXAMPLE 8   **Calculating a probability using the hypergeometric distribution**

An Internet-based company that sells discount accessories for cell phones often ships an excessive number of defective products. The company needs better control of quality. Suppose it has 20 identical car chargers on hand but that 5 are defective.

If the company decides to randomly select 10 of these items, what is the probability that 2 of the 10 will be defective?

Solution   Substituting $x = 2, n = 10, a = 5$, and $N = 20$ into the formula for the hypergeometric distribution, we get

$$h(2; 10, 5, 20) = \frac{\binom{5}{2}\binom{15}{8}}{\binom{20}{10}} = \frac{10 \times 6{,}435}{184{,}756} = 0.348 \qquad \blacksquare$$

In the preceding example, n was not small compared to N, and if we had made the mistake of using the binomial distribution with $n = 10$ and $p = \frac{5}{20} = 0.25$ to calculate the probability of two defectives, the result would have been 0.282, which is much too small. However, when n is small compared to N, less than $\frac{N}{10}$, the composition of the lot is not seriously affected by drawing the sample without replacement, and the binomial distribution with the parameters n and $p = \frac{a}{N}$ will yield a good approximation.

EXAMPLE 9   **A numerical comparison of the hypergeometric and binomial distributions**

Repeat the preceding example but with 100 car chargers, of which 25 are defective, by using

(a) the formula for the hypergeometric distribution;

(b) the formula for the binomial distribution as an approximation.

Solution   (a) Substituting $x = 2, n = 10, a = 25$, and $N = 100$ into the formula for the hypergeometric distribution, we get

$$h(2; 10, 25, 100) = \frac{\binom{25}{2}\binom{75}{8}}{\binom{100}{10}} = 0.292$$

(b) Substituting $x = 2, n = 10$, and $p = \frac{25}{100} = 0.25$ into the formula for the binomial distribution, we get

$$b(2; 10, 0.25) = \binom{10}{2}(0.25)^2(1 - 0.25)^{10-2}$$

$$= 0.282 \qquad \blacksquare$$

Observe that the difference between the two values is only 0.010. In general, it can be shown that $h(x; n, a, N)$ approaches $b(x; n, p)$ with $p = \frac{a}{N}$ when $N \to \infty$, and a good rule of thumb is to use the binomial distribution as an approximation to the hypergeometric distribution if $n \le \frac{N}{10}$.

Although we have introduced the hypergeometric distribution in connection with a problem of sampling inspection, it has many other applications. For instance, it can be used to find the probability that 3 of 12 homemakers prefer Brand A detergent to Brand B, if they are selected from among 200 homemakers among whom 40 actually prefer Brand A to Brand B. Also, it can be used in connection with a problem of selecting industrial diamonds, some of which have superior qualities and some of which do not, or in connection with a problem of sampling income tax returns, where a among N returns filed contain questionable deductions, and so on.

Exercises

4.1 Suppose that a probability of $\dfrac{1}{12}$ is assigned to each point of the sample space of part (a) of Exercise 3.1 on page 55. Find the probability distribution of the total number of lakes and streams that are contaminated.

4.2 An experiment consists of four tosses of a coin. Denoting the outcomes $HHTH, THTT, \ldots$ and assuming that all 16 outcomes are equally likely, find the probability distribution for the total number of heads.

4.3 Determine whether the following can be probability distributions of a random variable which can take on only the values 1, 2, 3, and 4.

(a) $f(1)=0.19,\quad f(2)=0.27,\quad f(3)=0.27,\quad$ and $f(4)=0.27$;

(b) $f(1)=0.24,\quad f(2)=0.24,\quad f(3)=0.24,\quad$ and $f(4)=0.24$;

(c) $f(1)=0.35,\quad f(2)=0.33,\quad f(3)=0.34,\quad$ and $f(4)=-0.02$.

4.4 Check whether the following can define probability distributions, and explain your answers.

(a) $f(x) = \dfrac{x}{12}$ for $x = 0, 1, 2, 3, 4$

(b) $f(x) = \dfrac{4 - x^2}{7}$ for $x = 0, 1, 2$

(c) $f(x) = \dfrac{1}{5}$ for $x = 4, 5, 6, 7, 8$

(d) $f(x) = \dfrac{3x + 1}{50}$ for $x = 1, 2, 3, 4, 5$

4.5 Given that $f(x) = \dfrac{k}{2^x}$ is a probability distribution for a random variable that can take on the values $x = 0, 1, 2, 3,$ and 4, find k.

4.6 With reference to Exercise 4.5, find an expression for the distribution function $F(x)$ of the random variable.

4.7 Prove that $b(x;\, n, p) = b(n - x;\, n, 1 - p)$.

4.8 Prove that $B(x;\, n, p) = 1 - B(n - x - 1;\, n, 1 - p)$.

4.9 Do the assumptions for Bernoulli trials appear to hold? Explain. If the assumptions hold, identify success and the probability of interest.

(a) A TV ratings company will use their electronic equipment to check a sample of homes around the city to see whether or not each has a set tuned to the mayor's speech on the local channel.

(b) Among 6 nuclear power plants in a state, 2 have had serious violations in last five years. Two plants will be selected at random, one after the other, and the outcome of interest is a serious violation in the last five years.

4.10 What conditions for the binomial distribution, if any, fail to hold in the following situations?

(a) For each of a company's eight production facilities, record whether or not there was an accident in the past week. The largest facility has three times the number of production workers as the smallest facility.

(b) For each of three shifts, the number of units produced will be compared with the long-term average of 560 and it will be determined whether or not production exceeds 560 units. The second shift will know the result for the first shift before they start working, and the third shift will start with the knowledge of how the first two shifts performed.

4.11 Which conditions for the binomial distribution, if any, fail to hold in the following situations?

(a) The number of persons having a cold at a family reunion attended by 30 persons.

(b) Among 8 projectors in the department office, 2 do not work properly but are not marked defective. Two are selected and the number that do not work properly will be recorded.

4.12 Use Table 1, or software, to find

(a) $B(8;\, 16, 0.40)$;

(b) $b(8;\, 16, 0.40)$;

(c) $B(9;\, 12, 0.60)$;

(d) $b(9;\, 12, 0.60)$;

(e) $\displaystyle\sum_{k=6}^{20} b(k;\, 20, 0.15)$;

(f) $\displaystyle\sum_{k=6}^{9} b(k;\, 9, 0.70)$.

4.13 Use Table 1, or software, to find

(a) $B(7;\, 18, 0.45)$;

(b) $b(7;\, 18, 0.45)$;

(c) $B(8;\, 11, 0.95)$;

(d) $b(8;\, 11, 0.95)$;

(e) $\sum_{k=4}^{11} b(k; 11, 0.35)$;

(f) $\sum_{k=2}^{4} b(k; 10, 0.30)$.

4.14 Rework the decision problem in Example 7, supposing that only 3 of the 20 hard drives required repairs within the first year.

4.15 Human error is given as the reason for 75% of all accidents in a plant. Use the formula for the binomial distribution to find the probability that human error will be given as the reason for two of the next four accidents.

4.16 If the probability is 0.40 that steam will condense in a thin-walled aluminum tube at 10 atm pressure, use the formula for the binomial distribution to find the probability that, under the stated conditions, steam will condense in 4 of 12 such tubes.

4.17 During one stage in the manufacture of integrated circuit chips, a coating must be applied. If 70% of chips receive a thick enough coating, use Table 1 or software to find the probabilities that, among 15 chips:

(a) at least 12 will have thick enough coatings;

(b) at most 6 will have thick enough coatings;

(c) exactly 10 will have thick enough coatings.

4.18 The probability that the noise level of a wide-band amplifier will exceed 2 dB is 0.05. Use Table 1 or software to find the probabilities that among 12 such amplifiers the noise level of

(a) one will exceed 2 dB;

(b) at most two will exceed 2 dB;

(c) two or more will exceed 2 dB.

4.19 An agricultural cooperative claims that 90% of the watermelons shipped out are ripe and ready to eat. Find the probabilities that among 18 watermelons shipped out

(a) all 18 are ripe and ready to eat;

(b) at least 16 are ripe and ready to eat;

(c) at most 14 are ripe and ready to eat.

4.20 A quality-control engineer wants to check whether (in accordance with specifications) 95% of the electronic components shipped by his company are in good working condition. To this end, he randomly selects 15 from each large lot ready to be shipped and passes the lot if the selected components are all in good working condition; otherwise, each of the components in the lot is checked. Find the probabilities that the quality-control engineer will commit the error of

(a) holding a lot for further inspection even though 95% of the components are in good working condition;

(b) letting a lot pass through without further inspection even though only 90% of the components are in good working condition;

(c) letting a lot pass through without further inspection even though only 80% of the components are in good condition.

4.21 A food processor claims that at most 10% of her jars of instant coffee contain less coffee than claimed on the label. To test this claim, 16 jars of her instant coffee are randomly selected and the contents are weighed; her claim is accepted if fewer than 3 of the jars contain less coffee than claimed on the label. Find the probabilities that the food processor's claim will be accepted when the actual percentage of her jars containing less coffee than claimed on the label is

(a) 5%; (b) 10%; (c) 15%; (d) 20%.

4.22 Refer to Exercise 4.2.

(a) Determine the cumulative probability distribution $F(x)$.

(b) Graph the probability distribution of $f(x)$ as a bar chart and below it graph $F(x)$.

4.23 Four emergency radios are available for rescue workers but one does not work properly. Two randomly selected radios are taken on a rescue mission. Let X be the number that work properly between the two.

(a) Determine the probability distribution $f(x)$ of X.

(b) Determine the cumulative probability distribution $F(x)$ of X.

(c) Graph $f(x)$ as a bar chart and below it graph $F(x)$.

4.24 Suppose that, next month, the technology group will receive 24 requests for help with computer problems. Among these, 19 are software related and 5 are hardware problems. If a novice in the group is randomly assigned to 3 problems, what are the probabilities that she will address

(a) none of the hardware problems?

(b) only one of the hardware problems?

(c) at least 2 of the hardware problems

4.25 A maker of specialized instruments receives shipments of 24 circuit boards. Suppose one shipment contains 4 that are defective. An engineer selects a random sample of size 4. What are the probabilities that the sample will contain

(a) 0 defective circuit boards?

(b) 1 defective circuit board ?

(c) 2 or more defective circuit boards?

4.26 If 6 of 18 new buildings in a city violate the building code, what is the probability that a building inspector, who randomly selects 4 of the new buildings for inspection, will catch

(a) none of the buildings that violate the building code?

(b) 1 of the new buildings that violate the building code?

(c) 2 of the new buildings that violate the building code?

(d) at least 3 of the new buildings that violate the building code?

4.27 Among the 16 cities that a professional society is considering for its next 3 annual conventions, 7 are in the western part of the United States. To avoid arguments, the selection is left to chance. If none of the cities can be chosen more than once, what are the probabilities that

(a) none of the conventions will be held in the western part of the United States?

(b) all of the conventions will be held in the western part of the United States?

4.28 A shipment of 120 burglar alarms contains 5 that are defective. If 3 of these alarms are randomly selected and shipped to a customer, find the probability that the customer will get one bad unit by using

(a) the formula for the hypergeometric distribution;

(b) the formula for the binomial distribution as an approximation.

4.29 Refer to Exercise 4.24 but now suppose there will be 72 requests among which 57 are software related and 15 are hardware related problems. Find the probability that, among the three problems assigned to the novice, 1 will be a hardware problem and 2 will not, by using

(a) the hypergeometric distribution

(b) the binomial distribution as an approximation.

4.30 Binomial probabilities can be calculated using *MINITAB*.

> **Dialog box:**
> **Calc > Probability Distribution > Binomial**
> Choose **Probability.**
> Enter 7 in **Number of trials** and .33 in **Probability of success**.
> Choose **Input constant** and enter 2.
> Click **OK**.

Output:
Probability Density Function
Binomial with $n = 7$ and $p = 0.33$

x	$P(X = x)$
2	0.308760

Find the binomial probabilities for $x = 5, 10, 15$ and 20 when $n = 27$ and $p = 0.47$.

4.31 Cumulative binomial probabilities can be calculated using *MINITAB*.

> **Dialog Box:**
> **Calc > Probability Distribution > Binomial**
> Choose **Cumulative Distribution**.
> Enter 7 in **Number of trials** and .33 in **Probability of success**.
> Choose **Input constant** and enter 2.
> Click **OK**.

Output:
Cumulative Distribution Function
Binomial with $n = 7$ and $p = 0.33$

x	$P(X <= x)$
2	0.578326

Find the cumulative binomial probabilities $x = 5, 10, 15$ and 20 when $n = 27$ and $p = 0.47$.

4.4   The Mean and the Variance of a Probability Distribution

Besides the binomial and hypergeometric distributions, there are many other probability distributions that have important engineering applications. However, before we go any further, let us discuss some general characteristics of probability distributions.

One such characteristic, that of the symmetry or **skewness** of a probability distribution, was illustrated in Figure 4.4; two other characteristics are apparent in Figure 4.6, which shows the probability histograms of two binomial distributions. One of these binomial distributions has the parameters $n = 4$ and $p = 1/2$, and the other has the parameters $n = 16$ and $p = 1/2$. Essentially, these two probability distributions differ in two respects. The first probability distribution is centered about $x = 2$, whereas the other (whose histogram is shaded) is centered about $x = 8$, and we say

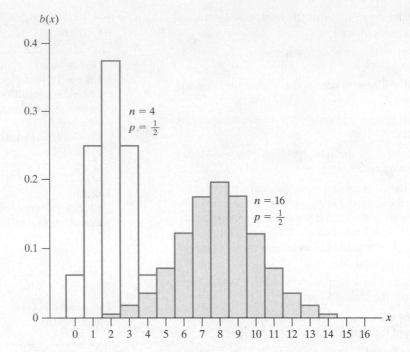

Figure 4.6
Probability histograms of two binomial distributions

that the two distributions differ in their **location**. Another distinction is that the histogram of the second distribution is broader, and we say that the two distributions differ in **variation**. To make such comparisons more specific, we shall introduce in this section two of the most important statistical measures, describing the location and the variation of a probability distribution—the **mean** and the **variance**, respectively.

The mean of a probability distribution is simply the mathematical expectation of a random variable having that distribution. If a random variable X takes on the values $x_1, x_2, \ldots,$ or x_k, with the probabilities $f(x_1), f(x_2), \ldots,$ and $f(x_k)$, its mathematical expectation or expected value is

$$x_1 \cdot f(x_1) + x_2 \cdot f(x_2) + \cdots + x_k \cdot f(x_k) = \sum (\text{value}) \times (\text{probability})$$

using the $\sum$ notation.

The mean of a probability distribution is denoted by the Greek letter μ (mu).

Alternatively, the **mean** of a random variable X, or its probability distribution, is called its **expected value** and is denoted by $E(X)$. Both μ and $E(X)$ refer to the same quantity.

Mean of discrete probability distribution

$$\mu = E(X)$$
$$= \sum_{\text{all } x} x \cdot f(x)$$

The mean of a probability distribution measures its center in the sense of an average, or by analogy to physics, in the sense of a center of gravity. Note that the above formula for μ is, in fact, that for the **first moment about the origin** of a discrete system of masses $f(x)$ arranged on a weightless straight line at distances x

from the origin. We do not have to divide here by

$$\sum_{\text{all } x} f(x)$$

as we do in the usual formula for the x-coordinate of the center of gravity, since the sum equals 1 by definition.

EXAMPLE 10   **The mean number of heads in three tosses of a fair coin**

Find the mean of the probability distribution of the number of heads obtained in 3 flips of a balanced coin.

Solution   The probabilities for 0, 1, 2, or 3 heads are $\frac{1}{8}, \frac{3}{8}, \frac{3}{8}$, and $\frac{1}{8}$ as can easily be verified by counting equally likely possibilities or by using the formula for the binomial distribution with $n = 3$ and $p = \frac{1}{2}$. Thus,

$$\mu = 0 \cdot \frac{1}{8} + 1 \cdot \frac{3}{8} + 2 \cdot \frac{3}{8} + 3 \cdot \frac{1}{8} = \frac{3}{2} \qquad \blacksquare$$

EXAMPLE 11   **The mean number of preferred used car attributes**

With reference to the used car example and the probabilities given on page 85, find the mean of the probability distribution of the number of preferred attributes.

Solution   Substituting $x = 0, 1, 2$, and 3 and the corresponding probabilities into the formula for μ, we get

$$\mu = 0 \, (0.18) + 1 \, (0.50) + 2 \, (0.29) + 3 \, (0.03)$$
$$= 1.17 \qquad \blacksquare$$

Returning to the second probability distribution of Figure 4.6, we could find its mean by calculating all the necessary probabilities (or by looking them up in Table 1) and substituting them into the formula for μ. However, if we reflect for a moment, we might argue that there is a 50-50 chance for a success on each trial, there are 16 trials, and it would seem reasonable to expect 8 heads and 8 tails (in the sense of a mathematical expectation). Similarly, we might argue that if a binomial distribution has the parameters $n = 200$ and $p = 0.20$, we can expect a success 20% of the time and, hence, on the average $200(0.20) = 40$ successes in 200 trials. These two values are, indeed, correct, and it can be shown in general that

Mean of binomial distribution

$$\mu = n \cdot p$$

for the mean of a binomial distribution. To prove this formula, we substitute the expression that defines $b(x; n, p)$ into the formula for μ, and we get

$$\mu = \sum_{x=0}^{n} x \cdot \frac{n!}{x!(n-x)!} \, p^x (1-p)^{n-x}$$

Then, making use of the fact that

$$\frac{x}{x!} = \frac{1}{(x-1)!}$$

and $n! = n(n - 1)!$, we factor out n and p to obtain

$$\mu = np \sum_{x=1}^{n} \frac{(n-1)!}{(x-1)!(n-x)!} \, p^{x-1}(1-p)^{n-x}$$

where the summation starts with $x = 1$ since the original summand is zero for $x = 0$. If we now let $y = x - 1$ and $m = n - 1$, we obtain

$$\mu = np \sum_{y=0}^{m} \frac{m!}{y!(m-y)!} \, p^{y}(1-p)^{m-y}$$

and this last sum can easily be recognized as that of all the terms of the binomial distribution with the parameters m and p. Hence, this sum equals 1 and it follows that $\mu = np$.

EXAMPLE 12   **Using $\mu = np$ to find the mean number of heads in three tosses**

Find the mean of the probability distribution of the number of heads obtained in 3 flips of a balanced coin.

Solution   For a binomial distribution with $n = 3$ and $p = \frac{1}{2}$, we get $\mu = 3 \cdot \frac{1}{2} = \frac{3}{2}$, and this agrees with the result obtained on page 99.   ∎

The formula $\mu = np$ applies, of course, only to binomial distributions. For other special distributions, we can express the mean in terms of their parameters. For instance, for the mean of the hypergeometric distribution with the parameters n, a, and N, we can write

Mean of hypergeometric distribution

$$\mu = n \cdot \frac{a}{N}$$

In Exercise 4.43, the reader will be asked to derive this formula by a method similar to the one we used to derive the formula for the mean of a binomial distribution.

EXAMPLE 13   **Using the formula for the mean of a hypergeometric distribution**

With reference to Example 8 in which 5 of 20 cell phone chargers are defective, find the mean of the probability distribution of the number of defectives in a sample of 10 randomly chosen for inspection.

Solution   Substituting $n = 10$, $a = 5$, and $N = 20$ into the above formula for μ, we get

$$\mu = 10 \cdot \frac{5}{20} = 2.5$$

In other words, if we inspect 10 of the chargers, we can expect 2.5 defectives, where *expect* is to be interpreted in the sense, it represents the long-run average number of defectives if 10 chargers are repeatedly selected from 20 chargers of which 5 are defective.   ∎

To study the second of the two properties of probability distributions mentioned on page 98, their variation, let us refer again to the two probability distributions of Figure 4.6. For the one where $n = 4$, there is a high probability of getting values close to the mean, but for the one where $n = 16$, there is a high probability of getting values scattered over considerable distances away from the mean. Using this property, it may seem reasonable to measure the variation of a probability distribution with the quantity

$$\sum_{\text{all } x} (x - \mu) \cdot f(x)$$

namely, the average amount by which the values of the random variable deviate from the mean. Unfortunately,

$$\sum_{\text{all } x} (x - \mu) \cdot f(x) = \sum_{\text{all } x} x \cdot f(x) - \sum_{\text{all } x} \mu \cdot f(x)$$

$$= \mu - \mu \cdot \sum_{\text{all } x} f(x) = \mu - \mu = 0$$

so that this expression is always equal to zero. However, since we are really interested in the magnitude of the deviations $x - \mu$ and not in their signs, it suggests itself that we average the absolute values of these deviations from the mean. This would, indeed, provide a measure of variation, but on purely theoretical grounds we prefer to work instead with the squares of the deviations from the mean. These quantities are also nonnegative, and their average is indicative of the spread or dispersion of a probability distribution. We thus define the **variance** of a probability distribution $f(x)$, or that of the random variable X which has that probability distribution, as

Variance of probability distribution

$$\sigma^2 = \sum_{\text{all } x} (x - \mu)^2 \cdot f(x)$$

where σ is the lowercase Greek letter for s. This measure is not in the same units (or dimension) as the values of the random variable, but we can adjust for this by taking the square root. This results in a measure of variation that is expressed in the same units in which the random variable is expressed. The **standard deviation** is defined as

Standard deviation of probability distribution

$$\sigma = \sqrt{\sum_{\text{all } x} (x - \mu)^2 \cdot f(x)}$$

EXAMPLE 14   **Calculating the standard deviations of two probability distributions**

Compare the standard deviations of the two probability distributions of Figure 4.6, on page 98.

Solution   Since $\mu = 4 \cdot \dfrac{1}{2} = 2$ for the binomial distribution with $n = 4$ and $p = \dfrac{1}{2}$, we find that the variance of this probability distribution is

$$\sigma^2 = (0 - 2)^2 \cdot \frac{1}{16} + (1 - 2)^2 \cdot \frac{4}{16} + (2 - 2)^2 \cdot \frac{6}{16}$$

$$+ (3 - 2)^2 \cdot \frac{4}{16} + (4 - 2)^2 \cdot \frac{1}{16} = 1$$

and, hence, that its standard deviation is $\sigma = 1$. Similarly, it can be shown that for the other distribution $\sigma = 2$, and we find that the second (shaded) distribution with the greater spread also has the greater standard deviation.   ■

An alternative formula for variance, that the reader is asked to verify in Exercise 4.49, sometimes simplifies the calculation of variance.

Computing formula for variance

$$\sigma^2 = \sum_{\text{all } x} x^2 \cdot f(x) - \mu^2$$

$$= E[X^2] - \mu^2$$

where $E[X^2]$ is defined as $\sum_{\text{all } x} x^2 \cdot f(x)$.

EXAMPLE 15 **Calculating variance using the alternative computing formula**

Use the preceding computing formula to determine the variance of the probability distribution of the number of points rolled with a balanced die.

Solution Since $f(x) = \dfrac{1}{6}$ for $x = 1, 2, 3, 4, 5$, and 6, we get

$$\mu = 1 \cdot \frac{1}{6} + 2 \cdot \frac{1}{6} + 3 \cdot \frac{1}{6} + 4 \cdot \frac{1}{6} + 5 \cdot \frac{1}{6} + 6 \cdot \frac{1}{6}$$

$$= \frac{7}{2}$$

$$E(X^2) = 1^2 \cdot \frac{1}{6} + 2^2 \cdot \frac{1}{6} + 3^2 \cdot \frac{1}{6} + 4^2 \cdot \frac{1}{6} + 5^2 \cdot \frac{1}{6} + 6^2 \cdot \frac{1}{6}$$

$$= \frac{91}{6}$$

and, hence,

$$\sigma^2 = \frac{91}{6} - \left(\frac{7}{2}\right)^2 = \frac{35}{12}$$

■

EXAMPLE 16 **The mean and variance of the number of incorrect addresses**

As part of a quality-improvement project focused on the delivery of mail at a department office within a large company, data were gathered on the number of different addresses that had to be changed so the mail could be redirected to the correct mail stop. The distribution, given in the first two columns of the table below, describes the number of redirects per delivery. Compute the mean and variance.

Solution We determine the columns $xf(x)$ and $x^2f(x)$

x	$f(x)$	$xf(x)$	$x^2f(x)$
0	.05	.0	0.0
1	.20	.2	0.2
2	.45	.9	1.8
3	.20	.6	1.8
4	.10	.4	1.6
Total		2.1	5.4

so $\mu = 2.1$ and $\sigma^2 = 5.4 - (2.1)^2 = 0.990$.

■

Given any probability distribution, we can always calculate σ^2 by substituting the corresponding probabilities $f(x)$ into the formula which defines the variance. As in the case of the mean, however, this work can be simplified to a considerable extent when we deal with special kinds of distributions. For instance, it can be shown that the variance of the binomial distribution with the parameters n and p is given by the formula

Variance of binomial distribution

$$\sigma^2 = n \cdot p \cdot (1 - p)$$

EXAMPLE 17 **Using the formula for variance of the binomial distribution**

Verify the result stated in the preceding example, that $\sigma = 2$ for the binomial distribution with $n = 16$ and $p = \dfrac{1}{2}$.

Solution Substituting $n = 16$ and $p = \dfrac{1}{2}$ into the formula for the variance of a binomial distribution, we get

$$\sigma^2 = 16 \cdot \frac{1}{2} \cdot \frac{1}{2} = 4$$

and, hence, $\sigma = \sqrt{4} = 2$. ∎

The variance of the hypergeometric distribution with the parameters n, a, and N is

Variance of hypergeometric distribution

$$\sigma^2 = n \frac{a}{N} \left(1 - \frac{a}{N} \right) \left(\frac{N - n}{N - 1} \right)$$

The factor $(N - n)/(N - 1)$ adjusts for the finite population.

EXAMPLE 18 **Using the formula for variance of the hypergeometric distribution**

With reference to Example 8 in which 5 of 20 cell phone chargers are defective, find the standard deviation of the probability distribution of the number of defectives in a sample of 10 randomly chosen for inspection.

Solution Substituting $n = 10$, $a = 5$, and $N = 20$ into the formula for the variance of a hypergeometric distribution, we get

$$\sigma^2 = 10 \frac{5}{20} \left(1 - \frac{5}{20} \right) \left(\frac{20 - 10}{20 - 1} \right) = \frac{75}{76}$$

and, hence, $\sigma = \sqrt{75/76} = 0.99$. ∎

When we first defined the variance of a probability distribution, it may have occurred to the reader that the formula looked exactly like the one which we use in physics to define second moments, or moments of inertia. Indeed, it is customary in statistics to define the **kth moment about the origin** as

$$\mu'_k = \sum_{\text{all } x} x^k \cdot f(x)$$

and the **kth moment about the mean** as

$$\mu_k = \sum_{\text{all } x} (x - \mu)^k \cdot f(x)$$

Thus, the mean μ is the first moment about the origin, and the variance σ^2 is the second moment about the mean. Higher moments are often used in statistics to give further descriptions of probability distributions. For instance, the third moment about the mean (divided by σ^3 to make this measure independent of the scale of measurement) is used to describe the symmetry or skewness of a distribution; the fourth moment about the mean (divided by σ^4) is, similarly, used to describe its "peakedness," or **kurtosis**. To determine moments about the mean, it is usually easiest to express moments about the mean in terms of moments about the origin and then to calculate the necessary moments about the mean. For the second moment about the mean we thus have the important formula $\sigma^2 = \mu_2' - \mu^2$.

4.5   Chebyshev's Theorem

Earlier in this chapter we used examples to show how the standard deviation measures the variation of a probability distribution, that is, how it reflects the concentration of probability in the neighborhood of the mean. If σ is large, there is a correspondingly higher probability of getting values farther away from the mean. Formally, the idea is expressed by the following theorem.

Chebyshev's theorem

> **Theorem 4.1**   If a probability distribution has mean μ and standard deviation σ, the probability of getting a value which deviates from μ by at least $k\sigma$ is at most $\dfrac{1}{k^2}$.

Symbolically,

$$P(\,|X - \mu| \geq k\sigma\,) \leq \frac{1}{k^2}$$

where $P(\,|X - \mu| \geq k\sigma\,)$ is the probability associated with the set of outcomes for which x, the value of a random variable having the given probability distribution, is such that $|x - \mu| \geq k\sigma$.

Thus, the probability that a random variable will take on a value which deviates (differs) from the mean by at least 2 standard deviations is at most $\frac{1}{4}$, the probability that it will take on a value which deviates from the mean by at least 5 standard deviations is at most $\frac{1}{25}$, and the probability that it will take on a value which deviates from the mean by 10 standard deviations or more is less than or equal to $\frac{1}{100}$.

To prove this theorem, consider any probability distribution $f(x)$, having mean μ, and variance σ^2. Dividing the sum defining the variance into three parts as indicated in Figure 4.7, we have

$$\sigma^2 = \sum_{\text{all } x} (x - \mu)^2 f(x)$$

$$= \sum_{R_1} (x - \mu)^2 f(x) + \sum_{R_2} (x - \mu)^2 f(x) + \sum_{R_3} (x - \mu)^2 f(x)$$

where R_1 is the region for which $x \leq \mu - k\sigma$, R_2 is the region for which $\mu - k\sigma < x < \mu + k\sigma$, and R_3 is the region for which $x \geq \mu + k\sigma$. Since $(x - \mu)^2 f(x)$ cannot

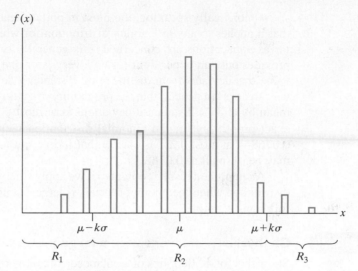

Figure 4.7
Diagram for proof of
Chebyshev's theorem

be negative, the above sum over R_2 is nonnegative, and without it the sum of the summations over R_1 and R_3 is less than or equal to σ^2; that is,

$$\sigma^2 \geq \sum_{R_1}(x - \mu)^2 f(x) + \sum_{R_3}(x - \mu)^2 f(x)$$

But $x - \mu \leq -k\sigma$ in the region R_1 and $x - \mu \geq k\sigma$ in the region R_3, so that in either case $|x - \mu| \geq k\sigma$. In both regions $(x - \mu)^2 \geq k^2\sigma^2$. If we now replace $(x - \mu)^2$ in each sum by $k^2\sigma^2$, a number less than or equal to $(x - \mu)^2$, we obtain the inequality

$$\sigma^2 \geq \sum_{R_1}k^2\sigma^2 f(x) + \sum_{R_1}k^2\sigma^2 f(x)$$

or

$$\frac{1}{k^2} \geq \sum_{R_1}f(x) + \sum_{R_3}f(x)$$

Since

$$\sum_{R_1}f(x) + \sum_{R_3}f(x)$$

represents the probability assigned to the region $R_1 \cup R_3$, namely, $P(|X - \mu| \geq k\sigma)$, this completes the proof of Theorem 4.1.

   To obtain an alternative form of Chebyshev's theorem, note that the event $|x - \mu| < k\sigma$ is the complement of the event $|x - \mu| \geq k\sigma$; hence, the probability of getting a value which deviates from μ by less than $k\sigma$ is at least $1 - \dfrac{1}{k^2}$.

EXAMPLE 19   **A probability bound using Chebyshev's theorem**

The number of customers who visit a car dealer's showroom on a Saturday morning is a random variable with $\mu = 18$ and $\sigma = 2.5$. With what probability can we assert that there will be more than 8 but fewer than 28 customers?

Solution   Let X be the number of customers. Since

$$k = \frac{28 - 18}{2.5} = \frac{18 - 8}{2.5} = 4$$

$$P(|X - \mu| < k\sigma) \geq 1 - \frac{1}{k^2} \quad \text{and} \quad P(8 < X < 28) \geq 1 - \frac{1}{4^2} = \frac{15}{16} \quad \blacksquare$$

Theoretically speaking, the most important feature of Chebyshev's theorem is that it applies to any probability distribution for which μ and σ exist. However, so far as applications are concerned, this generally is also its greatest weakness—it provides only an upper limit (often a very poor one) to the probability of getting a value that deviates from the mean by k standard deviations or more. For instance, we can assert in general that the probability of getting a value which differs from the mean by at least 2 standard deviations is at most 0.25, whereas the corresponding exact probability for the binomial distribution with $n = 16$ and $p = \frac{1}{2}$ is only 0.0768—"at most 0.25" is correct, but it does not tell us that the actual probability may be as small as 0.0768.

An important result is obtained if we apply Chebyshev's theorem to the binomial distribution when the number of trials is large. To illustrate this result, consider the following example.

EXAMPLE 20   **Chebyshev's theorem with a large number of Bernoulli trials**

Show that for 40,000 flips of a balanced coin, the probability is at least 0.99 that the proportion of heads will fall between 0.475 and 0.525.

Solution   Since

$$\mu = 40{,}000 \cdot \frac{1}{2} = 20{,}000 \qquad \sigma = \sqrt{40{,}000 \cdot \frac{1}{2} \cdot \frac{1}{2}} = 100$$

and

$$1 - \frac{1}{k^2} = 0.99$$

yields $k = 10$, the alternative form of Chebyshev's theorem tells us that the probability is at least 0.99 that we will get between $20{,}000 - 10(100) = 19{,}000$ and $20{,}000 + 10(100) = 21{,}000$ heads. Hence, the probability is at least 0.99 that the proportion of heads will fall between

$$\frac{19{,}000}{40{,}000} = 0.475 \qquad \text{and} \qquad \frac{21{,}000}{40{,}000} = 0.525 \qquad \blacksquare$$

Correspondingly, the reader will be asked to show in Exercise 4.47 that for 1,000,000 flips of a balanced coin the probability is at least 0.99 that the proportion of heads will fall between 0.495 and 0.505, and these results suggest that when n is large, the chances are that the proportion of heads will be very close to $p = \frac{1}{2}$.

When formulated for any binomial distribution with the parameters n and p, this result is referred to as the **law of large numbers**. Recall Figure 3.7 which demonstrates the stabilition of the long run relative frequency for the case $p = 0.6$. The law of large numbers guarantees this for all applications.

Exercises

4.32 Suppose that the probabilities are 0.4, 0.3, 0.2, and 0.1 that there will be 0, 1, 2, or 3 power failures in a certain city during the month of July. Use the formulas which define μ and σ^2 to find

(a) the mean of this probability distribution;

(b) the variance of this probability distribution.

4.33 Use the computing formula for σ^2 to rework part (b) of the preceding exercise.

4.34 The following table gives the probabilities that a certain computer will malfunction 0, 1, 2, 3, 4, 5, or 6 times on any one day:

Number of malfunctions: x	0	1	2	3	4	5	6
Probability: $f(x)$	0.17	0.29	0.27	0.16	0.07	0.03	0.01

Use the formulas which define μ and σ to find

(a) the mean of this probability distribution;

(b) the standard deviation of this probability distribution.

4.35 Use the computing formula for σ^2 to rework part (b) of the preceding exercise.

4.36 Find the mean and the variance of the uniform probability distribution given by

$$f(x) = \frac{1}{n} \qquad \text{for } x = 1, 2, 3, \ldots, n$$

[*Hint*: The sum of the first n positive integers is $n(n + 1)/2$, and the sum of their squares is $n(n + 1)$ $(2n + 1)/6$.]

4.37 Find the mean and the variance of the binomial distribution with $n = 4$ and $p = 0.70$ by using

(a) Table 1 and the formulas defining μ and σ^2;

(b) the special formulas for the mean and the variance of a binomial distribution.

4.38 As can easily be verified by means of the formula for the binomial distribution (or by listing all 32 possibilities), the probabilities of getting 0, 1, 2, 3, 4, or 5 heads in five flips of a balanced coin are

$$\frac{1}{32} \quad \frac{5}{32} \quad \frac{10}{32} \quad \frac{10}{32} \quad \frac{5}{32} \quad \frac{1}{32}$$

Find the mean of this probability distribution using

(a) the formula that defines μ;

(b) the special formula for the mean of a binomial distribution.

4.39 With reference to Exercise 4.38, find the variance of the probability distribution using

(a) the formula that defines σ^2;

(b) the computing formula for σ^2;

(c) the special formula for the variance of a binomial distribution.

4.40 If 95% of certain high-performance radial tires last at least 30,000 miles, find the mean and the standard deviation of the distribution of the number of these tires, among 20 selected at random, that last at least 30,000 miles, using

(a) Table 1, the formula which defines μ, and the computing formula for σ^2.

(b) the special formulas for the mean and the variance of a binomial distribution.

4.41 Find the mean and the standard deviation of the distribution of each of the following random variables (having binomial distributions):

(a) The number of heads obtained in 676 flips of a balanced coin.

(b) The number of 4's obtained in 720 rolls of a balanced die.

(c) The number of defectives in a sample of 600 parts made by a machine, when the probability is 0.04 that any one of the parts is defective.

(d) The number of students among 800 interviewed who do not like the food served at the university cafeteria, when the probability is 0.65 that any one of them does not like the food.

4.42 Find the mean and the standard deviation of the hypergeometric distribution with the parameters $n = 3$, $a = 4$, and $N = 8$

(a) by first calculating the necessary probabilities and then using the formulas which define μ and σ;

(b) by using the special formulas for the mean and the variance of a hypergeometric distribution.

4.43 Prove the formula for the mean of the hypergeometric distribution with the parameters n, a, and N, namely,

$$\mu = n \cdot \frac{a}{N}.$$

[*Hint*: Make use of the identity

$$\sum_{r=0}^{k} \binom{m}{r}\binom{s}{k-r} = \binom{m+s}{k}$$

which can be obtained by equating the coefficients of x^k in $(1 + x)^m(1 + x)^s$ and in $(1 + x)^{m+s}$.]

4.44 Construct a table showing the upper limits provided by Chebyshev's theorem for the probabilities of obtaining values differing from the mean by at least 1, 2, and 3 standard deviations and also the corresponding probabilities for the binomial distribution with $n = 16$ and $p = \frac{1}{2}$.

4.45 Over the range of cylindrical parts manufactured on a computer-controlled lathe, the standard deviation of the diameters is 0.002 millimeter.

(a) What does Chebyshev's theorem tell us about the probability that a new part will be within 0.006 unit of the mean μ for that run?

(b) If the 400 parts are made during the run, about what proportion do you expect will lie in the interval in part (a)?

4.46 In 1 out of 6 cases, material for bulletproof vests fails to meet puncture standards. If 405 specimens are tested, what does Chebyshev's theorem tell us about the probability of getting at most 30 or more than 105 cases that do not meet puncture standards?

4.47 Show that for 1 million flips of a balanced coin, the probability is at least 0.99 that the proportion of heads will fall between 0.495 and 0.505.

4.48 The time taken by students to fill out a loan request form has standard deviation 1.2 hours. What does Chebyshev's theorem tell us about the probability that a students' time will be within 4 hours of the mean μ for all potential loan applicants?

4.49 Prove that

(a) $\sigma^2 = E(X^2) - \mu^2$;

(b) $\mu_3 = \mu'_3 - 3\mu'_2 \cdot \mu + 2\mu^3$.

4.6 The Poisson Distribution and Rare Events

The **Poisson distribution** often serves as a model for counts which do not have a natural upper bound. It is an important probability distribution for describing the number of times an event randomly occurs in one unit of time or one unit of space. In one unit of time, each instant can be regarded as a potential trial in which the event may or may not occur. Although there are conceivably an infinite number of trials, usually only a few or moderate number of events take place.

The **Poisson distribution**, with mean λ (lambda), has probabilities given by

Poisson distribution

$$f(x; \lambda) = \frac{\lambda^x e^{-\lambda}}{x!} \qquad \text{for } x = 0, 1, 2, \ldots \qquad \lambda > 0$$

Using a method similar to that employed on page 99 to derive the formula for the mean of the binomial distribution, we can show that the mean and the variance of the Poisson distribution with the parameter λ are given by

Mean and variance of Poisson distribution

$$\mu = \lambda \qquad \text{and} \qquad \sigma^2 = \lambda$$

There is a different **Poisson distribution** for different values λ. They are all asymmetrical. If λ is an integer $f(\lambda - 1; \lambda) = f(\lambda; \lambda)$ and each is larger than any other probability. Otherwise, when λ is not an integer, the largest probability is assigned to the integer part of λ. When $0 < \lambda < 1$, the probability of 0 is the largest and the probabilities $f(x; \lambda)$ decrease as the value x increases as illustrated in Figure 4.8(a) for $\lambda = .7$. The distribution for $\lambda = 3$, representing $\lambda > 1$, has a more typical behavior. As illustrated in Figure 4.8 (b) the probabilities $f(x; \lambda)$ increase to $f(2; 3) = f(3; 3))$ and then decrease as x increases. This distribution has a long right-hand tail. As λ becomes large, the distribution becomes approximately symmetric.

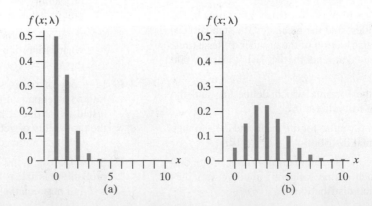

Figure 4.8
Two Poisson distributions
(a) $\lambda = .7$ and (b) $\lambda = 3$.

Since the Poisson distribution has many important applications, it has been extensively tabulated. Table 2W on the book's web site gives the values of the probabilities

$$F(x; \lambda) = \sum_{k=0}^{x} f(k; \lambda)$$

for values of λ in varying increments from 0.02 to 25, and its use is very similar to that of Table 1. Poisson probabilities are also calculated by many statistical software programs (see Exercises 4.70 and 4.71).

EXAMPLE 21

A Poisson distribution for counts of particles

For health reasons, homes need to be inspected for radon gas which decays and produces alpha particles. One device counts the number of alpha particles that hit its detector. To a good approximation, in one area, the count for the next week follows a Poisson distribution with mean 1.3. Determine

(a) the probability of exactly one particle next week.

(b) the probability of one or more particles next week.

(c) the probability of at least two but no more than four particles next week.

(d) The variance of the Poisson distribution.

Solution

Unlike the binomial case, there is no choice of a fixed Bernoulli trial here because one can always work with smaller intervals.

(a) $P(X = 1) = \dfrac{\lambda^1 e^{-\lambda}}{1!} = \dfrac{1.3\, e^{-1.3}}{1} = .3543$

Alternatively, using Table 2W, $F(1, 1.3) - F(0, 1.3) = 0.627 - 0.273 = 0.354$

(b) $P(X \geq 1) = 1 - P(X = 0) = 1 - e^{-1.3} = 0.727$

(c) $P(2 \leq X \leq 4) = F(4, 1.3) - F(1, 1.3) = 0.989 - 0.627 = 0.362$

This last calculation, depicted in Figure 4.9, visually demonstrates the subtraction of the cumulative probabilities for values below the upper limit, for the cumulative probabilities through the upper limit.

[Using **R**: (a) **dpois(1, 1.3)** (b) **1 - ppois(0, 1.3)** (c) **ppois(4, 1.3) - ppois(1, 1.3)**]

Figure 4.9
The calculation of
$P(2 \leq X \leq 4)$

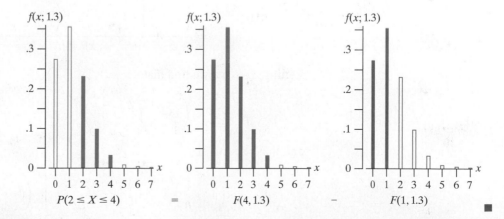

Let us point out that $x = 0, 1, 2, \ldots$ means that there is a countable infinity of possibilities, and this requires that we modify the third axiom of probability given on page 62. In its place we substitute the following axiom.

Modification of third axiom of probability

> **Axiom 3′**   If $A_1, A_2, A_3, \ldots$ is a finite or infinite sequence of mutually exclusive events in $\mathcal{S}$, then
>
> $$P(A_1 \cup A_2 \cup A_3 \cup \cdots) = P(A_1) + P(A_2) + P(A_3) + \cdots$$

The other postulates remain unchanged. To verify that $P(\mathcal{S}) = 1$ for this formula, we make use of Axiom 3′ and write

$$\sum_{x=0}^{\infty} f(x; \lambda) = \sum_{x=0}^{\infty} \frac{e^{-\lambda}\lambda^x}{x!} = e^{-\lambda} \sum_{x=0}^{\infty} \frac{\lambda^x}{x!}$$

Since the infinite series in the expression on the right is the Maclaurin's series for e^{λ}, it follows that

$$\sum_{x=0}^{\infty} f(x; \lambda) = e^{-\lambda} \cdot e^{\lambda} = 1$$

The Poisson Approximation to the Binomial Distribution

One interpretation of a rare event is one that occurs with a small probability in a single trial. When n is large and p is small, binomial probabilities are often approximated by means of the Poisson distribution with λ equal to the product np.

Let us now show that when $n \to \infty$ and $p \to 0$, while $np = \lambda$ remains constant, the limiting form of the binomial distribution is the Poisson distribution given above. First let us substitute $\dfrac{\lambda}{n}$ for p into the formula for the binomial distribution and simplify the resulting expression; thus, we get

$$b(x; n, p) = \frac{n!}{x!\,(n-x)!} \left(\frac{\lambda}{n}\right)^x \left(1 - \frac{\lambda}{n}\right)^{n-x}$$

$$= \frac{n(n-1)(n-2)\cdots(n-x+1)}{x!\,n^x} (\lambda)^x \left(1 - \frac{\lambda}{n}\right)^{n-x}$$

$$= \frac{\left(1 - \dfrac{1}{n}\right)\left(1 - \dfrac{2}{n}\right)\cdots\left(1 - \dfrac{x-1}{n}\right)}{x!} (\lambda)^x \left(1 - \frac{\lambda}{n}\right)^{n-x}$$

Letting $n \to \infty$, we find that

$$\left(1 - \frac{1}{n}\right)\left(1 - \frac{2}{n}\right)\cdots\left(1 - \frac{x-1}{n}\right) \to 1$$

and that

$$\left(1 - \frac{\lambda}{n}\right)^{n-x} = \left[\left(1 - \frac{\lambda}{n}\right)^{n/\lambda}\right]^{\lambda} \left(1 - \frac{\lambda}{n}\right)^{-x} \to e^{-\lambda}$$

Hence, the binomial distribution $b(x; n, p)$ approaches

$$\frac{\lambda^x e^{-\lambda}}{x!} \quad \text{for } x = 0, 1, 2, \ldots$$

This completes our proof; the distribution at which we arrived is called the **Poisson distribution**, as we already indicated on the page 108.

An acceptable rule of thumb is to use Poisson approximation of binomial probabilities if $n \geq 20$ and $p \leq 0.05$; if $n \geq 100$, the approximation is generally excellent so long as $np \leq 10$.

EXAMPLE 22   **Comparing Poisson and binomial probabilities**

It is known that 5% of the books bound at a certain bindery have defective bindings. Find the probability that 2 of 100 books bound by this bindery will have defective bindings using

(a) the formula for the binomial distribution;

(b) the Poisson approximation to the binomial distribution.

Solution   (a) Substituting $x = 2$, $n = 100$, and $p = 0.05$ into the formula for the binomial distribution, we get

$$b(2; 100, 0.05) = \binom{100}{2}(0.05)^2(0.95)^{98} = 0.081$$

(b) Substituting $x = 2$ and $\lambda = 100(0.05) = 5$ into the formula for the Poisson distribution, we get

$$f(2; 5) = \frac{5^2 \cdot e^{-5}}{2!} = 0.084$$

It is of interest to note that the difference between the two values we obtained (the error we would make by using the Poisson approximation) is only 0.003. [Had we used Table 2W instead of using a calculator to obtain e^{-5}, we would have obtained $f(2; 5) = F(2; 5) - F(1; 5) = 0.125 - 0.040 = 0.085$.]   ∎

EXAMPLE 23   **A Poisson approximation to binomial probabilities**

A heavy machinery manufacturer has 3,840 large generators in the field that are under warranty. If the probability is 1/1,200 that any one will fail during the given year, find the probabilities that $0, 1, 2, 3, 4, \ldots$ of the generators will fail during the given year.

Solution   The binomial distribution could be used when appropriate computer software is available. However, the expected number is small and the number of generators is large so the Poisson approximation is valid. We take

$$\lambda = 3,840 \cdot \frac{1}{1,200} = 3.2$$

Consulting Table 2W with $\lambda = 3.2$, and using the identity $f(x; \lambda) = F(x; \lambda) - F(x-1; \lambda)$, we obtain the results shown in the probability histogram of Figure 4.10.   ∎

In our justification of the Poisson approximation to the binomial distribution we let $\lambda = np$. For the variance we can write $\sigma^2 = np(1-p) = \lambda(1-p)$, which approaches λ as $p \to 0$. This matches the mean and variance of the Poisson distribution.

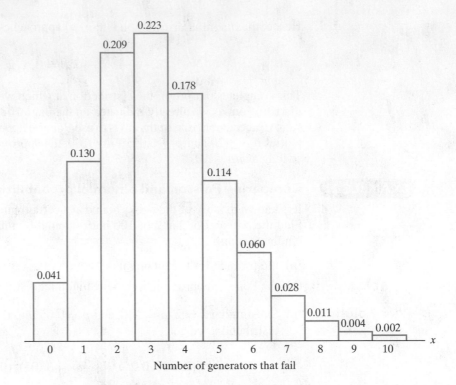

Figure 4.10
Probability histogram of
Poisson distribution with
$\lambda = 3.2$

4.7 Poisson Processes

In general, a **random process** is a physical process that is wholly or in part controlled by some sort of chance mechanism. It may be a sequence of repeated flips of a coin, measurements of the quality of manufactured products coming off an assembly line, the vibrations of airplane wings, the noise in a radio signal, or any one of numerous other phenomena. What characterizes such processes is their time dependence, namely, the fact that certain events do or do not take place (depending on chance) at regular intervals of time or throughout continuous intervals of time.

In this section we shall be concerned with processes taking place over continuous intervals of time or space, such as the occurrence of imperfections on a continuously produced bolt of cloth, the recording of radiation by means of a Geiger counter, the arrival of telephone calls at a virtual switchboard, or the passing by cars over an electronic counting device. We will now show that the mathematical model which we can use to describe many situations like these is that of the Poisson distribution. To find the probability of x successes during a time interval of length T, we divide the interval into n equal parts of length Δt, so that $T = n \cdot \Delta t$, and we assume that

1. The probability of a success during a very small interval of time Δt is given by $\alpha \cdot \Delta t$.
2. The probability of more than one success during such a small time interval Δt is negligible.
3. The probability of a success during such a time interval does not depend on what happened prior to that time.

This means that the assumptions underlying the binomial distribution are satisfied, and the probability of x successes in the time interval T is given by the binomial probability $b(x; n, p)$ with

$$n = \frac{T}{\Delta t} \qquad \text{and} \qquad p = \alpha \cdot \Delta t$$

Then, following the argument on page 110, we find that when $n \to \infty$ the probability of x successes during the time interval T is given by the corresponding Poisson probability with the parameter

$$\lambda = n \cdot p = \frac{T}{\Delta t} \cdot (\alpha \cdot \Delta t) = \alpha T$$

Since λ is the mean of this Poisson distribution, note that α is the average (mean) number of successes per unit time.

EXAMPLE 24   **Calculating probabilities concerning bad checks**

If a bank receives on the average $\alpha = 6$ bad checks per day, what are the probabilities that it will receive

(a) 4 bad checks on any given day?

(b) 10 bad checks over any 2 consecutive days?

Solution   (a) Substituting $x = 4$ and $\lambda = \alpha T = 6 \cdot 1 = 6$ into the formula for the Poisson distribution, we get

$$f(4; 6) = \frac{6^4 \cdot e^{-6}}{4!} = \frac{1{,}296(0.00248)}{24} = 0.134$$

(b) Here $\lambda = \alpha \cdot 2 = 12$ so we want to find $f(10; 12)$. We write

$$
\begin{aligned}
f(10; 12) &= F(10; 12) - F(9; 12) \\
&= 0.347 - 0.242 \\
&= 0.105
\end{aligned}
$$

where the values of $F(10; 12)$ and $F(9; 12)$ were obtained from Table 2W.   ∎

EXAMPLE 25   **Calculating the probabilities of internet interruptions**

A computing system manager states that the rate of interruptions to the internet service is 0.2 per week. Use the Poisson distribution to find the probability of

(a) one interruption in 3 weeks

(b) at least two interruptions in 5 weeks

(c) at most one interruption in 15 weeks.

Solution   Interruptions to the network occur randomly and the conditions for the Poisson distribution initially appear reasonable. We have $\lambda = 0.2$ for the expected number of interruptions in one week.

In terms of the cumulative probabilities,

(a) with $\lambda = (0.2) \cdot 3 = 0.6$, we get

$$
\begin{aligned}
F(1; 0.6) - F(0; 0.6) &= 0.878 - 0.549 \\
&= 0.329
\end{aligned}
$$

(b) With $\lambda = (0.2) \cdot 5 = 1.0$, we get

$$
\begin{aligned}
1 - F(1; 1.0) &= 1 - 0.736 \\
&= 0.264
\end{aligned}
$$

(c) With $\lambda = (0.2) \cdot 15 = 3.0$ we get

$$F(1; 3.0) = 0.199$$

■

The Poisson distribution has many important applications in *queuing* problems, where we may be interested, for example, in the number of customers arriving for service at a cafeteria, the number of ships or trucks arriving to be unloaded at a receiving dock, the number of aircraft arriving at an airport, and so forth. Thus, if, on average, 0.3 customer arrives per minute at a cafeteria, then the probability that exactly 3 customers will arrive during a 5-minute span is

$$F(3; 1.5) - F(2; 1.5) = 0.934 - 0.809 = 0.125$$

and if, on the average, 3 trucks arrive per hour to be unloaded at a warehouse, then the probability that at most 20 will arrive during an 8-hour day is

$$F(20; 24) = 0.243$$

4.8   The Geometric and Negative Binomial Distribution

On page 47 we indicated that a countably infinite sample space would be needed if we are interested in the number of cars persons have to inspect until they find one whose nitrogen oxide emission does not meet government standards. To treat this kind of problem in general, suppose that in a sequence of trials we are interested in the number of the trial on which the first success occurs. The three assumptions for Bernoulli trials are satisfied but the extra assumption underlying the binomial distribution is not. In other words, n is not fixed.

Clearly, if the first success is to come on the xth trial, it has to be preceded by $x - 1$ failures, and if the probability of a successes is p, the probability of $x - 1$ failures in $x - 1$ trials is $(1 - p)^{x-1}$. Then, if we multiply this expression by the probability p of a success on the xth trial, we find that the probability of getting the first success on the xth trial is given by

Geometric distribution

$$g(x; p) = p(1 - p)^{x-1} \qquad \text{for } x = 1, 2, 3, 4, \ldots$$

This probability distribution is called the **geometric distribution**. The reader will be asked to verify its mean and variance in Exercise 5.100.

Mean and variance of geometric distribution

$$\mu = \frac{1}{p} \qquad \sigma^2 = \frac{1 - p}{p^2}$$

EXAMPLE 26   **Calculating a probability using the geometric distribution**

If the probability is 0.05 that a certain kind of measuring device will show excessive drift, what is the probability that the sixth measuring device tested will be the first to show excessive drift?

Solution   Substituting $x = 6$ and $p = 0.05$ into the formula for the geometric distribution, we get

$$g(6; 0.05) = (0.05)(1 - 0.05)^{6-1}$$
$$= 0.039 \qquad \blacksquare$$

The **negative binomial distribution** describes the total number of Bernoulli trials, X, to obtain a specified number r successes. When $r = 1$, the negative binomial reduces to the geometric distribution.

If the rth success occurs at trial number x, it must be that $r - 1$ successes occurred in the first $x - 1$ trials and the last trial is a success. The probability distribution is then the product of the binomial probability $b(r - 1 ; x - 1, p)$ and p.

Negative binomial distribution

$$f(x) = \binom{x-1}{r-1} p^r (1-p)^{x-r} \quad \text{for } x = r, r+1, \ldots$$

The reader is asked, in Exercise 6.40, to show the mean and variance are given by

Mean and variance of negative binomial distribution

$$\mu = \frac{rp}{p} \qquad \sigma^2 = \frac{(1-p)r}{p^2}$$

Exercises

4.50 Prove that for the Poisson distribution
$$\frac{f(x+1; \lambda)}{f(x; \lambda)} = \frac{\lambda}{x+1}$$
for $x = 0, 1, 2, \ldots$.

4.51 Use the recursion formula of Exercise 4.50 to calculate the value of the Poisson distribution with $\lambda = 3$ for $x = 0, 1, 2, \ldots$, and 9, and draw the probability histogram of this distribution. Verify your results by referring to Table 2W or software.

4.52 Use Table 2W or software to find
(a) $F(4; 7)$; (b) $f(4; 7)$; (c) $\sum_{k=6}^{19} f(k; 8)$.

4.53 Use Table 2W or software to find
(a) $F(9; 12)$; (b) $f(9; 12)$; (c) $\sum_{k=3}^{12} f(k; 7.5)$.

4.54 Use the Poisson distribution to approximate the binomial probability $b(3; 100, 0.03)$.

4.55 In a given city, 6% of all drivers get at least one parking ticket per year. Use the Poisson approximation to the binomial distribution to determine the probabilities that among 80 drivers (randomly chosen in this city):
(a) 4 will get at least one parking ticket in any given year;
(b) at least 3 will get at least one parking ticket in any given year;
(c) anywhere from 3 to 6, inclusive, will get at least one parking ticket in any given year.

4.56 During inspection of the continuous process of making large rolls of floor coverings, 0.5 imperfections are spotted per minute on average. Use the Poison distribution to find the probabilities
(a) one imperfection in 4 minutes
(b) at least two in 8 minutes
(c) at most one in 10 minutes.

4.57 The number of gamma rays emitted per second by a certain radioactive substance is a random variable having the Poisson distribution with $\lambda = 5.8$. If a recording instrument becomes inoperative when there are more than 12 rays per second, what is the probability that this instrument becomes inoperative during any given second?

4.58 A consulting engineer receives, on average, 0.7 requests per week. If the number of requests follows a Poisson process, find the probability that
(a) in a given week, there will be at least 1 request;
(b) in a given 4-week period there will be at least 3 requests.

4.59 At a checkout counter customers arrive at an average of 1.5 per minute. Find the probabilities that
(a) at most 4 will arrive in any given minute;
(b) at least 3 will arrive during an interval of 2 minutes;
(c) at most 15 will arrive during an interval of 6 minutes.

4.60 Environmental engineers, concerned about the effects of releasing warm water from a power plants' cooling system into a Great Lake, decided to sample many organisms both inside and outside of a warm water plume. For the zoo-plankton Cyclops, they collect 100 cc of water and count the number of Cyclops. The expected number is 1.7 per 100 cc.

Use the Poisson distribution to find the probability of

(a) 1 Cyclops in a 100 cc sample

(b) less than or equal to 6 but more than one in a 100cc sample

(c) exactly 4 Cyclops in a sample of size 200cc.

(d) 2 or more Cyclops in a sample of size 200cc.

4.61 In a "torture test," a light switch is turned on and off until it fails. If the probability that the switch will fail any time it is turned on or off is 0.001, what is the probability that the switch will fail *after* it has been turned on or off 1,200 times? Assume that the conditions underlying the geometric distribution are met. [*Hint*: Use the formula for the value of an infinite geometric progression.]

4.62 An automated weight monitor can detect underfilled cans of beverages with probability 0.98. What is the probability it fails to detect an underfilled can for the first time when it encounters the 10th underfilled can?

4.63 A company fabricates special-purpose robots, and records show that the probability is 0.10 that one of its new robots will require repairs during confirmation tests. What is the probability that the eighth robot it builds in a month is the first one to require repairs?

4.64 Referring to Exercise 4.63, find the probability that the 12th robot built in a month is the second to require repairs.

4.65 During an assembly process, parts arrive just as they are needed. However, at one station, the probability is 0.01 that a defective part will arrive in a one-hour period. Find the probability that

(a) exactly 1 defective part arrives in a 4-hour span;

(b) 1 or more defective parts arrive in a 4-hour span;

(c) exactly 1 defective part arrives in a 4-hour span and exactly 1 defective part arrives in the next 4-hour span.

4.66 The arrival of trucks at a receiving dock is a Poisson process with a mean arrival rate of 2 per hour.

(a) Find the probability that exactly 5 trucks arrive in a two-hour period.

(b) Find the probability that 8 or more trucks arrive in a two-hour period.

(c) Find the probability that exactly 2 trucks arrive in a one-hour period and exactly 3 trucks arrive in the next one-hour period.

4.67 The number of flaws in a fiber optic cable follows a Poisson process with an average of 0.6 per 100 feet.

(a) Find the probability of exactly 2 flaws in a 200-foot cable.

(b) Find the probability of exactly 1 flaw in the first 100 feet and exactly 1 flaw in the second 100 feet.

4.68 Differentiating with respect to p on both sides of the equation

$$\sum_{x=1}^{\infty} p(1-p)^{x-1} = 1$$

show that the geometric distribution

$$f(x) = p(1-p)^{x-1} \qquad \text{for } x = 1, 2, 3, \ldots$$

has the mean $1/p$.

4.69 Use the formulas defining μ and σ^2 to show that the mean and the variance of the Poisson distribution are both equal to λ.

4.70 Poisson probabilities can be calculated using *MINITAB*.

Dialog box:
Calc > Probability Distribution > Poisson
Choose **Probability.**
Choose **Input constant** and enter 2. Type 1.64 in **Mean.**
Click **OK.**

Output:
Poisson with mean = 1.64

x	$P(X = x)$
2	0.260864

Find the Poisson probabilities for $x = 2$ and $x = 3$ when

(a) $\lambda = 2.73$; (b) $\lambda = 4.33$.

4.71 Cumulative Poisson probabilities can be calculated using *MINITAB*.

Dialog box:
Calc > Probability Distribution > Poisson
Choose **Cumulative Distribution.**
Choose **Input constant** and enter 2. Type 1.64 in **Mean.**
Click **OK.**

Output:
Poisson with mean = 1.64

x	$P(X <= x)$
2	0.772972

Find the cumulative Poisson probabilities for $x = 2$ and $x = 3$ when

(a) $\lambda = 2.73$; (b) $\lambda = 4.33$.

4.9   The Multinomial Distribution

An immediate generalization of the binomial distribution arises when each trial can have more than two possible outcomes. This happens, for example, when a manufactured product is classified as superior, average, or poor, when a student's performance is graded as an A, B, C, D, or F, or when an experiment is judged successful, unsuccessful, or inconclusive. To treat this kind of problem in general, let us consider the case where there are n independent trials, with each trial permitting k mutually exclusive outcomes whose respective probabilities are

$$p_1, p_2, \ldots, p_k \qquad \text{with} \quad \sum_{i=1}^{k} p_i = 1$$

Referring to the outcomes as being of the first kind, the second kind, $\ldots$, and the kth kind, we shall be interested in the probability $f(x_1, x_2, \ldots, x_k)$ of getting x_1 outcomes of the first kind, x_2 outcomes of the second kind, $\ldots$, and x_k outcomes of the kth kind, with

$$\sum_{i=1}^{k} x_i = n$$

Using arguments similar to those which we employed in deriving the equation for the binomial distribution in Section 4.2, it can be shown that the desired probability is given by

Multinomial distribution

$$f(x_1, x_2, \ldots, x_k) = \frac{n!}{x_1! x_2! \ldots x_k!} \, p_1^{x_1} p_2^{x_2} \cdots p_k^{x_k}$$

for $x_i = 0, 1, \ldots, n$ for each i, but with the x_i subject to the restriction

$$\sum_{i=1}^{k} x_i = n$$

The **joint probability distribution** whose values are given by these probabilities is called the **multinomial distribution**; it owes its name to the fact that for the various values of the x_i the probabilities are given by the corresponding terms of the multinomial expansion of $(p_1 + p_2 + \cdots + p_k)^n$.

EXAMPLE 27   **Calculating a probability using the multinomial distribution**

The probabilities that the light bulb of a certain kind of projector will last fewer than 40 hours of continuous use, anywhere from 40 to 80 hours of continuous use, or more than 80 hours of continuous use are 0.30, 0.50, and 0.20. Find the probability that among eight such bulbs 2 will last fewer than 40 hours, 5 will last anywhere from 40 to 80 hours, and 1 will last more than 80 hours.

Solution   Substituting $n = 8$, $x_1 = 2$, $x_2 = 5$, $x_3 = 1$, $p_1 = 0.30$, $p_2 = 0.50$, and $p_3 = 0.20$ into the formula, we get

$$f(2, 5, 1) = \frac{8!}{2!5!1!} (0.30)^2 (0.50)^5 (0.20)^1$$
$$= 0.0945 \qquad \blacksquare$$

Exercises

4.72 Suppose that the probabilities are, respectively, 0.40, 0.40, and 0.20 that in city driving a certain kind of imported car will average less than 22 miles per gallon, anywhere from 22 to 25 miles per gallon, or more than 25 miles per gallon. Find the probability that among 12 such cars tested, 4 will average less than 22 miles per gallon, 6 will average anywhere from 22 to 25 miles per gallon, and 2 will average more than 25 miles per gallon.

4.73 As can easily be shown, the probabilities of getting 0, 1, or 2 heads with a pair of balanced coins are $\frac{1}{4}$, $\frac{1}{2}$, and $\frac{1}{4}$. What is the probability of getting 2 tails twice, 1 head and 1 tail 3 times, and 2 heads once in 6 tosses of a pair of balanced coins?

4.74 Suppose the probabilities are 0.89, 0.09, and 0.02 that the finish on a new car will be rated acceptable, easily repairable, or unacceptable. Find the probability that, among 20 cars painted one morning, 17 have acceptable finishes, 2 have repairable finishes, and 1 finish is unacceptable.

4.75 Using the same sort of reasoning as in the derivation of the formula for the hypergeometric distribution, we can derive a formula which is analogous to the multi-nomial distribution but applies to sampling without replacement. A set of N objects contains a_1 objects of the first kind, a_2 objects of the second kind, . . . , and a_k objects of the kth kind, so that $a_1 + a_2 + \cdots + a_k = N$. The number of ways in which we can select x_1 objects of the first kind, x_2 objects of the second kind, . . . , and x_k objects of the kth kind is given by the product of the number of ways in which we can select x_1 of the a_1 objects of the first kind, x_2 of the a_2 objects of the second kind, . . . , and x_k of the a_k objects of the kth kind. Thus, the probability of getting that many objects of each kind is simply this product divided by the total number of ways in which $x_1 + x_2 + \cdots + x_k = n$ objects can be selected from the whole set of N objects.

(a) Write a formula for the probability of obtaining x_1 objects of the first kind, x_2 objects of the second kind, . . . and x_k objects of the kth kind.

(b) If 20 defective glass bricks include 10 that have cracks but no discoloration, 7 that are discolored but have no cracks, and 3 that have cracks and discoloration, what is the probability that among 6 of the bricks chosen at random for further checks 3 will have cracks only, 2 will only be discolored, and 1 will have cracks as well as discoloration?

4.10   Simulation

In recent years, simulation techniques have been applied to many problems in the various sciences. If the processes being simulated involve an element of chance, these techniques are referred to as **Monte Carlo methods**. Very often, the use of Monte Carlo simulation eliminates the cost of building and operating expensive equipment. It is used, for instance, in the study of collisions of photons with electrons, the scattering of neutrons, and similar complicated phenomena. Monte Carlo methods are also useful in situations where direct experimentation is impossible—say, in studies of the spread of cholera epidemics, which, of course, cannot be induced experimentally on human populations. In addition, Monte Carlo techniques are sometimes applied to the solution of mathematical problems which cannot be solved by direct means, or where a direct solution is too costly or requires too much time.

A classical example of the use of Monte Carlo methods in the solution of a problem of pure mathematics is the determination of π (the ratio of the circumference of a circle to its diameter) by probabilistic means. Early in the eighteenth century, George de Buffon, a French naturalist, proved that if a very fine needle of length a is thrown at random on a board ruled with equidistant parallel lines, the probability that the needle will intersect one of the lines is $2a/\pi b$, where b is the distance between the parallel lines. What is remarkable about this fact is that it involves the constant $\pi = 3.1415926 \ldots$, which in elementary geometry is approximated by the circumferences of regular polygons enclosed in a circle of radius $\frac{1}{2}$. Buffon's result implies that if such a needle is actually tossed a great many times, the proportion of the time it crosses one of the lines gives an estimate of $2a/\pi b$ and, hence, an estimate of π since a and b are known. Early experiments of this kind yielded an estimate of 3.1596 (based on 5,000 trials) and an estimate of 3.155 (based on 3,204 trials) in the middle of the nineteenth century.

Although Monte Carlo methods are sometimes based on actual gambling devices (for example, the needle tossing in the estimation of π), it is usually expedient to use so-called **random digits** or **random numbers** generated by computer software. We will illustrate an application using a table of random numbers that consists of many pages on which the digits of 0, 1, 2, ..., and 9 are set down in a "random" fashion, much as they would appear if they were generated one at a time by a gambling device giving each digit an equal probability of being selected. Actually, we could also construct such tables ourselves—say, by repeatedly drawing numbered slips out of a hat or by using a perfectly constructed spinner—but in practice such tables are usually generated by means of computers.

Although tables of random numbers are constructed so that the digits can be looked upon as values of a random variable having the discrete uniform distribution $f(x) = \dfrac{1}{10}$ for $x = 0, 1, 2, \ldots$, or 9, they can be used to simulate values of any discrete random variable, and even continuous random variables.

To illustrate the use of a table of random numbers, let us simulate, say, tossing three balanced coins. The distribution for the number of heads is

Number of Heads	Probability
0	$1/8 = 0.125$
1	$3/8 = 0.375$
2	$3/8 = 0.375$
3	$1/8 = 0.125$

Since the probabilities in this distribution are given to three decimal places, we use three-digit random numbers. Our scheme is to allocate 125 (or one-eighth) of the 1,000 random numbers from 000 to 999 to 0 heads, 375 (or three-eighths) to 1 head, 375 (or three-eighths) to 2 heads, and 125 (or one-eighth) to 3 heads.

We use the following scheme:

Number of Heads	Probability	Cumulative Probability	Random Numbers
0	0.125	0.125	000–124
1	0.375	0.500	125–449
2	0.375	0.875	500–874
3	0.125	1.000	875–999

The column of **cumulative probabilities** was added to facilitate the assignment of the random numbers. Observe that in each case the last random digit is one less than the number formed by the three decimal digits of the corresponding cumulative probability.

With this scheme, if we arbitrarily use the twenty-second, twenty-third, and twenty-fourth columns of the first page of Table 7W, starting with the sixth row and going down the page, we get 197, 365, 157, 520, 946, 951, 948, 568, 586, and 089, and we interpret this as 1, 1, 1, 2, 3, 3, 3, 2, 2, and 0 heads.

The method we have illustrated here with reference to a game of chance can be used to simulate observations of any random variable with a given probability distribution.

However, in practice it is much more efficient to use common computer software based on this scheme.

EXAMPLE 28

Simulation of arrival of cars at toll booth

Suppose that the probabilities are 0.082, 0.205, 0.256, 0.214, 0.134, 0.067, 0.028, 0.010, 0.003, and 0.001 that 0, 1, 2, 3, . . . , or 9 cars will arrive at a toll booth of a turnpike during any one-minute interval in the early afternoon.

Use computer software to simulate the arrival of cars at the toll booth during 20 one-minute intervals in the early afternoon.

Solution   We illustrate using *MINITAB* with the values set in C1 and the probabilities in C2.

Data:
C1: 0, 1, . . . , 9
C2: 0.082, 0.205, . . . , 0.001

Dialog box:
Calc > Random Data > Discrete
Type 20 after **Generate**. Type *C*3 below **Store**. Type *C*1
in **Values in:**. Type *C*2 in **Probabilities in** Click **OK**.

Output:

$$4 \quad 1 \quad 5 \quad 4 \quad 1 \quad 2 \quad 5 \quad 0 \quad 1 \quad 4$$
$$3 \quad 3 \quad 1 \quad 0 \quad 1 \quad 1 \quad 2 \quad 5 \quad 1 \quad 2$$

Suppose we are interested in a somewhat complex event, say, 11 or more cars arrive in at least one three-minute interval among the 20 one-minute intervals. It is a simple manner to repeat the simulation of 20 one-minute periods 100 times. The probability that 11 or more cars arrive in at least one three-minute interval is estimated by the proportion of times that event occurs. In the single sample of size 20 here, that event does not occur.   ∎

Exercises

4.76 Simulate tossing a coin.

  (a) For a balanced coin, generate 100 flips.

  (b) For a coin with probability of heads 0.8, generate 100 flips.

4.77 The probabilities that a computer software salesperson will make 0, 1, 2, 3, 4, or 5 sales on any one day are 0.14, 0.28, 0.27, 0.18, 0.09, and 0.04.

  (a) Simulate the salesperson's sales on 25 days.

  (b) Repeat the simulation of sales on 25 days a total of 100 times. Estimate the probability that the sales on three consecutive days are greater than 9.

4.78 Depending on the availability of parts, a company can manufacture 3, 4, 5, or 6 units of a certain item per week with corresponding probabilities of 0.10, 0.40, 0.30, and 0.20. The probabilities that there will be a weekly demand for 0, 1, 2, 3, …, or 8 units are, respectively, 0.05, 0.10, 0.30, 0.30, 0.10, 0.05, 0.05, 0.04, and 0.01. If a unit is sold during the week that it is made, it will yield a profit of $100; this profit is reduced by $20 for each week that a unit has to be stored. Simulate the operation of this company for 50 consecutive weeks and estimate its expected weekly profit.

Do's and Don'ts

Do's

1. Keep in mind that any scheme for assigning a numerical value to each possible outcome should quantify a feature of the outcome that is important to the scientist. That is, any random variable should convey pertinent information about the outcome.

2. Describe the chance behavior of a discrete random variable X by its probability distribution function

$$f(x) = P[X = x] \qquad \text{for each possible value } x$$

3. Summarize a probability distribution, or the random variable, by its

$$\text{mean: } \mu = \sum_{\text{all } x} x \cdot f(x) \qquad\qquad \text{variance: } \sigma^2 = \sum_{\text{all } x} (x - \mu)^2 \cdot f(x)$$

$$\text{standard deviation: } \sigma = \sqrt{\sum_{\text{all } x} (x - \mu)^2 \cdot f(x)}$$

4. Use a special family of distributions, for instance the binomial distribution

$$b(x;\, n,\, p) = \binom{n}{x} p^x (1 - p)^{n-x} \qquad \text{for } x = 0, 1, \ldots, n$$

having mean np and variance $np(1 - p)$, if the underlying assumptions are reasonable. The hypergeometric distribution might be entertained when sampling without replacement from a finite collection of units each of which is one of two possible types. It will be well approximated by the binomial when the sample size n is a small fraction of the population size N.

5. For counts whose possible values do not have a specified upper limit, consider the Poisson distribution

$$f(x;\, \lambda) = \frac{\lambda^x e^{-\lambda}}{x!} \qquad \text{for } x = 0, 1, 2, \ldots \qquad \lambda > 0$$

having mean λ and variance λ. You do need to check that the Poisson distribution is reasonable. The sample mean and variance should be about the same size.

Don'ts

1. Never apply the binomial distribution to counts without first checking that the conditions hold for Bernoulli trials: independent trials with the same probability of success for each trial. If the conditions are satisfied, then the binomial distribution is appropriate for the number of successes in a fixed number of trials.

2. Never use the formula $np(1 - p)$ for the variance of a count of successes without checking that the trials are independent.

Review Exercises

4.79 A manufacturer of smart phones has the following probability distribution for the number of defects per phone:

x	f(x)
0	.89
1	.07
2	.03
3	.01

(a) Determine the probability of 2 or more defects.

(b) Is a randomly selected phone more likely to have 0 defects or 1 or more defects?

4.80 Upon reviewing recent use of conference rooms at an engineering consulting firm, an industrial engineer determined the following probability distribution for the number of requests for a conference room per half-day:

x	f(x)
0	.07
1	.15
2	.45
3	.25
4	.08

(a) Currently, the building has two conference rooms. What is the probability that the number of requests will exceed the number of rooms for a given half-day?

(b) What is the probability that the two conference rooms will not be fully utilized on a given half-day?

(c) How many additional conference rooms are required so that the probability of denying a request is not more than 0.10?

4.81 Refer to Exercise 4.80 and obtain the
(a) mean; (b) variance; (c) standard deviation for the number of requests for conference rooms.

4.82 Determine whether the following can be probability distributions of a random variable that can take on only the values of 0, 1, and 2:

(a) $f(0) = 0.34$ $f(1) = 0.34$ and $f(2) = 0.34$.

(b) $f(0) = 0.2$ $f(1) = 0.6$ and $f(2) = 0.2$.

(c) $f(0) = 0.7$ $f(1) = 0.4$ and $f(2) = -0.1$.

4.83 Check whether the following can define probability distributions, and explain your answers.

(a) $f(x) = \dfrac{x}{10}$, for $x = 0, 1, 2, 3, 4$.

(b) $f(x) = \dfrac{1}{3}$, for $x = -1, 0, 1$.

(c) $f(x) = \dfrac{(x-1)^2}{4}$, for $x = 0, 1, 2, 3$.

4.84 A basketball player makes 90% of her free throws. What is the probability she will miss for the first time on the seventh shot?

4.85 If the probability is 0.20 that a downtime of an automated production process will exceed 2 minutes, find the probability that 3 of 8 downtimes of the process will exceed 2 minutes using (a) the formula for the binomial distribution; (b) Table 1 or software.

4.86 If the probability is 0.85 that a fully charged digital camera battery will take 150 or more pictures, find the probabilities that among 18 such batteries

(a) 16 will take 150 pictures or more;

(b) at least 14 will take 150 pictures or more;

(c) at most two will not take 150 pictures or more.

4.87 In 16 experiments studying the electrical behavior of single cells, 12 use micro-electrodes made of metal and the other 4 use micro-electrodes made from glass tubing. If 2 of the experiments are to be terminated for financial reasons, and they are selected at random, what are the probabilities that

(a) neither uses micro-electrodes made from glass tubing?

(b) only one uses micro-electrodes made from glass tubing?

(c) both use micro-electrodes made from glass tubing?

4.88 As can be easily verified by means of the formula for the binomial distribution, the probabilities of getting 0, 1, 2, or 3 heads in 3 flips of a coin whose probability of heads is 0.4 are 0.216, 0.432, 0.288, and 0.064. Find the mean of this probability distribution using

(a) the formula that defines μ;

(b) the special formula for the mean of a binomial distribution.

4.89 With reference to Exercise 4.88, find the variance of the probability distribution using

(a) the formula that defines σ^2;

(b) the special formula for the variance of a binomial distribution.

4.90 Find the mean and the standard deviation of the distribution of each of the following random variables (having binomial distributions):

(a) The number of heads in 440 flips of a balanced coin.

(b) The number of 6's in 300 rolls of a balanced die.

(c) The number of defectives in a sample of 700 parts made by a machine, when the probability is 0.03 that any one of the parts is defective.

4.91 Use the Poisson distribution to approximate the binomial probability $b(1; 100, 0.02)$.

4.92 With reference to Exercise 4.87, find the mean and the variance of the distribution of the number of microelectrodes made from glass tubing using

(a) the probabilities obtained in that exercise;

(b) the special formulas for the mean and the variance of a hypergeometric distribution.

4.93 The daily number of orders filled by the parts department of a repair shop is a random variable with $\mu = 142$ and $\sigma = 12$. According to Chebyshev's theorem, with what probability can we assert that on any one day it will fill between 82 and 202 orders?

4.94 Records show that the probability is 0.00004 that a car will have a flat tire while driving through a certain tunnel. Use the formula for the Poisson distribution to approximate the probability that at least 2 of 10,000 cars passing through the tunnel will have a flat tire.

4.95 The number of weekly breakdowns of a computer is a random variable having a Poisson distribution with $\lambda = 0.2$. What is the probability that the computer will operate without a breakdown for 3 consecutive weeks?

4.96 A manufacturer determines that a big screen HDTV set had probabilities of 0.8, 0.15, 0.05, respectively, of being placed in the categories acceptable, minor defect, or major defect. If 3 HDTVs are inspected,

(a) find the probability that 2 are acceptable and 1 is a minor defect;

(b) find the marginal distribution of the number in minor defect;

(c) compare your answer in part (b) with the binomial probabilities $b(x; 3, 0.15)$. Comment.

4.97 Suppose that the probabilities are 0.2466, 0.3452, 0.2417, 0.1128, 0.0395, 0.0111, 0.0026, and 0.0005 that there will be 0, 1, 2, 3, 4, 5, 6, or 7 polluting spills in the Great Lakes on any one day. Simulate the numbers of polluting spills in the Great Lakes in 30 days.

4.98 A candidate invited for a visit has probability 0.6 of being hired. Let X be the number of candidates that visit before 2 are hired. Find

(a) $P(X \le 4)$;

(b) $P(X \ge 5)$.

Key Terms

CHAPTER

5

PROBABILITY DENSITIES

ontinuous sample spaces and continuous random variables rise when we deal with quantities that are measured on a continuous scale. For instance, when we can measure the speed of a car, the amount of alcohol in a person's blood, the efficiency of a solar collector, or the tensile strength of a new alloy.

In this chapter we learn how to determine and work with probabilities relating to continuous sample spaces and continuous random variables. We first introduce probability densities in Section 5.1. The discussion expands to the normal distribution in Sections 5.2 and 5.3 and various other special probability densities in Sections 5.4 through 5.9. Problems involving more than one random variable are discussed in Section 5.10. Section 5.11 presents the moment generating function method, a tool for finding the distribution of the sum of independent random variables. A method for checking whether a data set appears to be generated by a normal distribution is introduced in Section 5.12.

5.1 Continuous Random Variables

When we first introduced the concept of a random variable in Chapter 4, we presented it as a real-valued function defined over the sample space of an experiment. We illustrated this idea with the random variable giving the number of preferred attributes possessed by a used car, assigning the numbers 0, 1, 2, or 3 (whichever was appropriate) to the 18 possible outcomes of the experiment. In the continuous case, where random variables can assume values on a continuous scale, the procedure is very much the same. The outcomes of an experiment are represented by the points on a line segment or a line. Then, a random variable is created by appropriately assigning a number to each point by means of some rule or equation.

When the value of a random variable is given directly by a measurement or observation, we usually do not bother to differentiate among the value of the random variable, the measurement which we obtain, and the outcome of the experiment, which is the corresponding point on the real axis. If an experiment consists of determining what force is required to break a given tensile-test specimen, the result itself, say, 138.4 pounds, is the value of the random variable, X, with which we are concerned. There is no real need in that case to add that the sample space of the experiment consists of all (or part of) the points on the positive real axis.

In general, we write $P(a \leq X \leq b)$ for the probability associated with the points of the sample space for which the value of a random variable falls on the interval from a to b. The problem of defining probabilities in connection with continuous sample spaces and continuous random variables involves some complications. To illustrate the nature of these complications, let us consider the following situation.

Suppose we want to know the probability that if an accident occurs on a freeway whose length is 200 miles, it will happen at some given location or, perhaps, some particular stretch of the road. The outcomes of this experiment can be looked upon as a continuum of points. Namely, those on the continuous interval from 0 to 200.

Next, suppose the probability that the accident occurs on any interval of length L is $L/200$, with L measured in miles. Note that this arbitrary assignment of probability is consistent with Axioms 1 and 2 on page 60, since the probabilities are all nonnegative and less than or equal to 1, and $P(S) = \dfrac{200}{200} = 1$.

So far, we are considering only events represented by intervals which form part of the line segment from 0 to 200. Using Axiom $3'$ on page 110, we can also obtain probabilities of events that are not intervals but which can be represented by the union of finitely many or countably many intervals. Thus, for two nonoverlapping intervals of length L_1 and L_2 we have a probability of

$$\frac{L_1 + L_2}{200}$$

and for an infinite sequence of nonoverlapping intervals of length $L_1, L_2, L_3, \ldots$, we have a probability of

$$\frac{L_1 + L_2 + L_3 + \cdots}{200}$$

Note that the probability that the accident occurs at any given point is equal to zero because we can look upon a point as an interval of zero length. However, the probability that the accident occurs in a very short interval is positive; for instance, for an interval of length 1 foot the probability is $(5{,}280 \times 200)^{-1} = 9.5 \times 10^{-7}$.

Thus, in extending the concept of probability to the continuous case, we again use Axioms 1, 2, and $3'$, but we shall have to restrict the meaning of the term *event*. So far as practical considerations are concerned, this restriction is of no consequence. We simply do not assign probabilities to some rather abstruse point sets, which cannot be expressed as the unions or intersections of finitely many or countably many intervals.

The way in which we assigned probabilities in the preceding example is, of course, very special; it is similar in nature to the way in which we assign equal probabilities to the six faces of a die, heads and tails, the 52 cards in a standard deck, and so forth. To treat the problem of associating probabilities with continuous random variables generally, suppose we are interested in the probability that a given random variable will take on a value on the interval from a to b, where a and b are constants with $a \leq b$. Suppose, furthermore, that we divide the interval from a to b into m equal subintervals of width Δx containing, respectively, the points $x_1, x_2, \ldots, x_m$, and that the probability that the random variable will take on a value in the subinterval containing x_i is given by $f(x_i) \cdot \Delta x$. Then the probability that the random variable with which we are concerned will take on a value in the interval from a to b is given by

$$P(a \leq X \leq b) = \sum_{i=1}^{m} f(x_i) \cdot \Delta x$$

When f is an integrable function defined for all values of the random variable with which we are concerned, we shall define the probability that the value of the random variable falls between a and b by letting $\Delta x \to 0$. Namely,

$$P(a \leq X \leq b) = \int_{a}^{b} f(x)\, dx$$

As illustrated in Figure 5.1, this definition of probability in the continuous case presupposes the existence of an appropriate function f which, integrated from any constant a to any constant b (with $a \leq b$), gives the probability that the corresponding random variable takes on a value on the interval from a to b. Note that the value

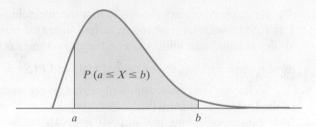

Figure 5.1
Probability as area under f

$f(x)$ does not give the probability that the corresponding random variable takes on the value x. In the continuous case, probabilities are given by integrals and not by the values $f(x)$.

To obtain the probability that a random variable will actually take on a given value x, we might first determine the probability that it will take on a value on the interval from $x - \Delta x$ to $x + \Delta x$, and then let $\Delta x \to 0$. However, if we did this it would become apparent that the result is always zero. The fact that the probability is always zero that a continuous random variable will take on any given value x should not be disturbing. Indeed, our definition of probability for the continuous case provides a remarkably good model for dealing with measurements or observations. Owing to the limits of our ability to measure, experimental data never seem to come from a continuous sample space. Thus, while temperatures are productively thought of as points on a continuous scale, if we report a temperature measurement of 74.8 degrees centigrade, we really mean that the temperature lies in the interval from 74.75 to 74.85 degrees centigrade, and not that it is exactly 74.800 . . . .

It is important to add that when we say that there is a zero probability that a random variable will take on any given value x, this does not mean that it is impossible that the random variable will take on the value x. In the continuous case, a zero probability does not imply logical impossibility, but the whole matter is largely academic since, owing to the limitations of our ability to measure and observe, we are always interested in probabilities connected with intervals and not with isolated points.

As an immediate consequence of the fact that in the continuous case probabilities associated with individual points are always zero, we find that if we speak of the probability associated with the interval from a to b, it does not matter whether either endpoint is included. Symbolically,

$$P(a \leq X \leq b) = P(a \leq X < b) = P(a < X \leq b) = P(a < X < b)$$

Drawing an analogy with the concept of a density function in physics, we call the functions f, whose existence we stipulated in extending our definition of probability to the continuous cases, **probability density functions**, or simply **probability densities**. Whereas density functions are integrated to obtain weights, probability density functions are integrated to obtain probabilities. We will follow the common practice of also calling $f(x)$ the probability density function with the understanding that we are referring to the function f which assigns the value $f(x)$ to x, for each x that is a possible value for the random variable X.

Since a probability density, integrated between any two constants a and b, gives the probability that a random variable assumes a value between these limits, f cannot be just any real-valued integrable function. However, imposing the conditions that

$$f(x) \geq 0 \quad \text{for all } x$$

and

$$\int_{-\infty}^{\infty} f(x)\, dx = 1$$

insures that the axioms of probability (with the modification about events discussed on page 125) are satisfied. Note the similarity between these conditions and those for probability distributions given on page 86.

As in the discrete case, we let $F(x)$ be the probability that a random variable with the probability density $f(x)$ takes on a value less than or equal to x. We again refer to the corresponding function F as the **cumulative distribution function** or just the **distribution function** of the random variable. Thus, for any value x, $F(x) = P(X \leq x)$ is the area under the probability density function over the interval $-\infty$ to x. In the usual calculus notation for the integral,

$$F(x) = \int_{-\infty}^{x} f(t)\, dt$$

Consequently, the probability that the random variable will take on a value on the interval from a to b is $F(b) - F(a)$, and according to the fundamental theorem of integral calculus it follows that

$$\frac{dF(x)}{dx} = f(x)$$

wherever this derivative exists.

EXAMPLE 1 **Calculating probabilities from the probability density function**
If a random variable has the probability density

$$f(x) = \begin{cases} 2\,e^{-2x} & \text{for } x > 0 \\ 0 & \text{for } x \leq 0 \end{cases}$$

find the probabilities that it will take on a value

(a) between 1 and 3;

(b) greater than 0.5.

Solution Evaluating the necessary integrals, we get

(a)
$$\int_{1}^{3} 2\,e^{-2x}\, dx = e^{-2} - e^{-6} = 0.133$$

(b)
$$\int_{0.5}^{\infty} 2\,e^{-2x}\, dx = e^{-1} = 0.368 \quad \blacksquare$$

Note that in the preceding example we make the domain of f include all the real numbers even though the probability is zero that x will be negative. This is a practice we shall follow throughout the book. It is also apparent from the graph of this function in Figure 5.2 that it has a discontinuity at $x = 0$; indeed, a probability density need not be everywhere continuous, as long as it is integrable between any two limits a and b (with $a \leq b$).

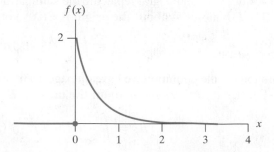

Figure 5.2
Graph of probability density
$f(x) = 2\,e^{-2x},\ x > 0$

EXAMPLE 2   **Determining a distribution function from its density function**

With reference to the preceding example, find the distribution function and use it to determine the probability that the random variable will take on a value less than or equal to 1.

Solution   Performing the necessary integrations, we get

$$F(x) = \begin{cases} 0 & \text{for } x \le 0 \\ \int_0^x 2\,e^{-2t}\,dt = 1 - e^{-2x} & \text{for } x > 0 \end{cases}$$

and substitution of $x = 1$ yields

$$F(1) = 1 - e^{-2} = 0.865$$ ∎

Note that the distribution function of this example is nondecreasing and that $F(-\infty) = 0$ and $F(\infty) = 1$. Indeed, it follows by definition that these properties are shared by all distribution functions.

EXAMPLE 3   **A probability density function assigns probability one to $(-\infty, \infty)$**

Find k so that the following can serve as the probability density of a random variable:

$$f(x) = \begin{cases} 0 & \text{for } x \le 0 \\ kxe^{-4x^2} & \text{for } x > 0 \end{cases}$$

Solution   To satisfy the first of the two conditions on page 126, k must be nonnegative, and to satisfy the second condition we must have

$$\int_{-\infty}^{\infty} f(x)\,dx = \int_0^{\infty} kxe^{-4x^2}\,dx = \int_0^{\infty} \frac{k}{8} \cdot e^{-u}\,du = \frac{k}{8} = 1$$

so that $k = 8$.   ∎

To describe probability densities, we define statistical measures that are very similar the ones that describe probability distributions. The first moment about the origin is again called the **mean**, and it is denoted by μ. Alternatively, it is also called the expected value of a random variable having the probability density $f(x)$ and denoted by $E(X)$.

Mean of a probability density

$$\mu = E(X) = \int_{-\infty}^{\infty} xf(x)\,dx$$

This expected value is analogous to that for the discrete case introduced in Section 4.4 but with an integral replacing the summation.

The **kth moment about the origin** is $E(X^k)$ or

$$\mu_k' = \int_{-\infty}^{\infty} x^k \cdot f(x)\,dx$$

analogous to the definition we gave on page 103.

Further, the **kth moment about the mean** is $E(X - \mu)^k$, or

$$\mu_k = \int_{-\infty}^{\infty} (x - \mu)^k \cdot f(x)\,dx$$

In particular, the second moment about the mean is again referred to as the **variance** and it is written as σ^2. As before, it measures the spread of a probability density in the sense that it gives the expected value of the squared deviation from the mean.

Variance of a probability density

$$\sigma^2 = \int_{-\infty}^{\infty} (x - \mu)^2 f(x)\, dx = \int_{-\infty}^{\infty} x^2 f(x)\, dx - \mu^2$$

Alternately, $\sigma^2 = E(X - \mu)^2 = E(X^2) - \mu^2$

Again, σ is referred to as the **standard deviation**.

EXAMPLE 4 **Determining the mean and variance using the probability density function**

With reference to Example 1, find the mean and the variance of the given probability density.

Solution Performing the necessary integrations, using integrations by parts, we get

$$\mu = \int_{-\infty}^{\infty} x f(x)\, dx = \int_0^{\infty} x \cdot 2 e^{-2x}\, dx = \frac{1}{2}$$

Alternatively, the expectation of x is $E(X) = 0.5$

$$\sigma^2 = \int_{-\infty}^{\infty} (x - \mu)^2 f(x)\, dx = \int_0^{\infty} \left(x - \frac{1}{2} \right)^2 \cdot 2 e^{-2x}\, dx = \frac{1}{4} \qquad \blacksquare$$

Exercises

5.1 Verify that the function of Example 1 is, in fact, a probability density.

5.2 If the probability density of a random variable is given by

$$f(x) = \begin{cases} k x^2 & 0 < x < 1 \\ 0 & \text{elsewhere} \end{cases}$$

find the value k and the probability that the random variable takes on a value

(a) between $\frac{1}{4}$ and $\frac{3}{4}$; (b) greater than $\frac{2}{3}$.

5.3 With reference to the preceding exercise, find the corresponding distribution function and use it to determine the probabilities that a random variable having this distribution function will take on a value

(a) greater than 0.8; (b) between 0.2 and 0.4.

5.4 If the probability density of a random variable is given by

$$f(x) = \begin{cases} x & \text{for } 0 < x < 1 \\ 2 - x & \text{for } 1 \leq x < 2 \\ 0 & \text{elsewhere} \end{cases}$$

find the probabilities that a random variable having this probability density will take on a value

(a) between 0.2 and 0.8; (b) between 0.6 and 1.2.

5.5 With reference to the preceding exercise, find the corresponding distribution function, and use it to determine the probabilities that a random variable having the distribution function will take on a value

(a) greater than 1.8;

(b) between 0.4 and 1.6.

5.6 Given the probability density $f(x) = \dfrac{k}{1 + x^2}$ for $-\infty < x < \infty$, find k.

5.7 If the distribution function of a random variable is given by

$$F(x) = \begin{cases} 1 - \dfrac{4}{x^2} & \text{for } x > 2 \\ 0 & \text{for } x \leq 2 \end{cases}$$

find the probabilities that this random variable will take on a value

(a) less than 3; (b) between 4 and 5.

5.8 Find the probability density that corresponds to the distribution function of Exercise 5.7. Are there any points at which it is undefined? Also sketch the graphs of the distribution function and the probability density.

5.9 Let the phase error in a tracking device have probability density

$$f(x) = \begin{cases} \cos x & 0 < x < \pi/2 \\ 0 & \text{elsewhere} \end{cases}$$

Find the probability that the phase error is

(a) between 0 and $\pi/4$; (b) greater than $\pi/3$.

5.10 The length of satisfactory service (years) provided by a certain model of laptop computer is a random variable having the probability density

$$f(x) = \begin{cases} \dfrac{1}{4.5}\, e^{-x/4.5} & \text{for } x > 0 \\ 0 & \text{for } x \leq 0 \end{cases}$$

Find the probabilities that one of these laptops will provide satisfactory service for

(a) at most 2.5 years;

(b) anywhere from 4 to 6 years;

(c) at least 6.75 years.

5.11 In a certain city, the daily consumption of electric power (in millions of kilowatt-hours) is a random variable having the probability density

$$f(x) = \begin{cases} \dfrac{1}{9}\, x e^{-x/3} & \text{for } x > 0 \\ 0 & \text{for } x \leq 0 \end{cases}$$

If the city's power plant has a daily capacity of 12 million kilowatt-hours, what is the probability that this power supply will be inadequate on any given day?

5.12 Prove that the identity $\sigma^2 = \mu_2' - \mu^2$ holds for any probability density for which these moments exist.

5.13 Find μ and σ^2 for the probability density of Exercise 5.2.

5.14 Find μ and σ^2 for the probability density of Exercise 5.4.

5.15 Find μ and σ for the probability density obtained in Exercise 5.8.

5.16 Find μ and σ for the distribution of the phase error of Exercise 5.9.

5.17 Find μ for the distribution of the satisfactory service of Exercise 5.10.

5.18 Show that μ_2' and, hence, σ^2 do not exist for the probability density of Exercise 5.6.

5.2 The Normal Distribution

Among the special probability densities we study in this chapter, the **normal probability density**, usually referred to simply as the **normal distribution**, is by far the most important.[1] It was studied first in the eighteenth century when scientists observed an astonishing degree of regularity in errors of measurement. They found that the patterns (distributions) they observed were closely approximated by a continuous distribution, which they referred to as the "normal curve of errors" and attributed to the laws of chance. The equation of the normal probability density, whose graph (shaped like the cross section of a bell) is shown in Figure 5.3, is

Normal distribution

$$f(x; \mu, \sigma^2) = \frac{1}{\sqrt{2\pi}\,\sigma}\, e^{-(x-\mu)^2/2\sigma^2} \qquad -\infty < x < \infty$$

In Exercises 5.42 and 5.43, the reader will be asked to verify that its parameters μ and σ are indeed its mean and its standard deviation.

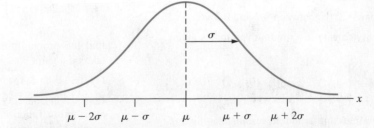

Figure 5.3
Graph of normal probability density

---

[1]The words *density* and *distribution* are often used interchangeably in the literature of applied statistics.

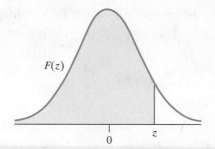

Figure 5.4
The standard normal
probabilities $F(z) = P(Z \leq z)$

Since the normal probability density cannot be integrated in closed form between every pair of limits a and b, probabilities relating to normal distributions are usually obtained from special tables, such as Table 3 at the back endpapers of this book. This table pertains to the **standard normal distribution**, namely, the normal distribution with $\mu = 0$ and $\sigma = 1$, and its entries are the values of

$$F(z) = \frac{1}{\sqrt{2\pi}} \int_{-\infty}^{z} e^{-t^2/2}\, dt = P(Z \leq z)$$

for positive or negative $z = 0.00, 0.01, 0.02, \ldots, 3.49$, and also $z = 3.50$, $z = 4.00$, and $z = 5.00$. The cumulative probabilities $F(z)$ correspond to the area under the standard normal density to the left of z, as shown by the shaded area in Figure 5.4.

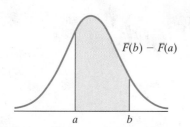

Figure 5.5
The standard normal
probability $F(b) - F(a) =$
$P(a < Z \leq b)$

To find the probability that a random variable having the standard normal distribution will take on a value between a and b, we use the equation

$$P(a < Z \leq b) = F(b) - F(a)$$

as shown by the shaded area in Figure 5.5. We also sometimes make use of the identity $F(-z) = 1 - F(z)$, which holds for all symmetric distributions centered around 0. The reader is asked to verify this in Exercise 5.41.

Given access to statistical software or a statistical calculator, that approach is preferable to looking in tables. The solution to Example 5 includes the R commands (see Appendix C on R and Exercise 5.44 for MINITAB).

> **EXAMPLE 5**   **Calculating some standard normal probabilities**
>
> Find the probabilities that a random variable having the standard normal distribution will take on a value

(a) between 0.87 and 1.28;

(b) between −0.34 and 0.62;

(c) greater than 0.85;

(d) greater than −0.65.

Solution   It is helpful to first indicate the area of interest in a graph as in Figure 5.6.

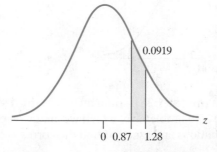

Figure 5.6
$P(0.87 < Z < 1.28)$

Looking up the necessary values in Table 3, for part (a) we get

$$F(1.28) - F(0.87) = 0.8997 - 0.8078$$
$$= 0.0919$$

As indicated in Figure 5.7 for part (b),

$$F(0.62) - F(-0.34) = 0.7324 - 0.3669$$
$$= 0.3655$$

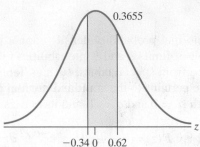

Figure 5.7
$P(-0.34 < Z < 0.62)$

As indicated in Figure 5.8 for part (c),

$$1 - F(0.85) = 1 - 0.8023$$
$$= 0.1977$$

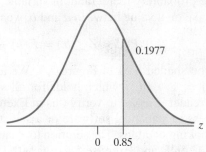

Figure 5.8
$P(Z > 0.85)$

As indicated in Figure 5.9 for part (d)

$$1 - F(-0.65) = 1 - 0.2578 = 0.7422$$

or, alternatively,

$$1 - F(-0.65) = 1 - [1 - F(0.65)]$$
$$= F(0.65)$$
$$= 0.7422$$

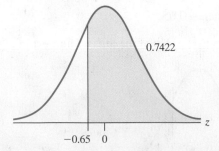

Figure 5.9
$P(Z > -0.65)$

∎

[Using **R**: (a) **pnorm(1.28) - pnorm(.87)**   (b) **1 - pnorm(.85)**]

There are also problems in which we are given probabilities relating to standard normal distributions and asked to find the corresponding values of z.

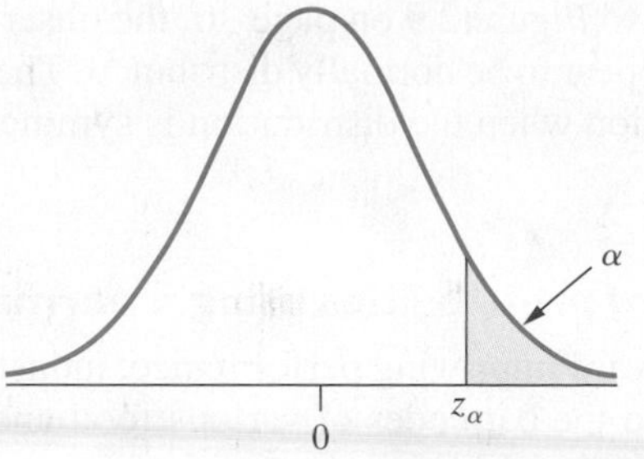

Figure 5.10
The z_α notation for a standard normal distribution

Let z_α be such that the probability is α that it will be exceeded by a random variable having the standard normal distribution. That is, $\alpha = P(Z > z_\alpha)$ as illustrated in Figure 5.10.

The results of the next example are used extensively in subsequent chapters.

EXAMPLE 6 **Two important values for z_α**

Find (a) $z_{0.01}$; (b) $z_{0.05}$.

Solution (a) Since $F(z_{0.01}) = 0.99$, we look for the entry in Table 3 which is closest to 0.99 and get 0.9901 corresponding to $z = 2.33$. Thus $z_{0.01} = 2.33$.

(b) Since $F(z_{0.05}) = 0.95$, we look for the entry in Table 3 which is closest to 0.95 and get 0.9495 and 0.9505 corresponding to $z = 1.64$ and $z = 1.65$. Thus, by interpolation, $z_{0.05} = 1.645$. ∎

[Using **R**: (a) **qnorm(.99)** (b) **qnorm(.95)**]

To use Table 3 in connection with a random variable X which has a normal distribution with the mean μ and the variance σ^2, we refer to the corresponding **standardized random variable**,

$$Z = \frac{X - \mu}{\sigma}$$

which can be shown to have the standard normal distribution. Thus, to find the probability that the original random variable will take on a value less than or equal to a, in Table 3 we look up

$$F\left(\frac{a - \mu}{\sigma}\right)$$

Also, to find the probability that a random variable having the normal distribution with the mean μ and the variance σ^2 will take on a value between a and b, we have only to calculate the probability that a random variable having the standard normal distribution will take on a value between

$$\frac{a - \mu}{\sigma} \quad \text{and} \quad \frac{b - \mu}{\sigma}$$

That is, to find probabilities concerning X, we convert its values to z **scores** using

$$Z = \frac{X - \mu}{\sigma}$$

Normal probabilities

When X has the normal distribution with mean μ and standard deviation σ.

$$P(a < X \leq b) = F\left(\frac{b - \mu}{\sigma}\right) - F\left(\frac{a - \mu}{\sigma}\right)$$

According to Figure 2.9 on page 20, the observations on the strength of an aluminum alloy appear to be normally distributed. The normal distribution is often used to model variation when the distribution is symmetric and has a single mode.

According to Figure 2.9 on page 20,

EXAMPLE 7   **Calculation of probabilities using a normal distribution**

With an eye toward improving performance, industrial engineers study the ability of scanners to read the bar codes of various food and household products. The maximum reduction in power, occurring just before the scanner cannot read the bar code at a fixed distance, is called the maximum attenuation. This quantity, measured in decibels, varies from product to product. After collecting considerable data, the engineers decided to model the variation in maximum attenuation as a normal distribution with mean 10.1 dB and standard deviation 2.7 dB.

(a) For the next food or product, what is the probability that its maximum attenuation is between 8.5 dB and 13.0 dB?

(b) According to the normal model, what proportion of the products have maximum attenuation between 8.5 dB and 13.0 dB?

(c) What proportion of the products have maximum attenuation greater than 15.1 dB?

Solution   **(a)** We treat the maximum attenuation of the next product, X, as a random selection for the normal distribution with $\mu = 10.1$ and $\sigma = 2.7$. Consequently, $Z = (X - 10.1)/2.7$ and, from Table 3, we get

$$F\left(\frac{13.0 - 10.1}{2.7}\right) - F\left(\frac{8.5 - 10.1}{2.7}\right) = F(1.07) - F(-0.59)$$
$$= 0.8577 - 0.2776$$
$$= 0.5801$$

as illustrated in Figure 5.11.

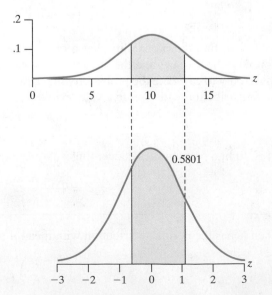

Figure 5.11
$P(8.5 < X < 13.0) =$
$P(-0.59 < Z < 1.07)$

(b) The variation in maximum attenuation for the vast, but finite, collection of all different products is modeled by a normal distribution. The proportion of products having maximum attenuation between 8.5 and 13.0 dB corresponds to the probability in part (a). When we consider the even larger infinite population of all existing products and those that could have been made, we still refer to 0.5801 as the proportion having maximum attenuation between 8.5 and 13.0 dB.

(c) Looking up the necessary value in Table 3,

$$1 - F\left(\frac{15.1 - 10.1}{2.7}\right) = 1 - F(1.85)$$
$$= 1 - 0.9678$$
$$= 0.0322$$

corresponding to the shaded area in Figure 5.12.

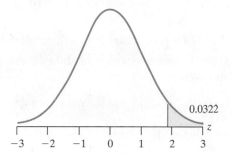

Figure 5.12
$P(X \geq 15.1)$

[Using **R**: (a) **pnorm(13.0, 10.1, 2.7) − pnorm(8.5, 10.1, 2.7)**
(c) **1 − pnorm(15.1, 10.1, 2.7)**]

EXAMPLE 8   **Normal Distribution as a Population Distribution**

A major manufacturer of processed meats monitors the amount of each ingredient. The weight(lb) of cheese per run is measured on $n = 80$ occasions. (courtesy of David Brauch))

```
72.2 67.8 78.0 64.4 76.3 72.3 73.1 71.7 66.2 63.3 85.4 67.4
66.3 76.3 57.7 50.3 77.4 63.1 73.9 67.4 74.7 68.2 87.4 86.4
69.4 58.0 63.3 72.7 73.6 68.8 63.3 63.3 73.0 64.8 73.1 70.9
85.9 74.4 75.9 72.3 84.3 61.8 79.2 64.3 65.4 66.7 77.2 50.0
70.3 90.4 63.9 62.1 68.2 55.1 52.6 68.5 55.2 73.5 53.7 61.7
47.9 72.3 61.1 71.8 83.1 71.2 58.8 61.8 86.8 64.5 52.3 58.3
65.9 80.2 75.1 59.9 62.3 48.8 64.3 75.4
```

Figure 5.13 suggests that the histogram, and therefore the population distribution, is well approximated by a normal distribution with mean $\mu = 68.4$ and standard deviation $\sigma = 9.6$ pounds. You are asked to examine the assumption of a normal distribution more closely in Exercise 5.102.

Using the normal population distribution,

(a) Find the probability of using 80 or more pounds of cheese.

(b) Set a limit so that only 10 % of production runs have less than L pounds of cheese.

(c) Determine a new mean for the distribution so that only 5 % of the runs have less than L pounds.

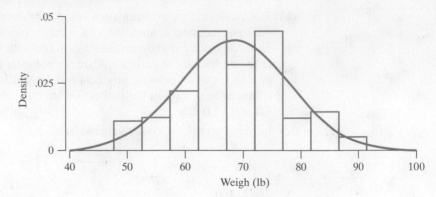

Figure 5.13
A normal distribution models
weight of cheese.

Solution   **(a)** $Z = (X - 68.4)/9.6$ and, from Table 3, we get

$$1 - F\left(\frac{80 - 68.4}{9.6}\right) = 1 - F(1.208) = 1 - .8865 = .1135$$

About 1 out of 9 production runs will result in more than 80 pounds of cheese.

(b) From Table 3, the entry with probability closest to .1000 is $z_{0.10} = 1.28$. The limit L is given by

$$L = \mu - \sigma z_{0.10} = 68.4 - 9.6 \times 1.28 = 56.1 \text{ pounds}$$

(c) The new value of the mean μ must satisfy

$$-z_{.05} = \frac{L - \mu}{9.6}$$

where $z_{0.05} = 1.645$ so

$$\mu = L + 9.6 \times z_{0.05} = 56.1 + 9.6 \times 1.645 = 71.9 \text{ pounds}$$

The mean must be increased by 3.5 pounds to decrease the percentage of units below the limit L from 10% to 5%.   ∎

[Using **R**:   (a) **1-pnorm(80,68.4,9.6)**   (b) **L = 68.4+9.6*qnorm(.10)**
                                                 (c) **L+9.6*qnorm(.95)**]

EXAMPLE 9   **Calculating probabilities when ln X has a normal distribution**

After collecting a large number of assays of the gold content in rocks from an open pit mine, a mining engineer postulates that the natural log of the gold content (oz/st gold) follows a normal distribution with mean -4.6 and variance 1.21. Under this distribution, would it be unusual to get 0.0015 oz/st gold or less in an assay?

Solution   Because it is ln X that has a normal distribution, the question concerns the standardized value

$$\frac{\ln(0.0015) - (-4.6)}{\sqrt{1.21}} = -1.729$$

The standard normal probability of obtaining this value or smaller (see Figure 5.14) is

$$F\left(\frac{\ln(0.0015) - (-4.6)}{\sqrt{1.21}}\right) = F(-1.73) = 0.0419$$

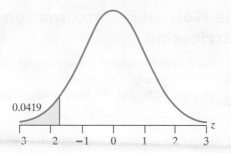

Figure 5.14
$P(Z \le -1.73) =$
$P(X \le 0.0015)$

This probability is small, so we suspect that the postulated normal distribution with mean -4.6 does not apply. An assay with this small amount of gold content could suggest that the specimen was collected outside of the vein. ∎

Although the normal distribution applies to continuous random variables, it is often used to approximate distributions of discrete random variables. Quite often, this yields satisfactory results, provided that we make the **continuity correction** illustrated in the following example.

EXAMPLE 10   **A continuity correction to improve the normal approximation to a count variable**

In a certain city, the number of power outages per month is a random variable, having a distribution with $\mu = 11.6$ and $\sigma = 3.3$. If this distribution can be approximated closely with a normal distribution, what is the probability that there will be at least 8 outages in any one month?

Solution   The answer is given by the area of the shaded region of Figure 5.15—the area to the right of 7.5, not 8. The reason for this is that the number of outages is a discrete random variable, and if we want to approximate its distribution with a normal distribution, we must "spread" its values over a continuous scale. We do this by representing each integer k by the interval from $k - \frac{1}{2}$ to $k + \frac{1}{2}$. For instance, 3 is represented by the interval from 2.5 to 3.5, 10 is represented by the interval from 9.5 to 10.5, and "at least 8" is represented by the interval to the right of 7.5. Thus the desired probability is approximated by

$$1 - F\left(\frac{7.5 - 11.6}{3.3}\right) = 1 - F(-1.24)$$
$$= F(1.24)$$
$$= 0.8925$$

∎

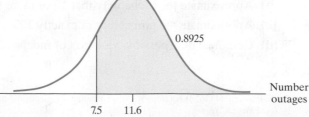

Figure 5.15
Diagram for example dealing
with power outages

5.3   The Normal Approximation to the Binomial Distribution

Unlike the Poisson approximation that applies when p is small, the normal distribution approximates the binomial distribution when n is large and p, the probability of a success, is not close to 0 or 1. We state, without proof, the following theorem:

Normal approximation to binomial distribution

> **Theorem 5.1**   If X is a random variable having the binomial distribution with the parameters n and p, the limiting form of the distribution function of the standardized random variable
>
> $$Z = \frac{X - np}{\sqrt{np(1 - p)}}$$
>
> as $n \to \infty$, is given by the standard normal distribution
>
> $$F(z) = \int_{-\infty}^{z} \frac{1}{\sqrt{2\pi}} e^{-t^2/2} \, dt \quad -\infty < z < \infty$$

Note that although X takes on only the values $0, 1, 2, \ldots, n$, in the limit as $n \to \infty$, the distribution of the corresponding standardized random variable is continuous, and the corresponding probability density is the standard normal density.

A good rule of thumb for the normal approximation

> Use the normal approximation to the binomial distribution only when np and $n(1 - p)$ are both greater than 15.

Note that in Example 11, which is an application of Theorem 5.1, we use again the continuity correction given on page 137.

EXAMPLE 11   The current consensus is that there are three types of neutrinos and each is accompanied by an antimatter version. Further, any single neutrino can change (oscillate) from one type to another. When one type of antimatter neutron, called an electron antineutrino, travels two kilometers from a reactor to the detector it will disappear if it interacts with an electron neutrino and changes into another type. At one site, physicists have performed a path breaking experiment[2] that measured an important constant for this change. At a specific detector, with electron antineutrinos of average energy, this constant translates into probability .056 of disappearing.

Consider the outcomes of the next 300 electron antineutrinos leaving the reactor and heading toward the detector. Assuming that the conditions for Bernoulli trials hold,

(a) find the mean and standard deviation of the number which will disappear.

(b) Approximate the probability that 12 or more will disappear.

(c) Approximate the probability of exactly 12.

(d) Comment on a possible violation of independence.

[2]F. P. An et. al (2013) An improved measurement of Electron Antineutrino disappearances at Day Bay, *Chin.Phys.* C37

Figure 5.16
Normal approximation to
Binomial (b) $P(X \geq 12)$
(c) $P(X = 12)$.

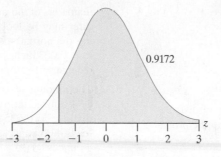

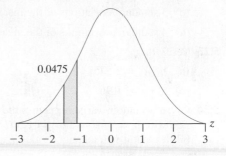

Solution   We take the probability $p = 0.056$ which is the value estimated from the physics experiment.

(a) Using the formulas for mean and standard deviation, we find

$$\text{Mean} = np = 300 \times 0.056 = 16.80$$
$$\text{Standard deviation} = \sqrt{np(1-p)} = \sqrt{300 \times 0.056 \times 0.944} = 3.982$$

(b) Since $np > 15$, the normal distribution provides a good approximation to the probability

$$1 - F\left(\frac{11.5 - 16.80}{3.982}\right) = 0.9084$$

as illustrated in Figure 5.16. Over ninety percent of the time there will be 12 or more disappearances among the 300.

The exact value .9142 is obtained using **1 - pbinom(11,300,.056)** in R.

(c) $F\left(\dfrac{12.5 - 16.80}{3.982}\right) - F\left(\dfrac{11.5 - 16.80}{3.982}\right) = 0.1401 - 0.0916 = 0.0485$

(d) If two or more electron antineutrinos are so close that they interfere with each other, or even collide, independence is violated.

The exact calculation is always preferrable when p is given but the approximation is important for inference when it is not.   ∎

Exercises

5.19 If a random variable has the standard normal distribution, find the probability that it will take on a value

(a) less than 1.75;

(b) less than -1.25;

(c) greater than 2.06;

(d) greater than -1.82.

5.20 If a random variable has the standard normal distribution, find the probability that it will take on a value

(a) between 0 and 2.3;

(b) between 1.22 and 2.43;

(c) between -1.45 and -0.45;

(d) between -1.70 and 1.35.

5.21 Recycled concrete materials must undergo a strength test before it can be incorporated in the base of a highway. The strength measurement, for each sample of

material, can be modeled by a normal distribution with mean 171 and standard deviation 20.5 MPa.

(a) If engineering specifications require the sample to have tensile strength greater than 140 MPa, what is the probability that a sample will fail to meet the specifications?

(b) In the long run, what proportion of the samples will fail? Explain your answer.

(c) The mean strength can be increased by mixing other materials with the recycled materials. What new mean is required, when the standard deviation is 20.5, to reduce the probability of not meeting specifications to 0.02?

5.22 If a random variable has a normal distribution, what are the probabilities that it will take on a value within

(a) 1 standard deviation of the mean;

(b) 2 standard deviations of the mean;

(c) 3 standard deviations of the mean;

(d) 4 standard deviations of the mean?

5.23 Verify that

(a) $z_{0.005} = 2.575$;

(b) $z_{0.025} = 1.96$.

5.24 Given a random variable having the normal distribution with $\mu = 16.2$ and $\sigma^2 = 1.5625$, find the probabilities that it will take on a value

(a) greater than 16.8;

(b) less than 14.9;

(c) between 13.6 and 18.8;

(d) between 16.5 and 16.7.

5.25 The time for a super glue to set can be treated as a random variable having a normal distribution with mean 30 seconds. Find its standard deviation if the probability is 0.20 that it will take on a value greater than 39.2 seconds.

5.26 Artificial silk fibers made from spider silk proteins have high tensile strengths. Tensile strength can be modeled by a normal distribution with mean 150 MJ/m^3 and standard deviation 10. Find the probability that a fiber will have tensile strength

(a) at least 142;

(b) between 141.5 and 159.5 MJ/m^3 ?

5.27 Refer to Exercise 5.26 but suppose a potential large contract contains the specification that at most 5% have tensile strength less than 135 MJ/m^3. If the manufacturing process is improved to meet this specification, determine

(a) the new mean μ if the standard deviation is 10 MJ/m^3;

(b) the new standard deviation if the mean is 150 MJ/m^3.

5.28 Find the *quartiles*

$$-z_{0.25} \quad z_{0.50} \quad z_{0.25}$$

of the standard normal distribution.

5.29 The daily high temperature in a computer server room at the university can be modeled by a normal distribution with mean $68.7°F$ and standard deviation $1.2°F$. Find the probability that, on a given day, the high temperature will be

(a) between 68.3 and $70.3°F$

(b) greater than $71.5°F$.

5.30 With reference to the preceding exercise, for which temperature is the probability 0.05 that it will be exceeded during one day?

5.31 Specifications for a certain job call for washers with an inside diameter of 0.300 ± 0.005 inch. If the inside

diameters of the washers supplied by a given manufacturer may be looked upon as a random variable having the normal distribution with $\mu = 0.302$ inch and $\sigma = 0.003$ inch, what percentage of these washers will meet specifications?

5.32 The amount of coffee, that a filling machine puts in a "4 ounce" jar, follows a normal distribution. Verify that if $\sigma = 0.025$ ounce and the mean amount of coffee is 4.05 ounces, 98% of the jars will contain at least 4 ounces.

5.33 A stamping machine produces can tops whose diameters are normally distributed with a standard deviation of 0.01 inch. At what "normal" (mean) diameter should the machine be set so that no more than 5% of the can tops produced have diameters exceeding 3 inches?

5.34 Extruded plastic rods are automatically cut into nominal lengths of 6 inches. Actual lengths are normally distributed about a mean of 6 inches and their standard deviation is 0.06 inch.

(a) What proportion of the rods have lengths that are outside the tolerance limits of 5.9 inches to 6.1 inches?

(b) To what value does the standard deviation need to be reduced if 99% of the rods must be within tolerance?

5.35 If a random variable has the binomial distribution with $n = 40$ and $p = 0.40$, use the normal approximation to determine the probabilities that it will take on

(a) the value 22;

(b) a value less than 8.

5.36 A manufacturer knows that, on average, 2% of the electric toasters that he makes will require repairs within 90 days after they are sold. Use the normal approximation to the binomial distribution to determine the probability that among 1,200 of these toasters at least 30 will require repairs within the first 90 days after they are sold.

5.37 The probability that an electronic component will fail in less than 1,000 hours of continuous use is 0.25. Use the normal approximation to find the probability that among 200 such components fewer than 45 will fail in less than 1,000 hours of continuous use.

5.38 Smartphone owners have a high risk of being victims of identity fraud. In a recent year about 7% of smartphone owners were victims. Assume the same rate of identity fraud holds today. Use the normal distribution to approximate the probability that, out of random sample of 230 smartphone owners, the number that are victims of identity fraud this year will be

(a) 9 or fewer.

(b) 18 or more.

5.39 Refer to Example 11 concerning the experiment that confirms electron antineutrinos change type. Suppose instead that there are 400 electron antineutrinos leaving the reactor. Repeat parts (a)–(c) of the example.

5.40 To illustrate the law of large numbers mentioned on page 106, find the probabilities that the proportion of heads will be anywhere from 0.49 to 0.51 when a balanced coin is flipped

(a) 1,000 times;

(b) 10,000 times.

5.41 Verify the identity $F(-z) = 1 - F(z)$ given on page 131.

5.42 Verify that the parameter μ in the expression for the normal density on page 130, is, in fact, its mean.

5.43 Verify that the parameter σ^2 in the expression for the normal density on page 130 is, in fact, its variance.

5.44 Normal probabilities can be calculated using *MINITAB*. Let X have a normal distribution with mean

11.3 and standard deviation 5.7. The following steps yield the cumulative probability of 9.31 or smaller, or $P(X \leq 9.31)$.

> **Dialog box:**
> **Calc > Probability Distribution > Normal**
> Choose **Cumulative Distribution.** Choose **Input constant** and enter 9.31.
> Type 11.3 in **Mean** and 5.7 in **standard deviation.**
> Click **OK.**

Output: Normal with mean = 11.3000 and standard deviation = 5.70000

x	$P(X <= x)$
9.3100	0.3635

For this same normal distribution, find the probability

(a) of 8.493 or smaller;

(b) of 16.074 or smaller.

5.4 Other Probability Densities

In the application of statistics to problems in engineering and physical science, we encounter many probability densities other than the normal distribution. These include the t, F, and chi square distributions; the fundamental sampling distributions that we introduce in Chapter 6. We also treat the exponential and Weibull distributions, which we apply to problems of reliability and life testing in Chapter 16.

In the remainder of this chapter we shall discuss five continuous distributions, the *uniform distribution*, the *log-normal distribution*, the *gamma distribution*, the *beta distribution*, and the *Weibull distribution*, for the twofold purpose of widening your familiarity with well known probability densities and to lay the foundation for future applications.

5.5 The Uniform Distribution

The **uniform distribution**, with the parameters α and β, has probability density function

Uniform distribution

$$f(x) = \begin{cases} \dfrac{1}{\beta - \alpha} & \text{for } \alpha < x < \beta \\ 0 & \text{elsewhere} \end{cases}$$

whose graph is shown in Figure 5.17. Note that all values of x from α to β are "equally likely" in the sense that the probability that x lies in an interval of width Δx entirely contained in the interval from α to β is equal to $\Delta x/(\beta - \alpha)$, regardless of the exact location of the interval.

To illustrate how a physical situation might give rise to a uniform distribution, suppose that a wheel of a locomotive has the radius r and that x is the location of a point on its circumference measured along the circumference from some reference point 0. When the brakes are applied, some point will make sliding contact with the

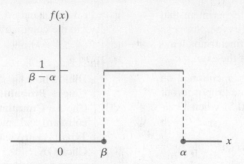

Figure 5.17
Graph of uniform probability density

rail, and heavy wear will occur at that point. For repeated application of the brakes, it would seem reasonable to assume that x is a value of a random variable having the uniform distribution with $\alpha = 0$ and $\beta = 2\pi r$. If this assumption were incorrect, that is, if some set of points on the wheel made contact more often than others, the wheel would eventually exhibit "flat spots" or wear out of round.

To determine the mean and the variance of the uniform distribution, we first evaluate the two integrals

$$\mu = \int_\alpha^\beta x \cdot \frac{1}{\beta - \alpha} \, dx = \frac{\alpha + \beta}{2}$$

and

$$\mu_2' = \int_\alpha^\beta x^2 \cdot \frac{1}{\beta - \alpha} \, dx = \frac{\alpha^2 + \alpha\beta + \beta^2}{3}$$

Thus,

Mean of uniform distribution

$$\mu = \frac{\alpha + \beta}{2}$$

and, making use of the formula $\sigma^2 = \mu_2' - \mu^2$, we find that

Variance of uniform distribution

$$\sigma^2 = \frac{1}{12}(\beta - \alpha)^2$$

5.6 The Log-Normal Distribution

The **log-normal distribution** occurs in practice whenever we encounter a random variable which is such that its logarithm has a normal distribution. Its probability density is given by

Log-normal distribution

$$f(x) = \begin{cases} \dfrac{1}{\sqrt{2\pi}\,\beta}\, x^{-1}\, e^{-(\ln x - \alpha)^2/2\beta^2} & \text{for } x > 0, \ \beta > 0 \\ 0 & \text{elsewhere} \end{cases}$$

where $\ln x$ is the natural logarithm of x. A graph of the log-normal distribution with $\alpha = 0$ and $\beta = 1$ is shown in Figure 5.18. It can be seen from the figure that this distribution is positively skewed, that is, it has a long right-hand tail.

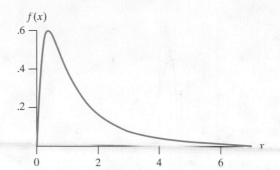

Figure 5.18
Graph of log-normal
probability density, $\alpha = 0$,
$\beta = 1$.

To find the probability that a random variable having the log-normal distribution will take on a value between a and b $(0 < a < b)$, we must evaluate the integral

$$\int_a^b \frac{1}{\sqrt{2\pi}\,\beta} x^{-1}\, e^{-(\ln x - \alpha)^2/2\beta^2}\, dx$$

Changing variable by letting $y = \ln x$ and identifying the integrand as the normal density with $\mu = \alpha$ and $\sigma = \beta$, we find that the desired probability is given by

$$\int_{\ln a}^{\ln b} \frac{1}{\sqrt{2\pi}\,\beta}\, e^{-(y-\alpha)^2/2\beta^2}\, dy = F\left(\frac{\ln b - \alpha}{\beta}\right) - F\left(\frac{\ln a - \alpha}{\beta}\right)$$

where F is the distribution function of the standard normal distribution.

EXAMPLE 12   **Calculating a log-normal probability**

The current gain of certain transistors is measured in units which make it equal to the logarithm of I_o/I_i, the ratio of the output to the input current. If this logarithm is normally distributed with $\mu = 2$ and $\sigma = 0.1$, find the probability that I_o/I_i will take on a value between 6.1 and 8.2.

Solution   Since $\alpha = 2$ and $\beta = 0.1$, we get

$$F\left(\frac{\ln 8.2 - 2}{0.1}\right) - F\left(\frac{\ln 6.1 - 2}{0.1}\right) = F(1.0) - F(-1.92)$$

$$= 0.8139 \qquad\blacksquare$$

EXAMPLE 13   **Graphing a probability density function on top of a density histogram to help assess fit**

Make a density histogram of the interrequest times on page 19 and relate it to a log-normal distribution.

Solution   Figure 5.19 gives the density histogram. To accurately portray the pattern, shorter intervals are used for the smaller times. We have also plotted the log-normal density with $\alpha = 8.85$ and $\beta = 1.03$. The log-normal fit is explored further in Section 5.12. (See also Exercise 5.103.) $\blacksquare$

EXAMPLE 14   **A log-normal probability calculation for a risk analysis**

As part of a risk analysis concerning a nuclear power plant, engineers must model the strength of steam generator supports in terms of their ability to withstand the peak acceleration caused by earthquakes. Expert opinion suggests that ln (strength)

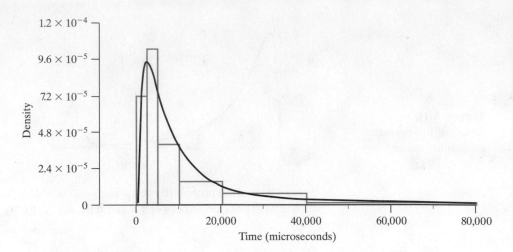

Figure 5.19
Density histogram of
interrequest time

is normally distributed with $\mu = 4.0$ and $\sigma^2 = 0.09$. Find the probability that the supports will survive a peak acceleration of 33.

Solution   Since $\alpha = 4.0$ and $\beta = 0.30$, we find

$$1 - F\left(\frac{\ln(33) - 4.0}{0.30}\right) = 1 - F(-1.68) = 0.9535$$ ∎

To find a formula for the mean of the log-normal distribution, we write

$$\mu = \frac{1}{\sqrt{2\pi}\,\beta} \int_0^\infty x \cdot x^{-1} \, e^{-(\ln x - \alpha)^2/2\beta^2} \, dx$$

and let $y = \ln x$, so

$$\mu = \frac{1}{\sqrt{2\pi}\,\beta} \int_{-\infty}^\infty e^y \, e^{-(y-\alpha)^2/2\beta^2} \, dy$$

This integral is evaluated by completing the square on the exponent $y - (y - \alpha)^2/2\beta^2$, to produce an integrand in the form of a normal density. The final result, which the reader will be asked to verify in Exercise 5.48, is

Mean of log-normal distribution

$$\mu = e^{\alpha + \beta^2/2}$$

Similar, but more lengthy, calculations yield

Variance of log-normal distribution

$$\sigma^2 = e^{2\alpha + \beta^2}\left(e^{\beta^2} - 1\right)$$

EXAMPLE 15   **Calculating the mean and variance for a log-normal distribution**

With reference to Example 12, find the mean and the variance of the distribution of the ratio of the output to the input current.

Solution   Substituting $\alpha = 2$ and $\beta = 0.1$ into the above formulas, we get

$$\mu = e^{2 + (0.1)^2 / 2} = 7.43$$

and

$$\sigma^2 = e^{4 + (0.1)^2} (e^{(0.1)^2} - 1) = 0.55$$ ∎

5.7 The Gamma Distribution

Several important probability densities whose applications will be discussed later are special cases of the **gamma distribution**. This distribution has probability density

Gamma distribution

$$f(x) = \begin{cases} \dfrac{1}{\beta^{\alpha} \Gamma(\alpha)} x^{\alpha - 1} e^{-x/\beta} & \text{for } x > 0, \ \alpha > 0, \ \beta > 0 \\ 0 & \text{elsewhere} \end{cases}$$

where $\Gamma(\alpha)$ is a value of the **gamma function**, defined by

$$\Gamma(\alpha) = \int_0^{\infty} x^{\alpha - 1} e^{-x} \, dx$$

Integration by parts shows that

$$\Gamma(\alpha) = (\alpha - 1)\Gamma(\alpha - 1)$$

for any $\alpha > 1$ and, hence, that $\Gamma(\alpha) = (\alpha - 1)!$ when α is a positive integer. Graphs of several gamma distributions are shown in Figure 5.20 and they exhibit the fact that these distributions are positively skewed. In fact, the skewness decreases as α increases for any fixed value of β.

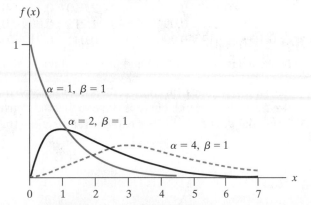

Figure 5.20
Graph of some gamma
probability density functions

The mean and the variance of the gamma distribution are obtained by making use of the gamma function and its special properties mentioned above. The mean

$$\mu = \frac{1}{\beta^{\alpha} \Gamma(\alpha)} \int_0^{\infty} x \cdot x^{\alpha - 1} e^{-x/\beta} \, dx$$

and, after letting $y = x/\beta$, we get

$$\mu = \frac{\beta}{\Gamma(\alpha)} \int_0^{\infty} y^{\alpha} e^{-y} \, dy = \frac{\beta \Gamma(\alpha + 1)}{\Gamma(\alpha)}$$

Then, using the identity $\Gamma(\alpha + 1) = \alpha \cdot \Gamma(\alpha)$, we arrive at the result.

Mean of gamma distribution

$$\mu = \alpha \beta$$

Using similar methods, it can also be shown that the variance of the gamma distribution is given by

Variance of gamma distribution

$$\sigma^2 = \alpha \beta^2$$

In the special case where $\alpha = 1$, we get the **exponential distribution**, whose probability density is thus

Exponential distribution

$$f(x) = \begin{cases} \dfrac{1}{\beta}\, e^{-x/\beta} & \text{for } x > 0, \ \beta > 0 \\ 0 & \text{elsewhere} \end{cases}$$

and whose mean and variance are $\mu = \beta$ and $\sigma^2 = \beta^2$. Note that the distribution of Example 1 is an exponential distribution with $\beta = \frac{1}{2}$.

EXAMPLE 16   **An exponential density function on top of a density histogram**

An engineer observing a nuclear reaction measures the time intervals between the emissions of beta particles. (Courtesy of consulting client)

0.894	0.991	0.261	0.186	0.311	0.817	2.267	0.091	0.139	0.083
0.235	0.424	0.216	0.579	0.429	0.612	0.143	0.055	0.752	0.188
0.071	0.159	0.082	1.653	2.010	0.158	0.527	1.033	2.863	0.365
0.459	0.431	0.092	0.830	1.718	0.099	0.162	0.076	0.107	0.278
0.100	0.919	0.900	0.093	0.041	0.712	0.994	0.149	0.866	0.054

Make a density histogram and plot an exponential density as an approximation.

Solution   These decay times (in milliseconds) are presented as a density histogram in Figure 5.21. The smooth curve is the exponential density with $\beta = 0.55$. Fit to an exponential density is further explored in Section 16.3. The exponential density has mean $\beta = 0.55$ and standard deviation 0.55.   ∎

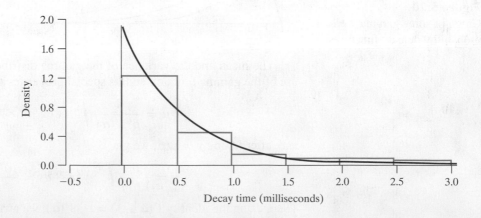

Figure 5.21
Density histogram of decay times

The exponential distribution has many important applications. For instance, it can be shown that in connection with Poisson processes (see Section 4.7) the **waiting time** between successive arrivals (successes) has an exponential distribution. More specifically, it can be shown that if in a Poisson process the mean arrival rate (average number of arrivals per unit time) is α, the time until the first arrival, or the waiting time between successive arrivals, has an exponential distribution with $\beta = \dfrac{1}{\alpha}$ (see Exercise 5.62).

EXAMPLE 17 **Probability calculations using the exponential distribution**

With reference to the example on page 114, where on the average three trucks arrived per hour to be unloaded at a warehouse, what are the probabilities that the time between the arrival of successive trucks will be

(a) less than 5 minutes? **(b)** at least 45 minutes?

Solution   Assuming the arrivals follow a Poisson process with $\alpha = 3$, then $\beta = \frac{1}{3}$ and we get

(a)
$$\int_{0}^{1/12} 3\,e^{-3x}\,dx = 1 - e^{-1/4} = 0.221$$

(b)
$$\int_{3/4}^{\infty} 3\,e^{-3x}\,dx = e^{-9/4} = 0.105$$

∎

5.8   The Beta Distribution

When a random variable takes on values on the interval from 0 to 1, one choice of a probability density is the **beta distribution** whose probability density is

Beta distribution

$$f(x) = \begin{cases} \dfrac{\Gamma(\alpha + \beta)}{\Gamma(\alpha) \cdot \Gamma(\beta)}\, x^{\alpha-1}(1-x)^{\beta-1} & \text{for } 0 < x < 1, \alpha > 0, \beta > 0 \\[2ex] 0 & \text{elsewhere} \end{cases}$$

The mean and the variance of this distribution are given by

Mean and variance of beta distribution

$$\mu = \frac{\alpha}{\alpha + \beta} \quad \text{and} \quad \sigma^2 = \frac{\alpha\,\beta}{(\alpha + \beta)^2\,(\alpha + \beta + 1)}$$

Note that for $\alpha = 1$ and $\beta = 1$ we obtain as a special case the uniform distribution of Section 5.5 defined on the interval from 0 to 1. The following example, pertaining to a proportion, illustrates a typical application of the beta distribution.

EXAMPLE 18 **Probability calculations using a beta distribution**

In a certain county, the proportion of highway sections requiring repairs in any given year is a random variable having the beta distribution with $\alpha = 3$ and $\beta = 2$ (shown in Figure 5.22).

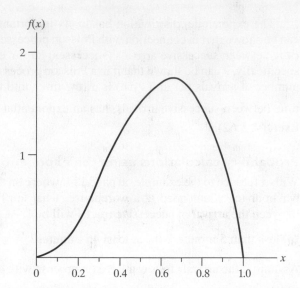

Figure 5.22
Graph of the beta density with
$\alpha = 3$ and $\beta = 2$

(a) On the average, what percentage of the highway sections require repairs in any given year?

(b) Find the probability that at most half of the highway sections will require repairs in any given year.

Solution   (a) $\mu = \dfrac{3}{3+2} = 0.60,$

which means that on the average 60% of the highway sections require repairs in any given year.

(b) Substituting $\alpha = 3$ and $\beta = 2$ into the formula for the beta distribution and making use of the fact that $\Gamma(5) = 4! = 24$, $\Gamma(3) = 2! = 2$, and $\Gamma(2) = 1! = 1$, we get

$$f(x) = \begin{cases} 12x^2(1-x) & \text{for } 0 < x < 1 \\ 0 & \text{elsewhere} \end{cases}$$

Thus, the desired probability is given by

$$\int_0^{1/2} 12x^2(1-x)\,dx = \frac{5}{16} \qquad\blacksquare$$

In most realistically complex situations, probabilities relating to gamma and beta distributions are obtained from computer programs.

5.9 The Weibull Distribution

Closely related to the exponential distribution is the **Weibull distribution**, whose probability density is given by

Weibull distribution

$$f(x) = \begin{cases} \alpha\,\beta x^{\beta-1}\,e^{-\alpha x^{\beta}} & \text{for } x > 0,\ \alpha > 0,\ \beta > 0 \\ 0 & \text{elsewhere} \end{cases}$$

The Weibull distribution has cumulative distribution function

$$F(x) = 1 - e^{-\alpha x^{\beta}} \qquad x > 0$$

which is obtained from

$$F(x) = \int_{0}^{x} \alpha \beta w^{\beta-1} e^{-\alpha w^{\beta}} \, dw$$

by making the change of variable $y = w^{\beta}$. Then

$$F(x) = \int_{0}^{x^{\beta}} \alpha e^{-\alpha y} \, dy = 1 - e^{-\alpha x^{\beta}}.$$

If X has the Weibull distribution and $Y = X^{\beta}$, then

$$P(X^{\beta} \leq y) = P(X \leq y^{1/\beta}) = 1 - e^{-\alpha(y^{1/\beta})^{\beta}} = 1 - e^{-\alpha y}$$

which is the cumulative distribution of the exponential distribution. That is, when X has the Weibull distribution then $Y = X^{\beta}$ has an exponential distribution. The graphs of several Weibull distributions with $\alpha = 1$ and $\beta = \frac{1}{2}$, 1, and 2 are shown in Figure 5.23.

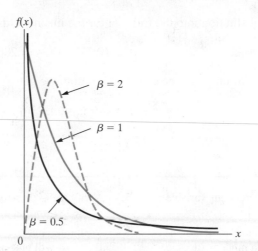

Figure 5.23
Graphs of Weibull densities with $\alpha = 1$ and $\beta = \frac{1}{2}$, 1, and 2

The mean of the Weibull distribution having the parameters α and β may be obtained by evaluating the integral

$$\mu = \int_{0}^{\infty} x \cdot \alpha \beta x^{\beta-1} e^{-\alpha x^{\beta}} \, dx$$

Making the change of variable $u = \alpha x^{\beta}$, we get

$$\mu = \alpha^{-1/\beta} \int_{0}^{\infty} u^{1/\beta} e^{-u} \, du$$

Recognizing the integral as

$$\Gamma\left(1 + \frac{1}{\beta}\right)$$

where the gamma function is defined on page 145, we obtain the mean of the Weibull distribution.

Mean of Weibull distribution	$$\mu = \alpha^{-1/\beta}\, \Gamma\left(1 + \frac{1}{\beta}\right)$$

Using a similar method to determine first μ_2', the reader will be asked to show in Exercise 5.70 that the variance of this distribution is given by

Variance of Weibull distribution	$$\sigma^2 = \alpha^{-2/\beta}\left\{\Gamma\left(1 + \frac{2}{\beta}\right) - \left[\Gamma\left(1 + \frac{1}{\beta}\right)\right]^2\right\}$$

EXAMPLE 19   **Probability calculations using a Weibull distribution**

Suppose that the lifetime of a certain kind of an emergency backup battery (in hours) is a random variable X having the Weibull distribution with $\alpha = 0.1$ and $\beta = 0.5$. Find

(a) the mean lifetime of these batteries;

(b) the probability that such a battery will last more than 300 hours.

Solution   (a) Substitution into the formula for the mean yields

$$\mu = (0.1)^{-2}\, \Gamma(3) = 200 \text{ hours}$$

(b) Performing the necessary integration, we get

$$\int_{300}^{\infty} (0.05)x^{-0.5}\, e^{-0.1x^{0.5}}\, dx = e^{-0.1(300)^{0.5}}$$
$$= 0.177 \quad \blacksquare$$

Exercises

5.45 Find the distribution function of a random variable having a uniform distribution on $(0, 1)$.

5.46 In certain experiments, the error made in determining the solubility of a substance is a random variable having the uniform density with $\alpha = -0.025$ and $\beta = 0.025$. What are the probabilities that such an error will be

(a) between 0.010 and 0.015;

(b) between -0.012 and 0.012?

5.47 From experience Mr. Harris has found that the low bid on a construction job can be regarded as a random variable having the uniform density

$$f(x) = \begin{cases} \dfrac{3}{4C} & \text{for } \dfrac{2C}{3} < x < 2C \\[2mm] 0 & \text{elsewhere} \end{cases}$$

where C is his own estimate of the cost of the job. What percentage should Mr. Harris add to his cost estimate when submitting bids to maximize his expected profit?

5.48 Verify the expression given on page 144 for the mean of the log-normal distribution.

5.49 With reference to the Example 12, find the probability that I_o/I_i will take on a value between 7.0 and 7.5.

5.50 If a random variable has the log-normal distribution with $\alpha = -1$ and $\beta = 2$, find its mean and its standard deviation.

5.51 With reference to the preceding exercise, find the probabilities that the random variable will take on a value

(a) between 3.2 and 8.4;

(b) greater than 5.0.

5.52 If a random variable has the gamma distribution with $\alpha = 2$ and $\beta = 3$, find the mean and the standard deviation of this distribution.

5.53 With reference to Exercise 5.52, find the probability that the random variable will take on a value less than 5.

5.54 In a certain city, the daily consumption of electric power (in millions of kilowatt-hours) can be treated as a random variable having a gamma distribution with $\alpha = 3$ and $\beta = 2$. If the power plant of this city has a daily capacity of 12 million kilowatt-hours, what is the probability that this power supply will be inadequate on any given day?

5.55 With reference to the Example 14, suppose the expert opinion is in error. Calculate the probability that the supports will survive if

(a) $\mu = 3.0$ and $\sigma^2 = 0.09$;

(b) $\mu = 4.0$ and $\sigma^2 = 0.25$.

5.56 Verify the expression for the variance of the gamma distribution given on page 146.

5.57 Show that when $\alpha > 1$, the graph of the gamma density has a relative maximum at $x = \beta(\alpha - 1)$. What happens when $0 < \alpha < 1$ and when $\alpha = 1$?

5.58 The amount of time that a surveillance camera will run without having to be reset is a random variable having the exponential distribution with $\beta = 50$ days. Find the probabilities that such a camera will

(a) have to be reset in less than 20 days;

(b) not have to be reset in at least 60 days.

5.59 With reference to Exercise 4.95, find the percent of the time that the interval between breakdowns of the computer will be

(a) less than 1 week;

(b) at least 5 weeks.

5.60 With reference to Exercise 4.58, find the probabilities that the time between successive requests for consulting will be

(a) less than 0.5 week;

(b) more than 3 weeks.

5.61 Given a Poisson process with on the average α arrivals per unit time, find the probability that there will be no arrivals during a time interval of length t, namely, the probability that the waiting times between successive arrivals will be at least of length t.

5.62 Use the result of Exercise 5.61 to find an expression for the probability density of the waiting time between successive arrivals.

5.63 Verify for $\alpha = 3$ and $\beta = 3$ that the integral of the beta density, from 0 to 1, is equal to 1.

5.64 If the annual proportion of erroneous income tax returns filed with the IRS can be looked upon as a random variable having a beta distribution with $\alpha = 2$ and $\beta = 9$, what is the probability that in any given year there will be fewer than 10% erroneous returns?

5.65 Suppose that the proportion of defectives shipped by a vendor, which varies somewhat from shipment to shipment, may be looked upon as a random variable having the beta distribution with $\alpha = 1$ and $\beta = 4$.

(a) Find the mean of this beta distribution, namely, the average proportion of defectives in a shipment from this vendor.

(b) Find the probability that a shipment from this vendor will contain 25% or more defectives.

5.66 Show that when $\alpha > 1$ and $\beta > 1$, the beta density has a relative maximum at

$$x = \frac{\alpha - 1}{\alpha + \beta - 2}$$

5.67 With reference to the Example 19, find the probability that such a battery will not last 100 hours.

5.68 Suppose that the time to failure (in minutes) of certain electronic components subjected to continuous vibrations may be looked upon as a random variable having the Weibull distribution with $\alpha = \frac{1}{5}$ and $\beta = \frac{1}{3}$.

(a) How long can such a component be expected to last?

(b) What is the probability that such a component will fail in less than 5 hours?

5.69 Suppose that the service life (in hours) of a nano scale semiconductor device is a random variable having the Weibull distribution with $\alpha = 0.025$ and $\beta = 0.500$. What is the probability that such a device will still be in operating condition after 4,000 hours?

5.70 Verify the formula for the variance of the Weibull distribution given on page 150.

5.10 Joint Distributions—Discrete and Continuous

Discrete Variables

Often, experiments are conducted where two random variables are observed simultaneously in order to determine not only their individual behavior but also the degree of relationship between them.

For two discrete random variables X_1 and X_2, we write the probability that X_1 will take the value x_1 and X_2 will take the value x_2 as $P(X_1 = x_1, X_2 = x_2)$. Consequently, $P(X_1 = x_1, X_2 = x_2)$ is the probability of the intersection of the events

$X_1 = x_1$ and $X_2 = x_2$. The distribution of probability is specified by listing the probabilities associated with all possible pairs of values x_1 and x_2, either by formula or in a table. We refer to the function $f(x_1, x_2) = P(X_1 = x_1, X_2 = x_2)$ and the corresponding possible values (x_1, x_2) as the **joint probability distribution** of X_1 and X_2.

EXAMPLE 20

Calculating probabilities from a discrete joint probability distribution

Let X_1 and X_2 have the joint probability distribution in the table below.

Joint Probability Distribution $f(x_1, x_2)$ of X_1 and X_2			
		x_1	
	0	1	2
x_2 0	0.1	0.4	0.1
1	0.2	0.2	0

(a) Find $P(X_1 + X_2 > 1)$.

(b) Find the probability distribution $f_1(x_1) = P(X_1 = x_1)$ of the individual random variable X_1.

Solution

(a) The event $X_1 + X_2 > 1$ is composed of the pairs of values $(1, 1)$, $(2, 0)$, and $(2,1)$. Adding their corresponding probabilities

$$P(X_1 + X_2 > 1) = f(1, 1) + f(2, 0) + f(2, 1) = 0.2 + 0.1 + 0 = 0.3$$

(b) Since the event $X_1 = 0$ is composed of the two pairs of values $(0, 0)$ and $(0, 1)$, we add their corresponding probabilities to obtain

$$P(X_1 = 0) = f(0, 0) + f(0, 1) = 0.1 + 0.2 = 0.3$$

Continuing, we obtain $P(X_1 = 1) = 0.6$ and $P(X_1 = 2) = 0.1$. In summary, $f_1(0) = 0.3$, $f_1(1) = 0.6$, and $f_1(2) = 0.1$ is the probability distribution of X_1.

Rewriting the frequency table but including the row and column totals,

Joint Probability Distribution $f(x_1, x_2)$ of X_1 and X_2 with Marginal Distributions				
		x_1		Total
	0	1	2	$f_2(x_2)$
x_2 0	0.1	0.4	0.1	0.6
1	0.2	0.2	0	0.4
Total $f_1(x_1)$	0.3	0.6	0.1	1.0

Note that the probability distribution $f_1(x_1)$ of X_1 appears in the lower margin of this enlarged table. The probability distribution $f_2(x_2)$ of X_2 appears in the right-hand margin of the table. Consequently, the individual distributions are called **marginal probability distributions**.

From the example, we see that for each fixed value x_1 of X_1, the marginal probability distribution is obtained as

$$P(X_1 = x_1) = f_1(x_1) = \sum_{\text{values } x_2} f(x_1, x_2)$$

where the sum is over all possible values of the second variable with x_1 fixed.

Consistent with the definition of conditional probability of events when A is the event $X_1 = x_1$ and B is the event $X_2 = x_2$, the **conditional probability distribution** of X_1 given $X_2 = x_2$ is defined as

$$f_1(x_1 \mid x_2) = \frac{f(x_1, x_2)}{f_2(x_2)} \qquad \text{for all } x_1 \text{ provided } f_2(x_2) \neq 0$$

If $f_1(x_1 \mid x_2) = f_1(x_1)$ for all x_1 and x_2, so the conditional probability distribution is free of x_2, or, equivalently, if

$$f(x_1, x_2) = f_1(x_1)f_2(x_2) \qquad \text{for all } x_1, x_2$$

the two random variables are **independent**.

EXAMPLE 21   **A conditional probability distribution**

With reference to the previous example, find the conditional probability distribution of X_1 given $X_2 = 1$. Are X_1 and X_2 independent?

Solution $f_1(0 \mid 1) = \dfrac{f(0, 1)}{f_2(1)} = \dfrac{0.2}{0.4} = 0.5, \quad f_1(1 \mid 1) = \dfrac{f(1, 1)}{f_2(1)} = \dfrac{0.2}{0.4} = 0.5,$ and

$$f_1(2 \mid 1) = \frac{f(2, 1)}{f_2(1)} = \frac{0}{0.4} = 0$$

Since $f_1(0 \mid 1) = 0.5 \neq 0.3 = f_1(0)$, the conditional probability distribution is not free of the value x_2. Equivalently, $f(0, 1) = 0.2 \neq (0.3)(0.4) = f_1(0)f_2(1)$ so X_1 and X_2 are *dependent*. ∎

Suppose that instead we are concerned with k random variables X_1, $X_2, \ldots, X_k$. Let x_1 be a possible value for the first random variable X_1, x_2 be a possible value for the second random variable X_2, and so on with x_k a possible value for the kth random variable. Then the probabilities

$$P(X_1 = x_1, X_2 = x_2, \ldots, X_k = x_k) = f(x_1, x_2, \ldots, x_k)$$

need to be specified. We refer to the function f and the corresponding k-tuples of possible values $(x_1, x_2, \ldots, x_k)$ as the **joint probability distribution** of these discrete random variables.

The probability distribution $f_i(x_i)$ of the individual variable X_i is called the **marginal probability distribution** of the ith random variable

$$f_i(x_i) = \sum_{x_1} \cdots \sum_{x_{i-1}} \sum_{x_{i+1}} \cdots \sum_{x_k} f(x_1, x_2, \ldots, x_k)$$

where the summation is over all possible k-tuples where the ith component is held fixed at the specified value x_i.

Continuous Variables

There are many situations in which we describe an outcome by giving the values of several continuous random variables. For instance, we may measure the weight and the hardness of a rock; the volume, pressure, and temperature of a gas; or the thickness, compressive strength, and potassium content of a piece of glass. If $X_1, X_2, \ldots, X_k$ are k continuous random variables, we shall refer to $f(x_1, x_2, \ldots, x_k)$ as the **joint probability density** of these random variables, if the probability that $a_1 \leq X_1 \leq b_1, a_2 \leq X_2 \leq b_2, \ldots$, and $a_k \leq X_k \leq b_k$ is given by the multiple integral

$$\int_{a_k}^{b_k} \cdots \int_{a_2}^{b_2} \int_{a_1}^{b_1} f(x_1, x_2, \ldots, x_k) \, dx_1 \, dx_2 \ldots dx_k$$

Thus, not every function $f(x_1, x_2, \ldots, x_k)$ can serve as a joint probability density, but if

$$f(x_1, x_2, \ldots, x_k) \geq 0$$

for all values of $x_1, x_2, \ldots, x_k$, and

$$\int_{-\infty}^{\infty} \int_{-\infty}^{\infty} \cdots \int_{-\infty}^{\infty} f(x_1, x_2, \ldots, x_k) \, dx_1 \, dx_2 \ldots dx_k = 1$$

it can be shown that the axioms of probability (with the modification of the definition of "event" discussed in Section 5.1) are satisfied.

To extend the concept of a cumulative distribution function to the k-variable case, we write as $F(x_1, x_2, \ldots, x_k)$ the probability that the first random variable will take on a value less than or equal to x_1, the second random variable will take on a value less than or equal to $x_2, \ldots$, and the kth random variable will take on a value less than or equal to x_k, and we refer to the corresponding function F as the **joint cumulative distribution function** of the k random variables.

EXAMPLE 22 **Calculating probabilities from a joint probability density function**

If the joint probability density of two random variables is given by

$$f(x_1, x_2) = \begin{cases} 6e^{-2x_1-3x_2} & \text{for } x_1 > 0, \ x_2 > 0 \\ 0 & \text{elsewhere} \end{cases}$$

find the probabilities that

(a) the first random variable will take on a value between 1 and 2 and the second random variable will take on a value between 2 and 3;

(b) the first random variable will take on a value less than 2 and the second random variable will take on a value greater than 2.

Solution Performing the necessary integrations, we get

(a)
$$\int_2^3 \int_1^2 6e^{-2x_1-3x_2} \, dx_1 \, dx_2 = (e^{-2} - e^{-4})(e^{-6} - e^{-9})$$
$$= 0.0003$$

(b)
$$\int_2^\infty \int_0^2 6e^{-2x_1-3x_2} \, dx_1 \, dx_2 = (1 - e^{-4})e^{-6}$$
$$= 0.0024$$

EXAMPLE 23   **Determining a joint cumulative distribution function**

Find the joint cumulative distribution function of the two random variables of the preceding exercise, and use it to find the probability that both random variables will take on values less than 1.

Solution   By definition,

$$F(x_1, x_2) = \begin{cases} \int_0^{x_2} \int_0^{x_1} 6 e^{-2u-3v} \, du \, dv & \text{for } x_1 > 0, \ x_2 > 0 \\ 0 & \text{elsewhere} \end{cases}$$

so that

$$F(x_1, x_2) = \begin{cases} (1 - e^{-2x_1})(1 - e^{-3x_2}) & \text{for } x_1 > 0, \ x_2 > 0 \\ 0 & \text{elsewhere} \end{cases}$$

and, hence,

$$F(1, 1) = (1 - e^{-2})(1 - e^{-3})$$
$$= 0.8216$$

∎

Given the joint probability density of k random variables, the probability density of the ith random variable can be obtained by integrating out the other variables; symbolically,

Marginal density

$$f_i(x_i) = \int_{-\infty}^{\infty} \cdots \int_{-\infty}^{\infty} f(x_1, x_2, \ldots, x_k) \, dx_1 \ldots dx_{i-1} \, dx_{i+1} \ldots dx_k$$

and, in this context, the function f_i is called the **marginal density** of the ith random variable. Integrating out only some of the k random variables, we can similarly define **joint marginal densities** of any two, three, or more of the k random variables.

EXAMPLE 24   **Determining a marginal density from a joint density**

With reference to Example 22. find the marginal density of the first random variable.

Solution   Integrating out x_2, we get

$$f_1(x_1) = \begin{cases} \int_0^{\infty} 6 e^{-2x_1 - 3x_2} \, dx_2 & \text{for } x_1 > 0 \\ 0 & \text{elsewhere} \end{cases}$$

or

$$f_1(x_1) = \begin{cases} 2 e^{-2x_1} & \text{for } x_1 > 0 \\ 0 & \text{elsewhere} \end{cases}$$

∎

To explain what we mean by the **independence** of continuous random variables, we could proceed as with discrete random variables and define *conditional probability densities* first; however, it will be easier to say that

Independent random variables

> k random variables $X_1, \ldots, X_k$ are **independent** if and only if
>
> $$F(x_1, x_2, \ldots, x_k) = F_1(x_1) \cdot F_2(x_2) \cdots F_k(x_k)$$
>
> for all values $x_1, x_2, \ldots, x_k$ of these random variables.

In this notation $F(x_1, x_2, \ldots, x_k)$ is, as before, the joint distribution function of the k random variables, while $F_i(x_i)$ for $i = 1, 2, \ldots, k$ are the corresponding individual distribution function of the respective random variables. The same condition applies for discrete random variables.

EXAMPLE 25   **Checking independence via the joint cumulative distribution**

With reference to Example 23, check whether the two random variables are independent.

Solution   As we already saw in Example 23, the joint distribution function of the two random variables is given by

$$F(x_1, x_2) = \begin{cases} (1 - e^{-2x_1})(1 - e^{-3x_2}) & \text{for } x_1 > 0 \text{ and } x_2 > 0 \\ 0 & \text{elsewhere} \end{cases}$$

Now, since $F_1(x_1) = F(x_1, \infty)$ and $F_2(x_2) = F(\infty, x_2)$, it follows that

$$F_1(x_1) = \begin{cases} 1 - e^{-2x_1} & \text{for } x_1 > 0 \\ 0 & \text{elsewhere} \end{cases}$$

and

$$F_2(x_2) = \begin{cases} 1 - e^{-3x_2} & \text{for } x_2 > 0 \\ 0 & \text{elsewhere} \end{cases}$$

Thus, $F(x_1, x_2) = F_1(x_1) \cdot F_2(x_2)$ for all (x_1, x_2) and the two random variables are independent.   ∎

When k random variables have a joint probability density, the k **random variables are independent** if and only if their joint probability density equals the product of the corresponding values of the marginal densities of the k random variables; symbolically,

$$f(x_1, x_2, \ldots, x_k) = f_1(x_1) \cdot f_2(x_2) \cdots f_k(x_k) \qquad \text{for all } (x_1, \ldots, x_k).$$

EXAMPLE 26   **Establishing independence by factoring the joint probability density**

With reference to Example 22, verify that

$$f(x_1, x_2) = f_1(x_1) \cdot f_2(x_2)$$

Solution   Example 24 shows that

$$f_1(x_1) = \begin{cases} 2 e^{-2x_1} & \text{for } x_1 > 0 \\ 0 & \text{elsewhere} \end{cases}$$

and in the same way,

$$f_2(x_2) = \begin{cases} 3\,e^{-3x_2} & \text{for } x_2 > 0 \\ 0 & \text{elsewhere} \end{cases}$$

Thus,

$$f_1(x_1) \cdot f_2(x_2) = \begin{cases} 6\,e^{-2x_1 - 3x_2} & \text{for } x_1 > 0 \text{ and } x_2 > 0 \\ 0 & \text{elsewhere} \end{cases}$$

and it can be seen that $f_1(x_1) \cdot f_2(x_2) = f(x_1, x_2)$ for all (x_1, x_2). ∎

Given two continuous random variables X_1 and X_2, we define the **conditional probability density** of the first given that the second takes on the value x_2 as

Conditional probability density

$$f_1(x_1 \mid x_2) - \frac{f(x_1, x_2)}{f_2(x_2)} \quad \text{provided } f_2(x_2) \neq 0$$

where $f(x_1, x_2)$ and $f_2(x_2)$ are, as before, the joint density of the two random variables and the marginal density of the second. Note that this definition parallels that of the conditional probability distribution on page 153. Also, the joint probability density is the product

$$f(x_1, x_2) = f_1(x_1 \mid x_2)\, f_2(x_2).$$

EXAMPLE 27 **Determining a conditional probability density**
If two random variables have the joint probability density

$$f(x_1, x_2) = \begin{cases} \dfrac{2}{3}(x_1 + 2x_2) & \text{for } 0 < x_1 < 1,\, 0 < x_2 < 1 \\ 0 & \text{elsewhere} \end{cases}$$

find the conditional density of the first given that the second takes on the value x_2.

Solution First we find the marginal density of the second random variable by integrating out x_1, and we get

$$f_2(x_2) = \int_0^1 \frac{2}{3}(x_1 + 2x_2)\,dx_1 = \frac{1}{3}(1 + 4x_2) \quad \text{for } 0 < x_2 < 1$$

and $f_2(x_2) = 0$ elsewhere. Hence, by definition, the conditional density of the first random variable given that the second takes on the value x_2 is given by

$$f_1(x_1 \mid x_2) = \frac{\dfrac{2}{3}(x_1 + 2x_2)}{\dfrac{1}{3}(1 + 4x_2)} = \frac{2x_1 + 4x_2}{1 + 4x_2} \quad \text{for } 0 < x_1 < 1,\, 0 < x_2 < 1$$

and $f_1(x_1 \mid x_2) = 0$ for $x_1 \leq 0$ or $x_1 \geq 1$ and $0 < x_2 < 1$. ∎

Properties of Expectation

Consider a function $g(X)$ of a single random variable X. For instance, if X is an oven temperature in degrees centigrade, then

$$g(X) = \frac{9}{5}X + 32$$

is the same temperature in degrees Fahrenheit.

The **expectation** of the function $g(X)$ is again the *sum* of the products *value* × *probability*.

Expected value of $g(X)$

> In the discrete case, where X has probability distribution $f(x)$
>
> $$E[g(X)] = \sum_{x_i} g(x_i)f(x_i)$$
>
> In the continuous case, where X has probability density function $f(x)$
>
> $$E[g(X)] = \int_{-\infty}^{\infty} g(x)f(x)\,dx$$

If X has mean $\mu = E(X)$, then taking $g(x) = (x - \mu)^2$, we have $E[g(X)] = E(X - \mu)^2$, which is just the variance σ^2 of X.

For any random variable Y, let $E(Y)$ denote its expectation, which is also its mean μ_Y. Its variance is $Var(Y)$ which is also written as σ_Y^2.

When $g(x) = ax + b$, for given constants a and b, then random variable $g(X)$ has expectation

$$E(aX + b) = \int_{-\infty}^{\infty} (ax + b)f(x)\,dx = a\int_{-\infty}^{\infty} xf(x)\,dx + b\int_{-\infty}^{\infty} f(x)\,dx$$
$$= aE(X) + b$$

and variance

$$Var(aX + b) = \int_{-\infty}^{\infty} (ax + b - a\mu_X - b)^2 f(x)\,dx$$
$$= a^2 \int_{-\infty}^{\infty} (x - \mu_X)^2 f(x)\,dx = a^2 Var(X)$$

To summarize,

> For given constants a and b
> $$E(aX + b) = aE(X) + b \quad \text{and} \quad Var(aX + b) = a^2 Var(X)$$

EXAMPLE 28   **The mean and standard deviation of a standardized random variable**
Let X have mean μ and standard deviation σ. Use the properties of expectation to show that the standardized random variable

$$Z = \frac{X - \mu}{\sigma}$$

has mean 0 and standard deviation 1.

Solution   Since Z is of the form

$$Z = \frac{X - \mu}{\sigma} = \frac{1}{\sigma}X - \frac{\mu}{\sigma} = aX + b$$

where $a = 1/\sigma$ and $b = -\mu/\sigma$,

$$E(Z) = \frac{1}{\sigma} E(X) - \frac{\mu}{\sigma} = \frac{\mu}{\sigma} - \frac{\mu}{\sigma} = 0$$

and the variance of Z is

$$\left(\frac{1}{\sigma}\right)^2 Var(X) = \frac{\sigma^2}{\sigma^2} = 1$$

because $E(X) = \mu$ and $Var(X) = \sigma^2$. ∎

EXAMPLE 29   **Determining the mean and variance of 20 X**

Suppose the daily amount of electricity X required for a plating process has mean 10 and standard deviation 3 kilowatt-hours. If the cost of electricity is 20 dollars per kilowatt hour, find the mean, variance, and standard deviation of the daily cost of electricity.

Solution   The daily cost of electricity, $g(X) = 20X$, has mean $20 E(X) = 20 \times 10 = 200$ dollars and variance $(20)^2 Var(X) = (20)^2 3^2 = 3{,}600$. Its standard deviation is $\sqrt{3{,}600} = 60$ dollars. ∎

Given any collection of k random variables, the function $Y = g(X_1, X_2, \ldots, X_k)$ is also a random variable. Examples include $Y = X_1 - X_2$ when $g(x_1, x_2) = x_1 - x_2$ and $Y = 2X_1 + 3X_2$ when $g(x_1, x_2) = 2x_1 + 3x_2$. The random variable $g(X_1, X_2, \ldots, X_k)$ has expected value, or mean, which is the sum of the products *value × probability*.

Expected value of $g(X_1, X_2, \ldots, X_k)$

In the discrete case,

$$E[g(X_1, X_2, \ldots, X_k)] = \sum_{x_1} \sum_{x_2} \cdots \sum_{x_k} g(x_1, x_2, \ldots, x_k) f(x_1, x_2, \ldots, x_k)$$

In the continuous case,

$$E[g(X_1, X_2, \ldots, X_k)]$$
$$= \int_{-\infty}^{\infty} \int_{-\infty}^{\infty} \cdots \int_{-\infty}^{\infty} g(x_1, x_2, \ldots, x_k) f(x_1, x_2, \ldots, x_k) \, dx_1 \, dx_2 \cdots dx_k$$

Several important properties of expectation can be deduced from this definition. Taking $g(x_1, x_2) = (x_1 - \mu_1)(x_2 - \mu_2)$, we see that the product $(x_1 - \mu_1)(x_2 - \mu_2)$ will be positive if both values x_1 and x_2 are above their respective means or both are below their respective means. Otherwise it will be negative. The expected value $E[(X_1 - \mu_1)(X_2 - \mu_2)]$ will tend to be positive when large X_1 and X_2 tend to occur together and small X_1 and X_2 tend to occur together, with high probability. This measure $E[(X_1 - \mu_1)(X_2 - \mu_2)]$ of joint variation is called the population **covariance** of X_1 and X_2.

If X_1 and X_2 are independent so $f(x_1, x_2) = f_1(x_1) f_2(x_2)$,

$$\int_{-\infty}^{\infty} \int_{-\infty}^{\infty} (x_1 - \mu_1)(x_2 - \mu_2) f(x_1, x_2) \, dx_1 \, dx_2$$

$$= \int_{-\infty}^{\infty} (x_1 - \mu_1) f_1(x_1) \, dx_1 \cdot \int_{-\infty}^{\infty} (x_2 - \mu_2) f_2(x_2) \, dx_2 = 0$$

This result concerning zero covariance can be stated as

**Independence implies that
the covariance is zero**

When X_1 and X_2 are independent, their covariance

$$E[(X_1 - \mu_1)(X_2 - \mu_2)] = 0$$

Further, the expectation of a linear combination of two independent random variables $Y = a_1X_1 + a_2X_2$ is

$$\mu_Y = E(Y) = E(a_1X_1 + a_2X_2)$$

$$= \int_{-\infty}^{\infty} \int_{-\infty}^{\infty} (a_1x_1 + a_2x_2)f_1(x)f_2(x_2)\,dx_1\,dx_2$$

$$= a_1 \int_{-\infty}^{\infty} x_1 f_1(x_1)\,dx_1 \int_{-\infty}^{\infty} f_2(x_2)\,dx_2$$

$$+ a_2 \int_{-\infty}^{\infty} f_1(x_1)\,dx_1 \int_{-\infty}^{\infty} x_2 f_2(x_2)\,dx_2$$

$$= a_1 E(X_1) + a_2 E(X_2)$$

This result holds even if the two random variables are not independent. Also,

$$Var(Y) = E(Y - \mu_Y)^2 = E[(a_1X_1 + a_2X_2 - a_1\mu_1 - a_2\mu_2)^2]$$

$$= E[(a_1(X_1 - \mu_1) + a_2(X_2 - \mu_2))^2]$$

$$= E[a_1^2(X_1 - \mu_1)^2 + a_2^2(X_2 - \mu_2)^2 + 2a_1a_2(X_1 - \mu_1)(X_2 - \mu_2)]$$

$$= a_1^2 E[(X_1 - \mu_1)^2] + a_2^2 E[(X_2 - \mu_2)^2] + 2a_1a_2 E[(X_1 - \mu_1)(X_2 - \mu_2)]$$

$$= a_1^2 Var(X_1) + a_2^2 Var(X_2)$$

since the third term is zero because we assumed X_1 and X_2 are independent.

These properties hold for any number of random variables whether they are continuous or discrete.

**The mean and variance of
linear combinations**

Let X_i have mean μ_i and variance σ_i^2 for $i = 1, 2, \ldots, k$. The linear combination $Y = a_1X_1 + a_2X_2 + \cdots + a_kX_k$ has

$$E(a_1X_1 + a_2X_2 + \cdots + a_kX_k) = a_1 E(X_1) + a_2 E(X_2) + \cdots + a_k E(X_k)$$

or

$$\mu_Y = \sum_{i=1}^{k} a_i \mu_i$$

When the random variables are independent,

$$Var(a_1X_1 + a_2X_2 + \cdots + a_kX_k) = a_1^2 Var(X_1)$$
$$+ a_2^2 Var(X_2) + \cdots + a_k^2 Var(X_k)$$

or

$$\sigma_Y^2 = \sum_{i=1}^{k} a_i^2 \sigma_i^2$$

EXAMPLE 30   **Variances of $X_1 - X_2$ and $X_1 + X_2$ when X_1 and X_2 are independent**

Let X_1 have mean μ_1 and variance σ_1^2 and let X_2 have mean μ_2 and variance σ_2^2. Find the mean and variance of

(a) $X_1 - X_2$ and **(b)** $X_1 + X_2$

if X_1 and X_2 are independent.

Solution   **(a)** Note that $X_1 - X_2$ is of the form $a_1X_1 + a_2X_2$ with $a_1 = -a_2 = 1$ so it has mean

$$1 \cdot \mu_1 + (-1)\mu_2 = \mu_1 - \mu_2$$

and variance

$$(1)^2 \cdot \sigma_1^2 + (-1)^2 \sigma_2^2 = \sigma_1^2 + \sigma_2^2$$

(b) Since $X_1 + X_2$ corresponds to the case with $a_1 = a_2 = 1$, it has mean

$$1 \cdot \mu_1 + 1 \cdot \mu_2 = \mu_1 + \mu_2$$

and variance

$$1^2 \cdot \sigma_1^2 + 1^2 \cdot \sigma_2^2 = \sigma_1^2 + \sigma_2^2 \qquad \blacksquare$$

EXAMPLE 31   **Finding the mean and variance of $2X_1 + X_2 - 5$**

If X_1 has mean 4 and variance 9 while X_2 has mean -2 and variance 6, and the two are independent, find

(a) $E\left(2X_1 + X_2 - 5\right)$
(b) $Var\left(2X_1 + X_2 - 5\right)$

Solution   According to the properties of expectation, the constant -5 is added to the expectation of $2X_1 + X_2$ but the variance is unchanged.

(a)
$$E(2X_1 + X_2 - 5) = E(2X_1 + X_2) - 5$$
$$= 2E(X_1) + E(X_2) - 5 = 2(4) + (-2) - 5 = 1$$

(b)
$$Var(2X_1 + X_2 - 5) = Var(2X_1 + X_2)$$
$$= 2^2 Var(X_1) + Var(X_2) = 2^2(9) + 6 = 42 \qquad \blacksquare$$

EXAMPLE 32   **The mean and variance of total time to coat and rinse**

The time to complete a coating process, X_1, has mean 35 minutes and variance 11, while the time to rinse, X_2, has mean 8 minutes and variance 5. Find the mean and standard deviation of the total time to coat and rinse.

Solution   According to properties of expectation, the total time $X_1 + X_2$ has mean $35 + 8 = 43$ minutes. Treating the coating and rinsing times as independent, the variance is $11 + 5 = 16$, so that the standard deviation is 4. $\qquad \blacksquare$

EXAMPLE 33   **The mean and variance of the sample mean $\overline{X}$**

Let the n random variables $X_1, X_2, \ldots, X_n$ be independent and each have the same distribution with mean μ and variance σ^2. Use the properties of expectation to show that the sample mean $\overline{X}$ has

(a) mean: $\mu_{\overline{X}} = E(\overline{X}) = \mu$

(b) variance: $\sigma^2_{\overline{X}} = Var(\overline{X}) = \dfrac{\sigma^2}{n}$

Solution   (a) The sample mean

$$\overline{X} = \frac{X_1 + X_2 + \cdots + X_n}{n} = \frac{1}{n}X_1 + \frac{1}{n}X_2 + \cdots + \frac{1}{n}X_n$$

is a linear combination with constants $a_i = 1/n$ for $i = 1, 2, \ldots, n$. Consequently,

$$E(\overline{X}) = \frac{1}{n}E(X_1) + \frac{1}{n}E(X_2) + \cdots + \frac{1}{n}E(X_n) = \sum_{i=1}^{n}\frac{1}{n}\mu = \frac{1}{n}n\mu = \mu$$

so the expected value or mean of $\overline{X}$ is the same as the mean of each observation.

(b) The variance of $\overline{X}$ is

$$Var(\overline{X}) = \left(\frac{1}{n}\right)^2 Var(X_1) + \left(\frac{1}{n}\right)^2 Var(X_2) + \cdots + \left(\frac{1}{n}\right)^2 Var(X_n)$$

$$= \sum_{i=1}^{n}\left(\frac{1}{n}\right)^2 \sigma^2 = \left(\frac{1}{n}\right)^2 n\sigma^2 = \frac{\sigma^2}{n}$$

so the variance of $\overline{X}$ equals the variance of a single observation divided by n. ∎

EXAMPLE 34   **The expected value of the sample variance**

Let the n random variables $X_1, X_2, \ldots, X_n$ be independent and each have the same distribution with mean μ and variance σ^2. Use the properties of expectation to show that σ^2 is the mean, or expectation, of the sample variance

$$\sum_{i=1}^{n}(X_i - \overline{X})^2/(n-1)$$

Solution   We write $(X_i - \overline{X})^2 = (X_i - \mu + \mu - \overline{X})^2 = (X_i - \mu)^2 + (\mu - \overline{X})^2 + 2(X_i - \mu)(\mu - \overline{X})$ so the numerator of the sample variance is

$$\sum_{i=1}^{n}(X_i - \overline{X})^2 = \sum_{i=1}^{n}(X_i - \mu)^2 + \sum_{i=1}^{n}(\mu - \overline{X})^2 + 2\sum_{i=1}^{n}(X_i - \mu)(\mu - \overline{X})$$

and the last term equals $-2(\overline{X} - \mu)\sum_{i=1}^{n}(X_i - \mu) = -2n(\overline{X} - \mu)^2$. Consequently,

$$\sum_{i=1}^{n}(X_i - \overline{X})^2 = \sum_{i=1}^{n}(X_i - \mu)^2 - n(\overline{X} - \mu)^2$$

Now $E(X_i - \mu)^2 = Var(X_i) = \sigma^2$ and, by Example 34, $E(\overline{X}) = \mu$ and $E(\overline{X} - \mu)^2 = Var(\overline{X}) = \sigma^2/n$. Taking expectation term by term and summing,

$$E\left[\sum_{i=1}^{n}(X_i - \overline{X})^2\right] = \sum_{i=1}^{n}\sigma^2 - n\frac{\sigma^2}{n} = (n-1)\sigma^2$$

Dividing both sides by $n-1$, we conclude that σ^2 is the expected value of the sample variance. ∎

Exercises

5.71 Two scanners are needed for an experiment. Of the five available, two have electronic defects, another one has a defect in the memory, and two are in good working order. Two units are selected at random.

(a) Find the joint probability distribution of $X_1 = $ the number with electronic defects, and $X_2 = $ the number with a defect in memory.

(b) Find the probability of 0 or 1 total defects among the two selected.

(c) Find the marginal probability distribution of X_1.

(d) Find the conditional probability distribution of X_1 given $X_2 = 0$.

5.72 Two random variables are independent and each has a binomial distribution with success probability 0.4 and 2 trials.

(a) Find the joint probability distribution.

(b) Find the probability that the second random variable is greater than the first.

5.73 If two random variables have the joint density

$$f(x_1, x_2) = \begin{cases} x_1 x_2 & \text{for } 0 < x_1 < 2,\ 0 < x_2 < 1 \\ 0 & \text{elsewhere} \end{cases}$$

find the probabilities that

(a) both random variables will take on values less than 1;

(b) the sum of the values taken on by the two random variables will be less than 1.

5.74 With reference to the preceding exercise, find the marginal densities of the two random variables.

5.75 With reference to Exercise 5.73, find the joint cumulative distribution function of the two random variables, the cumulative distribution functions of the individual random variables, and check whether the two random variables are independent.

5.76 If two random variables have the joint density

$$f(x, y) = \begin{cases} \frac{6}{5}(x + y^2) & \text{for } 0 < x < 1,\ 0 < y < 1 \\ 0 & \text{elsewhere} \end{cases}$$

find the probability that $0.2 < X < 0.5$ and $0.4 < Y < 0.6$.

5.77 With reference to the preceding exercise, find the joint cumulative distribution function of the two random variables and use it to verify the value obtained for the probability.

5.78 With reference to Exercise 5.76, find both marginal densities and use them to find the probabilities that

(a) $X > 0.8$;

(b) $Y < 0.5$.

5.79 With reference to Exercise 5.76, find

(a) an expression for $f_1(x \mid y)$ for $0 < y < 1$;

(b) an expression for $f_1(x \mid 0.5)$;

(c) the mean of the conditional density of the first random variable when the second takes on the value 0.5.

5.80 With reference to Example 27, find expressions for

(a) the conditional density of the first random variable when the second takes on the value $x_2 = 0.25$;

(b) the conditional density of the second random variable when the first takes on the value x_1.

5.81 If three random variables have the joint density

$$f(x, y, z) = \begin{cases} k(x + y)e^{-z} & \text{for } 0 < x < 2, \\ & \quad 0 < y < 1, z > 0 \\ 0 & \text{elsewhere} \end{cases}$$

find

(a) the value of k;

(b) the probability that $X > Y$ and $Z > 1$.

5.82 With reference to the preceding exercise, check whether

(a) the three random variables are independent;

(b) any two of the three random variables are pairwise independent.

5.83 A pair of random variables has the **circular normal distribution** if their joint density is given by

$$f(x_1, x_2)$$
$$= \frac{1}{2\pi\sigma^2} e^{-[(x_1 - \mu_1)^2 + (x_2 - \mu_2)^2]/2\sigma^2}$$

for $-\infty < x_1 < \infty$ and $-\infty < x_2 < \infty$.

(a) If $\mu_1 = 2$ and $\mu_2 = -2$, and $\sigma = 10$, use Table 3 to find the probability that $-8 < X_1 < 14$ and $-9 < X_2 < 3$.

(b) If $\mu_1 = \mu_2 = 0$ and $\sigma = 3$, find the probability that (X_1, X_2) is contained in the region between the two circles $x_1^2 + x_2^2 = 9$ and $x_1^2 + x_2^2 = 36$.

5.84 A precision drill positioned over a target point will make an acceptable hole if it is within 5 microns of the target. Using the target as the origin of a rectangular system of coordinates, assume that the coordinates (x, y) of the point of contact are values of a pair of random variables having the circular normal distribution (see Exercise 5.83) with $\mu_1 = \mu_2 = 0$ and $\sigma = 2$. What is the probability that the hole will be acceptable?

5.85 With reference to Exercise 5.73, find the expected value of the random variable whose values are given by $g(x_1, x_2) = x_1 + x_2$.

5.86 With reference to Exercise 5.76, find the expected value of the random variable whose values are given by $g(x, y) = x^2 y$.

5.87 If measurements of the length and the width of a rectangle have the joint density

$$f(x, y) = \begin{cases} \dfrac{1}{ab} & \text{for } L - \dfrac{a}{2} < x < L + \dfrac{a}{2}, \\ & W - \dfrac{b}{2} < y < W + \dfrac{b}{2} \\ 0 & \text{elsewhere} \end{cases}$$

find the mean and the variance of the corresponding distribution of the area of the rectangle.

5.88 Establish a relationship between $f_1(x_1 | x_2)$, $f_2(x_2 | x_1)$, $f_1(x_1)$, and $f_2(x_2)$.

5.89 If X_1 has mean 1 and variance 5 while X_2 has mean -1 and variance 5, and the two are independent, find

(a) $E(X_1 + X_2)$;

(b) $Var(X_1 + X_2)$.

5.90 If X_1 has mean -4 and variance 3 while X_2 has mean 5 and variance 4, and the two are independent, find

(a) $E(X_1 - X_2)$;

(b) $Var(X_1 - X_2)$.

5.91 If X_1 has mean 1 and variance 3 while X_2 has mean -2 and variance 5, and the two are independent, find

(a) $E(X_1 + 2X_2 - 3)$;

(b) $Var(X_1 + 2X_2 - 3)$.

5.92 The time for an older machine to complete a check on a computer chip, X_1, has mean 65 milliseconds and variance 16, while the time for a newer model, X_2, has mean 45 milliseconds and variance 9. Find the expected time savings using the newer model when

(a) checking a single chip;

(b) checking 200 chips.

(c) Find the standard deviations in parts (a) and (b), assuming all of the checking times are independent.

5.93 Let $X_1, X_2, \ldots, X_{20}$ be independent and let each have the same marginal distribution with mean 10 and variance 3. Find

(a) $E(X_1 + X_2 + \cdots + X_{20})$;

(b) $Var(X_1 + X_2 + \cdots + X_{20})$.

5.11 Moment Generating Functions*

An alternative to a probability distribution can sometimes greatly simplify the calculation of moments. The **moment generating function** (mgf) of a random variable X, or its probability distribution, is the function defined by

$$M(t) = E(e^{tX})$$

which is the expectation of the exponential function e^{tX}. In the discrete case,

$$M(t) = E(e^{tX}) = \sum_{\text{all } x_i} e^{tx_i} f(x_i)$$

and

$$M(t) = E(e^{tX}) = \int_{-\infty}^{\infty} e^{tx} f(x)\, dx$$

*This section may be skipped on first reading.

in the continuous case. For each fixed t, the integrand is a positive function of x so $M(t)$ is either finite or infinite. Note that $M(0) = E(e^{0X}) = E(1) = 1$ always exists, but we require $M(t)$ to exist for an interval of values of t.

The probability distribution, or random variable X, is said to **possess a moment generating function** $M(t)$ if this function is finite for t in some interval containing zero, say $|t| \le T$ for some $T > 0$.

Under the condition that $M(t)$ is finite for $|t| \le T$, for some $T > 0$, we can obtain successive derivatives by differentiating under the integral or summation sign. In the continuous case, we obtain

$$M'(t) = \frac{d}{dt}M(t) = \int_{-\infty}^{\infty} \frac{d}{dt}e^{tx}f(x)\,dx = \int_{-\infty}^{\infty} x\,e^{tx}f(x)\,dx$$

$$M''(t) = \frac{d^2}{dt^2}M(t) = \int_{-\infty}^{\infty} \frac{d}{dt}xe^{tx}f(x)\,dx = \int_{-\infty}^{\infty} x^2 e^{tx}f(x)\,dx$$

For either the continuous or discrete case,

$$M^{(k)}(t) = \frac{d^k}{dt^k}M(t) = E(X^k e^{tX}) \qquad \text{for } k = 1, 2, \ldots$$

Setting $t = 0$, we obtain the moments about the origin

$$M'(0) = E(X)$$
$$M''(0) = E(X^2)$$

Differentiating k times, the kth derivative is related to the kth moment

$$M^{(k)}(0) = E(X^k) \qquad \text{for } k = 1, 2, \ldots$$

Basic properties of moment generating functions

> **Theorem 5.2**   If the moment generating function is finite for $|t| \le T$, for some $T > 0$, it uniquely determines the probability distribution.
>
> Then, all moments exist and can be obtained from the relation
>
> $$M^{(k)}(0) = E(X^k)$$

From the definition of expectation, the moments of any random variable must be obtained by integration or summation of a series. When the moment generating function is available, this process can be replaced by a straightforward differentiation. We find the moment generating functions and illustrate the calculation of mean and variance for several common distributions in the next examples.

EXAMPLE 35   **Moment generating function for binomial distribution**

Let X have the binomial distribution with probability distribution

$$b(x \mid n, p) = \binom{n}{x} p^x (1 - p)^{n-x} \qquad \text{for } x = 0, 1, \ldots, n$$

Show that

(a) $M(t) = (1 - p + pe^t)^n$ for all t

(b) $E(X) = np$ and $Var(X) = np(1 - p)$

Solution   **(a)** By definition of the moment generating function

$$M(t) = \sum_{x=0}^{n} e^{tx} \binom{n}{x} p^x (1-p)^{n-x}$$

$$= \sum_{x=0}^{n} \binom{n}{x} (e^t p)^x (1-p)^{n-x}$$

$$= (pe^t + 1 - p)^n \qquad \text{for all } t$$

where we have used the binomial formula

$$(a+b)^n = \sum_{x=0}^{n} \binom{n}{x} a^x b^{n-x}$$

(b) Differentiating $M(t)$, we find

$$M'(t) = npe^t (pe^t + 1 - p)^{n-1}$$
$$M''(t) = (n-1)np^2 e^{2t}(pe^t + 1 - p)^{n-2} + npe^t(pe^t + 1 - p)^{n-1}$$

Evaluating these derivatives at $t = 0$, we obtain the moments

$$E(X) = np$$
$$E(X^2) = (n-1)np^2 + np$$

Also, the variance is

$$Var(X) = E(X^2) - [E(X)]^2 = np(1-p)$$ ∎

EXAMPLE 36   **Moment generating function for Poisson distribution**

Let X have the Poisson distribution with probability distribution

$$f(x) = \frac{\lambda^x}{x!} e^{-\lambda} \qquad \text{for } x = 0, 1, \ldots, \infty$$

Show that

(a) $M(t) = e^{\lambda(e^t - 1)}$ for all t

(b) $E(X) = \lambda$ and $Var(X) = \lambda$

The mean and variance of the Poisson distribution are equal.

Solution   **(a)** By definition of the moment generating function

$$M(t) = \sum_{x=0}^{\infty} e^{tx} \frac{\lambda^x}{x!} e^{-\lambda} = \sum_{x=0}^{\infty} \frac{(\lambda e^t)^x}{x!} e^{-\lambda}$$

$$= e^{-\lambda} e^{\lambda e^t} = e^{\lambda(e^t - 1)} \qquad \text{for } -\infty < t < \infty$$

where we have used the series $e^y = \sum_{k=0}^{\infty} \frac{y^k}{k!}$

(b) Differentiating $M(t)$, we find

$$M'(t) = \lambda e^t e^{\lambda(e^t - 1)}$$
$$M''(t) = \lambda e^t e^{\lambda(e^t - 1)} + \lambda^2 e^{2t} e^{\lambda(e^t - 1)}$$

Evaluating these derivatives at $t = 0$, we obtain the moments

$$E(X) = \lambda$$
$$E(X^2) = \lambda + \lambda^2$$

Also, the variance is

$$Var(X) = E(X^2) - [E(X)]^2 = \lambda$$

∎

EXAMPLE 37   **Moment generating function for gamma distribution**

The gamma distribution has probability density function

$$f(x) = \frac{1}{\beta^\alpha \Gamma(\alpha)} x^{\alpha-1} e^{-x/\beta} \qquad \text{for } x > 0$$

Show that its moment generating function is

$$M(t) = \frac{1}{(1 - \beta t)^\alpha}$$

and verify the mean and variance.

Solution

$$M(t) = \int_0^\infty e^{tx} \frac{1}{\beta^\alpha \Gamma(\alpha)} x^{\alpha-1} e^{-x/\beta} dx$$

$$= \frac{1}{\beta^\alpha \Gamma(\alpha)} \int_0^\infty x^{\alpha-1} e^{-x(1-\beta t)/\beta} dx$$

This last integral is finite for all $t < 1/\beta$ and can be evaluated by multiplying and dividing by $(1 - \beta t)^\alpha$ to obtain a gamma density with parameters α and $\beta/(1 - \beta t)$. We conclude that

$$M(t) = \frac{1}{\beta^\alpha \Gamma(\alpha)} \Gamma(\alpha) \frac{\beta^\alpha}{(1 - \beta t)^\alpha} = \frac{1}{(1 - \beta t)^\alpha}$$

To obtain the moments, we differentiate and find

$$M'(t) = \alpha \frac{1}{(1 - \beta t)^{\alpha+1}} \beta$$

$$M''(t) = (\alpha + 1)\alpha \frac{1}{(1 - \beta t)^{\alpha+2}} \beta^2$$

Setting $t = 0$,

$$E(X) = M'(0) = \alpha\beta \qquad \text{and} \qquad E(X^2) = M''(0) = \alpha(\alpha + 1)\beta^2$$

so $Var(X) = \alpha\beta^2$.

∎

EXAMPLE 38   **Moment generating function for chi square distribution**

The gamma distribution having $\alpha = \nu/2$ and $\beta = 2$ is called the chi square distribution with ν degrees of freedom. Show that the moment generating function is

$$M(t) = \frac{1}{(1 - 2t)^{\nu/2}}$$

and that $E(X) = \nu$ and $Var(X) = 2\nu$.

Solution These results follow from the previous example because this is a gamma distribution with $\alpha = \nu/2$ and $\beta = 2$. ■

EXAMPLE 39 **Moment generating function for normal distribution**

Show that the normal distribution, whose probability density function

$$f(x) = \frac{1}{\sqrt{2\pi}\sigma} e^{-(x-\mu)^2/2\sigma^2} \quad \text{has} \quad M(t) = e^{t\mu + \frac{1}{2}t^2\sigma^2}$$

which exists for all t. Also, verify the first two moments.

Solution To obtain the moment generating function, we use the identity

$$tx - \frac{1}{2}\frac{(x-\mu)^2}{\sigma^2} = -\frac{1}{2}\frac{[x-(t\sigma^2+\mu)]^2}{\sigma^2} + t\mu + \frac{1}{2}t^2\sigma^2$$

obtained by completing the square. Then

$$M(t) = E(e^{tX}) = \int_{-\infty}^{\infty} e^{tx} \frac{1}{\sqrt{2\pi}\sigma} e^{-(x-\mu)^2/2\sigma^2} dx$$

$$= \int_{-\infty}^{\infty} \frac{1}{\sqrt{2\pi}\sigma} e^{-\frac{1}{2}[x-(t\sigma^2+\mu)]^2/\sigma^2} dx \times e^{t\mu + \frac{1}{2}t^2\sigma^2} = e^{t\mu+\frac{1}{2}t^2\sigma^2}$$

To obtain the moments of the normal, we differentiate once to obtain

$$M'(t) = e^{t\mu + \frac{1}{2}t^2\sigma^2}(\mu + t\sigma^2)$$

and a second time to get

$$M''(t) = e^{t\mu + \frac{1}{2}t^2\sigma^2}[(\mu + t\sigma^2)^2 + \sigma^2].$$

Setting $t = 0$,

$$E[X] = M'(0) = \mu \quad \text{and} \quad E(X^2) = M''(0) = \sigma^2 + \mu^2$$

so $Var(X) = \sigma^2$ as the notation suggests. ■

A basic property relates the moment generating function of $a + bX$ to that of X.

Moment generating function of $a + bX$

Theorem 5.3 Let X have moment generating function $M(t)$ and let a and b be constants. Then

$$M_{a+bX}(t) = E(e^{(a+bX)t}) = e^{at} \cdot M(bt)$$

For instance, the moment generating function of $X - \mu$, corresponding to $b = 1$ and $a = -\mu$, is

$$M_{X-\mu}(t) = e^{-\mu t} \cdot M_X(t)$$

EXAMPLE 40 **Converting to the standard normal distribution**

Let X be distributed as normal with mean μ and variance σ^2. Use moment generating functions to show that

$$Z = \frac{X-\mu}{\sigma}$$

has a standard normal distribution.

Solution   By the previous example, X has moment generating function

$$M(t) = e^{t\mu + \frac{1}{2}t^2\sigma^2}$$

Further

$$Z = \frac{X - \mu}{\sigma} = a + bX \qquad \text{with} \qquad a = \frac{-\mu}{\sigma} \qquad \text{and} \qquad b = \frac{1}{\sigma}$$

Therefore, the moment generating function of Z is

$$M_Z(t) = e^{at} \cdot M(bt) = e^{-(\mu/\sigma)t} \, e^{(t/\sigma)\mu + \frac{1}{2}(t/\sigma)^2\sigma^2} = e^{\frac{1}{2}t^2}$$

This last expression is the moment generating function of a normal distribution having mean 0 and variance 1. It exists for all t so the moment generating function uniquely determines the distribution.   ∎

One of most useful properties of moment generating functions is a multiplication property for independent random variables.

Moment generating function of sum under independence

> **Theorem 5.4**   Let X and Y be independent random variables with moment generating functions M_X and M_Y. The sum $Z = X + Y$ has moment generating function
>
> $$M_Z(t) = M_X(t) M_Y(t)$$
>
> on the interval of t where $M_X(t)$ and $M_Y(t)$ exist.

Proof

$$M_Z(t) = E(e^{tZ}) = E(e^{t(X+Y)}) = E(e^{tX} e^{tY})$$

Then, from the assumption of independence,

$$= E(e^{tX} e^{tY}) = E(e^{tX})E(e^{tY}) = M_X(t) M_Y(t)$$

EXAMPLE 41   **Sum of two independent normal random variables is normal**

Let X and Y be independent normal random variables. Let X have mean μ_X and variance σ_X^2 while Y has mean μ_Y and variance σ_Y^2. Use moment generating functions to show that

(a) $X + Y$ has a normal distribution with mean $\mu_X + \mu_Y$ and variance $\sigma_X^2 + \sigma_Y^2$.

(b) $X - Y$ has a normal distribution with mean $\mu_X - \mu_Y$ and variance $\sigma_X^2 + \sigma_Y^2$.

Solution   **(a)** From a previous example, the two moment generating functions are

$$M_X(t) = e^{t\mu_X + \frac{1}{2}t^2\sigma_X^2}$$
$$M_Y(t) = e^{t\mu_Y + \frac{1}{2}t^2\sigma_Y^2}$$

Their product is

$$M_X(t)M_Y(t) = e^{t\mu_X + \frac{1}{2}t^2\sigma_X^2} \, e^{t\mu_Y + \frac{1}{2}t^2\sigma_Y^2}$$
$$= e^{t(\mu_X + \mu_Y) + \frac{1}{2}t^2(\sigma_X^2 + \sigma_Y^2)}$$

which we identify as the moment generating function of a normal random variable having mean $\mu_X + \mu_Y$ and variance $\sigma_X^2 + \sigma_Y^2$.

(b) Since X and $-Y$ are independent and $-Y$ has moment generating function $M_Y(-t)$, the result follows (see Exercise 5.98).   ∎

Exercises

5.94 Let $f(x) = 0.2$ for $x = 0, 1, 2, 3, 4$.

(a) Find the moment generating function.

(b) Obtain $E(X)$ and $E(X^2)$ by differentiating the moment generating function.

5.95 Let

$$f(x) = 0.25 \binom{2}{x} \quad \text{for } x = 0, 1, 2$$

(a) Find the moment generating function.

(b) Obtain $E(X)$ and $E(X^2)$ by differentiating the moment generating function.

5.96 Let Z have a normal distribution with mean 0 and variance 1.

(a) Find the moment generating function of Z^2.

(b) Identify the distribution of Z^2 by recognizing the form of the moment generating function.

5.97 Let X be a continuous random variable having probability density function

$$f(x) = \begin{cases} 2e^{-2x} & \text{for } x > 0 \\ 0 & \text{elsewhere} \end{cases}$$

(a) Find the moment generating function.

(b) Obtain $E(X)$ and $E(X^2)$ by differentiating the moment generating function.

5.98 Establish the result in Example 41 concerning the difference of two independent normal random variables, X and Y.

5.99 Let X and Y be independent normal random variables with

$$E(X) = 2 \quad \text{and} \quad \sigma_X^2 = 4$$
$$E(Y) = 1 \quad \text{and} \quad \sigma_Y^2 = 9$$

(a) Use moment generating functions to show that $2X - 3Y + 5$ has a normal distribution.

(b) Find the mean and variance of the random variable in part (a).

5.100 Let X have the geometric distribution

$$f(x) = p(1-p)^{x-1} \quad \text{for } x = 1, 2, \ldots$$

(a) Obtain the moment generating function for

$$t < -\ln(1-p)$$

[*Hint:* Recall that $\sum_{k=0}^{\infty} r^k = \dfrac{1}{1-r}$ for $|r| < 1$.]

(b) Obtain $E(X)$ and $E(X^2)$ by differentiating the moment generating function.

5.12   Checking If the Data Are Normal

In many instances, an experimenter needs to check whether a data set appears to be generated by a normally distributed random variable. As indicated in Figure 2.8, the normal distribution can serve to model variation in some quantities. Further, many commonly used statistical procedures, which we describe in later chapters, require that the probability distribution be nearly normal. Consequently, in a great number of applications it is prudent to check the assumption that the data are normal.

Although they involve an element of subjective judgment, graphical procedures are the most helpful for detecting serious departures from normality. Histograms can be checked for lack of symmetry. A single long tail certainly contradicts the assumption of a normal distribution. However, another special graph, called a **normal scores plot** or **normal quantile plot**, is even more effective in detecting departures from normality. To introduce such a plot, we consider a sample of size 4. In practice, we need a minimum of 15–20 observations in order to evaluate the agreement with normality.

The term **normal scores** refers to an idealized sample from the standard normal distribution. It consists of the values of z that divide the axes into equal probability intervals. For sample size $n = 4$, the normal scores are

$$m_1 = -z_{0.20} = -0.84$$
$$m_2 = -z_{0.40} = -0.25$$
$$m_3 = z_{0.40} = 0.25$$
$$m_4 = z_{0.20} = 0.84$$

as illustrated in Figure 5.24.

Figure 5.24
The standard normal
distribution and the normal
scores for $n = 4$

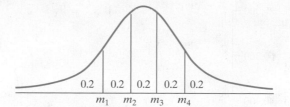

To construct a normal scores plot,

1. Order the data from smallest to largest;
2. Obtain the normal scores;
3. Plot the ith largest observation, versus the ith normal score m_i, for all i.

EXAMPLE 42 **A simple normal scores plot**

Suppose the four observations are 67, 48, 76, 81. Construct a normal scores plot.

Solution The ordered observations are 48, 67, 76, 81. Above, we found that $m_1 = -z_{0.20} = -0.84$, so we plot the pair $(-0.84, 48)$. Continuing, we obtain Figure 5.25.

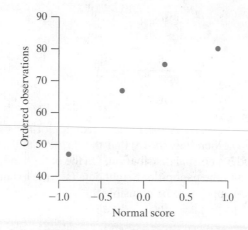

Figure 5.25
The normal scores plot

If the data were from a standard normal distribution, we would expect the ith largest observation to approximate the ith normal score so that the normal scores plot would resemble a 45° line through the origin. When the distribution is normal with an unspecified μ and σ,

$$z = \frac{x - \mu}{\sigma}$$

so the idealized z values can be converted to idealized x values through the relation $x = \mu + \sigma z$. Because the idealized values have this linear relation, it is sufficient to plot the ordered observations versus the normal scores obtained from the standard normal distribution. If the normal distribution prevails, the pattern should still be a straight line. But the line need not pass through the origin or have slope 1.

The construction of normal scores plots by hand is a difficult task at best. Fortunately, they can be treated easily with most statistical programs. (See Exercise 5.102.) Many slight variants are used in the calculation of the normal scores but the plots are very similar if more than 20 observations are plotted. Whichever computer program you use, if a normal distribution is plausible, the plot will have a straight-line appearance.

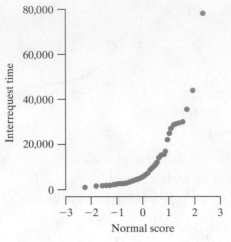

Figure 5.26
The normal scores plot of the interrequest times

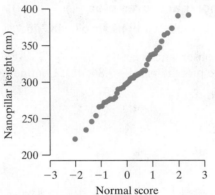

Figure 5.27
Normal scores plot of nanopillar heights

Figure 5.26 shows the normal scores plot for the interrequest times given on page 19. The bending shows that the largest values are larger than would be expected under a normal distribution. On the other hand, Figure 5.27 exhibits a normal scores plot of the nanopillar height data (see the example on page 15), and a normal distribution appears to be plausible.

5.13   Transforming Observations to Near Normality

When the histogram or normal scores plot indicate that the assumption of a normal distribution is invalid, transformations of the data can often improve the agreement with normality. Scientists regularly express their observations in natural logs. We consider a few other transformations, as indicated in Table 5.1.

Table 5.1 Some useful transformations					
Make Large Values Smaller :				**Make Large Values Larger :**	
$-\dfrac{1}{x}$ $\ln x$ $x^{1/4}$ $\sqrt{x}$				x^2 x^3	

If the observations are positive and the distribution has a long tail or a few stragglers on the right, then the $\ln x$ or $\sqrt{x}$ transformations will pull the large values down farther than they pull the central or small values. If the transformed observations have a nearly straight line normal scores plot, it is usually advantageous to use the normality of this new scale to perform any statistical analysis. Further, the validity of many of the powerful statistical methods described in later chapters rests on the assumption that the probability distribution is nearly normal. By choosing a transformation that leads to nearly normal data, the investigator can greatly extend the range of validity of these techniques.

EXAMPLE 43   **A transformation to better approximate a normal distribution**

Transform the interrequest times in the example on page 19 to better approximate a normal distribution.

Solution   On a computer, we calculate $\sqrt{x}$, take the square root again to obtain $x^{1/4}$, and take the natural logarithm $\ln x$ of all 50 values. The transformation $\ln x$ appears to work best. The histogram and normal scores plot are shown in Figure 5.28 for both the original and transformed data. The quality of the fit further confirms the log-normal model. ∎

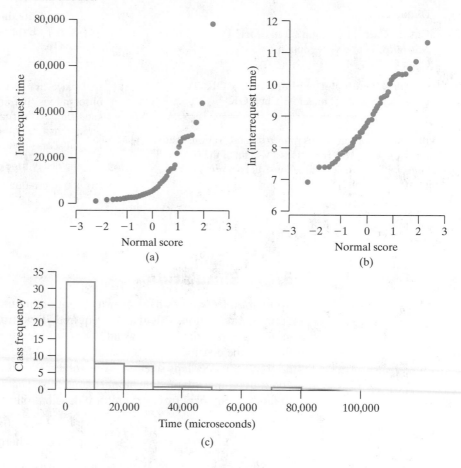

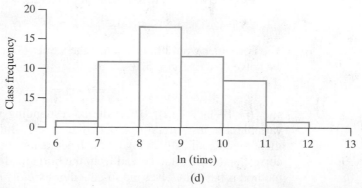

Figure 5.28
(a) The normal scores plot of interrequest time (b) The normal scores plot of ln (interrequest time)
(c) Histogram of interrequest time (d) Histogram of ln (interrequest time)

Exercises

5.101 For any 11 observations,

(a) Use software or Table 3 to verify the normal scores

$$-1.38 \;\; -0.97 \;\; -0.67 \;\; -0.43 \;\; -0.21 \;\; 0 \;\; 0.21 \;\; 0.43 \;\; 0.67 \;\; 0.97 \;\; 1.38$$

(b) Construct a normal scores plot using the observations on the times between neutrinos in Exercise 2.7.

5.102 (Normal scores plots) The *MINITAB* commands

> **Dialog box:**
>
> **Calc > Calculator** Type $C2$ in **Store**. Type *NSCOR(C1)* in **Expression**.
> Click **OK**.
>
> **Graph > Scatteplot > Simple**. Click **OK**. Type $C1$ under **Y** and $C2$ under **X**. Click **OK**.

will create a normal scores plot from observations that were set in C1. (*MINITAB* uses a variant of the normal scores, m_i, that we defined.) Construct a normal scores plot of

(a) the cheese data of Example 8,

(b) the decay time data on page 146.

5.103 (Transformations) The *MINITAB* commands

> **Dialog box:**
>
> **Calc > Calculator** Type $C2$ in **Store**. Type *LOGE(C1)* in **Expression**.
> Click **OK**.
>
> **Calc > Calculator** Type $C3$ in **Store**. Type *SQRT(C1)* **Expression**.
> Click **OK**.
>
> **Calc > Calculator** Type $C4$ in **Store**. Type *SQRT(C3)* **Expression**.
> Click **OK**.

will place $\ln x$ in C2, $\sqrt{x}$ in C3, and $x^{1/4}$ in C4 for observations that are set in C1. Normal scores plots can then be constructed as in Exercise 5.102. Try these three transformations and construct the corresponding normal scores plots for

(a) the decay time data on page 146;

(b) the interrequest time data on page 19.

5.14   Simulation

Simulation techniques have grown up with computers. They are ideally suited for doing the repetitive calculations required. To **simulate** the observation of continuous random variables, we usually start with uniform random numbers and relate these to the distribution function of interest. We could use two- or three-digit random integers, perhaps selected from Table 7W, but most software programs have a continuous uniform random number generator. That is, they produce approximations to random numbers from the uniform distribution

$$f(x) = \begin{cases} 1 & 0 < x < 1 \\ 0 & \text{elsewhere} \end{cases}$$

Suppose we wish to simulate an observation from the exponential distribution

$$F(x) = 1 - e^{-0.3x}, \quad 0 < x < \infty$$

The computer would first produce the value u from the uniform distribution. Then we solve (see Exercise 5.104)

$$u = F(x) = 1 - e^{-0.3x}$$

so $x = [-\ln(1-u)]/0.3$ is the corresponding value of an exponential random variable. For instance, if $u = 0.45$, then $x = [-\ln(1-0.45)]/0.3 = 1.993$. This is illustrated graphically in Figure 5.29, where u is located on the vertical scale and the corresponding x value is read from the horizontal scale. (The theory on which this method is based involves the so-called *probability integral transformation*, which is

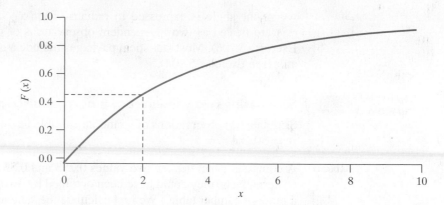

Figure 5.29
Exponential cumulative
distribution with mean $\frac{10}{3}$

presented in Example 15, Chapter 6.) If we wish to simulate a sample from F, the preceding process is repeated with a different u for each new observation x.

A similar procedure applies to the simulation of observations from a Weibull distribution. Starting with the value of a uniform variable u, we now solve (see Exercise 5.105)

$$u = F(x) = 1 - e^{-\alpha x^{\beta}}$$

for

$$x = \left[-\frac{1}{\alpha} \ln(1-u) \right]^{1/\beta}$$

which is the corresponding value of a Weibull random variable.

EXAMPLE 44   **Simulating five values from a Weibull distribution**

Simulate five observations of a random variable having the Weibull distribution with $\alpha = 0.05$ and $\beta = 2.0$.

Solution   A computer generates the five values 0.57, 0.74, 0.26, 0.77, 0.12. (Alternatively, we could read two digits at a time from a random number table.) We calculate

$$x = [-20.0 \ln(1 - 0.57)]^{1/2} = 4.108$$
$$x = [-20.0 \ln(1 - 0.74)]^{1/2} = 5.191$$

The reader can show that the last three uniform numbers yield $x = 2.454, 5.422, 1.599$. ∎

Suppose we need to simulate values from the normal distribution with a specified μ and σ^2. By the relation

$$z = \frac{x - \mu}{\sigma}$$

it follows that $x = \mu + \sigma z$, so a value x can be calculated from the value of a standard normal variable z. Although z can be obtained from the value for a uniform variable u by numerically solving $u = F(z)$, another approach called the Box-Muller-Marsaglia method is almost universally preferred. It starts with a pair of independent uniform variables (u_1, u_2) and produces two standard normal variables

$$z_1 = \sqrt{-2 \ln(u_2)} \, \cos(2\pi u_1)$$
$$z_2 = \sqrt{-2 \ln(u_2)} \, \sin(2\pi u_1)$$

where the angle is expressed in radians. Then $x_1 = \mu + \sigma z_1$ and $x_2 = \mu + \sigma z_2$ are treated as two independent observations of normal random variables (see Exercise 5.106). Most statistical packages include a normal random number generator (see Exercise 5.108).

EXAMPLE 45   **Simulating two values from a normal distribution**

Simulate two observations of a random variable having the normal distribution with $\mu = 50$ and $\sigma = 5$.

Solution   A computer generates the two values 0.253 and 0.531 from a uniform distribution. (Alternatively, they could have been obtained by reading three digits at a time from a random number table.) We first calculate the standard normal values

$$z_1 = \sqrt{-2\ln(0.531)}\,\cos(2\pi \cdot 0.253) = -0.021$$
$$z_2 = \sqrt{-2\ln(0.531)}\,\sin(2\pi \cdot 0.253) = 1.125$$

and then the normal values

$$x_1 = 50 + 5z_1 = 50 + 5(-0.021) = 49.895$$
$$x_2 = 50 + 5z_2 = 50 + 5(1.125) = 55.625$$

∎

Exercises

5.104 Verify that

(a) the exponential density $0.3\,e^{-0.3x}$, $x > 0$ corresponds to the distribution function $F(x) = 1 - e^{-0.3x}$, $x > 0$;

(b) the solution of $u = F(x)$ is given by $x = [-\ln(1-u)]/0.3$.

5.105 Verify that

(a) the Weibull density $\alpha\beta x^{\beta-1} e^{-ax^{\beta}}$, $x > 0$, corresponds to the distribution function $F(x) = 1 - e^{-ax^{\beta}}$, $x > 0$;

(b) the solution of $u = F(x)$ is given by $x = \left[-\dfrac{1}{\alpha}\ln(1-u)\right]^{1/\beta}$.

5.106 Consider two independent standard normal variables whose joint probability density is

$$\frac{1}{2\pi}\,e^{-(z_1^2 + z_2^2)/2}$$

Under a change to polar coordinates, $z_1 = r\cos(\theta)$, $z_2 = r\sin(\theta)$, we have $r^2 = z_1^2 + z_2^2$ and $dz_1\,dz_2 = r\,dr\,d\theta$, so the joint density of r and θ is

$$r e^{-r^2/2}\,\frac{1}{2\pi}, \quad 0 < \theta < 2\pi,\ r > 0$$

Show that

(a) r and θ are independent and that θ has a uniform distribution on the interval from 0 to 2π;

(b) $u_1 = \theta/2\pi$ and $u_2 = 1 - e^{-r^2/2}$ have independent uniform distributions;

(c) the relations between (u_1, u_2) and (z_1, z_2) on page 175 hold [note that $1 - u_2$ also has a uniform distribution, so $\ln(u_2)$ can be used in place of $\ln(1 - u_2)$].

5.107 The statistical package *MINITAB* has a random number generator. To simulate 5 values from an exponential distribution having mean $\beta = 0.05$, choose

---

Dialog Box:

Calc > Random Data > Exponential
Type 5 after **Generate**, $C1$ in **Column** and 0.05 in **Mean**.
Then click **OK**.

Output:
One call produced the output

0.031949 0.004643 0.030029 0.112834 0.064642

Generate 8 values from the exponential distribution with $\beta = 0.2$.

5.108 The statistical package *MINITAB* has a normal random number generator. To simulate 5 values from a normal distribution having mean 7 and standard deviation 4, and place them in C1, use the commands

Dialog Box:

Calc > Random Sample > Normal
Type 5 after **Generate**, C1 in **Column**, 7 in **Mean** and 4 in **standard deviation**
Click **OK**.

Output:
One call produced the output

5.42137   6.98061   9.41352   7.05932   5.87297

Generate 8 values for a normal variable with $\mu = 123$ and $\sigma^2 = 23.5$.

Do's and Don'ts

Do's

1. Describe the behavior of a continuous random variable X by specifying its probability density function which satisfies

$$f(x) \geq 0 \quad \text{for all } x \quad \text{and} \quad \int_{-\infty}^{\infty} f(x)\,dx = 1$$

2. Remember that it is only meaningful to talk about the probability that a continuous random variable X lies in an interval. It is always the case that $P(X = x) = 0$ for every possible value x.

3. Obtain the probability that the value of X will lie in an interval by finding the area under the curve f over the interval.

$$P(X \leq b) = \int_{-\infty}^{b} f(x)\,dx$$

$$= \text{area under the density function to the left of } x = b$$

$$P(a < X \leq b) = \int_{a}^{b} f(x)\,dx$$

$$= \text{area under the density function between}$$
$$x = a \quad \text{and} \quad x = b$$

4. Summarize a probability density of the continuous random variable X by its

$$\text{mean:} \quad \mu = \int_{-\infty}^{\infty} x f(x)\,dx \qquad \text{variance:} \quad \sigma^2 = \int_{-\infty}^{\infty} (x - \mu)^2 f(x)\,dx$$

$$\text{standard deviation:} \quad \sigma = \sqrt{\int_{\infty}^{\infty} (x - \mu)^2 f(x)\,dx}$$

5. When X has a normal distribution with mean μ and variance σ^2, obtain the probability of an interval $P(X \leq b)$ by converting the limit b to the standardized value $(b - \mu)/\sigma = z$ and obtaining the probability

$$P\left(Z \leq \frac{b - \mu}{\sigma}\right) = P(X \leq b)$$

from the standard normal table.

6. Use the properties of expectation and variance

$$E(aX+b)=aE(X)+b \quad \text{and} \quad Var(aX+b)=a^2\,Var(X)$$

More generally,

$$E(a_1X_1+a_2X_2+b)=a_1E(X_1)+a_2E(X_2)+b$$

and, if X_1 and X_2 are independent,

$$Var(a_1X_1+a_2X_2+b)=a_1^2\,Var(X_1)+a_2^2\,Var(X_2)$$

Don'ts

1. Never apply the normal approximation to the binomial

$$Z=\frac{X-np}{\sqrt{np(1-p)}}$$

when the expected number of successes (or failures) is too small. That is, when either

$$np \quad \text{or} \quad n(1-p) \text{ is 15 or less}$$

2. Don't add variances according to

$$Var(X_1+X_2)=Var(X_1)+Var(X_2)$$

unless the two random variables are independent or have zero covariance.

3. Don't just assume that data come from a normal distribution. When there are at least 20 to 25 observations, it is good practice to construct a normal scores plot to check this assumption.

Review Exercises

5.109 If the probability density of a random variable is given by

$$f(x)=\begin{cases} k(1-x^2) & \text{for } 0<x<1 \\ 0 & \text{elsewhere} \end{cases}$$

find the value of k and the probabilities that a random variable having this probability density will take on a value

(a) between 0.1 and 0.2;

(b) greater than 0.5.

(c) Find μ and σ^2.

5.110 With reference to the preceding exercise, find the corresponding distribution function and use it to determine the probabilities that a random variable having this distribution function will take on a value

(a) less than 0.3;

(b) between 0.4 and 0.6.

5.111 In certain experiments, the error made in determining the density of a silicon compound is a random variable having the probability density

$$f(x)=\begin{cases} 25 & \text{for } -0.02<x<0.02 \\ 0 & \text{elsewhere} \end{cases}$$

Find the probabilities that such an error will be

(a) between -0.03 and 0.04;

(b) between -0.005 and 0.005.

5.112 A thermocouple measures the temperature of reactions. The temperature reading can be modeled by a normal distribution having mean μ and standard deviation $0.001°C$ where μ is the true temperature. Find the probability that the reading will differ from the true temperature by

(a) less than $0.0015°C$

(b) more than $0.0018°C$?

5.113 Refer to Exercise 5.112. Suppose the design of thermocouple can be improved and the standard deviation decreased. Determine the new value for the standard deviation that would restrict the probability of an error greater than $0.0018°C$ to be less than 0.02.

5.114 The burning time of an experimental rocket is a random variable having the normal distribution with $\mu = 4.76$ seconds and $\sigma = 0.04$ second. What is the probability that this kind of rocket will burn

   (a) less than 4.66 seconds;

   (b) more than 4.80 seconds;

   (c) anywhere from 4.70 to 4.82 seconds?

5.115 Verify that

   (a) $z_{0.10} = 1.28$;

   (b) $z_{0.001} = 3.09$.

5.116 Referring to Exercise 5.28, find the *quartiles* of the normal distribution with $\mu = 102$ and $\sigma = 27$.

5.117 The probability density shown in Figure 5.19 is the log-normal distribution with $\alpha = 8.85$ and $\beta = 1.03$. Find the probability that

   (a) the interrequest time is more than 200 microseconds;

   (b) the interrequest time is less than 300 microseconds.

5.118 The probability density shown in Figure 5.21 is the exponential distribution

$$f(x) = \begin{cases} 0.55\, e^{-0.55x} & 0 < x \\ 0 & \text{elsewhere} \end{cases}$$

Find the probability that

   (a) the time to observe a particle is more than 200 microseconds;

   (b) the time to observe a particle is less than 10 microseconds.

5.119 Referring to the normal scores in Exercise 5.101, construct a normal scores plot of the suspended solids data in Exercise 2.68.

5.120 A change is made to one product page on the retail companies' web site. To determine if the change does improve the efficiency of that product page, data must be collected on the proportion of visitors to the new page that ultimately purchase the product. It is known that 3.2% of visitors, to the original page, make purchases. Assume that this proportion holds for the next 500 visitors to the new page. Use the normal distribution to approximate the probability that, among these 500 visitors, the number who purchase will be

   (a) 11 or fewer.

   (b) 21 or more.

5.121 If n salespeople are employed in a door-to-door selling campaign, the gross sales volume in thousands of dollars may be regarded as a random variable having the gamma distribution with $\alpha = 100\sqrt{n}$ and $\beta = \frac{1}{2}$. If the sales costs are $5,000 per salesperson, how many salespeople should be employed to maximize the expected profit?

5.122 A mechanical engineer models the bending strength of a support beam in a transmission tower as a random variable having the Weibull distribution with $\alpha = 0.02$ and $\beta = 3.0$. What is the probability that the beam can support a load of 4.5?

5.123 Let the times to breakdown for the processors of a parallel processing machine have joint density

$$f(x, y) = \begin{cases} 0.04\, e^{-0.2x - 0.2y} & \text{for } x > 0,\ y > 0 \\ 0 & \text{elsewhere} \end{cases}$$

where x is the time for the first processor and y is the time for the second. Find

   (a) the marginal distributions and their means;

   (b) the expected value of the random variable whose values are given by $g(x, y) = x + y$.

   (c) Verify in this example that the mean of a sum is the sum of the means.

5.124 Two random variables are independent and each has a binomial distribution with success probability 0.6 and 2 trials.

   (a) Find the joint probability distribution.

   (b) Find the probability that the second random variable is greater than the first.

5.125 If X_1 has mean -5 and variance 3 while X_2 has mean 1 and variance 4, and the two are independent, find

   (a) $E(3X_1 + 5X_2 + 2)$;

   (b) $Var(3X_1 + 5X_2 + 2)$.

5.126 Let $X_1, X_2, \ldots, X_{30}$ be independent and let each have the same marginal distribution with mean -3 and variance 4. Find

   (a) $E(X_1 + X_2 + \cdots + X_{30})$;

   (b) $Var(X_1 + X_2 + \cdots + X_{30})$.

5.127 Refer to Example 7 concerning scanners. The maximum attenuation has a normal distribution with mean 10.1 dB and standard deviation 2.7 dB.

   (a) What proportion of the products has maximum attenuation less than 6 dB?

   (b) What proportion of the products has maximum attenuation between 6 dB and 14 dB?

5.128 Refer to the heights of pillars in the example on page 15. The variation in the population of heights of pillars can be modeled as a normal distribution with mean 306.6 nm and standard deviation 37.0 nm.

(a) For a pillar selected at random, what is the probability that its height is greater than 350 nm?

(b) According to the normal model, what proportion of all existing pillars has heights greater than 350 nm? Explain your answer.

(c) What proportion of the pillars has heights between 270 nm and 350 nm?

Summary of Distributions

The formulas for the discrete and continuous distributions, together with their means, variances, and moment generating functions, are given in Table 5.2(a) and (b).

Key Terms

Table 5.2(a) Discrete distributions

Distribution	Probability Distribution $f(x)$	Mean	Variance	Moment Generating Function
Binomial $b(x; n, p)$	$\binom{n}{x}p^x(1-p)^{n-x}$, $\;x=0,1,\ldots,n$	np	$np(1-p)$	$(pe^t+1-p)^n$
Geometric $g(x; p)$	$p(1-p)^{x-1}$, $\;x=1,2,\ldots$	$\dfrac{1}{p}$	$\dfrac{1-p}{p^2}$	$\dfrac{pe^t}{1-(1-p)e^t}$
Hypergeometric $h(x; n, a, N)$	$\binom{a}{x}\binom{N-a}{n-x}\bigg/\binom{N}{n}$ $x=0,1,2,\ldots,\min(N-a,n)$	$n\dfrac{a}{N}$ $p=a/N$	$n\dfrac{a}{N}\left(1-\dfrac{a}{N}\right)\left(\dfrac{N-n}{N-1}\right)$	complicated
Poisson $f(x; \lambda)$	$\dfrac{\lambda^x e^{-\lambda}}{x!}$, $\;x=0,1,\ldots$	λ	λ	$e^{\lambda(e^t-1)}$
Negative binomial	$\binom{x-1}{r-1}p^r q^{x-r}$, $\;x=r,r+1,\ldots$	r/p	$r(1-p)/p^2$	$\left(\dfrac{pe^t}{1-(1-p)e^t}\right)^r$

Table 5.2(b) Continuous distributions

Distribution	Probability Density Function $f(x)$	Mean	Variance	Moment Generating Function
Uniform	$\dfrac{1}{\beta-\alpha}, \quad \alpha < x < \beta$	$\dfrac{\alpha+\beta}{2}$	$\dfrac{(\beta-\alpha)^2}{12}$	$\dfrac{e^{\beta t}-e^{\alpha t}}{(\beta-\alpha)t}$
Normal	$\dfrac{1}{\sqrt{2\pi}\,\sigma}\,e^{-\frac{1}{2\sigma^2}(x-\mu)^2}, \quad -\infty < x < \infty$	μ	σ^2	$e^{\mu t+\frac{1}{2}\sigma^2 t^2}$
Exponential	$\dfrac{1}{\beta}e^{-x/\beta}, \quad 0 \le x < \infty$	β	β^2	$\dfrac{1}{1-\beta t}, \quad t < \dfrac{1}{\beta}$
Lognormal	$\dfrac{1}{\sqrt{2\pi}\,\beta}x^{-1}e^{-(\ln x - \alpha)^2/2\beta^2}, \quad 0 < x < \infty$	$e^{\alpha+\beta^2/2}$	$(e^{\beta^2}-1)e^{2\alpha+\beta^2}$	complicated
Gamma	$\dfrac{1}{\Gamma(\alpha)\,\beta^\alpha}x^{\alpha-1}e^{-x/\beta}, \quad 0 < x < \infty$	$\alpha\beta$	$\alpha\beta^2$	$\dfrac{1}{(1-\beta t)^\alpha}, \quad t < \dfrac{1}{\beta}$
Beta	$\dfrac{1}{B(\alpha,\beta)}x^{\alpha-1}(1-x)^{\beta-1}, \quad 0 < x < 1$	$\dfrac{\alpha}{\alpha+\beta}$	$\dfrac{\alpha\beta}{(\alpha+\beta)^2(\alpha+\beta+1)}$	complicated
Weibull	$\alpha\beta x^{\beta-1}e^{-\alpha x^\beta}, \quad 0 < x$	$\alpha^{-1/\beta}\Gamma\left(1+\dfrac{1}{\beta}\right)$	$\alpha^{-2/\beta}\left\{\Gamma\left(1+\dfrac{2}{\beta}\right)-\left[\Gamma\left(1+\dfrac{1}{\beta}\right)\right]^2\right\}$	complicated

SAMPLING DISTRIBUTIONS

I n most of the inference methods we shall study in this book, it will be assumed that we are dealing with a particular kind of sample called a *random sample*. This attention to random samples, which we discuss in Section 6.1, is due to their permitting valid, or logical, generalizations from sample data. Then, in Sections 6.2 through 6.4, we see how certain statistics (that is, certain quantities determined from samples) can be expected to vary from sample to sample. Section 6.5 connects the sampling distributions arising from normal distributions. Techniques for deriving sampling distributions are described in Sections 6.6 and 6.7. The concept of a sampling distribution—the distribution of a statistic calculated on the basis of a random sample—is basic to all of statistical inference.

6.1 Populations and Samples

Usage of the term **population** in statistics is a carryover from the days when statistics was applied mainly to sociological and economic phenomena. Recall from Chapter 1 that today the term population of units applies to sets or collections of objects, actual or conceptual. In contrast, statistical population, or just population, refers to sets of numbers, measurements, or observations under investigation. For example, if we are interested in determining the average number of television sets per household in the United States, the totality of these numbers of sets, one for each household, constitutes the population for this study. Similarly, the population from which inspectors draw a sample to determine some quality characteristic of a manufactured product may be the corresponding measurements for all units in a given lot; depending on the objectives of the inspection, it may also consist of the corresponding measurements for all units that may conceivably be manufactured.

In some cases, such as the one above concerning the number of television sets per household, the population is **finite**. In other cases, such as the determination of some characteristic of all units past, present, and future that might conceivably be manufactured by a given process, it is convenient to think of the population as **infinite**. Similarly, we look upon the results obtained in an unending series of flips of a coin as a sample from the hypothetically infinite population consisting of all conceivably possible flips of the coin.

Populations are often described by the distribution of their values. It is common practice to refer to a population in terms of its corresponding probability distribution or density function. For example, we may refer to a fixed number of flips of a coin as a sample from a "binomial population" or to certain measurements as a sample from a "normal population." Hereafter, when referring to a "population $f(x)$" we shall mean a population described by a probability distribution or a density $f(x)$.

If a population is infinite, it is impossible to observe all its values, and even if it is finite it may be impractical or uneconomical to observe it in its entirety. Thus, it is usually necessary to use a **sample**, a part of a population, and infer from it results pertaining to the entire population. Clearly, such results can be useful only if the sample is in some way "representative" of the population. It would be unreasonable, for instance, to expect useful generalizations about the population of family incomes in the United States in the year 2015 on the basis of data pertaining to home owners only. Similarly, we can hardly expect reasonable generalizations about the performance of a tire if it is tested only on smooth roads.

To assure that a sample is representative of the population from which it is obtained, and to provide a framework for the application of probability theory to problems of sampling, we shall limit our discussion to **random samples**. Before a random sample of size n is selected, the observations are modeled as the random variables $X_1, X_2, \ldots, X_n$. For sampling from finite populations, random samples are defined as follows:

Random sample (finite population)

> A set of observations $X_1, X_2, \ldots, X_n$ constitutes a random sample of size n from a finite population of size N, if its values are chosen so that each subset of n of the N elements of the population has the same probability of being selected.

Note that this definition of randomness pertains essentially to the manner in which the sample values are selected. This holds also for the following definition of a random sample from an infinite population:

Random sample (infinite population)

> A set of observations $X_1, X_2, \ldots, X_n$ constitutes a random sample of size n from the infinite population $f(x)$ if
>
> 1. Each X_i is a random variable whose distribution is given by $f(x)$.
> 2. These n random variables are independent.

We also apply the term **random sample** to the set of observed values $x_1, x_2, \ldots, x_n$ of the random variables. The lower case distinguishes the realization of a random sample from the upper case, which represents the random variables before they are observed.

There are several ways of assuring the selection of a sample that is at least approximately random. When dealing with a finite population, we can serially number the elements of the population and then select a sample with the aid of a random number generator or a table of random digits (see discussion on page 8). For instance, if a population has $N = 500$ elements and we wish to select a random sample of size $n = 10$, we can use three arbitrarily selected columns of Table 7W to obtain 10 different three-digit numbers less than or equal to 500, which will then serve as the serial numbers of the elements to be included in the sample.

When the population size is large, the use of random numbers can become very laborious and at times practically impossible. For instance, if a sample of five cartons of canned peaches is to be chosen for inspection from among the many thousands stored in a warehouse, one can hardly expect to number all the cartons, make a selection with the use of random numbers, and then pull out the ones that were chosen. In a situation like this, one really has very little choice but to make the selection relatively haphazard, hoping that this will not seriously violate the assumption of randomness which is basic to most statistical theory.

When dealing with infinite populations, the situation is somewhat different since we cannot physically number the elements of the population; but efforts should be made to approach conditions of randomness by the use of artificial devices. For example, in selecting a sample from a production line we may be able to approximate conditions of randomness by choosing one unit each half hour; when tossing a coin we can try to flip it in such a way that neither side is intentionally favored; and so forth. The proper use of artificial or mechanical devices for selecting random samples is always preferable to human judgment, as it is extremely difficult to avoid unconscious biases when making almost any kind of selection.

Even with the careful choice of artificial devices, it is all too easy to commit gross errors in the selection of a random sample. To illustrate some of these pitfalls, suppose we have the task of selecting logs being fed into a sawmill by a constant-speed conveyor belt, for the purpose of obtaining a random sample of their lengths. One sampling device, which at first sight would seem to assure randomness, consists of measuring the logs which pass a given point at the end of a certain number of 10-minute intervals. However, further thought reveals that this method of selection favors the longer logs, since they require more time to pass the given point. Thus, the sample is not random since the longer logs have a better chance of being included.

Another common mistake in selecting a sample is that of sampling from the wrong population or from a poorly specified population. As we have pointed out earlier, we would hardly get a sample from which we could generalize about family incomes in the United States if we limited our sample to home owners. Similarly, if we wanted to determine the effect of vibrations on a structural member, we should be careful to delineate the frequency band of vibrations that is of relevance, and to vibrate test specimens only at frequencies selected randomly from this band.

EXAMPLE 46   **Selecting where to sample in an area that may be contaminated**

In many environmental cleanup studies, engineers are faced with the problem of evaluating the status of land areas or bodies of water. It is not always easy to collect a representative sample where the observations can be treated as independent and from the same distribution. To illustrate some of the key issues, consider sampling from a contaminated area in City C that covers a city block. Locations must be selected for taking the soil samples that will then be analyzed for the presence of heavy metals.

Solution   One recommended approach for homogeneous land areas is to sample according to a rectangular grid as shown in Figure 6.1(a). This grid could even be randomly placed over the area.

However, more was known about this area. At the time the pollution occurred, a smelter was located in the position indicted by the shaded area in Figure 6.1(b). The soil contamination by heavy metals is definitely not homogeneous! Materials from the smelter flowed toward the north side of the smelter. This area is definitely a "hot spot." The average amount of heavy metals, obtained by averaging the measurements from all locations, is not representative of the contamination problem. Using the average from the whole area as a summary description downplays the seriousness of the contamination around the smelter.

The smelter runoff area should be treated as a separate population. Soil should be collected from at least two locations within the area of this suspected hot spot.

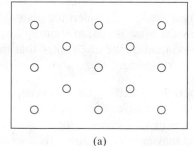

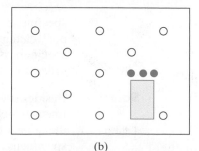

Figure 6.1
Choosing where to sample   (a)   (b)

We suggest three sites indicated by solid circles. The corresponding measurements of heavy metals should be used to describe only that area.

While sampling the "hot spot" separately might seem like the obvious approach, the owners of the site in the situation on which the example is based wanted to collect data from more locations throughout the whole block. This, they hoped, would lower the average amount of heavy metals enough so that only simple cleanup tactics would be required rather than resorting to the expensive solution of trucking out the soil. ∎

EXAMPLE 47 **Always replicate at least one key measurement**

Refer to the environmental cleanup study in the previous example. Explain why soil samples should be taken in at least two locations within the hot spot, rather than just doing two chemical analyses on essentially a single soil sample.

Solution One major aspect of understanding the contamination problem is to evaluate the condition of the known hot spot. Even within the hot spot, the amount of contamination could vary considerably because of, for instance, any particular location's position relative to the old stream of runoff from the smelter. We strongly recommend that soil be collected from at least two soil samples—not just that two chemical analyses be performed on soil from a single location. This approach will provide measurements that can also be used to estimate the total amount of variability in the measurements from the "hot spot."

By repeating the chemical analysis on soil from a single location, we could estimate the variability in the chemical analysis and possibly identify an outlier that might suggest a faulty chemical analysis. However, no matter how many times we repeated the chemical analysis, we would not know how much variation to expect if we sampled soil from the same hot spot but several feet from the first sample. That is, nearby locations are likely to be very much alike but those farther apart are less alike. We need to sample from at least two different locations to determine the degree of homogeneity within the hot spot. ∎

The purpose of most statistical investigations is to generalize from information contained in random samples about the population from which the samples were obtained. In particular, we are usually concerned with the problem of making inferences about the **parameters** of populations, such as the mean μ or the standard deviation σ. In making such inferences, we use **statistics** such as $\bar{x}$ and s, namely quantities calculated on the basis of sample observations. In practice, the term *statistic* is also applied to the corresponding random variables.

EXAMPLE 48 **Sample-to-sample variation must be understood to accurately assess total water quality**

The quality of water leaving a plant must be maintained. It is monitored by taking a tiny volume of water called a test specimen. If the quality of the water in the specimen is bad, action may be taken. The action could range from taking more specimens, making a phone call to alert the plant operators, preparing a written report, changing how the plant is run, to shutting down the plant. If the water in the specimen is of good quality, we often infer that the total volume of discharge is satisfactory. Discuss sampling.

Solution Besides sound laboratory practice, judgments depend crucially on getting test specimens which are representative of the total volume of effluent discharged. Because the quality of water will vary over the total effluent at any one time, the actual test specimens selected may or may not correctly convey the water quality.

The key idea in this discussion is that of sampling variability. Not all choices of actual test specimens will produce the same values for water quality or even a correct appraisal of quality. This variability can be overcome, to a greater or lesser extent, by taking a large enough number of test specimens. The results from examining enough test specimens should then quite accurately reflect water quality most of the time. ■

Since the selection of a random sample is controlled largely by chance, so are the values we obtain for statistics. The remainder of this chapter will be devoted to sampling distributions, namely, to distributions which describe the chance fluctuations of statistics calculated on the basis of random samples.

6.2   The Sampling Distribution of the Mean (σ known)

Suppose that a random sample of n observations, from some population, leads to the observed value $\bar{x}$ as an estimate of the population mean. It should be clear that if we took a second random sample of size n from the population, it would be quite unreasonable to expect the identical value for $\bar{x}$, and if we took several more samples, probably no two of the $\bar{x}$'s would be alike. The differences among such x's are generally attributed to chance. This raises important questions concerning their distribution, specifically concerning the extent of their chance fluctuations.

To approach this question experimentally, suppose that 50 random samples of size $n = 10$ are to be taken from a population having the **discrete uniform distribution**

$$f(x) = \begin{cases} \dfrac{1}{10} & \text{for } x = 0, 1, 2, \ldots, 9 \\ 0 & \text{elsewhere} \end{cases}$$

Sampling is with replacement, so to speak, so that we are sampling from an infinite population. A convenient way of obtaining these samples is to use a table of random digits letting each sample consist of 10 consecutive digits in arbitrarily chosen rows or columns. Actually proceeding in this way, we get 50 samples whose means are

4.4	3.2	5.0	3.5	4.1	4.4	3.6	6.5	5.3	4.4
3.1	5.3	3.8	4.3	3.3	5.0	4.9	4.8	3.1	5.3
3.0	3.0	4.6	5.8	4.6	4.0	3.7	5.2	3.7	3.8
5.3	5.5	4.8	6.4	4.9	6.5	3.5	4.5	4.9	5.3
3.6	2.7	4.0	5.0	2.6	4.2	4.4	5.6	4.7	4.3

Grouping these means into a distribution with the classes [2.0, 3.0), [3.0, 4.0), [4.0, 5.0), [5.0, 6.0), and [6.0, 7.0), where the left endpoint is included, we get

$\bar{x}$	Frequency
[2.0, 3.0)	2
[3.0, 4.0)	14
[4.0, 5.0)	19
[5.0, 6.0)	12
[6.0, 7.0)	3
	50

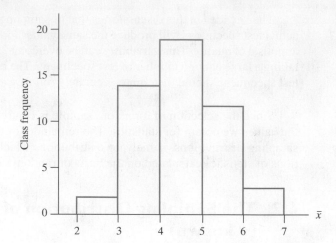

Figure 6.2
Experimental sampling
distribution of the mean

and it is apparent from this distribution as well as its histogram shown in Figure 6.2 that the distribution of the means is fairly **bell-shaped**, even though the population itself has a uniform distribution. This raises the question whether our result is typical of what we might expect; that is, whether we would get similar distributions if we repeated the experiment again and again.

To answer this kind of question, we shall have to investigate the **theoretical sampling distribution** of the mean which, for the given example, provides us with the probabilities of getting means in the interval [2.0, 3.0) or in [3.0, 4.0), ..., [6.0, 7.0) and perhaps values less than 2.0 or greater than or equal to 7.0. Although we could evaluate these probabilities for this particular example, it is usually preferable to refer to some general theorems concerning sampling distributions. The first of these gives expressions for the mean $\mu_{\overline{X}}$ and the variance $\sigma_{\overline{X}}^2$ of sampling distributions of the mean $\overline{X}$.

Formulas for $\mu_{\overline{X}}$ and $\sigma_{\overline{X}}^2$

> **Theorem 6.1**   If a random sample of size n is taken from a population having the mean μ and the variance σ^2, then $\overline{X}$ is a random variable whose distribution has the mean μ.
>
> For samples from infinite populations the variance of this distribution is $\dfrac{\sigma^2}{n}$.
>
> For samples from a finite population of size N the variance is $\dfrac{\sigma^2}{n} \cdot \dfrac{N-n}{N-1}$.

The result for infinite populations was established in Example 33 on page 162 using the properties of expectation. Alternatively, we now prove that $\mu_{\overline{X}} = \mu$ for the continuous case directly starting from the definition on page 159.

$$\mu_{\overline{X}} = \int_{-\infty}^{\infty} \int_{-\infty}^{\infty} \cdots \int_{-\infty}^{\infty} \sum_{i=1}^{n} \frac{x_i}{n} f(x_1, x_2, \ldots, x_n) \, dx_1 \, dx_2 \ldots dx_n$$

$$= \frac{1}{n} \sum_{i=1}^{n} \int_{-\infty}^{\infty} \int_{-\infty}^{\infty} \cdots \int_{-\infty}^{\infty} x_i \, f(x_1, x_2, \ldots, x_n) \, dx_1 \, dx_2 \ldots dx_n$$

where $f(x_1, x_2, \ldots, x_n)$ is the joint density of the random variables which constitute the random sample. Using the assumption of a random sample, we can write

$$f(x_1, x_2, \ldots, x_n) = f(x_1) f(x_2) \ldots f(x_n)$$

and we now have

$$\mu_{\overline{X}} = \frac{1}{n} \sum_{i=1}^{n} \int_{-\infty}^{\infty} f(x_1)\, dx_1 \ldots \int_{-\infty}^{\infty} x_i f(x_i)\, dx_i \ldots \int_{-\infty}^{\infty} f(x_n)\, dx_n$$

Since each integral except the one with the integrand $x_i f(x_i)$ equals 1, and the one with the integrand $x_i f(x_i)$ equals μ, we finally obtain

$$\mu_{\overline{X}} = \frac{1}{n} \sum_{i=1}^{n} \mu = \mu$$

and this completes the proof. (For the discrete case the proof follows the same steps, with integral signs replaced by $\sum$'s.)

To prove that $\sigma_{\overline{X}}^2 = \sigma^2/n$ for the continuous case, we shall make the simplifying assumption that $\mu = 0$, which does not involve any loss of generality as the reader will be asked to show in Exercise 6.18. Using the definition on page 159, we thus have

$$\sigma_{\overline{X}}^2 = \int_{-\infty}^{\infty} \int_{-\infty}^{\infty} \cdots \int_{-\infty}^{\infty} \overline{x}^2 f(x_1, x_2, \ldots, x_n)\, dx_1\, dx_2 \ldots dx_n$$

and making use of the fact that

$$\overline{x}^2 = \frac{1}{n^2} \left(\sum_{i=1}^{n} x_i \right)^2 = \frac{1}{n^2} \left(\sum_{i=1}^{n} x_i^2 + \sum \sum_{i \neq j} x_i x_j \right)$$

we obtain

$$\sigma_{\overline{X}}^2 = \frac{1}{n^2} \sum_{i=1}^{n} \int_{-\infty}^{\infty} \int_{-\infty}^{\infty} \cdots \int_{-\infty}^{\infty} x_i^2 f(x_1, x_2, \ldots, x_n)\, dx_1\, dx_2 \ldots dx_n$$

$$+ \frac{1}{n^2} \sum \sum_{i \neq j} \int_{-\infty}^{\infty} \int_{-\infty}^{\infty} \cdots \int_{-\infty}^{\infty} x_i x_j f(x_1, x_2, \ldots, x_n)\, dx_1\, dx_2 \ldots dx_n$$

where $\displaystyle\sum \sum_{i \neq j}$ extends over all i and j from 1 to n, not including the terms where $i = j$. Again using the fact that

$$f(x_1, x_2, \ldots, x_n) = f(x_1)\, f(x_2) \ldots f(x_n)$$

we can write each of the preceding multiple integrals as a product of simple integrals, where each integral with integrand $f(x)$ equals 1. We thus obtain

$$\sigma_{\overline{X}}^2 = \frac{1}{n^2} \sum_{i=1}^{n} \int_{-\infty}^{\infty} x_i^2 f(x_i)\, dx_i + \frac{1}{n^2} \sum \sum_{i \neq j} \int_{-\infty}^{\infty} x_i f(x_i)\, dx_i$$

$$\times \int_{-\infty}^{\infty} x_j f(x_j)\, dx_j$$

and since each integral in the first sum equals σ^2 while each integral in the second sum equals 0, we finally have

$$\sigma_{\overline{X}}^2 = \frac{1}{n^2} \sum_{i=1}^{n} \sigma^2 = \frac{\sigma^2}{n}$$

This completes the proof of the second part of the theorem. We shall not prove the corresponding result for random samples from finite populations. But it should be noted that in the resulting formula for $\sigma_{\overline{X}}^2$ the factor

$$\frac{N-n}{N-1}$$

often called the **finite population correction factor**, is close to 1 (and can be omitted for most practical purposes) unless the sample constitutes a substantial portion of the population.

EXAMPLE 49   **Calculating a finite population correction factor**

Find the value of the finite population correction factor for $n = 10$ and $N = 1,000$.

Solution
$$\frac{1,000-10}{1,000-1} = 0.991$$
∎

Although it should not come as a surprise that $\mu_{\overline{X}} = \mu$, the fact that $\sigma_{\overline{X}}^2 = \sigma^2/n$ for random samples from infinite populations is interesting and important. To point out its implications, let us apply Chebyshev's theorem to the sampling distribution of the mean, substituting $\overline{X}$ for X and $\sigma/\sqrt{n}$ for σ in the formula for the alternate form of the theorem (see page 104). We thus obtain

$$P\left(|\overline{X} - \mu| < \frac{k\sigma}{\sqrt{n}}\right) \geq 1 - \frac{1}{k^2}$$

and, letting $k\sigma/\sqrt{n} = \varepsilon$, we get

$$P(|\overline{X} - \mu| < \varepsilon) \geq 1 - \frac{\sigma^2}{n\varepsilon^2}$$

Thus, for any given $\varepsilon > 0$, the probability that $\overline{X}$ differs from μ by less than ε can be made arbitrarily close to 1 by choosing n sufficiently large. In less rigorous language, the larger the sample size, the closer we can expect $\overline{X}$ to be to the mean of the population. In this sense we can say that the mean becomes more and more reliable as an estimate of μ as the sample size is increased. This result—that $\overline{X}$ becomes arbitrarily close to μ with arbitrarily high probability—is called the **law of large numbers**.

Theorem 6.2   Let $X_1, X_2, \ldots, X_n$ be independent random variables each having the same mean μ and variance σ^2. Then, for any positive ε,

$$P(|\overline{X} - \mu| > \varepsilon) \rightarrow 0 \quad \text{as } n \rightarrow \infty$$

Law of large numbers

As the sample size increases, unboundedly, the probability that the sample mean differs from the population mean μ, by more than an arbitrary amount ε, converges to zero.

EXAMPLE 50   **Law of large numbers and long-run relative frequency**

Consider an experiment where a specified event A has probability p of occurring. Suppose that, when the experiment is repeated n times, outcomes from different trials are independent. Show that

$$\text{relative frequency of } A = \frac{\text{number of times } A \text{ occurs in } n \text{ trials}}{n}$$

becomes arbitrarily close to p, with arbitrarily high probability, as the number of times the experiment is repeated grows unboundedly.

Solution   We can define n random variables $X_1, X_2, \ldots, X_n$ where $X_i = 1$ if A occurs on the ith trial and $X_i = 0$ otherwise. The X_i are independent and identically distributed with mean $\mu = p$ and variance $\sigma^2 = p(1-p)$ since $E(X_1^2) = 1^2 \cdot p + 0^2(1-p) = p$. Then $X_1 + \cdots + X_n$ is the number of times that A occurs in n trials of the experiment and $\overline{X}$ is the relative frequency of A.

We apply the law of large numbers and conclude that, for an arbitrary positive amount ε,

$$P(\,|\,\text{relative frequency of } A - p\,| > \varepsilon\,) = P(\,|\overline{X} - p\,| > \varepsilon\,) \to 0 \quad \text{as } n \to \infty$$

Beginning with the axioms of probability, we are led to a theorem that determines the long-run relative frequency of an event.   ∎

The reliability of the mean as an estimate of μ is often measured by $\sigma_{\overline{X}} = \sigma/\sqrt{n}$, also called the **standard error of the mean**. Note that this measure of the reliability of the mean decreases in proportion to the square root of n; for instance, it is necessary to quadruple the size of the sample in order to halve the standard deviation of the sampling distribution of the mean. This also indicates what might be called a "law of diminishing returns" so far as increasing the sample size is concerned. Usually it does not pay to take excessively large samples since the extra labor and expense is not accompanied by a proportional gain in reliability. For instance, if we increase the size of a sample from 25 to 2,500, the errors to which we are exposed are reduced only by a factor of 10.

Let us now return to the experimental sampling distribution on page 187, and let us check how closely its mean and variance correspond to the values we should expect in accordance with Theorem 6.1. Since the population from which the 50 samples of size $n = 10$ were obtained has the mean

$$\mu = \sum_{x=0}^{9} x \cdot \frac{1}{10} = 4.5$$

and the variance

$$\sigma^2 = \sum_{x=0}^{9} (x - 4.5)^2 \frac{1}{10} = 8.25$$

Theorem 6.1 leads us to expect a mean of $\mu_{\overline{X}} = 4.5$ and a variance of $\sigma_{\overline{X}}^2 = 8.25/10 = 0.825$. Calculating the mean and the variance from the 50 sample means on page 187, we get $\overline{x}_{\overline{x}} = 4.43$ and $s_{\overline{x}}^2 = 0.930$, which are reasonably close to the theoretical values.

Theorem 6.1 provides only partial information about the theoretical sampling distribution of the mean. In general, it is impossible to determine such a distribution exactly without knowledge of the actual form of the population. Even then, it can be quite difficult. But it is possible to find the limiting distribution as $n \to \infty$ of a random variable whose values are closely related to $\overline{X}$, assuming only that the population has a finite variance σ^2. The random variable we are referring to here is the **standardized sample mean**

$$Z = \frac{\overline{X} - \mu}{\sigma/\sqrt{n}}$$

whose values are given by the difference between $\overline{x}$ and μ divided by the standard error of the mean. With reference to this random variable, we can now state the following theorem, called the **central limit theorem:**

Central limit theorem

> **Theorem 6.3**   If $\overline{X}$ is the mean of a random sample of size n taken from a population having the mean μ and the finite variance σ^2, then
>
> $$Z = \frac{\overline{X} - \mu}{\sigma / \sqrt{n}}$$
>
> is a random variable whose distribution function approaches that of the standard normal distributions as $n \to \infty$.

The central limit theorem provides a normal distribution that allows us to assign probabilities to intervals of values for $\overline{X}$. Regardless of the form of the population distribution, the distribution of $\overline{X}$ is approximately normal with mean μ and variance σ^2/n whenever n is large. This tendency toward normality is illustrated in Figure 6.3 for a uniform population distribution and an exponential population distribution.

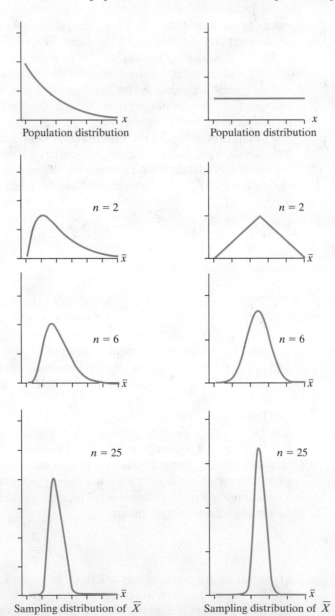

Figure 6.3
An illustration of the approach toward normality for the sampling distribution of $\overline{X}$ as sample size increases

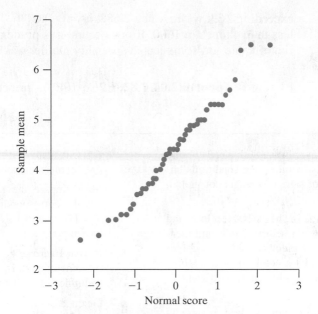

Figure 6.4
Experimental verification of
the central limit theorem

Although proving the central limit theorem is beyond the scope of this text, we can obtain experimental verification by constructing a normal scores plot of the 50 sample means on page 187, which were obtained by sampling with replacement from a discrete uniform population. In Figure 6.4, the points fall close to a straight line. It seems that, even for $n = 10$, the sampling distribution of the mean for this example exhibits a pattern that generally resembles that of a normal distribution.

In practice, **the normal distribution provides an excellent approximation to the sampling distribution of the mean $\overline{X}$ for n as small as 25 or 30**, with hardly any restrictions on the shape of the population. As we see in our example, the sampling distribution of the mean has the general shape of a normal distribution even for samples of size $n = 10$ from a discrete uniform distribution.

A stronger result holds for normal populations.

When the random samples come from a normal population, the sampling distribution of the mean is normal regardless of the size of the random sample.

EXAMPLE 51   **A probability calculation based on the central limit theorem concerns operator time**

Car mufflers are constructed by nearly automatic machines. One manufacturer finds that, for any type of car muffler, the time for a person to set up and complete a production run has a normal distribution with mean 1.82 hours and standard deviation 1.20. What is the probability that the sample mean of the next 40 runs will be from 1.65 to 2.04 hours.

Solution   Theorem 6.3, the central limit theorem, applies whatever the form of the population distribution. We need only find the normal curve area between

$$z = \frac{1.65 - 1.82}{1.20/\sqrt{40}} = -0.896 \quad \text{and} \quad z = \frac{2.04 - 1.82}{1.20/\sqrt{40}} = 1.16$$

From Table 3, we obtain the probability 0.6917.

If it turns out that $\bar{x}$ is 2.33 hours, serious doubt will be cast on whether the sample came from a population having $\mu = 1.82$ and $\sigma = 1.20$. The probability of exceeding 2.33, with z-value 2.688, is only 0.0036. These large values would occur less than 4 times in 1000. If one occurs, it is prudent to look for a cause. Maybe a run of a rare and complicated speciality muffler was required.   ∎

[Using **R**:   **pnorm(2.04, 1.82, 1.2/sqrt(40)) - pnorm(1.65, 1.82, 1.2/sqrt(40))**]

Exercises

6.1 An inspector examines every twentieth piece coming off an assembly line. List some of the conditions under which this method of sampling might not yield a random sample.

6.2 Clear plastic wrap is made in sheets 100 feet long and wound in rolls. An inspector selects 7 rolls and measures the clarity of a small piece taken from a corner at the end of each roll. List a condition under which this method of sampling might not yield a random sample.

6.3 Explain why the following will not lead to random samples from the desired populations.

(a) To determine what the average person spends on a vacation, a market researcher interviews passengers on a luxury cruise.

(b) To determine the average income of its graduates 10 years after graduation, the alumni office of a university sent questionnaires in 2016 to all the members of the class of 2006 and based its estimate on the questionnaires returned.

(c) To determine public sentiment about certain import restrictions, an interviewer asks voters: "Do you feel that this unfair practice should be stopped?"

6.4 A market research organization wants to try a new product in 8 of 50 states. Use Table 7W or software to make this selection.

6.5 How many different samples of size $n = 2$ can be chosen from a finite population of size

(a) $N = 9$;

(b) $N = 23$?

6.6 With reference to Exercise 6.5, what is the probability of each sample in part (a) and the probability of each sample in part (b), if the samples are to be random?

6.7 Take 30 slips of paper and label five each -4 and 4, four each -3 and 3, three each -2 and 2, and two each $-1, 0$ and 1.

(a) If each slip of paper has the same probability of being drawn, find the probability of getting $-4, -3, -2, -1, 0, 1, 2, 3, 4$ and find the mean and the variance of this distribution.

(b) Draw 50 samples of size 10 from this population, each sample being drawn without replacement, and calculate their means.

(c) Calculate the mean and the variance of the 50 means obtained in part (b).

(d) Compare the results obtained in part (c) with the corresponding values expected according to Theorem 6.1. [Note that μ and σ^2 were obtained in part (a).]

6.8 Repeat Exercise 6.7, but select each sample with replacement; that is, replace each slip of paper and reshuffle before the next one is drawn.

6.9 Given the infinite population whose distribution is given by

x	$f(x)$
1	0.25
2	0.25
3	0.25
4	0.25

list the 16 possible samples of size 2 and use this list to construct the distribution of $\overline{X}$ for random samples of size 2 from the given population. Verify that the mean and the variance of this sampling distribution are identical with corresponding values expected according to Theorem 6.1.

6.10 Suppose that we convert the 50 samples referred to on page 187 into 25 samples of size $n = 20$ by combining the first two, the next two, and so on. Find the means of these samples and calculate their mean and their standard deviation. Compare this mean and this standard deviation with the corresponding values expected in accordance with Theorem 6.1.

6.11 When we sample from an infinite population, what happens to the standard error of the mean if the sample size is

(a) increased from 50 to 200;

(b) increased from 400 to 900;

(c) decreased from 225 to 25;

(d) decreased from 640 to 40?

6.12 What is the value of the finite population correction factor in the formula for $\sigma_{\overline{X}}^2$ when

(a) $n = 5$ and $N = 250$;

(b) $n = 10$ and $N = 500$;

(c) $n = 100$ and $N = 5,000$?

6.13 For large sample size n, verify that there is a 50-50 chance that the mean of a random sample from an infinite population with the standard deviation σ will differ from μ by less than $0.6745 \cdot \sigma/\sqrt{n}$. It has been the custom to refer to this quantity as the **probable error of the mean**.

6.14 The mean of a random sample of size $n = 25$ is used to estimate the mean of an infinite population that has standard deviation $\sigma = 2.4$. What can we assert about the probability that the error will be less than 1.2, if we use

(a) Chebyshev's theorem;

(b) the central limit theorem?

6.15 Hard disks for computers must spin evenly, and one departure from level is called roll. The roll for any disk can be modeled as a random variable having mean 0.2250 mm and standard deviation 0.0042 mm. The sample mean roll $\overline{X}$ will be obtained from a random

sample of 40 disks. What is the probability that $\overline{X}$ will lie between 0.2245 and 0.2260 mm?

6.16 A wire-bonding process is said to be in control if the mean pull strength is 10 pounds. It is known that the pull-strength measurements are normally distributed with a standard deviation of 1.5 pounds. Periodic random samples of size 4 are taken from this process and the process is said to be "out of control" if a sample mean is less than 7.75 pounds. Comment.

6.17 If the distribution of the weights of all men traveling by air between Dallas and El Paso has a mean of 163 pounds and a standard deviation of 18 pounds, what is the probability that the combined gross weight of 36 men traveling on a plane between these two cities is more than 6,000 pounds?

6.18 If X is a continuous random variable and $Y = X - \mu$, show that $\sigma_Y^2 = \sigma_X^2$.

6.19 Prove that $\mu_{\overline{X}} = \mu$ for random samples from discrete (finite or countably infinite) populations.

6.3 The Sampling Distribution of the Mean (σ unknown)

Application of the theory of the preceding section requires knowledge of the population standard deviation σ. If n is large, this does not pose any problems even when σ is unknown, as it is reasonable in that case to substitute for it the sample standard deviation s. However, when it comes to the random variable whose values are given by

$$\frac{\overline{x} - \mu}{s/\sqrt{n}}$$

very little is known about its exact sampling distribution for small values of n unless we make the assumption that the sample comes from a normal population. Under this assumption, one can prove the following:

A random variable having the t distribution

> **Theorem 6.4** If $\overline{X}$ is the mean of a random sample of size n taken from a normal population having the mean μ and the variance σ^2, and $S^2 = \sum_{i=1}^{n} \frac{(X_i - \overline{X})^2}{n-1}$, then
> $$t = \frac{\overline{X} - \mu}{S/\sqrt{n}}$$
> is a random variable having the t distribution with the parameter $\nu = n - 1$.

The lowercase t notation helps differentiate this important statistic from others. This theorem is more general than Theorem 6.3 in the sense that it does not require knowledge of σ; on the other hand, it is less general than Theorem 6.3 in the sense that it requires the assumption of a normal population.

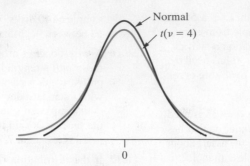

Figure 6.5
t distribution and standard normal distribution

As can be seen from Figure 6.5, the overall shape of a *t* **distribution** is similar to that of a normal distribution—both are bell-shaped and symmetrical about the mean. Like the standard normal distribution, the *t* distribution has the mean 0, but its variance depends on the parameter v (*nu*), called the number of **degrees of freedom**. The variance of the *t* distribution exceeds 1, but it approaches 1 as $n \rightarrow \infty$. In fact, it can be shown that the *t* distribution with v degrees of freedom approaches the standard normal distribution as $v \rightarrow \infty$.

Table 4 in Appendix B contains selected values of t_α for various values of v, where t_α is such that the area under the *t* distribution to its right is equal to α. In this table the left-hand column contains values of v, the column headings are areas α in the right-hand tail of the *t* distribution, and the entries are values of t_α. (See also Figure 6.6.) It is not necessary to tabulate values of t_α for $\alpha > 0.50$, as it follows from the symmetry of the *t* distribution that $t_{1-\alpha} = -t_\alpha$. Thus, the value of *t* that corresponds to a left-hand tail area of α is $-t_\alpha$.

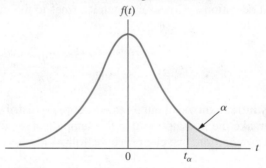

Figure 6.6
Tabulated value of t_α

Note that in the bottom row of Table 4 the entries correspond to the values of *z* that cut off right-hand tails of area α under the standard normal curve. Using the notation z_α for such a value of *z*, it can be seen, for example, that $z_{0.025} = 1.96 = t_{0.025}$ for $v = \infty$. In fact, observing that the values of t_α for 29 or more degrees of freedom are close to the corresponding values of z_α, we conclude that **the standard normal distribution provides a good approximation to the *t* distribution for samples of size 30 or more**.

EXAMPLE 52   **Using a probability calculation from the *t* distribution to refute a claim**

A treatment plant that sends effluent into the river claims the mean suspended solids is never above 40 mg/l. Measurements of the suspended solids in river water on

$n = 14$ Monday mornings yield $\bar{x} = 46$ and $s = 9.4$ mg/l. Based on data collected over a period of many years, it is reasonable to assume that the individual measurements follow a normal distribution.

Do the data support or refute the treatment plant's claim?

Solution   We calculate

$$t = \frac{\bar{X} - 40}{S/\sqrt{n}} = \frac{46 - 40}{9.4/\sqrt{20}} = 2.855$$

which is a value of a random variable having a t distribution with $\nu = 20 - 1 = 19$ degrees of freedom provided the mean is 40. Since the probability that t will exceed $2.531 = t_{0.01}$ is 0.01, the probability of observing a value as large or larger than 2.855 is even smaller. We conclude that the data strongly refute the treatment plant's claim. In all likelihood the mean suspended solids is more than 40 mg/l.   ∎

[Using **R**: For the upper tail, use .99 in **qt(.99,19)** and **1 - pt(2.855,19)**]

The assumption that the sample must come from a normal population is not so severe a restriction as it may seem. Studies have shown that the distribution of random variable

$$\frac{\bar{X} - \mu}{S/\sqrt{n}}$$

is fairly close to a t distribution even for samples from certain nonnormal populations. In practice, it is necessary to make sure primarily that the population from which we are sampling is approximately bell-shaped and not too skewed. A practical way of checking this assumption is to construct a normal scores plot, as described on page 170. (If such a plot shows a distinct curve rather than a straight line, it may be possible to "straighten it out" by transforming the data—say, by taking their logarithms or their square roots, as discussed in Chapter 5, Section 5.13.)

6.4   The Sampling Distribution of the Variance

So far we have discussed only the sampling distribution of the mean. If we take the medians or the standard deviations of the 50 samples on page 187, we would similarly obtain experimental sampling distributions of these statistics. In this section we shall be concerned with the theoretical sampling distribution of the sample variance for random samples from normal populations. Since S^2 cannot be negative, we should suspect that this sampling distribution is not a normal curve; in fact, it is related to the gamma distribution (see page 145) with $\alpha = \nu/2$ and $\beta = 2$, called the **chi square distribution**. Specifically, using the square of the Greek letter χ (chi), we have the following theorem.

A random variable having the chi square distribution

> **Theorem 6.5**   If S^2 is the variance of a random sample of size n taken from a normal population having the variance σ^2, then
>
> $$\chi^2 = \frac{(n-1)S^2}{\sigma^2} = \frac{\sum_{i=1}^{n}(X_i - \bar{X})^2}{\sigma^2}$$
>
> is a random variable having the chi square distribution with the parameter $\nu = n - 1$.

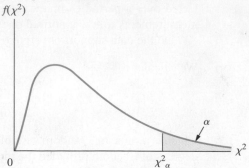

χ^2 distribution with ν degrees of freedom

Figure 6.7
Tabulated values of chi square

Table 5W on the book website contains selected values of χ_{α}^2 for various values of ν, again called the number of **degrees of freedom**, where χ_{α}^2 is such that the area under the chi square distribution to its right is equal to α. In this table the left-hand column contains values of ν, the column headings are areas α in the right-hand tail of the chi square distribution, and the entries are values of χ_{α}^2. (See also Figure 6.7.) Unlike the t distribution, it is necessary to tabulate values of χ_{α}^2 for $\alpha > 0.50$, because the chi square distribution is not symmetrical.

EXAMPLE 53   **A probability calculation based on the χ^2 helps monitor variability**

Plastic sheeting produced by a machine must be periodically monitored for possible fluctuations in thickness. Uncontrollable variation in the viscosity of the liquid mold produces some variation in thickness. Based on experience with a great many samples, when the machine is working well, an observation on thickness has a normal distribution with standard deviation $\sigma = 1.35$ mm.

Samples of 20 thickness measurements are collected regularly. A value of the sample standard deviation exceeding 1.4 mm signals concern about the product. Find the probability that, when $\sigma = 1.35$, the next sample will signal concern about the product.

Solution   The chi square statistic

$$\chi^2 = \frac{(n-1)s^2}{\sigma^2} = \frac{19 \cdot 1.4^2}{1.2^2} = 30.6$$

From Table 5W, for 19 degrees of freedom, $\chi_{0.05}^2 = 30.1$. The probability of a false signal of concern is less than 0.05. In the long run, a false signal will occur less than 5 times in 100 samples.   ∎

[Using **R**:   For the upper tail, use $1 - .05 = .95$ in **qchisq(39.6,19)** and **1 - pchisq(30.6, 19)**]

A problem closely related to that of finding the distribution of the sample variance is that a finding the distribution of the ratio of the variances of two independent random samples. This problem is important because it arises in tests in which we want to determine whether two samples come from populations having equal variances. If they do, the two sample variances should be nearly the same; that is, their ratio should be close to 1. To determine whether the ratio of two sample variances is too small or too large, we use the F distribution.

> **Theorem 6.6** If S_1^2 and S_2^2 are the variances of independent random samples of size n_1 and n_2, respectively, taken from two normal populations having the same variance, then
>
> $$F = \frac{S_1^2}{S_2^2}$$
>
> is a random variable having the F distribution with the parameters $\nu_1 = n_1 - 1$ and $\nu_2 = n_2 - 1$.

A random variable having the F distribution

The **F distribution** is related to the beta distribution (page 147), and its two parameters, ν_1 and ν_2, are called the **numerator** and **denominator degrees of freedom**. As it would require too large a table to give values of F_α corresponding to many different right-hand tail probabilities α, and since $\alpha = 0.05$ and $\alpha = 0.01$ are most commonly used in practice, Table 6W contains only values $F_{0.05}$ and $F_{0.01}$ for various combinations of values of ν_1 and ν_2. (See also Figure 6.8.)

F distribution with ν_1 and ν_2 degrees of freedom

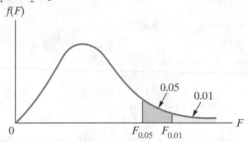

Figure 6.8
Tabulated values of F

EXAMPLE 54   **Using the F distribution, Table 6W, to evaluate a probability**

If two independent random samples of size $n_1 = 7$ and $n_2 = 13$ are taken from a normal population, what is the probability that the variance of the first sample will be at least three times as large as that of the second sample?

Solution   From Table 6W we find that $F_{0.05} = 3.00$ for $\nu_1 = 7 - 1 = 6$ and $\nu_2 = 13 - 1 = 12$; thus, the desired probability is 0.05. ∎

[Using **R**: For the upper tail, use $1 - .05 = .95$ in   **qf(.95, 6, 12)**]
    It is possible to use Table 6W also to find values of F corresponding to left-hand tail probabilities of 0.05 or 0.01. Writing $F_\alpha(\nu_1, \nu_2)$ for F_α with ν_1 and ν_2 degrees of freedom, we simply use the identity

$$F_{1-\alpha}(\nu_1, \nu_2) = \frac{1}{F_\alpha(\nu_2, \nu_1)}$$

EXAMPLE 55   **Using the F distribution, Table 6W, to find a left-hand tail probability**

Find the value of $F_{0.95}$ (corresponding to a left-hand tail probability of 0.05) for $\nu_1 = 10$ and $\nu_2 = 20$ degrees of freedom.

Solution   Making use of the identity and Table 6W, we get

$$F_{0.95}(10, 20) = \frac{1}{F_{0.05}(20, 10)} = \frac{1}{2.77} = 0.36$$ ∎

[Using **R**: For the lower tail, use $1 - .95 = .05$ in   **qf(.05,10,20)**]

Note that Theorems 6.4 and 6.5 require the assumption that we are sampling from normal populations. Unlike the situation with the t distribution, deviations from an underlying normal distribution, such as a long tail, may have a serious effect on these sampling distributions. Consequently, it is best to transform to near normality using the approach in Section 5.13 before invoking the sampling distributions in this section.

Exercises

6.20 The tensile strength (1,000 psi) of a new composite can be modeled as a normal distribution. A random sample of size 25 specimens has mean $\bar{x} = 45.3$ and standard deviation $s = 7.9$. Does this information tend to support or refute the claim that the mean of the population is 40.5?

6.21 The following are the times between 6 calls for an ambulance (in a certain city) and the patient's arrival at the hospital: 27, 15, 20, 32, 18, and 26 minutes. Use these figures to judge the reasonableness of the ambulance service's claim that it takes on the average of 20 minutes between the call for an ambulance and the patient's arrival at the hospital.

6.22 A process for making certain bearings is under control if the diameters of the bearings have a mean of 0.5000 cm. What can we say about this process if a sample of 10 of these bearings has a mean diameter of 0.5060 cm and a standard deviation of 0.0040 cm?

6.23 Hard disks for computers must spin evenly, and one departure from level is called pitch. Samples are regularly taken from production and each disk in the sample is placed in test equipment that yields a measurement of pitch. From many samples, it is concluded that the population is normal. The variance is $\sigma^2 = 0.065$ when the process is in control. A sample of size 10 is collected each week. The process will be declared out of control if the sample variance exceeds 0.122. What is the probability it will be declared out of control even though $\sigma^2 = 0.065$?

6.24 A random sample of 10 observations is taken from a normal population having the variance $\sigma^2 = 42.5$. Find the approximate probability of obtaining a sample standard deviation between 3.14 and 8.94.

6.25 If independent random samples of size $n_1 = n_2 = 8$ come from normal populations having the same variance, what is the probability that either sample variance will be at least 7 times as large as the other?

6.26 Find the values of

(a) $F_{0.95}$ for 15 and 12 degrees of freedom;

(b) $F_{0.99}$ for 5 and 20 degrees of freedom.

6.27 The chi square distribution with 4 degrees of freedom is given by

$$f(x) = \begin{cases} \dfrac{1}{4} \cdot x \cdot e^{-x/2} & x > 0 \\ 0 & x \le 0 \end{cases}$$

Find the probability that the variance of a random sample of size 5 from a normal population with $\sigma = 15$ will exceed 180.

6.28 The t distribution with 1 degree of freedom is given by

$$f(t) = \frac{1}{\pi}(1 + t^2)^{-1} \quad -\infty < t < \infty$$

Verify the value given for $t_{0.05}$ for $\nu = 1$ in Table 4.

6.29 The F distribution with 4 and 4 degrees of freedom is given by

$$f(F) = \begin{cases} 6F(1 + F)^{-4} & F > 0 \\ 0 & F \le 0 \end{cases}$$

If random samples of size 5 are taken from two normal populations having the same variance, find the probability that the ratio of the larger to the smaller sample variance will exceed 3.

6.5 Representations of the Normal Theory Distributions

The basic distributions of normal theory can all be defined in terms of independent standard normal random variables. The defining of a new random variable in terms of others is called a **representation**.

Let $Z, Z_1, Z_2, \ldots$, be independent standard normal random variables with mean 0 and variance 1. First, we define a chi square variable.

Representation of chi square random variable

> Let $Z_1, Z_2, \ldots, Z_\nu$ be independent standard normal random variables.
>
> $\chi^2 =$ sum of squares of ν independent standard normal variables
>
> $= \sum_{i=1}^{\nu} Z_i^2$ has a chi square distribution with ν degrees of freedom.

Consider two chi square random variables which have the representations $\chi_1^2 = \sum_{i=1}^{\nu_1} Z_i^2$ and $\chi_2^2 = \sum_{i=\nu_1+1}^{\nu_1+\nu_2} Z_i^2$. Since they depend on different sets of Z_i's, they are independent. Adding these two representations we conclude, as in Exercise 6.33, that the **sum of two independent chi square variables, $\chi_1^2 + \chi_2^2$, has a chi square distribution with degrees of freedom $\nu_1 + \nu_2$.**

Next, since $\chi^2 = \sum_{i=1}^{\nu} Z_i^2$ depends only on $Z_1, Z_2, \ldots, Z_\nu$ and they are independent of Z this χ^2 and Z are independent. We define a t random variable in terms of two independent random variables Z and χ^2.

Representation of t random variable

> Let the standard normal Z and chi square (χ^2), having ν degrees of freedom, be independent.
>
> $$t = \frac{\text{standard normal}}{\sqrt{\dfrac{\text{chi square}}{\text{degrees of freedom}}}} = \frac{Z}{\sqrt{\dfrac{\chi^2}{\nu}}} = \frac{Z}{\sqrt{\dfrac{\sum_{i=1}^{\nu} Z_i^2}{\nu}}}$$
>
> has a t distribution with ν degrees of freedom.

We define an F random variable in terms of two independent chi square variables χ_1^2 and χ_2^2 with ν_1 and ν_2 degrees of freedom, respectively.

Representation of F random variable

> Let the chi square variables χ_1^2, with ν_1 degrees of freedom, and χ_2^2, with ν_2 degrees of freedom, be independent.
>
> $$F_{\nu_1, \nu_2} = \frac{\dfrac{\text{chi square}}{\text{degrees of freedom}}}{\dfrac{\text{chi square}}{\text{degrees of freedom}}} = \frac{\dfrac{\chi_1^2}{\nu_1}}{\dfrac{\chi_2^2}{\nu_2}} = \frac{\dfrac{\sum_{i=1}^{\nu_1} Z_i^2}{\nu_1}}{\dfrac{\sum_{i=\nu_1+1}^{\nu_1+\nu_2} Z_i^2}{\nu_2}}$$
>
> has an F distribution with (ν_1, ν_2) degrees of freedom.

The basic case arises starting with n independent normal random variables $X_1, X_2, \ldots, X_n$ all having the same mean μ and standard deviation σ. Then

$$Z_i = \frac{X_i - \mu}{\sigma}$$

has a standard normal distribution for each i. It then holds that

$$\sqrt{n}\,\overline{Z} = \sqrt{n}\,\frac{1}{n}\sum_{i=1}^{n} Z_i = \sqrt{n}\left(\frac{\overline{X} - \mu}{\sigma}\right)$$

has a standard normal distribution.

Next,

$$\sum_{i=1}^{n} Z_i^2 = \sum_{i=1}^{n} (Z_i - \bar{Z} + \bar{Z})^2 = \sum_{i=1}^{n} (Z_i - \bar{Z})^2 + n\bar{Z}^2$$

The left-hand side has a chi square distribution with n degrees of freedom. The last term on the right is the square of the standard normal variable $\sqrt{n}\,\bar{Z}$ and so has a chi square distribution with 1 degree of freedom. It can be shown that the two terms on the right-hand side of the equation are independent and that

$$\sum_{i=1}^{n} (Z_i - \bar{Z})^2$$

has a chi square distribution with $n - 1$ degrees of freedom. Since

$$\frac{(n-1)S^2}{\sigma^2} = \sum_{i=1}^{n} \frac{(X_i - \bar{X})^2}{\sigma^2} = \sum_{i=1}^{n} (Z_i - \bar{Z})^2$$

we conclude that $(n - 1)S^2/\sigma^2$ has a chi square distribution with $n - 1$ degrees of freedom.

EXAMPLE 56 t^2 **has an F distribution**

Let t be distributed as a t distribution with ν degrees of freedom.

(a) Use the representation of t to show that t^2 has an F distribution with $(1, \nu)$ degrees of freedom.

(b) Use part (a) to show that $t_{\alpha/2} = F_\alpha(1, \nu)$.

Solution **(a)** Using the representation of a t random variable,

$$t^2 = \left(\frac{Z}{\sqrt{\dfrac{\chi^2}{\nu}}} \right)^2 = \frac{Z^2}{\dfrac{\chi^2}{\nu}}$$

Since, by the first representation above, Z^2 has a chi square distribution with 1 degree of freedom and it is independent of the denominator, we confirm the representation of the F distribution with $(1, \nu)$ degrees of freedom.

(b) $$1 - \alpha = P(-t_{\alpha/2} \le t \le t_{\alpha/2}) = P(t^2 \le t_{\alpha/2}^2)$$

By part (a), $t^2 = F$ so we have

$$1 - \alpha = P(F \le t_{\alpha/2}^2)$$

Because $t_{\alpha/2}^2$ satisfies the definition of $F_\alpha(1, \nu)$, the two must be equal. ∎

Exercises

6.30 Let $Z_1, \ldots, Z_5$ be independent and let each have a standard normal distribution.

(a) Specify the distribution of $Z_2^2 + Z_3^2 + Z_4^2 + Z_5^2$.

(b) Specify the distribution of

$$\frac{Z_1}{\sqrt{\dfrac{Z_2^2 + Z_3^2 + Z_4^2 + Z_5^2}{4}}}$$

6.31 Let $Z_1, \ldots, Z_6$ be independent and let each have a standard normal distribution. Specify the distribution of

$$\frac{Z_1 - Z_2}{\sqrt{\dfrac{Z_3^2 + Z_4^2 + Z_5^2 + Z_6^2}{8}}}$$

6.32 Let $Z_1, \ldots, Z_7$ be independent and let each have a standard normal distribution.

(a) Specify the distribution of $Z_1^2 + Z_2^2 + Z_3^2 + Z_4^2$.

(b) Specify the distribution of $Z_5^2 + Z_6^2 + Z_7^2$.

(c) Specify the distribution of the sum of variables in part (a) and part (b).

6.33 Let the chi square variables χ_1^2, with ν_1 degrees of freedom, and χ_2^2, with ν_2 degrees of freedom, be independent. Establish the result on page 201, that their sum is a chi square variable with $\nu_1 + \nu_2$ degrees of freedom.

6.6   The Moment Generating Function Method to Obtain Distributions*

The mgf method is a very convenient tool for obtaining the distribution function of the sum of independent random variables. Let X_1 have mgf $M_1(t)$, X_2 have mgf $M_2(t)$, X_3 have mgf $M_3(t)$, and so on. Then, by independence, the mgf of the sum $X_1 + X_2 + X_3$ is

$$M_{X_1+X_2+X_3}(t) = E(e^{t(X_1+X_2+X_3)}) = E(e^{tX_1}e^{tX_2}e^{tX_3})$$
$$= E(e^{tX_1}) \cdot E(e^{tX_2}) \cdot E(e^{tX_3})$$

or

$$M_{X_1+X_2+X_3}(t) = M_1(t) \cdot M_2(t) \cdot M_3(t)$$

For any number of independent random variables, we have the following result.

Moment generating function for sum of n independent random variables

> **Theorem 6.7**   Let $X_1, \ldots, X_n$ be independent random variables and let X_i have moment generating function $M_{X_i}(t)$ for $i = 1, \ldots, n$, where all moment generating functions exist for all $|t| \leq T$ some $T > 0$. Then the moment generating function of the sum exists for all $t \leq T$ and
>
> $$M_{X_1+X_2+\cdots+X_n}(t) = M_{X_1}(t) \cdot M_{X_2}(t) \cdots M_{X_n}(t)$$

The mgf of the sum of random variables is the product of the component mgf, under independence. When the product can be identified, we know the distribution of the sum. This argument is called the **moment generating function method**.

EXAMPLE 57   **Sum of n independent normal random variables is normal**

Let X_1 be $N(\mu_1, \sigma_1^2)$, X_2 be $N(\mu_2, \sigma_2^2)$, and X_3 be $N(\mu_3, \sigma_3^2)$, where the three random variables are independent.

(a) Find the distribution of $X_1 + X_2 + X_3$.

(b) Let X_i be $N(\mu_i, \sigma_i^2)$, for $i = 1, 2, \ldots, n$ and let the X_i be independent. Show that the distribution of their sum, $\sum_{i=1}^{n} X_i$, is normal with

$$\text{mean} = \sum_{i=1}^{n} \mu_i$$

$$\text{variance} = \sum_{i=1}^{n} \sigma_i^2$$

*This section may be skipped on first reading. Some key sampling distributions are verified.

Solution   **(a)** We know that X_1 has mgf

$$M_1(t) = e^{t\mu_1 + \frac{1}{2}t^2\sigma_1^2}$$

so

$$M_{X_1+X_2+X_3}(t) = e^{t\mu_1 + \frac{1}{2}t^2\sigma_1^2} \cdot e^{t\mu_2 + \frac{1}{2}t^2\sigma_2^2} \cdot e^{t\mu_3 + \frac{1}{2}t^2\sigma_3^2}$$

$$= e^{t(\mu_1+\mu_2+\mu_3) + \frac{1}{2}t^2(\sigma_1^2+\sigma_2^2+\sigma_3^2)}$$

This last form we identify as being an $N(\mu_1 + \mu_2 + \mu_3, \sigma_1^2 + \sigma_2^2 + \sigma_3^2)$. That is, the sum has a normal distribution where the mean is the sum of the component means and the variance is the sum of the variances.

(b)

$$M_{\sum_{i=1}^n X_i}(t) = \prod_{i=1}^n e^{t\mu_i + \frac{1}{2}t^2\sigma_i^2} = e^{t\left(\sum_{i=1}^n \mu_i\right) + \frac{1}{2}t^2\left(\sum_{i=1}^n \sigma_i^2\right)}$$

so the sum has a normal distribution with mean equal the sum of the component means and variance equal to the sum of variances. ∎

EXAMPLE 58   **Sum of independent Poisson random variables is Poisson**

Let X_i have a Poisson distribution with parameter λ_i, for $i = 1, 2, \ldots, n$ and let the X_i be independent. Show that the distribution of the their sum, $\sum_{i=1}^n X_i$, is Poisson with parameter

$$\lambda = \sum_{i=1}^n \lambda_i$$

Solution   We know that X_i has mgf

$$M_i(t) = e^{-\lambda_i + \lambda_i e^t}$$

Consequently,

$$M_{\sum_{i=1}^n X_i}(t) = \prod_{i=1}^n e^{-\lambda_i + \lambda_i e^t} = e^{\left(\sum_{i=1}^n \lambda_i\right) + \left(\sum_{i=1}^n \lambda_i\right)e^t}$$

This is the mgf of a Poisson distribution with parameter $\sum_{i=1}^n \lambda_i$. ∎

EXAMPLE 59   **Sum of chi square random variables is chi square**

Let X_i have a chi square distribution with ν_i degrees of freedom, for $i = 1, 2, \ldots, n$ and let the X_i be independent. Use the moment generating function method to show that the distribution of their sum, $\sum_{i=1}^n X_i$ is chi square with degrees of freedom $\sum_{i=1}^n \nu_i$.

Solution   **(a)** We know from Example 38, Chapter 5, that X_i has mgf $(1 - 2t)^{-\nu_i/2}$ so that $\sum_{i=1}^n X_i$ has mgf

$$M_{\sum_{i=1}^n X_i}(t) = \prod_{i=1}^n \frac{1}{(1-2t)^{\nu_i/2}} = \frac{1}{(1-2t)^{\sum_{i=1}^n \nu_i/2}}$$

which we identify as the mgf of a chi square distribution with $\sum_{i=1}^n \nu_i$ degrees of freedom. ∎

Exercises

6.34 Let $X_1, X_2, \ldots, X_8$ be 8 independent random variables. Find the moment generating function

$$M_{\sum X_i}(t) = E(e^{t(X_1+X_2+\cdots+X_8)})$$

of the sum when X_i has a Poisson distribution with mean

(a) $\lambda_i = 0.5$

(b) $\lambda_i = 0.04$

6.35 Let $X_1, X_2, \ldots, X_5$ be 5 independent random variables. Find the moment generating function

$$M_{\sum X_i}(t) = E(e^{t(X_1+X_2+\cdots+X_5)})$$

of the sum when X_i has a gamma distribution with $\alpha_i = 2i$ and $\beta_i = 2$.

6.36 Let $X_1, X_2,$ and X_3 be independent normal variables with

$$E(X_1) = 2 \quad \text{and} \quad \sigma_1^2 = 4$$
$$E(X_2) = 1 \quad \text{and} \quad \sigma_2^2 = 9$$
$$E(X_3) = -1 \quad \text{and} \quad \sigma_3^2 = 1$$

(a) Show that $X_1 + 2X_2 - 3X_3$ has a normal distribution.

(b) Find the mean and variance of the random variable in part (a).

6.37 Refer to Exercise 6.36.

(a) Show that $X_1 - 3X_2 + 2X_3 - 5$ has a normal distribution.

(b) Find the mean and variance of the random variable in part (a).

6.38 Let $X_1, X_2,$ and X_3 be independent normal variables with

$$E(X_1) = -4 \quad \text{and} \quad \sigma_1^2 = 1$$
$$E(X_2) = 0 \quad \text{and} \quad \sigma_2^2 = 4$$
$$E(X_3) = 3 \quad \text{and} \quad \sigma_3^2 = 1$$

(a) Show that $2X_1 - X_2 + 5X_3$ has a normal distribution.

(b) Find the mean and variance of the random variable in part (a).

6.39 Refer to Exercise 6.38.

(a) Show that $7X_1 + X_2 - 2X_3 + 7$ has a normal distribution.

(b) Find the mean and variance of the random variable in part (a).

6.40 Let $X_1, X_2, \ldots, X_r$ be r independent random variables each having the same geometric distribution.

(a) Show that the moment generating function $M_{\sum X_i}(t) = E(e^{t(X_1+X_2+\cdots+X_r)})$ of the sum is

$$[pe^t/(1-(1-p)e^t)]^r$$

(b) Relate the sum to the total number of trials to obtain r successes. This distribution, is given by

$$\binom{x-1}{r-1} p^r q^{x-r}, x = r, r+1, \cdots$$

(see page 115)

(c) Obtain the first two moments of this negative binomial by differentiating the mgf.

6.41 Refer to Exercise 6.40. Let $X_1, X_2, \ldots, X_n$ be n independent random variables each having a negative binomial distribution with success probability p but where X_i has parameter r_i.

(a) Show that the mgf $M_{\sum X_i}(t) = E(e^{t(X_1+X_2+\cdots+X_r)})$ of the sum $\sum X_i$ is

$$[pe^t/(1-(1-p)e^t)]^{\sum_{i=0}^n r_i}$$

(b) Identify the form of this mgf and specify the distribution of $\sum X_i$.

6.7 Transformation Methods to Obtain Distributions*

We briefly[1] introduce two further techniques for obtaining the probability distribution, or density, of a random variable that is a function of a random variable whose distribution is known. These are the **distribution function method** and the **transformation method**.

*This section may be skipped on first reading since the techniques are not used later in the book.

[1] An extended Section 6.7 is available at the Pearson Math & Stats Web site.

Distribution Function Method

The approach of the **distribution function method** is to first obtain the distribution function $G(y)$ of $Y = h(X)$, where X has known distribution function $F(x)$. The density, if needed, can be obtained by differentiation.

$$G(y) = P(Y \leq y) = P[h(X) \leq y]$$

Two examples will illustrate this method.

EXAMPLE 60   **The probability integral transformation**

Let X have distribution function $F(x)$ and density function $f(x)$ which is positive on an open interval and 0 elsewhere. Consider the **probability integral transformation** $Y = F(X)$ where the cumulative distribution distribution is evaluated at the random variable X. Show that $F(X)$ has a uniform distribution on (0, 1).

Solution   Choose any value y between 0 and 1. Since $F(x)$ has a positive derivative, there is a unique value x such that $F(x) = y$. This correspondence can be written as a function $x = w(y)$ and $F(w(y)) = y$ for all $0 < y < 1$. Then,

$$\begin{aligned} G(y) = P(Y \leq y) &= P(F(X) \leq y) \\ &= P(X \leq w(y)) \\ &= F(w(y)) = y \end{aligned}$$

for any $0 < y < 1$. The cumulative distribution function $G(y) = y$ is that of the uniform distribution. ∎

EXAMPLE 61   **Distribution function method applied to X^2**

Let X have distribution function $F(x)$ and density function $f(x)$.

(a) Show that its square, $Y = X^2$, has distribution function

$$G(y) = P(Y \leq y) = F(\sqrt{y}) - F(-\sqrt{y})$$

(b) If X has a standard normal distribution, show that its square has

$$g(y) = \frac{1}{\sqrt{2\pi}} y^{-1/2} e^{-y/2}$$

which is a chi square distribution with 1 degree of freedom

Solution   **(a)** We have

$$\begin{aligned} G(y) = P(Y \leq y) &= P(X^2 \leq y) \\ &= P(-\sqrt{y} \leq X \leq \sqrt{y}) \\ &= F(\sqrt{y}) - F(-\sqrt{y}) \end{aligned}$$

(b) Upon differentiating,

$$g(y) = f(\sqrt{y}) \frac{d\sqrt{y}}{dy} - f(-\sqrt{y}) \frac{d-\sqrt{y}}{dy} = f(\sqrt{y}) \frac{y^{-1/2}}{2} + f(-\sqrt{y}) \frac{y^{-1/2}}{2}$$

The reader is asked, in Exercise 6.42, to verify the stated density. ∎

For any differentiable strictly increasing function $h(x)$, with inverse function $w(y)$, we have $G(y) = F(w(y))$. By taking the derivative of both sides with respect to y, we obtain the expression for the density function of $h(X)$ presented next when discussing the transformation method.

Transformation Method

The **transformation method** expresses the probability density for a function of a random variable in terms of the density of the original variable. Consider $Y = h(X)$ where X has density $f(x)$. Initially, we assume that $h(x)$ is differentiable and either strictly increasing or strictly decreasing. Then, $y = h(x)$ can be solved for x. That is, $h(x)$ has an inverse $w(y) = x$.

Density function of h(X)

> **Theorem 6.8**   Let $Y = h(X)$ where X has density $f(x)$, and let $h(x)$ be differentiable and either strictly increasing or strictly decreasing on the range where $f(x) \neq 0$. The inverse function $w(y)$ exists and the density of Y is given by
>
> $$g(y) = \begin{cases} f(w(y))|w'(y)| & \text{where } w'(y) \neq 0 \\ 0 & \text{elsewhere} \end{cases}$$

EXAMPLE 62   **Transformation method: square root of chi square / degrees of freedom**

Let X have a chi square distribution with ν degrees of freedom. Apply the transformation to show that the density of $Y = \sqrt{X/\nu}$ is

$$\frac{\nu^{\nu/2}}{\Gamma(\frac{\nu}{2})2^{(\nu-2)/2}} y^{\nu-1} e^{-\nu y^2/2}, \; y > 0$$

Solution   The density of the chi square distribution with ν degrees of freedom is given by

$$f(x) = \frac{1}{\Gamma(\frac{\nu}{2})2^{\nu/2}} x^{\frac{\nu}{2}-1} e^{-x/2}$$

and $y = \sqrt{x/\nu} = h(x)$ has inverse $x = \nu y^2 = w(y)$. Since $w'(y) = 2\nu y$ is continuous and greater than 0 for $y > 0$,

$$g(y) = \frac{1}{\Gamma(\frac{\nu}{2})2^{\nu/2}} (\nu y^2)^{\frac{\nu}{2}-1} e^{-\nu y^2/2} 2\nu y$$

$$= \frac{\nu^{\nu/2}}{\Gamma(\frac{\nu}{2})2^{(\nu-2)/2}} y^{\nu-1} e^{-\nu y^2/2} \quad \blacksquare$$

We state two important transformations to obtain the sum or the ratio of two independent random variables.

> **Theorem 6.9**   Let X and Y be independent and let X have density $f_X(x)$ and Y have density $f_Y(y)$. Then the density of $Z = X + Y$ is given by the **convolution formula**
>
> **Convolution formula**
> $$f_{X+Y}(z) = \int_{-\infty}^{\infty} f_X(x)f_Y(z-x)\,dx \text{ for all } z$$
>
> The ratio of random variables $Z = Y/X$ has density
> $$f_{Y/X}(z) = \int_{-\infty}^{\infty} |x| f_X(x)f_Y(xz)\,dx \text{ for all } z$$

EXAMPLE 63   **Student's t distribution**

Let Y have a standard normal distribution and be independent of X which has a chi square distribution with v degrees of freedom. Apply the transformation technique to show that the density of

$$\frac{\text{standard normal}}{\sqrt{\dfrac{\text{chi square}}{\text{degrees of freedom}}}} = \frac{Y}{\sqrt{\dfrac{X}{v}}}$$

is given by

$$\frac{\Gamma\left(\frac{v+1}{2}\right)}{\sqrt{\pi v}\,\Gamma\left(\frac{v}{2}\right)}\left(1 + \frac{t^2}{v}\right)^{-\frac{v+1}{2}} \qquad \text{for } -\infty < t < \infty$$

This distribution is called the **student's t distribution**, or the **t distribution**.

Solution   Since Y has a standard normal distribution, using the conclusion from Example 17 but with Y replaced by X, the density of the ratio $T = Y/X$ is

$$\int_{-\infty}^{\infty} |x|\, f_X(x) f_Y(xt)\, dx$$

$$= \int_{0}^{\infty} |x|\, \frac{1}{\sqrt{2\pi}} e^{-(tx)^2/2}\, \frac{v^{v/2}}{\Gamma(\frac{v}{2}) 2^{(v-2)/2}} x^{v-1} e^{-vx^2/2}\, dx$$

$$= \int_{0}^{\infty} \frac{v^{v/2}}{\sqrt{\pi}\, \Gamma(\frac{v}{2}) 2^{(v-1)/2}} x^{v}\, e^{-\frac{1}{2}(v+t^2)x^2}\, dx$$

Making the change of variable $u = x^2(v + t^2)/2$, we obtain

$$(v + t^2)^{-\frac{v+1}{2}}\, \frac{v^{v/2}}{\sqrt{\pi}\, \Gamma(\frac{v}{2})} \int_{0}^{\infty} u^{\frac{v-1}{2}}\, e^{-u}\, du$$

and the result follows from the definition of $\Gamma\left(\frac{v+1}{2}\right)$.   ∎

Convolution Formula for Discrete Random Variables

There is also a convolution formula for the sum, $Z = X + Y$, of two independent discrete random variables X and Y. Let $f_X(x)$ denote the probability distribution of X and $f_Y(y)$ denote the probability distribution of Y. We restrict attention to cases where X and Y take on nonnegative integer values.

To find $f_Z(z) = P(Z = z)$, for each z, we recognize that the event $[Z = z]$ is the union of the disjoint events $[X = x$ and $Y = z - x]$ for $x = 0, 1, \ldots, z$. Consequently,

$$P(Z = z) = f_Z(z) = \sum_{x=0}^{z} P(X = x \text{ and } Y = z - x)$$

$$= \sum_{x=0}^{z} f_X(x) f_Y(z - x)$$

where the last step follows by independence. This last result is called the **convolution formula** for discrete random variables.

> **Theorem 6.10**   Let X and Y be non-negative and integer valued. The random variable $Z = X + Y$ has probability distribution $f_Z(z)$ given by
>
> **Discrete convolution formula**
>
> $$f_Z(z) = f_{X+Y}(z) = P(Z = z) = \sum_{x=0}^{z} f_X(x) f_Y(z - x) \text{ for } z = 0, 1, \dots$$

EXAMPLE 64   **Sum of two independent Poisson random variables**

Let X and Y be independent Poisson random variables where X has parameter λ_1 and Y has parameter λ_2. Show that the sum

$$X + Y \qquad \text{has a Poisson distribution with parameter} \qquad \lambda_1 + \lambda_2$$

Solution   By the discrete convolution formula, $Z = X + Y$ has probability distribution

$$f_Z(z) = \sum_{x=0}^{z} f_X(x) f_Y(z - x)$$

so

$$f_Z(z) = \sum_{x=0}^{z} \frac{\lambda_1^x}{x!} e^{-\lambda_1} \frac{\lambda_2^{z-x}}{(z - x)!} e^{-\lambda_2}$$

$$= e^{-(\lambda_1 + \lambda_2)} \sum_{x=0}^{z} \frac{\lambda_1^x}{x!} \frac{\lambda_2^{z-x}}{(z - x)!}$$

Using the binomial formula

$$(a + b)^m = \sum_{x=0}^{m} \binom{m}{x} a^x b^{m-x}$$

with $m = z$, $a = \lambda_1$, and $b = \lambda_2$, then multiplying and dividing by $z!$, we conclude that

$$\sum_{x=0}^{z} \frac{\lambda_1^x}{x!} \frac{\lambda_2^{z-x}}{(z - x)!} = \frac{(\lambda_1 + \lambda_2)^z}{z!}$$

and the result is established.

Remark: Note that the rate parameters λ_i add.   ∎

Exercises

6.42 Referring to Example 16, verify that

$$g(y) = \frac{1}{\sqrt{2\pi}} y^{-1/2} e^{-y/2}$$

6.43 Use the distribution function method to obtain the density of Z^3 when Z has a standard normal distribution.

6.44 Use the distribution function method to obtain the density of $1 - e^{-X}$ when X has the exponential distribution with $\beta = 1$.

6.45 Use the distribution function method to obtain the density of $\ln(X)$ when X has the exponential distribution with $\beta = 1$.

6.46 Use the transformation method to obtain the density of of X^2 when X has density $f(x) = 0.5x$ for $0 < x < 2$.

6.47 Use the transformation method to obtain the distribution of $-\ln(X)$ when X has the uniform distribution on $(0, 1)$.

6.48 Use the convolution formula, Theorem 6.9, to obtain the density of $X + Y$ when X and Y are independent and each has the exponential distribution with $\beta = 1$.

6.49 Use the transformation method, Theorem 6.9, to obtain the distribution of the ratio Y/X when when X and Y are independent and each has the same gamma distribution.

6.50 Use the discrete convolution formula, Theorem 6.10, to obtain the probability distribution of $X + Y$ when X and Y are independent and each has the uniform distribution on $\{0, 1, 2\}$.

Do's and Don'ts

Do's

1. Understand the concept of a sampling distribution. Each observation is the value of a random variable so a sample of n observations varies from one possible sample to another. Consequently, a statistic such as a sample mean varies from one possible sample to another. The probability distribution or density function which describes the chance behavior of the sample mean is called its *sampling distribution*.

2. When the underlying distribution has mean μ and variance σ^2, remember that the sampling distribution of $\overline{X}$ has

$$\text{mean of } \overline{X} = \mu = \text{population mean}$$
$$\text{variance of } \overline{X} = \frac{\sigma^2}{n} = \frac{\text{population variance}}{n}$$

3. When the underlying distribution is normal with mean μ and variance σ^2, calculate exact probabilities for $\overline{X}$ using the normal distribution with mean μ and variance $\dfrac{\sigma^2}{n}$

$$P(\overline{X} \le b) = P\left(Z \le \frac{b - \mu}{\sigma/\sqrt{n}}\right)$$

4. Apply the central limit theorem, when the sample size is large, to approximate the sampling distribution of $\overline{X}$ by a normal distribution with mean μ and variance $\dfrac{\sigma^2}{n}$. The probability $P(\overline{X} \le b)$ is approximately equal to the standard normal probability $P\left(Z \le \dfrac{b - \mu}{\sigma/\sqrt{n}}\right)$.

Don'ts

1. Don't confuse the population distribution, which describes the variation for a single random variable, with the sampling distribution of a statistic.

2. When sampling from a finite population of size N, don't use $\sigma/\sqrt{n}$ as the standard deviation of $\overline{X}$ unless the finite population correction factor is nearly 1.

3. When the population distribution is noticeably nonnormal, don't try to conclude that the sampling distribution of $\overline{X}$ is normal unless the sample size is at least moderately large, 30 or more.

Review Exercises

6.51 The panel for a national science fair wishes to select 10 states from which a student representative will be chosen at random from the students participating in the state science fair.

(a) Use Table 7W or software to select the 10 states.

(b) Does the total selection process give each student who participates in some state science fair an equal chance of being selected to be a representative at the national science fair?

6.52 How many different samples of size $n = 2$ can be chosen from a finite population of size

(a) $N = 12$;

(b) $N = 20$?

6.53 With reference to Exercise 6.52, what is the probability of choosing each sample in part (a) and the probability of choosing each sample in part (b), if the samples are to be random?

6.54 Referring to Exercise 6.52, find the value of the finite population correction factor in the formula for $\sigma_{\overline{X}}^2$ for part (a) and part (b).

6.55 The time to check out and process payment information at an office supplies Web site can be modeled as a random variable with mean $\mu = 63$ seconds and variance $\sigma^2 = 81$. If the sample mean $\overline{X}$ will be based on a random sample of $n = 36$ times, what can we assert about the probability of getting a sample mean greater than 66.75, if we use

(a) Chebyshev's theorem;

(b) the central limit theorem?

6.56 The number of pieces of mail that a department receives each day can be modeled by a distribution having mean 44 and standard deviation 8. For a random sample of 35 days, what can be said about the probability that the sample mean will be less than 40 or greater than 48 using

(a) Chebyshev's theorem;

(b) the central limit theorem?

6.57 If measurements of the specific gravity of a metal can be looked upon as a sample from a normal population having a standard deviation of 0.04, what is the probability that the mean of a random sample of size 25 will be "off" by at most 0.02?

6.58 Adding graphite to iron can improve its ductile qualities. If measurements of the diameter of graphite spheres within an iron matrix can be modeled as a normal distribution having standard deviation 0.16, what is the probability that the mean of a sample of size 36 will differ from the population mean by more than 0.06?

6.59 If 2 independent random samples of size $n_1 = 9$ and $n_2 = 16$ are taken from a normal population, what is the probability that the variance of the first sample will be at least 4 times as large as the variance of the second sample?

6.60 If 2 independent sample of sizes $n_1 = 26$ and $n_2 = 8$ are taken from a normal population, what is the probability that the variance of the second sample will be at least 2.4 times the variance of the first sample?

6.61 When we sample from an infinite population, what happens to the standard error of the mean if the sample size is

(a) increased from 100 to 200;

(b) increased from 200 to 300;

(c) decreased from 360 to 90?

6.62 A traffic engineer collects data on traffic flow at a busy intersection during the rush hour by recording the number of westbound cars that are waiting for a green light. The observations are made for each light change. Explain why this sampling technique will not lead to a random sample.

6.63 Explain why the following may not lead to random samples from the desired populations:

(a) To determine the smoothness of shafts, a manufacturer measures the roughness of the first piece made each morning.

(b) To determine the mix of cars, trucks, and buses in the rush hour, an engineer records the type of vehicle passing a fixed point at 1-minute intervals.

6.64 Several pickers are each asked to gather 30 ripe apples and put them in a bag.

(a) Would you expect all of the bags to weigh the same? For one bag, let X_1 be the weight of the first apple, X_2 the weight of the second apple, and so on. Relate the weight of this bag,

$$\sum_{i=1}^{30} X_i$$

to the approximate sampling distribution of $\overline{X}$.

(b) Explain how your answer to part (a) leads to the sampling distribution for the variation in bag weights.

(c) If the weight of an individual apple has mean $\mu = 0.2$ pound and standard deviation $\sigma = 0.03$ pound, find the probability that the total weight of the bag will exceed 6.2 pounds.

Key Terms

INFERENCES CONCERNING A MEAN

Recall from Chapter 6 that the purpose of most statistical investigations is to generalize from information contained in random samples about the populations from which the samples were obtained. In the classical approach the methods of statistical inference are divided into two major areas—estimation and tests of hypotheses. In Sections 7.2 and 7.3 we shall present some theory and some methods which pertain to the estimation of means. A general estimation procedure is introduced in Section 7.4. Sections 7.5, and 7.6 deal with the basic concepts of hypothesis testing, and Sections 7.7 and 7.8 deal with tests of hypotheses concerning a mean. Test performance, including power, is the subject of Section 7.9.

7.1 Statistical Approaches to Making Generalizations

To obtain new knowledge about a process or phenomena, relevant data must be collected. Usually, it is not possible to obtain a complete set of data but only a sample. Statistical inference arises whenever we need to make generalizations about a population on the basis of a sample. The main features of the sample can be described by the methods presented in Chapter 2. However, the central issue is not just the particular observed data but what can be said about the population that produced the sample. We call any generalization a **statistical inference** or just an **inference**.

The first step in making a statistical inference is to model the population by a probability distribution or density function that has a numerical feature of interest called a **parameter**. Earlier, we encountered parameters including μ and σ for normal distributions and p for binomial distributions. Next, a **statistic**, whose value can be calculated for every sample, serves as the source of information about a parameter. Any statistic, such as $\overline{X}$, S^2, or the sample median, is just a function of the sample.

Three points must be kept in mind when making inferences.

1. Because a sample is only part of the population, the numerical value of the statistic will not be the exact value of the parameter.
2. The observed value of the statistic depends on the particular sample selected.
3. Variability in the values of a statistic, over different samples, is unavoidable.

Statistical inferences are founded on an understanding of the manner is which variation in the population is transmitted, via sampling, to variation in a statistic.

How do we extract relevant information about the population by analyzing the sample? The two main classes of statistical inference are **estimation** of parameters and **testing hypotheses**. Estimation can be either a point estimator that gives a single number estimate of the value of the parameter or an interval estimate that specifies an interval of plausible values for the parameter. A test of hypotheses provides the answer to whether the data support or contradict an investigator's claim about the value of the parameter. Example 1 illustrates these three approaches to statistical inference.

EXAMPLE 1   **Types of Inference: point estimation, interval estimation, and testing hypotheses**

Refer to Example 14, Chapter 2, and the data on recycled concrete pavement. Green engineering practices require that their strength be accessed before using them in the base of roadways. Measurements of the resiliency modulus(MPa) on $n = 18$ specimens of recycled concrete aggregate produce the ordered values (Courtesy of Tuncer Edil).

136	143	147	151	158	160
161	163	165	167	173	174
181	181	185	188	190	205

The descriptive summary for the sample is

sample mean $\bar{x} = 168.2$ sample standard deviation $s = 18.10$
sample median 166   first quartile 158   third quartile 181

However, our goal here is not just the particular measurements recorded here but rather, it concerns the vast population of values for all possible recycled concrete pavements.

Discuss approaches for generalizing from this sample to the population.

Solution   We model the collection of values of the modulus, from all possible specimens of recycled concrete pavement, by a density function. The purpose of taking the sample is to learn about a feature of this unknown density function. The feature, or parameter, could be its mean μ or σ.

Concerning the parameter μ, we may wish to make one, two, or all three of following types of inference.

1. **Point estimation:** Estimate, by a single value, the unknown μ.
2. **Interval estimation:** Determine an interval of plausible values for μ.
3. **Testing hypotheses:** Determine whether or not the mean μ is 170 MPa, which is the mean value of an alternative material.   ∎

Logical deductions from the general to specific case are always correct. In contrast, when making statistical inferences, variability is unavoidable even when observations are made under the same, or nearly the same, conditions. Necessarily then, statistical inferences are based on a sample so they will sometimes be in error. An interval may not contain the value of parameter or the test of hypotheses may reach the wrong conclusion concerning the correctness of the hypothesis.

The realization that many highly variable observations can provide the basis for strong scientific evidence must be considered one the great intellectual advances of the twentieth century.

7.2   Point Estimation

Basically, **point estimation** concerns the choosing of a statistic, that is, a single number calculated from sample data. We should have some expectation, or assurance, that it is reasonably close to the parameter it is supposed to estimate. To explain what we mean here by reasonably close is not an easy task. First, the value of the parameter is unknown, and second, the value of the statistic is unknown until after the sample has been obtained. Thus, we can only ask whether, upon repeated sampling, the distribution of the statistic has certain desirable properties akin to closeness.

We know from Theorem 6.1 that the sampling distribution of the mean has the same mean as the population. This property suggests considering the sample mean $\overline{X}$ as a point estimator of the population mean μ. Closeness can then be expressed in terms of its standard deviation $\sigma/\sqrt{n}$. In the context of point estimation, we call this quantity the **standard error** and the value of its estimator $S/\sqrt{n}$ the **estimated standard error**.

Point estimation of a mean

Parameter: Population mean μ
Data: A random sample $X_1, \ldots, X_n$
Estimator: $\overline{X}$

Estimate of standard error: $\dfrac{S}{\sqrt{n}}$

EXAMPLE 2   **Point Estimation of the Stiffness of Recycled Road Material**

Refer to Example 1 and the data on recycled concrete pavement. Obtain a point estimate of μ, the mean resiliency modulus for recycled concrete. Also give the estimated standard error.

Solution   Our point estimator is $\overline{X}$ and its value for this sample

$$\overline{x} = 168.2 \text{ is the point estimate of } \mu.$$

The estimated standard error is

$$\frac{s}{\sqrt{n}} = \frac{18.10}{\sqrt{18}} = 4.27$$

where $s = 18.10$ MPa is given in Example 1.   ∎

Maximum Error of Estimate with High Probability

When we use a sample mean to estimate the mean of a population, we know that although we are using a method of estimation which has certain desirable properties, the chances are slim, virtually nonexistent, that the estimate will actually equal μ. Hence, it would seem desirable to accompany such a point estimate of μ with some statement as to how close we might reasonably expect the estimate to be. The error, $\overline{X} - \mu$, is the difference between the estimator and the quantity it is supposed to estimate.

To examine this error, let us make use of the fact that for large n

$$\frac{\overline{X} - \mu}{\sigma/\sqrt{n}}$$

is a random variable having approximately the standard normal distribution.

As illustrated in Figure 7.1, for any specified value of α

$$P\left(-z_{\alpha/2} \leq \frac{\overline{X} - \mu}{\sigma/\sqrt{n}} \leq z_{\alpha/2}\right) = 1 - \alpha$$

or, equivalently,

$$P\left(\frac{|\overline{X} - \mu|}{\sigma/\sqrt{n}} \leq z_{\alpha/2}\right) = 1 - \alpha$$

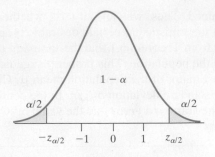

Figure 7.1
The sampling distribution
of $\dfrac{\overline{X} - \mu}{\sigma/\sqrt{n}}$

where $z_{\alpha/2}$ is such that the normal curve area to its right equals $\alpha/2$.

We now let E, called the **maximum error of estimate** stand for the maximum of these values of $|\overline{X} - \mu|$. Then, the error $|\overline{X} - \mu|$, will be less than

**Maximum error of
estimate**

$$E = z_{\alpha/2} \cdot \frac{\sigma}{\sqrt{n}}$$

with probability $1 - \alpha$. In other words, if we intend to estimate μ with the mean of a large ($n \geq 30$) random sample, we can assert with probability $1 - \alpha$ that the error, $|\overline{X} - \mu|$, will be at most $z_{\alpha/2} \cdot \dfrac{\sigma}{\sqrt{n}}$. The most widely used values for $1 - \alpha$ are 0.95 and 0.99, and the corresponding values of $z_{\alpha/2}$ are $z_{0.025} = 1.96$ and $z_{0.005} = 2.575$. (See Exercise 5.23 on page 140.)

EXAMPLE 3   **Specifying a high probability for the maximum error (σ known)**

An industrial engineer intends to use the mean of a random sample of size $n = 150$ to estimate the average mechanical aptitude (as measured by a certain test) of assembly line workers in a large industry. If, on the basis of experience, the engineer can assume that $\sigma = 6.2$ for such data, what can he assert with probability 0.99 about the maximum size of his error?

Solution   Substituting $n = 150$, $\sigma = 6.2$, and $z_{0.005} = 2.575$ into the preceding formula for E, we get

$$E = 2.575 \cdot \frac{6.2}{\sqrt{150}} = 1.30.$$

Thus, the engineer can assert with probability 0.99 that his error will be at most 1.30.   ∎

Suppose now that the engineer of this example collects his data and gets $\bar{x} = 69.5$. Can he still assert with probability 0.99 that the error is at most 1.30? First of all, $\bar{x} = 69.5$ either differs from the true average by at most 1.30 or it does not, and he does not know which. Consequently, it must be understood that the 0.99 probability applies to the method he used to determine the maximum error (getting the sample data and using the formula for E) and not directly to the parameter he is trying to estimate. To make this distinction, it has become the custom to use the word **confidence** here instead of *probability*. **In general, we make probability statements about future values of random variables (say, the potential error of an estimate) and confidence statements once the data have been obtained**. Accordingly, we would say in our example that the engineer can be 99% confident that the error of his estimate, $\bar{x} = 69.5$, is at most 1.30.

The methods discussed so far in this section require that σ be known or that it can be approximated with the sample standard deviation s, thus requiring that n be large. However, if it is reasonable to assume that we are sampling from a normal population, we can base our argument on Theorem 6.4 instead of Theorem 6.3, namely on the fact that

$$t = \frac{\overline{X} - \mu}{S/\sqrt{n}}$$

is a random variable having the t distribution with $n - 1$ degrees of freedom. Duplicating the steps on page 215, we thus arrive at the result that with probability $1 - \alpha$ the error we make in using $\overline{X}$ to estimate μ will be at most $t_{\alpha/2}S/\sqrt{n}$. Here $t_{\alpha/2}$ has probability $\alpha/2$ of being exceeded by a t random variable having $n - 1$ degrees of freedom. (See page 196.)

When $\overline{X}$ and S become available, we assert with $(1 - \alpha)100\%$ confidence that the error made in using $\bar{x}$ to estimate μ is at most

Maximum error of estimate, normal population (σ unknown)

$$E = t_{\alpha/2} \cdot \frac{s}{\sqrt{n}}$$

EXAMPLE 4   **A 98% confidence bound on the maximum error**

In six determinations of the melting point of an aluminum alloy, a chemist obtained a mean of 532.26 degrees Celsius with a standard deviation of 1.14 degree. If he uses this mean to estimate the actual melting point of the alloy, what can the chemist assert with 98% confidence about the maximum error?

Solution   Substituting $n = 6$, $s - 1.14$, and $t_{0.01} - 3.365$ (for $n - 1 = 5$ degrees of freedom) into the formula for E, we get

$$E = 3.365 \cdot \frac{1.14}{\sqrt{6}} = 4.24$$

Thus the chemist can assert with 98% confidence that his figure for the melting point of the aluminum alloy is off by at most 4.24 degrees.   ∎

Determination of Sample Size

The formula for E on page 216 can also be used to determine the sample size that is needed to attain a desired degree of precision. Suppose that we want to use the mean of a large random sample to estimate the mean of a population, and we want to be able to assert with probability $1 - \alpha$ that the error will be at most some prescribed quantity E [or assert later with $(1 - \alpha)100\%$ confidence that the error is at most E]. As before, we write

$$E = z_{\alpha/2} \cdot \frac{\sigma}{\sqrt{n}}$$

and upon solving this equation for n we get

Sample size determination

$$n = \left[\frac{z_{\alpha/2} \cdot \sigma}{E} \right]^2$$

To be able to use this formula we must know $1 - \alpha$, E, and σ, and for the latter we often substitute an estimate based on prior data of a similar kind (or, if necessary, a good guess).

EXAMPLE 5   **Selecting the sample size**

A research worker wants to determine the average time it takes a mechanic to rotate the tires of a car, and she wants to be able to assert with 95% confidence that the mean of her sample is off by at most 0.50 minute. If she can presume from past experience that $\sigma = 1.6$ minutes, how large a sample will she have to take?

Solution   Substituting $E = 0.50$, $\sigma = 1.6$, and $z_{0.025} = 1.96$ into the formula for n, we get

$$n = \left[\frac{1.96 \cdot 1.6}{0.50}\right]^2 = 39.3$$

or 40 rounded up to the nearest integer. Thus, the research worker will have to time 40 mechanics performing the task of rotating the tires of a car.   ∎

We know from Theorem 6.1 that the sampling distribution of the mean has the same mean as the population from which the sample is obtained. Hence, we can expect that the means of repeated random samples from a given population will center on the mean of this population and not about some other value.

To formulate this property more generally, let θ be the parameter of interest and $\widehat{\theta}$ be a statistic. The **hat notation** distinguishes the sample-based quantity from the parameter. We now make the following definition:

Unbiased estimator

> A statistic $\widehat{\theta}$ is said to be an **unbiased estimator**, or its value an unbiased estimate, if and only if the mean of the sampling distribution of the estimator $E(\widehat{\theta}) = \theta$, whatever the value of θ.

Thus, we call a statistic unbiased if "on the average" its values will equal the parameter it is supposed to estimate. Note that we have distinguished here between an estimator, a random variable, and an estimate, which is one of its values. Also, it is customary to apply the term *statistic* to both estimates and estimators.

It is a mathematical fact that $\overline{X}$ is an unbiased estimator of the population mean μ provided the observations are a random sample.

Generally speaking, the property of unbiasedness is one of the more desirable properties in point estimation, although it is by no means essential and it is sometimes out weighed by other factors. One shortcoming of the criterion of unbiasedness is that it will generally not provide a unique statistic for a given problem of estimation. For instance, it can be shown that for a random sample of size $n = 2$ the mean $\frac{X_1 + X_2}{2}$ as well as the weighted mean $\frac{aX_1 + bX_2}{a + b}$, where a and b are positive constants, are unbiased estimates of the mean of the population. If we further assume that the population is symmetric, so are the median and the midrange (the mean of the largest value and the smallest) for random samples of any size.

This suggests that we must seek a further criterion for deciding which of several unbiased estimators is best for estimating a given parameter. Such a criterion becomes evident when we compare the sampling distributions of the median and the mean for random samples of size n from the same normal population. The sampling distribution of the mean is normal and that of the median is nearly normal. Although these two sampling distributions have the same mean, the population mean μ, and although they are both symmetrical and bell-shaped, their variances differ. From Theorem 6.1, the variance of the sampling distribution of the mean for random samples from infinite populations is $\dfrac{\sigma^2}{n}$, and it can be shown that for random samples of the same size from normal populations the variance of the sampling distribution of the median is approximately $1.5708 \cdot \dfrac{\sigma^2}{n}$. Thus, it is more likely that the mean will be closer to μ than the median is to μ. Despite this long-run average property, given a particular sample we have no way of knowing which of the two is closest.

We formalize this important comparison of sampling distributions of statistics on the basis of their variances.

More efficient unbiased estimator

A statistic $\widehat{\theta}_1$ is said to be a more efficient unbiased estimator of the parameter θ than the statistic $\widehat{\theta}_2$ if

1. $\widehat{\theta}_1$ and $\widehat{\theta}_2$ are both unbiased estimators of θ;
2. the variance of the sampling distribution of the first estimator is no larger than that of the second and is smaller for at least one value of θ.

We have thus seen that for random samples from normal populations the mean $\overline{X}$ is more efficient than the median as an estimator of μ. In fact, it can be shown that in most practical situations where we estimate a population mean μ, the variance of the sampling distribution of no other unbiased statistic is less than that of the sampling distribution of the mean. In other words, in most practical situations the sample mean is an acceptable statistic for estimating a population mean μ. (There exist several other criteria for assigning the goodness of methods of point estimation, but we shall not discuss them in this book.)

7.3 Interval Estimation

Since point estimates cannot really be expected to coincide with the quantities they are intended to estimate, it is sometimes preferable to replace them with **interval estimates**. That is, with intervals for which we can assert with a reasonable degree of certainty that they will contain the parameter under consideration. To illustrate the construction of such an interval, suppose that we have a large ($n > 30$) random sample from a population with the unknown mean μ and the known variance σ^2.

Referring to the probability statement

$$P\left(-z_{\alpha/2} \leq \frac{\overline{X} - \mu}{\sigma/\sqrt{n}} \leq z_{\alpha/2}\right) = 1 - \alpha$$

shown on page 215, and rewriting the event as

$$\left[-z_{\alpha/2}\frac{\sigma}{\sqrt{n}} \leq \overline{X} - \mu \leq z_{\alpha/2}\frac{\sigma}{\sqrt{n}}\right] = \left[\overline{X} - z_{\alpha/2}\frac{\sigma}{\sqrt{n}} \leq \mu \leq \overline{X} + z_{\alpha/2}\frac{\sigma}{\sqrt{n}}\right]$$

we have

$$P\left(\overline{X} - z_{\alpha/2}\frac{\sigma}{\sqrt{n}} \leq \mu \leq \overline{X} + z_{\alpha/2}\frac{\sigma}{\sqrt{n}}\right) = 1 - \alpha$$

This last probability statement concerns a random interval covering the unknown parameter μ with probability $1 - \alpha$.

Before the data are obtained, we write the event as $\overline{X} - z_{a/2}\, \sigma/\sqrt{n} < \mu < \overline{X} + z_{a/2} \cdot \sigma/\sqrt{n}$. When the observed value $\bar{x}$ becomes available, we obtain

Large sample confidence interval for μ (σ known)

$$\bar{x} - z_{\alpha/2} \cdot \frac{\sigma}{\sqrt{n}} < \mu < \bar{x} + z_{\alpha/2} \cdot \frac{\sigma}{\sqrt{n}}$$

Thus, when a sample has been obtained and the value of $\bar{x}$ has been calculated, we can claim with $(1 - \alpha)100\%$ confidence that the interval from $\bar{x} - z_{\alpha/2} \cdot \dfrac{\sigma}{\sqrt{n}}$ to $\bar{x} + z_{\alpha/2} \cdot \dfrac{\sigma}{\sqrt{n}}$ contains μ. It is customary to refer to an interval of this kind as a **confidence interval** for μ having the **degree of confidence** $1 - \alpha$ or $(1 - \alpha)\,100\%$ and to its endpoints as the **confidence limits**.

EXAMPLE 6   **Calculating and interpreting a large sample confidence interval**

A random sample of size $n = 100$ is taken from a population with $\sigma = 5.1$. Given that the sample mean is $\bar{x} = 21.6$, construct a 95% confidence interval for the population mean μ.

Solution   Substituting the given values of $n, \bar{x}, \sigma$, and $z_{0.025} = 1.96$ into the confidence interval formula, we get

$$21.6 - 1.96 \cdot \frac{5.1}{\sqrt{100}} < \mu < 21.6 + 1.96 \cdot \frac{5.1}{\sqrt{100}}$$

or $20.6 < \mu < 22.6$. Of course, either the interval from 20.6 to 22.6 contains the population mean μ, or it does not, but we are 95% confident that it does. As was explained on page 216, this means that the method by which the interval was obtained "works" 95% of the time. In other words, in repeated applications of the confidence interval formula, 95% of the intervals can be expected to contain the means of the respective populations. ∎

The preceding confidence interval formula is exact only for random samples from normal populations, but for large samples it will generally provide good approximations. Since σ is unknown in most applications, we may have to make the further approximation of substituting the sample standard deviation s for σ.

Large sample confidence interval for μ

$$\bar{x} - z_{\alpha/2} \cdot \frac{s}{\sqrt{n}} < \mu < \bar{x} + z_{\alpha/2} \cdot \frac{s}{\sqrt{n}}$$

EXAMPLE 7   **A 99% confidence interval for the mean nanopillar height**

With reference to the nanopillar height data on page 15, for which we have $n = 50, \bar{x} = 305.58$ nm, and $s^2 = 1{,}366.86$ (hence, $s = 36.97$ nm), construct a 99% confidence interval for the population mean of all nanopillars.

Solution   Substituting into the confidence interval formula with $\bar{x} = 305.58$, $s = 36.97$, and $z_{0.005} = 2.575$, we get

$$305.58 - 2.575 \cdot \frac{36.97}{\sqrt{50}} < \mu < 305.58 + 2.575 \cdot \frac{36.97}{\sqrt{50}}$$

or $292.12 < \mu < 319.04$. We are 99% confident that the interval from 292.12 nm to 319.04 nm contains the true mean nanopillar height. ∎

For small samples ($n < 30$), we proceed as on page 215, provided it is reasonable to assume that we are sampling from a normal population. Thus, with $t_{\alpha/2}$ defined as on page 217, we get the $(1 - \alpha)100\%$ confidence interval formula

Small sample confidence interval for μ of normal population

$$\bar{x} - t_{\alpha/2} \cdot \frac{s}{\sqrt{n}} < \mu < \bar{x} + t_{\alpha/2} \cdot \frac{s}{\sqrt{n}}$$

This formula applies to samples from normal populations, but in accordance with the discussion on page 197, it may be used as long as the sample does not exhibit any pronounced departures from normality.

EXAMPLE 8

95% confidence interval for the mean of a normal population

We know that silk fibers are very tough but in short supply. Engineers are making breakthroughs to create synthetic silk fibers that can improve everything from car bumpers to bullet-proof vests or to make artificial blood vessels. One research group reports the summary statistics[1]

$$n = 18 \qquad \bar{x} = 22.6 \qquad s = 15.7$$

for the toughness (MJ/m^3) of processed fibers.

Construct a 95% confidence interval for the mean toughness of these fibers. Assume that the population is normal.

Solution The sample size is $n = 18$ and $t_{0.025} = 2.110$ for $n - 1 = 17$ degrees of freedom. The 95% confidence formula for μ becomes

$$22.6 - 2.110 \cdot \frac{15.7}{\sqrt{18}} < \mu < 22.6 + 2.110 \cdot \frac{15.7}{\sqrt{18}} \quad \text{or} \quad 14.79 < \mu < 30.41 \, MJ/m^3$$

We are 95% confident that the interval from 14.79 to 36.41 MJ/m^3 contains the mean toughness of all possible artificial fibers created by the current process.

The article does not give the original data but, since $n = 18$ is moderately large, the normal assumption is not critical unless an outlier exists. ∎

Because confidence intervals are an important way of making inferences, we review their interpretation in the context of 95% confidence intervals for μ.

Before the observations are made, $\overline{X}$ and S are random variables, so

1. The interval from $\overline{X} - t_{0.025} \dfrac{S}{\sqrt{n}}$ to $\overline{X} + t_{0.025} \dfrac{S}{\sqrt{n}}$ is a random interval. It is centered at $\overline{X}$ and its length is proportional to S.

2. The interval from $\overline{X} - t_{0.025} \dfrac{S}{\sqrt{n}}$ to $\overline{X} + t_{0.025} \dfrac{S}{\sqrt{n}}$ will cover the true (fixed) μ with probability 0.95.

Once the observations are made and we have the numerical values $\bar{x}$ and s.

[1] F. Teulé, et. al. (2012) Combining flagelliform and dragline spider silk motifs to produce tunable synthetic biopolymer fibers. *Biopolymers*, 97(6), 418–431.

3. The calculated interval from $\bar{x} - t_{0.025}\dfrac{s}{\sqrt{n}}$ to $\bar{x} + t_{0.025}\dfrac{s}{\sqrt{n}}$ is fixed. It is no longer possible to talk about the probability of covering μ. The interval either covers μ or it does not. Further, in any particular application, we have no way of knowing if μ is covered or not.

However, because 0.95 is the probability that we cover μ in each application, the long-run relative frequency interpretation of probability (or law of large numbers) promises that

$$\frac{\text{number of intervals that cover the true mean}}{\text{number of intervals calculated}} \to 0.95$$

when the intervals are calculated for a large number of different problems. This is what gives us 95% confidence! Over many different applications of the method, the proportion of intervals that cover μ should be nearly 0.95.

To emphasize these points, we simulated a sample of size $n = 10$ from a normal distribution with $\mu = 20$ and $\sigma = 5$. The 95% confidence interval was then calculated and graphed in Figure 7.2. This procedure was repeated 20 times. The different samples produce different values for $\bar{x}$ and, consequently, the intervals are centered at different points. The different values of the standard deviation s gave rise to intervals of different lengths. Unlike a real application, here we know that the true fixed mean is $\mu = 20$. The proportion of intervals that cover the true value of $\mu = 20$ should be near 0.95 and, in this instance, we happen to have exactly that proportion $19/20 = 0.95$.

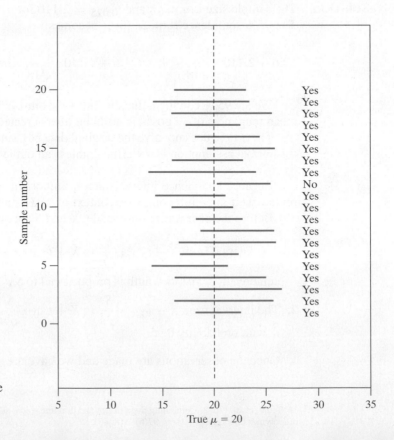

Figure 7.2

Interpretation of the confidence interval for population mean, true mean $\mu = 20$

Exercises

7.1 Civil engineers collected data from one area of Wisconsin on the amount of salt (tons) used to keep highways drivable during a snowstorm. The amount of salt for $n = 30$ storms

1111 2115 1573 2813 2815 2126 854 3965 1819 776
1484 2056 784 779 1373 1237 1701 1957 246 1730
2365 1902 2858 2236 1718 916 2830 2865 1574 1373

has $\bar{x} = 1798.4$ tons and $s^2 = 671{,}330.9$ so $s = 819.35$ tons. What can one assert with 95% confidence about the maximum error if $\bar{x} = 1798.4$ is used as a point estimate of the true population mean amount of salt required for a snowstorm?

7.2 With reference to the previous exercise, construct a 95% confidence interval for the true population mean amount of salt required for a snowstorm.

7.3 An industrial engineer collected data on the labor time required to produce an order of automobile mufflers using a heavy stamping machine. The data on times (hours) for $n = 52$ orders of different parts

2.15 2.27 0.99 0.63 2.45 1.30 2.63 2.20 0.99 1.00 1.05
3.44 0.49 0.93 2.52 1.05 1.39 1.22 3.17 0.85 1.18 2.27
1.52 0.48 1.33 4.20 1.37 2.70 0.63 1.13 3.81 0.20 1.08
2.92 2.87 2.62 1.03 2.76 0.97 0.78 4.68 5.20 1.90 0.55
1.00 2.95 0.45 0.70 2.43 3.65 4.55 0.33

has $\bar{x} = 1.865$ hours and $s^2 = 1.5623$ so $s = 1.250$ hours. What can one assert with 95% confidence about the maximum error if $\bar{x} = 1.865$ hours is used as a point estimate of the true population mean labor time required to run the heavy stamping machine?

7.4 With reference to the previous exercise, construct a 95% confidence interval for the true population mean labor time.

7.5 The manufacture of large liquid crystal displays (LCD's) is difficult. Some defects are minor and can be removed; others are unremovable. The number of unremovable defects, for each of $n = 45$ displays (Courtesy of Shiyu Zhou)

1	0	5	3	0	7	6	0	0	4	6	8
5	0	9	1	0	8	6	0	3	2	0	0
0	6	0	10	0	6	0	0	1	0	0	0
0	1	5	1	0	5	0	0	2			

has $\bar{x} = 2.467$ and $s = 3.057$ unremovable defects. What can one assert with 98% confidence about the maximum error if $\bar{x} = 2.467$ is used as a point estimate of the true population mean number of unremovable defects?

7.6 With reference to the previous exercise, construct a 98% confidence interval for the true population mean number of unremovable defects per display.

7.7 With reference to the $n = 50$ interrequest time observations in Example 6, Chapter 2, which have mean 11,795 and standard deviation 14,056, what can one assert with 95% confidence about the maximum error if $\bar{x} = 11{,}795$ is used as a point estimate of the true population mean interrequest time?

7.8 With reference to the previous exercise, construct a 95% confidence interval for the true mean interrequest time.

7.9 In a study of automobile collision insurance costs, a random sample of 80 body repair costs for a particular kind of damage had a mean of $472.36 and a standard deviation of $62.35. If $\bar{x} = \$472.36$ is used as a point estimate of the true average repair cost of this kind of damage, with what confidence can one assert that the error does not exceed $10?

7.10 Refer to Example 8. How large a sample will we need in order to assert with probability 0.95 that the sample mean will not differ from the true mean by more than 1.5. (replacing σ by s is reasonable here because the estimate is based on a sample of size eighteen.)

7.11 The dean of a college wants to use the mean of a random sample to estimate the average amount of time students take to get from one class to the next, and she wants to be able to assert with 99% confidence that the error is at most 0.25 minute. If it can be presumed from experience that $\sigma = 1.40$ minutes, how large a sample will she have to take?

7.12 One novel process of making green gasoline takes biomass in the form of sucrose and converts it into gasoline using catalytic reactions. At one step in a pilot plant process, a chemical engineer measures the output of carbon chains of length three. Nine runs with same catalyst produced the yields (gal)

0.63 2.64 1.85 1.68 1.09 1.67 0.73 1.04 0.68

What can the chemical engineer assert with 95% confidence about the maximum error if she uses the sample mean to estimate true mean yield?

7.13 With reference to the previous exercise, assume that yield has a normal distribution and obtain a 95% confidence interval for the true mean yield of the pilot plant process.

7.14 To monitor complex chemical processes, chemical engineers will consider key process indicators, which may be just yield but most often depend on several quantities. Before trying to improve a process, $n = 9$ measurements were made on a key performance indicator.

123 106 114 128 113 109 120 102 111

What can the engineer assert with 95% confidence about the maximum error if he uses the sample mean to estimate true mean value of the performance indicator?

7.15 With reference to the previous exercise, assume that the key performance indicator has a normal distribution and obtain a 95% confidence interval for the true value of the indicator.

7.16 Refer to Exercise 2.34, page 36, concerning material costs for rebuilding $n = 29$ traction motors. A computer calculation gives $\bar{x} = 1.4707$ and $s = 0.5235$ thousand dollars. Obtain a 90% confidence interval for the mean material costs to rebuild a motor.

7.17 Refer to the 2×4 lumber strength data in Exercise 2.58, page 38. According to the computer output, a sample of $n = 30$ specimens had $\bar{x} = 1908.8$ and $s = 327.1$. Find a 95% confidence interval for the population mean strength.

7.18 Refer to the data on page 40, on the number of defects per board for Product B. Obtain a 95% confidence interval for the population mean number of defects per board.

7.19 With reference to the thickness measurements in Exercise 2.41, page 37, obtain a 95% confidence interval for the mean thickness.

7.20 Ten bearings made by a certain process have a mean diameter of 0.5060 cm and a standard deviation of 0.0040 cm. Assuming that the data may be looked upon as a random sample from a normal population, construct a 95% confidence interval for the actual average diameter of bearings made by this process.

7.21 The freshness of produce at a mega-store is rated a scale of 1 to 5, with 5 being very fresh. From a random sample of 36 customers, the average score was 3.5 with a standard deviation of 0.8.

(a) Obtain a 90% confidence interval for the population mean, μ, or the mean score for all customers.

(b) Does μ lie in your interval obtained in part (a)? Explain.

(c) In long series of repeated experiments, with new random samples collected for each experiment, what proportion of the resulting confidence intervals will contain the true population mean? Explain your reasoning.

7.22 A copy shop records that in $n = 64$ cases, the cartridge for the copy machine lasted an average of 18,300 copies with a standard deviation of 2,800 copies.

(a) Obtain a 95% confidence interval for μ, the population mean number of copies before a new cartridge is needed for the copy machine.

(b) Does μ lie in your interval obtained in part (a)? Explain.

(c) In long series of repeated experiments, what proportion of the respective confidence intervals contain the true mean? Explain your reasoning.

7.23 Refer to Example 1 and the data on the resiliency modulus of recycled concrete.

(a) Obtain a 95% confidence interval for the population mean resiliency modulus μ.

(b) Is the population mean contained in your interval in part (a)? Explain.

(c) What did you assume about the population in your answer to part (a)?

(d) Why are you 95% confident about the interval in part (a)?

7.24 In an air-pollution study performed at an experiment station, the following amount of suspended benzene-soluble organic matter (in micrograms per cubic meter) was obtained for eight different samples of air:

$$2.2 \quad 1.8 \quad 3.1 \quad 2.0 \quad 2.4 \quad 2.0 \quad 2.1 \quad 1.2$$

Assuming that the population sampled is normal, construct a 95% confidence interval for the corresponding true mean.

7.25 Modify the formula for E on page 216 so that it applies to large samples which constitute substantial portions of finite populations, and use the resulting formula for the following problems:

(a) A sample of 50 scores on the admission test for a school of engineering is drawn at random from the scores of the 420 persons who applied to the school in 2015. If the sample mean and the standard deviation are $\bar{x} = 546$ and $s = 85$, what can we assert with 95% confidence about the maximum error if $\bar{x} = 546$ is used as an estimate of the mean score of all the applicants?

(b) A random sample of 40 drums of a chemical, drawn from among 200 such drums whose weights can be expected to have the standard deviation $\sigma = 12.2$ pounds, has a mean weight of 240.8 pounds. If we estimate the mean weight of all 200 drums as 240.8 pounds, what can we assert with 99% confidence about the maximum error?

7.26 Instead of the large sample confidence interval formula for μ on page 220, we could have given the alternative formula

$$\bar{x} - z_{\alpha/3} \cdot \frac{\sigma}{\sqrt{n}} < \mu < \bar{x} + z_{2\alpha/3} \cdot \frac{\sigma}{\sqrt{n}}$$

Explain why the one on page 220 is narrower, and hence preferable, to the one given here.

7.27 Suppose that we observe a random variable having the binomial distribution. Let X be the number of successes in n trials.

(a) Show that $\dfrac{X}{n}$ is an unbiased estimate of the binomial parameter p.

(b) Show that $\dfrac{X+1}{n+2}$ is not an unbiased estimate of
the binomial parameter p.

7.28 The statistical program *MINITAB* will calculate the small sample confidence
interval for μ. With the nanopillar height data in C1,

Dialog box:
Stat > Basic Statistics > 1-Sample t. Click on box and type C1. Choose **Options**.
Type 0.95 in **Confidence level** and choose **not equal**. Click **OK**. Click **OK**

---

produces the output

	N	Mean	StDev	SE Mean	95% CI
C1	50	305.580	36.971	5.229	(295.073, 316.087)

The R command **t.test(x,conf.level=.95)** produces similar results when the
data are in **x**.

(a) Obtain a 90% confidence interval for μ.

(b) Obtain a 95% confidence interval for μ with the aluminum alloy data on
page 19.

Alternatively, you can use the *MINITAB* commands
Stat > Basic statistics > Graphical summary
to produce the more complete output

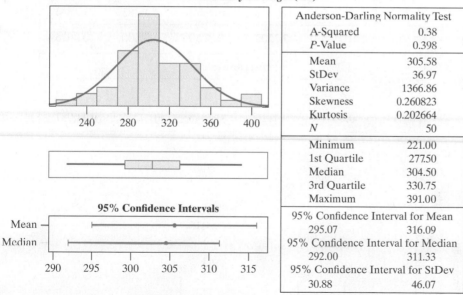

Summary for height (nm)

Anderson-Darling Normality Test	
A-Squared	0.38
P-Value	0.398
Mean	305.58
StDev	36.97
Variance	1366.86
Skewness	0.260823
Kurtosis	0.202664
N	50
Minimum	221.00
1st Quartile	277.50
Median	304.50
3rd Quartile	330.75
Maximum	391.00

95% Confidence Interval for Mean
295.07   316.09
95% Confidence Interval for Median
292.00   311.33
95% Confidence Interval for StDev
30.88   46.07

95% Confidence Intervals

Mean

Median

7.29 You can simulate the coverage of the small sample confidence intervals for μ by
generating 20 samples of size 10 from a normal distribution with $\mu = 20$ and $\sigma = 5$
and computing the 95% confidence intervals according to the formula on page 221.
Using *MINITAB*:

Calc > Random Data > Normal
Type 10 in **Generate**, $C1 - C20$ in **Store**, 20 in **Mean**, and 5 in **standard deviation**. Click **OK**.

Stat > Basic Statistics > 1-sample t, Click on box and Type $C1 - C20$. Click **OK**.

(a) From your output, determine the proportion of the 20 intervals that cover the true mean $\mu = 20$.

(b) Repeat with 20 samples of size 5.

7.4 Maximum Likelihood Estimation

Sometimes it is necessary to estimate parameters other than a mean or variance. A very general approach to estimation, proposed by R. A. Fisher, is called the *method of maximum likelihood*. To set the ideas, we begin with a special case. Suppose that one of just two distributions must prevail. For example, let X take the possible values 0, 1, 2, 3, or 4 with probabilities specified by distribution 1 or with probabilities specified by distribution 2 (see Table 7.1 and Figure 7.3).

The first is the binomial distribution with $p = 0.5$ and the second the binomial with $p = 0.3$, but this fact is not important to the argument.

Table 7.1 Two Possible Distributions for X					
Distribution 1					
x	0	1	2	3	4
$f(x)$	0.0625	0.2500	0.3750	0.2500	0.0625
Distribution 2					
x	0	1	2	3	4
$f(x)$	0.2401	0.4116	0.2646	0.0756	0.0081

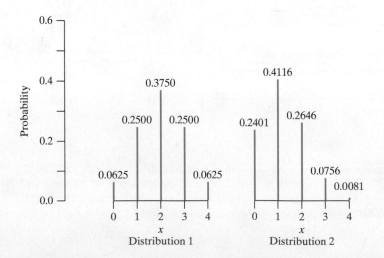

Figure 7.3
The two possible distributions for X

If we observe $X = 3$, should our estimate of the underlying distribution be distribution 1 or distribution 2? Suppose we take the attitude that we will select the distribution for which the observed value $x = 3$ has the highest probability of occurring. Because this calculation is done after the data are obtained, we use the terminology of maximizing *likelihood* rather than probability. For the first distribution, $P[X = 3] = 0.2500$ and for the second distribution $P[X = 3] = 0.0756$, so we estimate that the first distribution is the distribution that actually produced the observation 3.

If, instead, we observed $X = 1$, the estimate would be distribution 2 since 0.4116 is larger than 0.2500.

Let us take this example a step further and assume that X follows a binomial distribution with $n = 4$ but that $0 \leq p \leq 1$ is unknown. The count X then has the distribution

$$\binom{4}{x} p^x (1 - p)^{4-x} \quad \text{for} \quad x = 0, 1, 2, 3, 4$$

If we again observe $X = 3$, we evaluate the binomial distribution at $x = 3$ and obtain

$$4p^3 (1 - p)^{4-3} \quad \text{for} \quad 0 \leq p \leq 1$$

which is a function of p. We now vary p to best explain the observed result. This curve, $L(p)$, is shown in Figure 7.4.

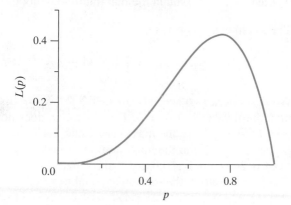

Figure 7.4
The likelihood curve
$L(p) = 4p^3(1 - p)$

To obtain the best explanation for what we did observe, we choose a value for the unknown p at which the maximum occurs. Using calculus, the maximum occurs at the value of p for which the derivative is zero.

$$\frac{d}{dp} 4p^3 (1 - p) = 4(3p^2 - 4p^3) = 0$$

Since the solution $p = 0$ yields a minimum, our estimate is $\widehat{p} = 0.75$. Note that this derivative is positive for $p < 3/4$ and negative for $p > 3/4$, confirming that $\widehat{p} = 0.75$ gives the global maximum. To review, this value maximizes the after-the-fact probability, or likelihood, of observing the value 3.

More generally, a random sample of size n is taken from a probability distribution, or density, $f(x; \theta)$ that depends on a parameter θ. The random sample produces n values $x_1, x_2, \ldots, x_n$, which we substitute into the joint probability distribution, or probability density function, and then study the resulting function of θ.

The function of θ that is obtained by substituting the observed values of the random sample $X_1 = x_1, \ldots, X_n = x_n$ into the joint probability distribution or the density function for $X_1, X_2, \ldots, X_n$

$$L(\theta|x_1, \ldots, x_n) = \prod_{i=1}^{n} f(x_i; \theta)$$

is called the **likelihood function** for θ.

We often simplify the notation and write $L(\theta)$ with the understanding that the likelihood function does depend on the values $x_1, x_2, \ldots, x_n$ from the random sample.

Given the values $x_1, x_2, \ldots, x_n$ from a random sample, one distinctive feature of the likelihood function is the value or values of θ at which it attains its maximum.

A statistic $\widehat{\theta}(X_1, \ldots, X_n)$ is a **maximum likelihood estimator** of θ if, for each sample $x_1, \ldots, x_n$, $\widehat{\theta}(x_1, \ldots, x_n)$ is a value for the parameter that maximizes the likelihood function $L(\theta|x_1, \ldots, x_n)$.

EXAMPLE 9   **The maximum likelihood estimator with Bernoulli trials**

Consider a characteristic that occurs in proportion p of a population. Let $X_1, \ldots, X_n$ be a random sample of size n so

$$P[X_i = 0] = 1 - p \quad \text{and} \quad P[X_i = 1] = p \quad \text{for} \quad i = 1, \ldots, n$$

where $0 \leq p \leq 1$. Obtain the maximum likelihood estimator of p.

Solution   The likelihood function is

$$L(p|x_1, x_2, \ldots, x_n) = \prod_{i=1}^{n} p^{x_i}(1 - p)^{1-x_i} = p^{\sum_{i=1}^{n} x_i}(1 - p)^{n - \sum_{i=1}^{n} x_i}$$

We first check two special cases. If $\sum_{i=1}^{n} x_i = 0$, then $L(p) = (1 - p)^n$ has the maximum value 1 at $p = 0$. The derivative does not vanish here. If $\sum_{i=1}^{n} x_i = n$, then $L(p) = p^n$ has the maximum value 1 at $p = 1$. Otherwise, $L(p)$ goes to 0 as p goes to 0 or 1 and the maximum must occur at a value of p where the derivative $L(p)$ is zero. Equivalently, we maximize the log-likelihood function $\ln L(p)$ over $0 < p < 1$. Setting the derivative equal to zero,

$$\frac{d}{dp} \ln L(p) = \frac{d}{dp} \left(\sum_{i=1}^{n} x_i \ln(p) + \left(n - \sum_{i=1}^{n} x_i \right) \ln(1 - p) \right)$$

$$= \frac{\sum_{i=1}^{n} x_i}{p} - \frac{n - \sum_{i=1}^{n} x_i}{1 - p} = 0$$

we obtain the maximum likelihood estimator $\widehat{p} = \sum_{i=1}^{n} x_i/n$. That is, $\widehat{p}$ is the fraction of persons in the sample that have the characteristic. Note that this definition of the estimator also includes the two special cases $\sum_{i=1}^{n} x_i = 0$ or $= n$ even though the derivative does not vanish is these cases.   ∎

EXAMPLE 10   **Maximum likelihood estimator: Poisson distribution**

Let $X_1, \ldots, X_n$ be a random sample of size n from the Poisson distribution

$$f(x|\lambda) = \frac{\lambda^x e^{-\lambda}}{x!},$$

where $0 \leq \lambda < \infty$. Obtain the maximum likelihood estimator of λ.

Solution   The likelihood function is

$$L(\lambda|x_1, \ldots, x_n) = \prod_{i=1}^{n} \frac{\lambda^{x_i} e^{-\lambda}}{x_i!} = \lambda^{\sum_{i=1}^{n} x_i} e^{-n\lambda} \times \frac{1}{\prod_{i=1}^{n} x_i!}$$

If $\sum_{i=1}^{n} x_i = 0$, then $L(\lambda) = e^{-n\lambda}$ has its maximum at $\widehat{\lambda} = 0$. Otherwise, we can maximize $\ln L(\lambda)$ by setting its derivative equal to zero.

$$\frac{\partial}{\partial \lambda} \ln L(\lambda) = \frac{\sum_{i=1}^{n} x_i}{\lambda} - n = 0$$

so the maximum likelihood estimator is $\widehat{\lambda} = \sum_{i=1}^{n} x_i/n = \bar{x}$. This same formula works for the special case $\sum_{i=1}^{n} x_i = 0$.   ∎

The method of maximum likelihood also applies to continuous distributions.

EXAMPLE 11   **Maximum likelihood estimator: normal distribution mean**

Let $X_1, \ldots, X_n$ be a random sample of size n from a normal distribution with known variance. Obtain the maximum likelihood estimator of μ.

Solution   Writing $(x_i - \mu)^2 = (x_i - \bar{x} + \bar{x} - \mu)^2$, the likelihood function is

$$L(\mu|x_1, \ldots, x_n) = \prod_{i=1}^{n} \frac{1}{\sqrt{2\pi\sigma^2}} e^{-(x_i-\mu)^2/2\sigma^2}$$

$$= e^{-n(\bar{x}-\mu)^2/2\sigma^2} \times \frac{1}{(2\pi\sigma^2)^{n/2}} e^{-\sum_{i=1}^{n}(x_i-\bar{x})^2/2\sigma^2}$$

This likelihood is maximized over all values of μ when the exponent $n(\bar{x}-\mu)^2/2\sigma^2$ is minimized. Therefore, the maximum likelihood estimator $\widehat{\mu} = \bar{x}$.   ∎

EXAMPLE 12   **Maximum likelihood estimator: normal distribution variance**

Let $X_1, \ldots, X_n$ be a random sample of size n from a normal distribution with known mean. Obtain the maximum likelihood estimator of σ^2.

Solution   The likelihood function is

$$L(\sigma^2|x_1, \ldots, x_n) = \prod_{i=1}^{n} \frac{1}{\sqrt{2\pi\sigma^2}} e^{-(x_i-\mu)^2/2\sigma^2} = \frac{1}{(2\pi\sigma^2)^{n/2}} e^{-\sum_{i=1}^{n}(x_i-\mu)^2/2\sigma^2}$$

The function $z^b e^{-cz}$ has a maximum at $z = b/c$ for $z \geq 0$ when b and c are positive. Taking $z = 1/\sigma^2$, $b = n/2$, and $c = \sum_{i=1}^{n}(x_i - \mu)^2/2$, we obtain the maximum likelihood estimator $\widehat{\sigma^2} = \sum_{i=1}^{n}(x_i - \mu)^2/n$, when μ is known.   ∎

Suppose that, in the example above, we were interested in estimating σ rather than σ^2. The likelihood is still

$$\frac{1}{(2\pi\sigma^2)^{n/2}} e^{-\sum_{i=1}^{n}(x_i-\mu)^2/2\sigma^2}$$

but is now considered to be a function of σ. Taking logarithms and differentiating, you may verify that, for $d, c > 0$, $z^d e^{-cz^2}$ has a maximum at $z = \sqrt{d/2c}$ when $z > 0$. Consequently, taking $z = 1/\sigma$, $d = n$, and $c = \sum_{i=1}^{n}(x_i - \mu)^2/2$, we obtain the maximum likelihood estimator

$$\widehat{\sigma} = \sqrt{\widehat{\sigma^2}} = \sqrt{\frac{\sum_{i=1}^{n}(x_i - \mu)^2}{n}}$$

This same argument can be extended to a more general **invariance** property for maximum likelihood estimators, which we state but do not prove. Specifically, if $g(\theta)$ is a continuous one-to-one function of θ and $\widehat{\theta}$ is the maximum likelihood estimator of θ, the maximum likelihood estimator of $g(\theta)$ is obtained by simple substitution.

$$\widehat{g(\theta)} = \text{maximum likelihood estimator of } g(\theta) = g(\widehat{\theta})$$

EXAMPLE 13   **Maximum likelihood estimator of function of λ**

The number of defective hard drives produced daily by a production line can be modeled as a Poisson distribution. The counts for ten days are

$$7 \quad 3 \quad 1 \quad 2 \quad 4 \quad 1 \quad 2 \quad 3 \quad 1 \quad 2$$

Obtain the maximum likelihood estimate of the probability of 0 or 1 defectives on one day.

Solution   From Example 10, the maximum likelihood estimate of λ is $\widehat{\lambda} = \bar{x} = 26/10 = 2.6$. Consequently, by the invariance property, the maximum likelihood estimate of

$$P(X = 0 \quad \text{or} \quad 1) = e^{-\lambda} + \frac{\lambda e^{-\lambda}}{1!}$$

is

$$e^{-\widehat{\lambda}} + \frac{\widehat{\lambda} e^{-\widehat{\lambda}}}{1!} = e^{-2.6} + \frac{2.6 \cdot e^{-2.6}}{1!} = 0.267$$

There will 1 or fewer defectives on just over one-quarter of the days. ∎

The method of maximum likelihood applies to more than one parameter.

EXAMPLE 14   **Maximum likelihood estimator: normal distributions**

Let $X_1, \ldots, X_n$ be a random sample of size n from a normal distribution. Obtain the maximum likelihood estimators of μ and σ^2. Also obtain the maximum likelihood estimator σ.

Solution   Using the expression in Example 11 for the joint probability density function, the likelihood function is

$$L(\mu, \sigma^2 | x_1, \ldots, x_n) = \prod_{i=1}^{n} \frac{1}{\sqrt{2\pi\sigma^2}} e^{-(x_i - \mu)^2/2\sigma^2}$$

$$= e^{-n(\bar{x} - \mu)^2/2\sigma^2} \times \frac{1}{(2\pi\sigma^2)^{n/2}} e^{-\sum_{i=1}^{n}(x_i - \bar{x})^2/2\sigma^2}$$

Only the first term contains μ, and it is maximized at $\widehat{\mu} = \bar{x}$ whatever the value of σ^2. The maximum of this first term is 1. Then, as in Example 12, the function $z^b e^{-cz}$ has a maximum at $z = b/c$. Taking $z = 1/\sigma^2$, $b = n/2$, and $c = \sum_{i=1}^{n}(x_i - \bar{x})^2/2$, we obtain the maximum likelihood estimator $\widehat{\sigma^2} = \sum_{i=1}^{n}(x_i - \bar{x})^2/n$. In summary $\widehat{\mu} = \bar{x}$ and $\widehat{\sigma^2} = \sum_{i=1}^{n}(x_i - \bar{x})^2/n$ are the maximum likelihood estimators of μ and σ^2.

Here $\widehat{\sigma^2}$ contains the divisor n, not $n-1$, so it is a biased estimate of σ^2.

Because the sample standard deviation σ is the square root of σ^2, the maximum likelihood estimator of σ is

$$\widehat{\sigma} = \sqrt{\widehat{\sigma^2}} = \sqrt{\sum_{i=1}^{n}(x_i - \bar{x})^2/n}$$

Consequently, we commonly write $\widehat{\sigma}^2$ for the maximum likelihood estimator of σ^2. ∎

EXAMPLE 15   **Estimation of yield for a green gas process**

One process of making green gasoline takes biomass in the form of sucrose and converts it into gasoline using catalytic reactions. At one step in a pilot plant process, the output includes carbon chains of length 3. Fifteen runs with same catalyst produced the yields (gal)

5.57	5.76	4.18	4.64	7.02	6.62	6.33	7.24
5.57	7.89	4.67	7.24	6.43	5.59	5.39	

Treating the yields as a random sample from a normal population,

(a) Obtain the maximum likelihood estimates of the mean yield and the variance.

(b) Obtain the maximum likelihood estimate of the coefficient of variation σ/μ.

Solution   **(a)** We calculate

$$\widehat{\mu} = \bar{x} = \frac{5.57 + 5.76 + \cdots + 5.39}{15} = \frac{90.14}{15} = 6.009 \text{ gal}$$

Recall that the maximum likelihood estimate of variance uses divisor n, not $n-1$.

$$\widehat{\sigma^2} = \frac{1}{n}\sum_{i=1}^{15}(x_i - \bar{x})^2 = \frac{1}{15}(16.2631) = 1.084$$

(b) The coefficient of variation is a function of μ and σ^2, so its maximum likelihood estimate is that same function of $\widehat{\mu}$ and $\widehat{\sigma}^2$.

$$\widehat{\left(\frac{\sigma}{\mu}\right)} = \frac{\widehat{\sigma}}{\widehat{\mu}} = \frac{\sqrt{1.084}}{6.009} = 0.173$$ ∎

Exercises

7.30 Refer to Example 13, Chapter 3, where 294 out of 300 ceramic insulators were able to survive a thermal shock.

    (a) Obtain the maximum likelihood estimate of the probability that a ceramic insulator will survive a thermal shock.

    (b) Suppose a device contains 3 ceramic insulators and all must survive the shock in order for the device to work. Find the maximum likelihood estimate of the probability that all three will survive a thermal shock.

7.31 Refer to Example 7, Chapter 10, where 48 of 60 transceivers passed inspection.

    (a) Obtain the maximum likelihood estimate of the probability that a transceiver will pass inspection.

    (b) Obtain the maximum likelihood estimate that the next two transceivers tested will pass inspection.

7.32 The daily number of accidental disconnects with a server follows a Poisson distribution. On five days

$$2 \quad 5 \quad 3 \quad 3 \quad 7$$

accidental disconnects are observed.

    (a) Obtain the maximum likelihood estimate of λ.

    (b) Find the maximum likelihood estimate of the probability that 3 or more accidental disconnects will occur.

7.33 In one area along the interstate, the number of dropped wireless phone connections per call follows a Poisson distribution. From four calls, the number of dropped connections is

$$2 \quad 0 \quad 3 \quad 1$$

(a) Find the maximum likelihood estimate of λ.

(b) Obtain the maximum likelihood estimate that the next two calls will be completed without any accidental drops.

7.34 Refer to Exercise 7.12.

(a) Obtain the maximum likelihood estimates of μ and σ.

(b) Find the maximum likelihood of the probability that the next run will have a yield greater than 2 gallons.

7.35 Refer to Exercise 7.14.

(a) Obtain the maximum likelihood estimates of μ and σ.

(b) Find the maximum likelihood of the coefficient of variation σ/μ.

7.36 Find the maximum likelihood estimator of p when

$$f(x; p) = p^x(1-p)^{1-x} \quad \text{for} \quad x = 0, 1$$

7.37 Let $x_1, \ldots, x_n$ be the observed values of a random sample of size n from the exponential distribution $f(x; \beta) = \beta^{-1}e^{-x/\beta}$ for $x > 0$.

(a) Find the maximum likelihood estimator of β.

(b) Obtain the maximum likelihood estimator of the probability that the next observation is greater than 1.

7.38 Let X have the negative binomial distribution

$$f(x) = \binom{x-1}{r-1} p^r(1-p)^{x-r} \quad \text{for} \quad x = r, r+1, \ldots$$

(a) Obtain the maximum likelihood estimator of p.

(b) For one engineering application, it is best to use components with a superior finish. Suppose $X = 27$ identical components are inspected, one at a time, before the $r = 3$rd component with superior finish is found. Find the maximum likelihood estimate of the probability that a component will have a superior finish.

7.5   Tests of Hypotheses

There are many problems in which, rather than estimate the value of a parameter, we must decide whether a statement concerning a parameter is true or false. Statistically speaking we test a hypothesis about a parameter. For example, in quality control work a random sample may serve to determine whether the "process mean" (for a given kind of measurement) has remained unchanged or whether it has changed to such an extent that the process has gone out of control and adjustments have to be made.

EXAMPLE 16   **Not all samples will lead to a correct assessment of water quality**

Refer to Example 1 of monitoring the quality of water leaving a plant. Why does evaluating a sample of specimens not always lead to correct conclusions regarding water quality?

Solution   The observed values of water quality will depend on the particular specimens in the sample. Because these values can vary from sample to sample, particular samples can produce misleading values and hence incorrect decisions. The possibility of making a mistake about water quality, on the basis of test specimens, cannot be completely eliminated unless the entire discharge could be accurately measured for the entire reporting period. This is, of course, technologically and economically infeasible.   ∎

We consider the problem of improving lithium car batteries. A research group is making great advances using a new type anode and they claim that the mean life is greater than 1600 recharge cycles. To support this claim, they create 36 new batteries and subject them to recharge cycles until they fail. The claim will be established if the sample mean lifetime is greater than 1660 cycles. Otherwise, the claim will not be established and further improvements are needed.

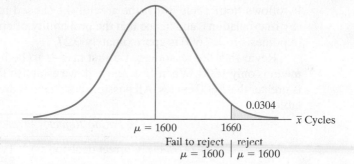

Figure 7.5
Probability of falsely rejecting claim

This provides a clear-cut criterion for accepting or not accepting the claim., but unfortunately, it is not infallible. Since the decision is based on a sample, the sample mean may exceed 1660 cycles even though the mean life is 1600 cycles. There is also the possibility that the sample mean will be less than 1660 cycles even though the mean life length is, say, 1680 cycles. Before adopting the criterion it is wise to investigate the chances that the criterion leads to wrong decisions.

To simplify this initial presentation of testing hypotheses, we assume that the standard deviation $\sigma = 192$ cycles is known. Let us first investigate the possibility that the sample mean will exceed 1660 even when the true mean life length is $\mu = 1600$. The probability this happens, purely due to chance, is the area of the shaded region in Figure 7.5. This area is determine by approximating the sampling distribution of $\overline{X}$ by a normal distribution. We have $\sigma_{\overline{X}} = \dfrac{192}{\sqrt{36}} = 32$. In standard units, the dividing line for the criterion is

$$z = \frac{1660 - 1600}{32} = 1.875$$

It follows from Table 3 that the area of the shaded region of Figure 7.5 is $1 - 0.9696 = 0.0304$ (by interpolation). Hence the probability of erroneously rejecting the hypothesis $\mu = 1600$ cycles is approximately 0.03.

Let us now consider the other possibility, where the procedure fails to detect that $\mu > 1600$ cycles. Suppose again, for the sake of argument, that the true mean is $\mu = 1680$ cycles, so that the probability of getting a sample mean less than or equal to 1660 cycles (and, hence, erroneously failing to reject $\mu = 1600$ cycles) is given by the area of the ruled region of Figure 7.6. As before, $\sigma_{\overline{X}} = 192$, so that the dividing line of the criterion, in standard units, is now

$$z = \frac{1660 - 1680}{32} = -0.625$$

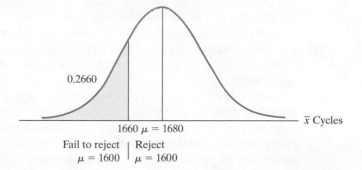

Figure 7.6
Probability of failing to reject claim

It follows from Table 3 that the area of the shaded region of Figure 7.6 is 0.2660 (by interpolation), and hence that the probability of erroneously failing to reject the hypothesis $\mu = 1600$ is approximately 0.27.

Reviewing the reasoning, we first take H to be hypothesis that the population mean is only 1600. When H is rejected, we establish the claim that mean life length is greater that 1600 cycles. All possible conclusions are summarized in the following table.

	Fail to Reject	Reject H
H is true	Correct decision	Type I error
H is false	Type II error	Correct decision

If hypothesis H is true and not rejected or false and rejected, the decision is in either case correct. If hypothesis H is true but rejected, it is rejected in error, and if hypothesis H is false but not rejected, this is also an error. The first of these errors is called a **Type I error**. The probability of committing it, when the hypothesis is true, is designated by the Greek letter α (alpha). The second error is called a **Type II error** and the probability of committing it is designated by the Greek letter β (beta). Thus, in our example we showed that for the given test criterion, $\alpha = 0.03$ when $\mu = 1600$ cycles, and $\beta = 0.27$ when $\mu = 1680$ cycles.

In calculating the probability of a Type II error in our example, we arbitrarily choose the alternative value $\mu = 1680$ cycles. However, in this problem, as in most others, there are infinitely many other alternatives, and for each one of them there is a positive probability β of erroneously accepting the hypotheses H. What to do about this will be discussed further in Section 7.9.

7.6   Null Hypotheses and Tests of Hypotheses

In the electric car battery example of the preceding section, we were able to calculate the probability of a Type I error because we formulated the hypothesis H as a single value for the parameter μ. That is, we formulated the hypothesis H so that μ was completely specified. Had we formulated instead $\mu \leq 1600$ cycles, where μ can take on more than one possible value, we would not have been able to calculate the probability of a Type I error without specifying by how much μ is less than 1600 cycles.

We often formulate hypotheses to be tested as a single value for a parameter; at least, we do this whenever possible. This usually requires that we hypothesize the opposite of what we hope to prove. For instance, if we want to show that one method of teaching computer programming is more efficient than another, we hypothesize that the two methods are equally effective. Similarly, if we want to show that one method of irrigating the soil is more expensive than another, we hypothesize that the two methods are equally expensive; and if we want to show that a new copper-bearing steel has a higher yield strength than ordinary steel, we hypothesize that the two yield strengths are the same. Since we hypothesize that there is no difference in the effectiveness of the two teaching methods, no difference in the cost of the two methods of irrigation, and no difference in the yield strength of the two kinds of steel, we call hypotheses like these **null hypotheses** and denote them H_0. Nowadays, the term *null hypothesis* is used for any hypothesis set up primarily to see whether it can be rejected.

The idea of setting up a null hypothesis is not an uncommon one, even in non-statistical thinking. In fact, this is exactly what is done in an American criminal court of law, where an accused person is assumed to be innocent unless he is proved guilty "beyond a reasonable doubt." The null hypothesis states that the accused is not guilty, and the probability expressed subjectively by the phrase "beyond reasonable doubt" reflects the probability α of risking a Type I error. Note that the "burden of proof" is always on the prosecution in the sense that the accused is found not guilty unless the null hypothesis of innocence is clearly disproved. This does not imply that the defendant has been proved innocent if found not guilty. It implies only that he has not been proved guilty. Of course, since we cannot legally "reserve judgment" if proof of guilt is not established, the accused is freed and we act as if the null hypothesis of innocence were accepted. Note that this is precisely what we may have to do in tests of statistical hypotheses, when we cannot afford the luxury of reserving judgment.

We develop procedures for hypothesis testing in the context of the electric car battery example. When the goal is to establish the long life claim, the alternative hypothesis should be $\mu > 1600$ cycles.

To approach problems of hypothesis testing systematically, it will help to proceed as outlined in the following five steps.

1. **Formulate a null hypothesis and an appropriate alternative hypothesis which we accept when the null hypothesis must be rejected.**[2]

In the electric car battery example, the null hypothesis is $\mu = 1600$ cycles and the alternative hypothesis is $\mu > 1600$ cycles. This kind of alternative, which is called a **one-sided alternative**, may also have the inequality going the other way. For instance, if we hope to be able to show that the average time required to do a certain job is less than 15 minutes, we would test the null hypothesis $\mu = 15$ against the alternative hypothesis $\mu < 15$.

The following is an example in which we would use the **two-sided alternative** $\mu \neq \mu_0$, where μ_0 is the value assumed under the null hypothesis: A food processor wants to check whether the average amount of coffee that goes into his 4-ounce jars is indeed 4 ounces. Since the food processor cannot afford to put much less than 4 ounces into each jar for fear of losing customer acceptance, nor can he afford to put much more than 4 ounces into each jar for fear of losing part of his profit, the appropriate alternative hypothesis is $\mu \neq 4$.

As in the examples of the two preceding paragraphs, **alternative hypotheses** usually specify that the population mean (or whatever other parameter may be of concern) is either not equal to, greater than, or less than the value assumed under the null hypothesis. For any given problem, the choice of an appropriate alternative depends mostly on what we hope to be able to show.

EXAMPLE 17 **Formulating the alternative hypothesis**

An appliance manufacturer is considering the purchase of a new machine for stamping out sheet metal parts. If μ_0 is the average number of good parts stamped out per hour by her old machine and μ is the corresponding average for the new machine, the manufacturer wants to test the null hypothesis $\mu = \mu_0$ against a suitable alternative. What should the alternative be if she does not want to buy the new machine unless it is more productive than the old one?

Solution The manufacturer should use the alternative hypothesis $\mu > \mu_0$ and purchase the new machine only if the null hypothesis can be rejected. ∎

Having formulated the null hypothesis and an alternative hypothesis, we proceed with the following step:

2. **Specify the probability of a Type I error. If possible, desired, or necessary, also specify the probabilities of Type II errors for particular alternatives.**

[2]See also the discussion on page 237.

The probability of a Type I error is also called the **level of significance**, and it is usually set at $\alpha = 0.05$ or $\alpha = 0.01$. Which value we choose in any given problem will have to depend on the risks, or consequences, of committing a Type I error. Observe, however, that we should not make the probability of a Type I error too small, because this will have the tendency to make the probabilities of serious Type II errors too large.

Step 2 can often be performed even when the null hypothesis specifies a range of values for the parameter. To illustrate, let us investigate briefly what might be done in the electric car battery example if we wanted to allow for the possibility that the mean battery life is less than 1600 cycles. That is, we test the null hypothesis $\mu \le 1600$ cycles against the alternative hypothesis $\mu > 1600$ cycles. Observe that if μ is less than 1600 cycles, the normal curve of Figure 7.5 on page 233 is shifted to the left, and the area under the curve to the right of 1660 becomes less than 0.0304. Thus, if the null hypothesis is $\mu \le 1600$ cycles, we can say that the probability of a Type I error is at most 0.0304, and we write $\alpha \le 0.0304$. In general, if the null hypothesis is of the form $\mu \le \mu_0$ or $\mu \ge \mu_0$, we need only specify the maximum probability of a Type I error. By performing the test as if the null hypothesis were $\mu = \mu_0$, we protect ourselves against the worst possibility. (See the example on page 241.)

After the null hypothesis, the alternative hypothesis, and the level of significance have been specified, the remaining steps are as follows:

3. **Based on the sampling distribution of an appropriate statistic, we construct a criterion for testing the null hypothesis against the given alternative.**
4. **We calculate from the data the value of the statistic on which the decision is to be based.**
5. **We decide whether to reject the null hypothesis or whether to fail to reject it.**

In the electric car battery example we studied the criterion using the normal-curve approximation to the sampling distribution of the mean. In general, step 3 depends not only on the statistic on which we want to base the decision and on its sampling distribution, but also on the alternative hypothesis we happen to choose. In the car battery example we used a **one-sided criterion (one-sided test** or **one-tailed test)** with the one-sided alternative $\mu > 1600$ cycles, rejecting the null hypothesis only for large values of the sample mean. In the example dealing with coffee jars, we choose a **two-sided criterion (two-sided test** or **two-tailed test)** to go with the two-sided alternative $\mu \ne 4$ ounces. In general, a test is said to be two-sided if the null hypothesis is rejected for values of the test statistic falling into either tail of its sampling distribution.

The purpose of this discussion has been to introduce some of the basic problems connected with the testing of statistical hypotheses. Although the methods we have presented are objective—that is, two experimenters analyzing the same data under the same conditions would arrive at identical results—their use does entail some arbitrary, or subjective, considerations. For instance, in the example on page 233 it was partially a subjective decision to draw the line between satisfactory and unsatisfactory values of μ at 1660 cycles. It is also partially a subjective decision to use a sample of 36 batteries, and to reject the manufacturer's claim for values of $\overline{X}$ exceeding 1660 cycles. Approaching the problem differently, the government agency investigating the manufacturer's claim could have specified values of α and β, thus controlling the risks to which they are willing to be exposed. The choice of α, the probability of a Type I error, could have been based on the consequences of making that kind of error, namely, the manufacturer's cost of having a good product condemned, the possible cost of subsequent litigation, the manufacturer's cost of unnecessarily adjusting his machinery, the cost to the public of not having the product available when needed, and so forth. The choice of β, the probability of a Type II error, could similarly have been based on the consequences of making that kind of error. Namely, the cost to the public of buying an inferior product, the manufacturer's savings in using inferior components but loss in good will, again the cost of possible ligation, and so forth. It should be obvious that it would be extremely difficult to put cash values on all these eventualities, but they must nevertheless be considered, at least indirectly, in choosing suitable criteria for testing statistical hypotheses.

In this text we shall discuss mainly the **Neyman-Pearson theory**, also called the **classical theory of testing hypothesis**. This means that cost factors and other considerations that are partly arbitrary and partly subjective only informally affect the choice of a sample size, the choice of an alternative hypothesis, the choice of α and β, and so forth.

In this approach, the maximum value of α is controlled over the null hypothesis. For instance, if the null hypothesis is $\mu \leq \mu_0$, then μ as well as σ is unspecified. This is a **composite hypothesis**. Otherwise, if all the parameters are completely specified, the hypothesis is **simple**. Even with a composite null hypothesis, we set the critical region so that the error probability is α on the boundary $\mu = \mu_0$. Then, the error probabilities will be even smaller under values of μ that are less than μ_0.

Because the probability of falsely rejecting the null hypothesis is controlled, the null hypothesis is retained unless the observations strongly contradict it. Consequently, if the goal of an experiment is to establish an assertion or hypothesis, that hypothesis must be taken as the alternative hypothesis.

Guideline for selecting the null hypothesis

When the goal of an experiment is to establish an assertion, the negation of the assertion should be taken as the null hypothesis. The assertion becomes the alternative hypothesis.

Similar reasoning suggests that even when costs of making a wrong decision are difficult to determine but the consequences of one error are much more serious than for the other, the hypotheses should be labeled so that the most serious error is the Type I error.

The alternative hypothesis is often denoted by H_1. In the electric car battery example, $H_1: \mu > 1600$. Recalling the notation H_0 for the null hypothesis, we summarize the concepts and notation for hypotheses, types of error, and the probability of error.

Notation for the hypotheses

H_1: The **alternative hypothesis** is the claim we wish to establish.
H_0: The **null hypothesis** is the negation of the claim.

The two kinds of error and their probabilities are

The errors and their probabilities

Type I error: Rejection of H_0 when H_0 is true.
Type II error: Nonrejection of H_0 when H_1 is true.

α = probability of making a Type I error (also called the **level of significance**)
β = probability of making a Type II error

It is important to understand that tests of hypotheses are structured to control the probability, α, of falsely rejecting the null hypothesis. To interpret this, suppose the null hypothesis prevails and the test is repeated many times with independent sets of data. When $\alpha = 0.05$, the law of large numbers tells us that

$$\frac{\text{number of times the null hypothesis is falsely rejected}}{\text{number of times a test is conducted}} \rightarrow \alpha$$

This is, of course, the long-run frequency interpretation of probability. In the long run only 1 in 20 times will the null hypothesis be falsely rejected.

Before we discuss various special tests about means in the remainder of this chapter, let us point out that the concepts we have introduced here apply equally well to hypotheses concerning proportions, standard deviations, the randomness of samples, and relationships among several variables.

Exercises

7.39 A computer manufacturer wants to establish that the average time to set up a new laptop computer is less than 2 hours.

(a) Formulate the null and alternative hypotheses.

(b) What error could be made if $\mu = 1.9$? Explain in the context of the problem.

(c) What error could be made if $\mu = 2.0$? Explain in the context of the problem.

7.40 A manufacturer of four-speed clutches for automobiles claims that the clutch will not fail until after 50,000 miles.

(a) Interpreting this as a statement about the mean, formulate a null and alternative hypothesis for verifying the claim.

(b) If the true mean is 55,000 miles, what error can be made? Explain your answer in the context of the problem.

(c) What error could be made if the true mean is 50,000 miles?

7.41 An airline claims that the typical flying time between two cities is 56 minutes.

(a) Formulate a test of hypotheses with the intent of establishing that the population mean flying time is different from the published time of 56 minutes.

(b) If the true mean is 50 minutes, what error can be made? Explain your answer in the context of the problem.

(c) What error could be made if the true mean is 56 minutes?

7.42 A company wants to establish that the mean life of its batteries, when used in a wireless mouse, is over 183 days. The data will consist of the life lengths of batteries in 64 different wireless mice.

(a) Formulate the null and alternative hypotheses.

(b) If the true mean is 190 days, what error can be made? Explain your answer in the context of the problem.

7.43 A statistical test of hypotheses includes the step of setting a maximum for the probability of falsely rejecting the null hypothesis. Engineers make many measurements on critical bridge components to decide if a bridge is safe or unsafe.

(a) Explain how you would formulate the null hypothesis.

(b) Would you prefer $\alpha = 0.05$ or $\alpha = 0.01$? Explain your reasoning.

7.44 Suppose you are scheduled to ride a space vehicle that will orbit the earth and return. A statistical test of hypotheses includes the step of setting a maximum for the probability of falsely rejecting the null hypothesis. Engineers need to make various measurements to decide if it is safe or unsafe to launch the vehicle.

(a) Explain how you would formulate the null hypothesis.

(b) Would you prefer $\alpha = 0.05$ or $\alpha = 0.01$? Explain your reasoning.

7.45 Suppose that an engineering firm is asked to check the safety of a dam. What type of error would it commit if it erroneously rejects the null hypothesis that the dam is safe? What type of error would it commit if it erroneously fails to reject the null hypothesis that the dam is safe? Would would the likely impact of these errors be?

7.46 Suppose that we want to test the null hypothesis that an antipollution device for cars is effective. Explain under what conditions we would commit a Type I error and under what conditions we would commit a Type II error.

7.47 If the criterion on page 232 is modified so that the manufacturer's claim is accepted for $\overline{X} > 1640$ cycles, find

(a) the probability of a Type I error;

(b) the probability of a Type II error when $\mu = 1680$ cycles.

7.48 Suppose that in the electric car battery example on page 232, n is changed from 36 to 50 while everything else remains the same. Find

(a) the probability of a Type I error;

(b) the probability of a Type II error when $\mu = 1680$ cycles.

7.49 It is desired to test the null hypothesis $\mu = 100$ pounds against the alternative hypothesis $\mu < 100$ pounds on the basis of the weights of a random sample of size $n = 40$ packages shipped by truck. The population has $\sigma = 12$ pounds. For what values of $\overline{X}$ must the null hypothesis be rejected if the probability of a Type I error is to be $\alpha = 0.01$?

7.50 Several square inches of gold leaf are required in the manufacture of a high-end component. Suppose that, the population of the amount of gold leaf has $\sigma = 8.4$ square inches. We want to test the null hypothesis $\mu = 80.0$ square inches against the alternative hypothesis $\mu < 80.0$ square inches on the basis of a random sample of size $n = 100$.

(a) If the null hypothesis is rejected for $\overline{X} < 78.0$ square inches and otherwise it is accepted, what is the probability of Type I error?

(b) What is the answer to part (a) if the null hypothesis is $\mu \geq 80.0$ square inches instead of $\mu = 80.0$ square inches?

7.51 A producer of extruded plastic products finds that his mean daily inventory is 1,250 pieces. A new marketing policy has been put into effect and it is desired to test the null hypothesis that the mean daily inventory is still the same. What alternative hypothesis should be used if

(a) it is desired to know whether or not the new policy changes the mean daily inventory;

(b) it is desired to demonstrate that the new policy actually reduces the mean daily inventory;

(c) the new policy will be retained so long as it cannot be shown that it actually increases the mean daily inventory?

7.52 Specify the null hypothesis and the alternative hypothesis in each of the following cases.

(a) An automobile manufacturer wants to confirm a supplier's claim that the maximum resistance in a wiring harness is less than 50 ohms.

(b) An investigator wants to establish the research department's claim that a new ballast will increase mean bulb life to more than 5,000 hours.

7.7 Hypotheses Concerning One Mean

The example concerning the long life electric car batteries illustrates the basic terminology and principles of hypothesis testing. Now let us see how we proceed in actual practice. Suppose, for instance, we want to establish that the thermal conductivity of a certain kind of cement brick differs from 0.340, the value claimed. We will test on the basis of $n = 35$ determinations and at the 0.05 level of significance. From information gathered in similar studies, we can expect that the variability of such determinations is given by $\sigma = 0.010$.

Following the outline of the preceding section, we begin with steps 1 and 2 by writing

1. *Null hypothesis*: $\mu = 0.340$

   Alternative hypothesis: $\mu \neq 0.340$

2. *Level of significance*: $\alpha = 0.05$

The alternative hypothesis is two-sided because we shall want to reject the null hypothesis if the mean of the determinations is significantly less than or significantly greater than 0.340.

Next, in step 3, we depart from the procedure used in the example of the preceding section and base the test on the standardized statistic

Statistic for test concerning mean (σ known)

$$Z = \frac{\overline{X} - \mu_0}{\sigma/\sqrt{n}}$$

instead of $\overline{X}$. The reason for working with standard units, or Z values, is that it enables us to formulate criteria which are applicable to a great variety of problems, not just one.

If z_α is, as before, such that the area under the standard normal curve to its right equals α, the **rejection regions** or **critical regions**, namely, the sets of values of Z for which we reject the null hypothesis $\mu = \mu_0$, can be expressed as in the following table:

Level α Rejection Regions for Testing $\mu = \mu_0$ (normal population and σ known)	
Alternative hypothesis	**Reject null hypothesis if:**
$\mu < \mu_0$	$Z < -z_\alpha$
$\mu > \mu_0$	$Z > z_\alpha$
$\mu \neq \mu_0$	$Z < -z_{\alpha/2}$ or $Z > z_{\alpha/2}$

If $\alpha = 0.05$, the dividing lines, or **critical values**, of the criteria are -1.645 and 1.645 for the one-sided alternatives, and -1.96 and 1.96 for the two-sided alternative. If $\alpha = 0.01$, the dividing lines of the criteria are -2.33 and 2.33 for the one-sided alternatives, and -2.575 and 2.575 for the two-sided alternative. These results come from Example 6, Chapter 5 and Exercise 5.23.

Returning now to the example dealing with the thermal conductivity of the cement bricks where $\alpha = 0.05$, suppose that the mean of the 35 determinations is 0.343. We continue by writing

3. *Criterion*: Reject the null hypothesis if $Z < -1.96$ or $Z > 1.96$, where

$$Z = \frac{\overline{X} - \mu_0}{\sigma/\sqrt{n}}$$

4. *Calculations*:

$$Z = \frac{0.343 - 0.340}{0.010/\sqrt{35}} = 1.77$$

5. *Decision*: Since $Z = 1.77$ falls on the interval from -1.96 to 1.96, the null hypothesis cannot be rejected; to put it another way, the difference between $\overline{x} = 0.343$ and $\mu = 0.340$ can be attributed to chance. We actually never establish that the null hypothesis holds. Instead, the conclusion is that we fail to reject the null hypothesis.

In problems like this, many research workers accompany the calculated value of Z with a corresponding **tail probability**, or *P-value*, which is the probability of getting a difference between $\overline{x}$ and μ_0 greater than or equal to that actually observed. Figure 7.7 illustrates the reasoning behind the P-value for a two-sided test. It shows the observed value of the test statistic, in the example above, as a solid dot at $z = 1.77$. It also shows the corresponding potential extreme value in the lower tail, the same distance from 0, as the dotted circle at $z = -1.77$. Because the alternative is

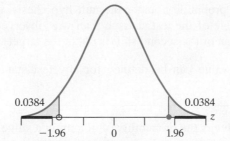

Figure 7.7
P-value $= 0.0384 + 0.0384 =$
0.0768 when $Z = 1.77$

two-sided, we must consider both of the tails to be more extreme than the observed value. The P-value is given by the total area under the standard normal curve to the left of -1.77 and to the right of 1.77, and it equals $2(1 - 0.9616) = 0.0768$. This P-value exceeds 0.05, which agrees with our earlier result. To review, Figure 7.7 shows that the rejection region, the value of Z, and the two tail areas.

EXAMPLE 18   **Calculating the P-value for a two-sided test**

A process for producing vinyl floor covering has been stable for a long period of time, and the surface hardness measurement of the flooring produced has a normal distribution with mean 4.5 and standard deviation $\sigma = 1.5$. A second shift has been hired and trained and their production needs to be monitored. Consider testing the hypothesis $H_0: \mu = 4.5$ versus $H_1: \mu \neq 4.5$. A random sample of hardness measurements is made of $n = 25$ vinyl specimens produced by the second shift. Calculate the P-value when using the test statistic

$$Z = \frac{\overline{X} - 4.5}{1.5/\sqrt{25}}$$

if $\overline{X} = 3.9$.

Solution   The observed value of the test statistic is

$$z = \frac{3.9 - 4.5}{1.5/\sqrt{25}} = -2.00$$

Figure 7.8 reviews the reasoning behind the P-value for a two-sided test. It shows the observed value of the test statistic as a solid dot at $z = -2.00$. It also shows the corresponding potential extreme value in the upper tail, at an equal distance from 0, as the dotted circle at $z = 2.00$. Because the alternative is two-sided, we must consider both large positive values of Z as well as large negative values to be more extreme than the observed value.

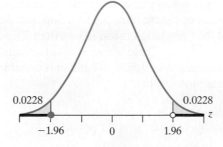

Figure 7.8
The P-value $= 0.0228 +$
$0.0228 = 0.0456$ for
$Z = -2.00$

From the normal table, $P(Z > 2.00) = 0.0228$. The probability that Z is smaller than -2.00 is also 0.0228. Consequently, the P-value is $0.0228 + 0.0228 = 0.0456$.

This is the probability, under the null hypothesis, of getting the same or a more extreme value of the test statistic than was observed. The small P-value suggests that the mean of the second shift is not at the target value of 4.5.   ■

The P-value can be defined for any test statistic in any hypotheses testing problem.

P-value for a given test statistic and null hypothesis

> The **P-value** is the probability of obtaining a value for the test statistic that is as extreme or more extreme than the value actually observed.
> Probability is calculated under the null hypothesis.

If the alternative hypothesis is right-sided—for instance, $H_1: \mu > \mu_0$—then only values greater than the observed value are more extreme. If the alternative hypothesis is left-sided, only values less than the observed value are more extreme. For two-sided alternatives, values in both tails need to be considered as in the example above.

Observe that giving a tail probability does not relieve us of the responsibility of specifying the level of significance before the test is actually performed.

The test we have described in this section is essentially an approximate large sample test. It is exact only when the population we are sampling is normal and σ is known. Typically, σ is unknown. If the sample size is large, we can approximate the original Z by substituting the sample standard deviation S for σ. This results in the **one-sample Z test**.

Statistic for large sample test concerning mean

$$Z = \frac{\overline{X} - \mu_0}{S/\sqrt{n}}$$

Level α Rejection Regions for Testing $\mu = \mu_0$ (large sample)	
Alternative hypothesis	**Reject null hypothesis if:**
$\mu < \mu_0$	$Z < -z_\alpha$
$\mu > \mu_0$	$Z > z_\alpha$
$\mu \neq \mu_0$	$Z < -z_{\alpha/2}$ or $Z > z_{\alpha/2}$

EXAMPLE 19   **A large sample test of the mean amount of cheese**

Refer to Example 8, Chapter 5, where the manufacturer of a pizza like product measures the amount of cheese used per run. Suppose that a consumer agency wishes to establish that the population mean is less than 71 pounds, the target amount established for this product. There are $n = 80$ observations and a computer calculation gives $\bar{x} = 68.45$ and $s = 9.583$. What can it conclude if the probability of a Type I error is to be at most 0.01?

Solution   1. *Null hypothesis*: $\mu \geq 71$ pounds
   Alternative hypothesis: $\mu < 71$ pounds

2. *Level of significance*: $\alpha \leq 0.01$

3. *Criterion*: Since the probability of a Type I error is greatest when $\mu = 71$ pounds, we proceed as if we were testing the null hypothesis $\mu = 71$ pounds

against the alternative hypothesis $\mu < 71$ pounds at the 0.01 level of significance. Thus, the null hypothesis must be rejected if $Z < -2.33$, where

$$Z = \frac{\overline{X} - \mu_0}{S/\sqrt{n}}$$

4. *Calculations*:

$$Z = \frac{68.45 - 71}{9.583/\sqrt{80}} = -2.38$$

5. *Decision*: Since $Z = -2.38$ is less than -2.33, the null hypothesis must be rejected at level of significance 0.01. In other words, the suspicion that $\mu < 71$ pounds is confirmed.

The small *P*-value 0.009, as shown in Figure 7.9, strengthens the conclusion. ■

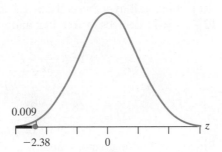

Figure 7.9
The *P*-value $= 0.009$ for $Z = -2.38$

0.009

-2.38 0 z

If the sample size is small and σ is unknown, the tests just described cannot be used. However, if the sample comes from a normal population (to within a reasonable degree of approximation), we can make use of the theory discussed in Section 6.3 and base the test of the null hypothesis $\mu = \mu_0$ on the statistic

Statistic for small sample test concerning mean (normal population)

$$t = \frac{\overline{X} - \mu_0}{S/\sqrt{n}}$$

which is a random variable having the *t* distribution with $n - 1$ degrees of freedom. The criteria for the **one sample *t* test** are like those for Z but are based on the *t* distribution.

Level α Rejection Regions for Testing $\mu = \mu_0$ (normal population and σ unknown) one sample *t* test	
Alternative hypothesis	Reject null hypothesis if:
$\mu < \mu_0$	$t < -t_\alpha$
$\mu > \mu_0$	$t > t_\alpha$
$\mu \neq \mu_0$	$t < -t_{\alpha/2}$ or $t > t_{\alpha/2}$

where t_α and $t_{\alpha/2}$ are based on $n - 1$ degrees of freedom.

EXAMPLE 20   **A t test of a normal population mean**

Scientists need to be able to detect small amounts of contaminants in the environment. As a check on current capabilities, measurements of lead content(μ g / L) are taken from twelve water specimens spiked with a known concentration. (courtesy of Paul Berthouex).

$$2.4 \quad 2.9 \quad 2.7 \quad 2.6 \quad 2.9 \quad 2.0 \quad 2.8 \quad 2.2 \quad 2.4 \quad 2.4 \quad 2.0 \quad 2.5$$

Test the null hypothesis $\mu = 2.25$ against the alternative hypothesis $\mu > 2.25$ at the 0.025 level of significance.

There are no outliers or other indications of non-normality. As a preliminary step we calculate $\bar{x} = 2.483$ and $s = 0.3129$.

Solution

1. *Null hypothesis*: $\mu = 2.25$ μg/L
   Alternative hypothesis: $\mu > 2.25$ μg/L

2. *Level of significance*: $\alpha = 0.025$

3. *Criterion*: Reject the null hypothesis if $t > 2.201$, where 2.201 is the value of $t_{0.025}$ for $12 - 1 = 11$ degrees of freedom and

$$t = \frac{\overline{X} - \mu_0}{S/\sqrt{n}}$$

4. *Calculations*:

$$t = \frac{2.483 - 2.25}{0.3129/\sqrt{12}} = 2.58$$

5. *Decision*: Since $t = 2.58$ is greater than 2.201, the null hypothesis must be rejected at level $\alpha = 0.025$. In other words, the mean lead content is above 2.25 μg/L. The exact tail probability, or *P*-value, cannot be determined from Table 4, but it is 0.013 (see Figure 7.10). The evidence against the mean lead content being 2.25 is even stronger than 0.025. Only about 13 in 1,000 times would we observe a value of t that is 2.58 or larger, if the mean really were 2.25 μg/L.

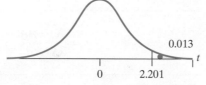

Figure 7.10
P-value = 0.013 for $t = 2.58$

Unfortunately, this statistical inference does not solve the original measurement capabilities question. The engineers did spike the samples but at level 1.25 μ g/L lead content. Either the laboratory procedure for determining lead content is producing high readings or the samples are contaminated. A new series of samples is needed to sort this out.

[Using **R**: data x   **t.test(x,mu=2.25, alt="greater",conf=.95)**
Drop **alt="greater",** for two-sided test or confidence interval]   ∎

Exercises

7.53 Refer to Exercise 7.1 where civil engineers recorded the amount of salt (tons) used to keep highways drivable during a snowstorm. The amount of salt for $n = 30$ storms has $\bar{x} = 1798.4$ tons and $s^2 = 671,330.9$, so $s = 819.35$ tons.

(a) Conduct a test of hypotheses with the intent of showing that the mean salt usage during a snowstorm is less than 2,000 tons. Take $\alpha = 0.05$.

(b) Based on your conclusion in part (a), what error could you have made? Explain in the context of the problem.

7.54 Refer to data in Exercise 7.3 on the labor time required to produce an order of automobile mufflers using a heavy stamping machine. The times (hours) for $n = 52$ orders of different parts has $\bar{x} = 1.865$ hours and $s^2 = 1.5623$, so $s = 1.250$ hours.

(a) Conduct a test of hypotheses with the intent of showing that the mean labor time is more than 1.5 hours. Take $\alpha = 0.05$.

(b) Based on your conclusion in part (a), what error could you have made? Explain in the context of the problem.

7.55 Refer to Exercise 7.5, where the number of unremovable defects, for each of $n = 45$ displays, has $\bar{x} = 2.467$ and $s = 3.057$ unremovable defects.

(a) Conduct a test of hypotheses with the intent of showing that the mean number of unremovable defects is less than 3.6. Take $\alpha = 0.025$.

(b) Based on your conclusion in part (a), what error could you have made? Explain in the context of the problem.

7.56 Refer to Exercise 7.12, where, at one step in a pilot plant process, the yield (gal) of carbon chains of length three was measured for $n = 9$ runs.

0.63 2.64 1.85 1.68 1.09 1.67 0.73 1.04 0.68

(a) Conduct a test of hypotheses with the intent of showing that the mean yield is less than 1.8. Take $\alpha = 0.05$ and assume a normal population.

(b) Based on your conclusion in part (a), what error could you have made? Explain in the context of the problem.

7.57 Refer to Exercise 7.14, where $n = 9$ measurements were made on a key performance indicator.

123 106 114 128 113 109 120 102 111

(a) Conduct a test of hypotheses with the intent of showing that the mean key performance indicator is different from 107. Take $\alpha = 0.05$ and assume a normal population.

(b) Based on your conclusion in part (a), what error could you have made? Explain in the context of the problem.

7.58 Refer to Exercise 7.22, where $n = 64$ cartridges for a copy machine produced a mean of 18,300 copies and a standard deviation of 2,800 copies.

(a) Conduct a test of hypotheses with the intent of showing that the mean number of copies is greater than 17,500 copies. Take $\alpha = 0.02$.

(b) Based on your conclusion in part (a), what error could you have made? Explain in the context of the problem.

7.59 Refer to Exercise 2.34, page 36, concerning material costs for rebuilding $n = 29$ traction motors. A computer calculation gives $\bar{x} = 1.4707$ and $s = 0.5235$ thousand dollars. At the 0.05 level of significance, conduct a test of hypotheses with the intent of showing the mean is greater than 1.3 thousand dollars.

7.60 In 64 randomly selected hours of production, the mean and the standard deviation of the number of acceptable pieces produced by a automatic stamping machine are $\bar{x} = 1,038$ and $s = 146$. At the 0.05 level of significance, does this enable us to reject the null hypothesis $\mu = 1,000$ against the alternative hypothesis $\mu > 1,000$?

7.61 With reference to the thickness measurements in Exercise 2.41, test the null hypothesis that $\mu = 30.0$ versus a two-sided alternative. Take $\alpha = 0.05$.

7.62 A random sample of 6 steel beams has a mean compressive strength of 58,392 psi (pounds per square inch) with a standard deviation of 648 psi. Use this information and the level of significance $\alpha = 0.05$ to test whether the true average compressive strength of the steel from which this sample came is 58,000 psi. Assume normality.

7.63 A manufacturer claims that the average tar content of a certain kind of cigarette is $\mu = 14.0$. In an attempt to show that it differs from this value, five measurements are made of the tar content (mg per cigarette):

14.5 14.2 14.4 14.3 14.6

Show that the difference between the mean of this sample, $\bar{x} = 14.4$, and the average tar claimed by the manufacturer, $\mu = 14.0$, is significant at $\alpha = 0.05$. Assume normality.

7.64 Suppose that in the preceding exercise the first measurement is recorded incorrectly as 16.0 instead of 14.5. Show that, even though the mean of the sample increases to $\bar{x} = 14.7$, the null hypothesis $H_0: \mu = 14.0$ is not rejected at level $\alpha = 0.05$. Explain the apparent paradox that even though the difference between observed $\bar{x}$ and μ has increased, the null hypothesis is no longer rejected.

7.65 The statistical program *MINITAB* will calculate t tests. With the nanopillar height data in C1,

Dialog box:
Stat > Basic Statistics > 1-Sample t. Click on box and type C1.
Click **Perform hypothesis test** and Type 300 in **Hypothesized mean**.
Choose **Options**. Type 0.95 in **Confidence level** and choose **not equal**. Click **OK**. Click **OK**

produces the output

Test of mu = 300 vs not = 300

	N	Mean	StDev	SE Mean	T	P
C1	50	305.580	36.971	5.229	1.07	0.291

You must compare your preselected α with the printed P-value in order to obtain the conclusion of your test. To perform a Z test, you need to find the P-value corresponding to the value 1.07 for the test statistic.

Here, with the two-sided alternative, we cannot reject $H_0: \mu = 300$ nm unless we take the significance level almost as large as 0.30.

(a) Test $H_0: \mu = 295$ with $\alpha = 0.05$.

(b) Test $H_0: \mu = 16.0$ for the speed of light data in Exercise 2.66 with $\alpha = 0.05$.

7.8 The Relation between Tests and Confidence Intervals

We now describe an important connection between tests for two-sided alternatives and confidence intervals. This relation provides the reason that most statisticians prefer the information available in a confidence interval statement as opposed to the information that the null hypothesis $\mu = \mu_0$ was or was not rejected.

To develop the relation, we consider the $(1 - \alpha)100\%$ confidence interval for μ given on page 221:

$$\bar{x} - t_{\alpha/2}\frac{s}{\sqrt{n}} < \mu < \bar{x} + t_{\alpha/2}\frac{s}{\sqrt{n}}$$

This interval is closely connected to a level α test of $H_0: \mu = \mu_0$ versus the two-sided alternative $H_1: \mu \neq \mu_0$. In terms of the values of $\bar{x}$ and s, this test has rejection region

$$\left|\frac{\bar{x} - \mu_0}{s/\sqrt{n}}\right| = |t| \geq t_{\alpha/2}$$

The acceptance region of this test is obtained by reversing the inequality to obtain all the values of $\bar{x}$ and s that do not lead to the rejection of $H_0: \mu = \mu_0$.

$$\text{acceptance region: } \left|\frac{\bar{x} - \mu_0}{s/\sqrt{n}}\right| < t_{\alpha/2}$$

The acceptance region can also be expressed as

$$\text{acceptance region: } \bar{x} - t_{\alpha/2}\frac{s}{\sqrt{n}} < \mu_0 < \bar{x} + t_{\alpha/2}\frac{s}{\sqrt{n}}$$

where the limits of the interval are identical to the preceding confidence interval. That is, the null hypothesis will not be rejected at level α if μ_0 lies within the $(1 - \alpha)100\%$ confidence interval for μ.

The $(1 - \alpha)100\%$ confidence interval gives the interval of plausible values for μ, so if μ_0 is contained in this interval, then it cannot be ruled out (i.e., cannot be rejected). The set of plausible values for μ, as determined by the $(1 - \alpha)100\%$ confidence interval, tells us at once about the outcome of all possible two-sided tests of hypothesis that specify a single value for μ.

EXAMPLE 21 **Illustrating the relation between tests and confidence intervals**

Referring to Example 20, $n = 12$ measurements of lead concentration yielded $\bar{x} = 2.483$ and $s = 0.3129$. Since $t_{0.025} = 2.201$ with 11 degrees of freedom, the 95% confidence interval is

$$\bar{x} - t_{\alpha/2}\frac{s}{\sqrt{n}} < \mu < \bar{x} + t_{\alpha/2}\frac{s}{\sqrt{n}}, \quad \text{or} \quad 2.28 < \mu < 2.68.$$

Use the relation between 95% confidence intervals and $\alpha = 0.05$ level tests to test the null hypothesis $\mu = 2.37$ versus two-sided the alternative hypothesis $\mu \neq 2.37$. Also test the null hypothesis $\mu = 2.26$ versus the alternative hypothesis $\mu \neq 2.26$.

Solution   In view of the relation just established, a test of $\mu = 2.37$ versus $\mu \neq 2.37$ would not reject H_0: $\mu = 2.37$, at the 5% level, since $\mu = 2.37$ falls within the confidence interval.

On the other hand $\mu = 2.26$ does not fall within the 95% confidence interval and hence that null hypothesis would be rejected at level $\alpha = 0.05$.   ∎

The relation between tests and confidence intervals holds quite generally. Suppose, for any value θ_0 of a parameter θ, we have a level α test of the null hypothesis $\theta = \theta_0$ versus the alternative $\theta \neq \theta_0$. Collect all the values θ_0 that would not be rejected. These form a $(1 - \alpha)100\%$ confidence interval for θ.

A confidence interval statement provides a more comprehensive inference than a statement concerning a two-sided test of a single null hypothesis. Consequently, we favor a confidence interval approach when one is available.

Exercises

7.66 Refer to the nonopillar height data on page 15. Using the 95% confidence interval, based on the t distribution, for the mean nanopillar height

N	Mean	StDev	SE Mean	95% CI
50	305.580	36.971	5.229	(295.073, 316.087)

   (a) decide whether or not to reject $H_0 : \mu = 320$ nm in favor of $H_1 : \mu \neq 320$ at $\alpha = 0.05$;

   (b) decide whether or not to reject $H_0 : \mu = 310$ nm in favor of $H_1 : \mu \neq 310$ at $\alpha = 0.05$.

   (c) What is your decision in part (b) if $\alpha = 0.02$? Explain.

7.67 Repeat Exercise 7.66 but replace the t test with the large sample Z test.

7.68 Refer to the green gas data on page 231. Using the 95% confidence interval, based on the t distribution for the mean yield

N	Mean	StDev	SE Mean	95% CI
15	6.00933	1.07780	0.27829	(5.41247, 6.60620)

   (a) decide whether or not to reject $H_0 : \mu = 5.5$ gal in favor of $H_0 : \mu \neq 5.5$ at $\alpha = 0.05$;

   (b) decide whether or not to reject $H_0 : \mu = 5.3$ gal in favor of $H_1 : \mu \neq 5.3$ at $\alpha = 0.05$.

   (c) Perform the t test for part (b) to verify your conclusion.

7.69 Refer to the labor time data in Exercise 7.3. Using the 90% confidence interval, based on the t distribution, for the mean labor time

N	Mean	StDev	SE Mean	90% CI
52	1.86462	1.24992	0.17333	(1.57423, 2.15500)

   (a) decide whether or not to reject $H_0 : \mu = 1.6$ in favor of $H_1 : \mu \neq 1.6$ at $\alpha = 0.10$;

   (b) decide whether or not to reject $H_0 : \mu = 2.2$ in favor of $H_1 : \mu \neq 2.2$ at $\alpha = 0.10$.

   (c) What is your decision in part (a) if $\alpha = 0.05$? Explain.

7.70 Repeat Exercise 7.69 but replace the t test with the large sample Z test.

7.9 Power, Sample Size, and Operating Characteristic Curves*

So far we have not paid much attention to Type II errors. In the electric car battery example of Section 7.6 we calculated one probability of a Type II error. Since the choice of the alternative hypothesis $\mu = 1680$ cycles in the electric car battery example was essentially arbitrary, it may be of interest to see how the testing procedure

*This section can be omitted on first reading.

will perform when other values of μ prevail. The actual mean battery life depends on the chemical reactions governing the battery life. Lacking a total scientific explanation, we do not know precisely what value to expect for this population mean. Consequently, we must investigate the probability of not rejecting (accepting) the null hypothesis under a range of possible values for μ. To this end, let

$$L(\mu) = \text{probability of accepting the null hypothesis when } \mu \text{ prevails}$$

Figure 7.11 presents the picture of a typical operating characteristic (OC) curve for the case where the alternative hypothesis is $\mu > \mu_0$. When the alternative hypothesis is $\mu < \mu_0$, the OC curve becomes the mirror image of that of Figure 7.11, reflected about the dashed vertical line through μ_0.

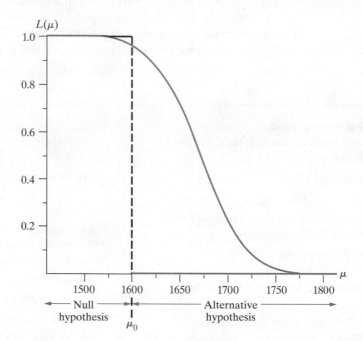

Figure 7.11
Operating characteristic curve

The function $L(\mu)$ completely characterizes the testing procedure whatever the value of population mean μ. If μ equals a value where the null hypothesis is true, then $1 - L(\mu)$ is the probability of the Type I error. When μ has a value where the alternative hypothesis is true, then $L(\mu)$ is the probability of a Type II error. That is, the function $L(\mu)$ carries complete information about the probabilities of both types of error.

To illustrate the calculation of $L(\mu)$, we continue with the electric car battery example on page 232, where we had $\mu_0 = 1600$, $\sigma = 192$, $n = 36$, and the dividing line of the criterion is $\bar{x} = 1660$. If the prevailing population mean is $\mu = 1640$, then $Z = \sqrt{n}(\bar{X} - 1640)/\sigma = 6(\bar{X} - 1640)/192$ is a standard normal variable. We reason that $L(1640)$ is the probability of observing

$$\bar{X} < 1660, \quad \text{or} \quad \frac{6(\bar{X} - 1640)}{192} < \frac{6(1660 - 1640)}{192} = 0.625$$

or $Z < 0.625$. Therefore, from Table 3, $L(1640) = 0.73$. Continuing with other possible values for μ, we obtain the results shown in the following table.

Values of μ	Probability of accepting null hypothesis
1560	0.999
1580	0.99
1600	0.97
1620	0.90
1640	0.74
1660	0.50
1680	0.27
1700	0.11
1720	0.03
1740	0.01
1760	0.001

Note that under the alternative hypothesis where $\mu > 1600$, the probability of committing a Type II error diminishes when μ is increased. Also the probability of not committing a Type I error approaches 1 when μ becomes much smaller than 1600 (and battery life is shorter).

The graph of $L(\mu)$ for various values of μ shown in Figure 7.11 is called the **operating characteristic curve**, or simply the **OC curve**, of the test.

In the context of sampling in order to decide whether or not to accept a shipment of electric car batteries on the basis of battery life, we would like to accept the shipment if the mean life is high and reject it if it is low. Based on the operating characteristic curve in Figure 7.11, the engineer can decide if the proposed procedure has small enough error probabilities at values of μ she deems important. Ideally we should want to reject the null hypothesis $\mu = \mu_0$ when actually μ exceeds μ_0, and to accept it when μ is less than or equal to μ_0. Thus, the ideal OC curve for our example would be given by the horizontal lines of Figure 7.11. In actual practice, OC curves can only approximate such ideal curves, with the approximation becoming better as the sample size is increased.

In contexts other than acceptance sampling plans, and for most statistical software packages, the performance of a test is expressed in terms of **power**:

$$\text{power} = \gamma(\mu) = P(\text{reject } H_0)$$

when μ is a value for the mean under the alternative hypothesis. Consequently, power and the operating characteristic are equivalent since $\gamma(\mu) = 1 - L(\mu)$.

For the one-sided alternative $H_0 : \mu > \mu_0$, with σ known, the rejection region for a level α Z test is

$$\sqrt{n}\,\frac{(\overline{X} - \mu_0)}{\sigma} > z_\alpha \quad \text{or} \quad \overline{X} > \frac{\sigma}{\sqrt{n}}\,z_\alpha + \mu_0$$

When the mean has the particular value μ_1 greater than μ_0, $\sqrt{n}\,(\overline{X} - \mu_1)/\sigma$ has a standard normal distribution and H_0 is rejected when

$$\overline{X} - \mu_1 > \frac{\sigma}{\sqrt{n}}\,z_\alpha + \mu_0 - \mu_1$$

The power at μ_1 is

$$\gamma(\mu_1) = P\left(\sqrt{n}\,\frac{(\overline{X} - \mu_1)}{\sigma} > z_\alpha + \sqrt{n}\,\frac{(\mu_0 - \mu_1)}{\sigma}\right)$$

$$= P\left(Z > z_\alpha + \sqrt{n}\,\frac{(\mu_0 - \mu_1)}{\sigma}\right)$$

when $H_1 : \mu > \mu_0$ and $\mu_1 > \mu_0$.

Similarly,

$$\gamma(\mu_1) = P\left(\sqrt{n}\,\frac{(\overline{X} - \mu_1)}{\sigma} < -z_\alpha + \sqrt{n}\,\frac{(\mu_0 - \mu_1)}{\sigma}\right)$$

$$= P\left(Z < -z_\alpha + \sqrt{n}\,\frac{(\mu_0 - \mu_1)}{\sigma}\right)$$

when $H_1\colon \mu < \mu_0$ and $\mu_1 < \mu_0$. Under a two-sided alternative,

$$\gamma(\mu_1) = P\left(Z < -z_{\alpha/2} + \sqrt{n}\,\frac{(\mu_0 - \mu_1)}{\sigma}\right) + P\left(Z > z_{\alpha/2} + \sqrt{n}\,\frac{(\mu_0 - \mu_1)}{\sigma}\right)$$

when $H_1\colon \mu \neq \mu_0$ and the population mean has value $\mu_1 \neq \mu_0$.

We observe that power and the Type II error probability β depend upon the

1. choice of significance level α. As α increases, power increases (β decreases).
2. difference between the null hypothesis value μ_0 and the particular value μ_1 under the alternative. Power increases (β decreases) as the difference between μ_0 and μ_1 increases.
3. value of the population standard deviation σ. Power decreases (β increases) as σ increases.
4. sample size n. Power increases (β decreases) as sample size increases.

Power calculations are most conveniently performed with statistical software (see Exercises 7.72 and 7.75).

EXAMPLE 22   **Determining the probability of a Type II error—one-sided test**

We want to investigate a claim about the average sound intensity of certain vacuum cleaners. Suppose the sound is a random variable having a normal distribution with a standard deviation of 3.6 dB. Specifically, we shall want to test the null hypothesis $\mu = 75.20$ against the alternative hypothesis $\mu > 75.20$ on the basis of measurements of the sound intensity of $n = 15$ of these machines. If the probability of a Type I error is to be $\alpha = 0.05$, what is the probability of a Type II error for $\mu = 77.00$ dB?

Solution   The test is one-sided, $\alpha = 0.05$, $z_{0.05} = 1.645$, and

$$z_\alpha + \sqrt{n}\,\frac{(\mu_0 - \mu_1)}{\sigma} = 1.645 + \sqrt{15}\,\frac{(75.20 - 77.0)}{3.6} = -0.291$$

so the power is $\gamma(77.0) = P(Z > -0.291) = 0.614$. The Type II error probability $\beta = 1 - 0.614 = 0.386$. ∎

EXAMPLE 23   **Determining the probability of a Type II error—two-sided test**

Suppose that the length of certain machine parts may be looked upon as a random variable having a normal distribution with a mean of 2.000 cm and a standard deviation of 0.050 cm. Specifically, we shall want to test the null hypothesis $\mu = 2.000$ against the alternative hypothesis $\mu \neq 2.000$ on the basis of the mean of a random sample of size $n = 30$. If the probability of a Type I error is to be $\alpha = 0.05$, what is the probability of a Type II error for $\mu = 2.010$?

Solution   The test is two-tailed, $\alpha = 0.05$, $\alpha/2 = 0.025$, $z_{0.025} = 1.96$, and

$$-z_{\alpha/2} + \sqrt{n}\,\frac{(\mu_0 - \mu_1)}{\sigma} = -1.96 + \sqrt{30}\,\frac{(2.000 - 2.010)}{0.050} = -3.055$$

$$z_{\alpha/2} + \sqrt{n}\,\frac{(\mu_0 - \mu_1)}{\sigma} = 1.96 + \sqrt{30}\,\frac{(2.000 - 2.010)}{0.050} = 0.865$$

The power is

$$\gamma(2.010) = P(Z < -3.055) + P(Z > .865) = 0.001 + 0.194 = 0.195$$

The Type II error probability $\beta = 1 - 0.195 = 0.805$.   ∎

When the alternative is one-sided, we can obtain an equation for the sample size required to give a specified power $\gamma = 1 - \beta$ at some value μ_1. For, $H_1 : \mu > \mu_0$ we require n so that

$$\gamma = 1 - \beta = P\left(Z > z_\alpha + \sqrt{n}\,\frac{(\mu_0 - \mu_1)}{\sigma}\right)$$

but then, by the definition z_β, we must have

$$-z_\beta = z_\alpha + \sqrt{n}\,\frac{(\mu_0 - \mu_1)}{\sigma}$$

Solving for n, we obtain the required sample size. The same expression for required sample size holds when $H_1 : \mu < \mu_0$.

Required sample size

When conducting a one-sided large sample Z test, the required sample size n must be at least as large as

$$n = \left(\sigma \frac{(z_\beta + z_\alpha)}{(\mu_0 - \mu_1)}\right)^2$$

For two-sided tests, the calculation of sample size is best relegated to computer software (see Exercise 7.72).

EXAMPLE 24   **Determining a sample size—one-sided test**

With reference to the electric car battery example on page 232, where we have $\mu_0 = 1600$ and $\sigma = 192$, when $\alpha = 0.05$, how large a sample do we need so that $\beta = 0.10$ for $\mu = 1680$?

Solution   The test is one-sided, $\mu_0 = 1600$, $\mu_1 = 1680$, and $\sigma = 192$. Also, $\alpha = 0.05$, $z_{0.05} = 1.645$, $\beta = 0.10$, $z_{0.10} = 1.28$. The required sample size is no smaller than

$$n = \left(\sigma \frac{(z_\beta + z_\alpha)}{(\mu_0 - \mu_1)}\right)^2 = \left(192 \frac{(1.28 + 1.645)}{(1600 - 1680)}\right)^2 = 49.2$$

so $n = 50$ is the required sample size.   ∎

Exercises

7.71 Refer to the example concerning average sound intensity on page 250. Calculate the power at $\mu_1 = 77$ when

(a) the level is changed to $\alpha = 0.03$.

(b) $\alpha = 0.05$ but the alternative is changed to the two-sided $H_1 : \mu \neq 75.2$.

7.72 *MINITAB* **calculation of power**
These calculations pertain to normal populations with known variance and provide an accurate approximation in the large sample case where σ is unknown. To calculate the power of the Z test at μ_1, you need to enter the

$$\text{difference} = \mu_1 - \mu_0.$$

Although $n = 15$ is not large, we illustrate with reference to the example concerning average sound intensity on page 250, where $\alpha = 0.05$, $\sigma = 3.6$, and $H_1 : \mu > 75.2$. We are given $\mu_1 = 77$, so the difference $= \mu_1 - \mu_0 = 77 - 75.2 = 1.80$.

Referring to the example of machine parts on page 250.

(a) Calculate the power at $\mu_1 = 2.020$.

(b) Repeat part (a) but take $\alpha = 0.03$.

Stat > Power and sample size > 1-Sample Z. Type 15 in **Sample sizes,** 1.8 in **differences** and 3.6 in **Standard deviation.** Click **Options** and choose **Greater than**. Type 0.05 in **Significance level.** Click **OK.** Click **OK.**

Notice that you have a choice of an alternative that is less than, not equal to or greater than.

Output: (partial)

Difference	Sample Size	Power
1.8	15	0.614718

7.73 Use computer software to repeat Exercise 7.71.

7.74 *MINITAB* **calculation of sample size**
Refer to Exercise 7.72, but this time leave **Sample size** blank but

Type *0.90* in **power**

to obtain the partial output concerning sample size

Difference	Sample Size	Target Power	Actual Power
1.8	35	0.9	0.905440

Refer to the example concerning sound intensity on page 250. Find the required sample size if power must be at least 0.96 at $\mu_1 = 77$.

7.75 *MINITAB* **calculation of power or OC curve**
Refer to the steps in Exercise 7.72, but enter a range of values for the difference. Here 0:3/.1 goes in steps from 0 to 3 in steps of .1 for Example 22.

Stat > Power and sample size > 1-Sample Z. Type 15 in **Sample sizes,** 0:3/.1 in **differences** and 3.6 in **Standard deviation**. Click **Options** and choose **Greater than**. Type 0.05 in **Significance level.** Click **OK.** Click **OK.**

With reference to the electric car battery example on page 232, use computer software to obtain the power curve for the $\alpha = 0.05$ one-sided test.

Do's and Don'ts

Do's

1. Calculate the estimated standard error $s/\sqrt{n}$ to accompany the point estimate $\bar{x}$ of a population mean.

2. Whatever the population, when the sample size is large, calculate the $100(1-\alpha)\%$ confidence interval for the mean

$$\bar{x} - z_{\alpha/2}\frac{s}{\sqrt{n}} < \mu < \bar{x} + z_{\alpha/2}\frac{s}{\sqrt{n}}$$

3. When the population is normal, calculate the $100(1-\alpha)\%$ confidence interval for the mean

$$\bar{x} - t_{\alpha/2}\frac{s}{\sqrt{n}} < \mu < \bar{x} + t_{\alpha/2}\frac{s}{\sqrt{n}}$$

where $t_{\alpha/2}$ is obtained from the t distribution with $n-1$ degrees of freedom.

4. Understand the interpretation of a $100(1-\alpha)\%$ confidence interval. When the population is normal, before the data are collected,

$$\left(\bar{X} - t_{\alpha/2}\frac{S}{\sqrt{n}}, \bar{X} + t_{\alpha/2}\frac{S}{\sqrt{n}}\right)$$

is a random interval that will cover the fixed unknown mean with probability $1-\alpha$. In many repeated applications of this method, about proportion $1-\alpha$ of the times the interval will cover the respective population mean.

5. When conducting a test of hypothesis, formulate the assertion that the experiment seeks to confirm as the alternative hypothesis.

6. When the sample size is large, base a test of the null hypothesis $H_0:\mu = \mu_0$ on the test statistic

$$\frac{\bar{X} - \mu_0}{S/\sqrt{n}}$$

which has, approximately, a standard normal distribution. When the population is normal, the same statistic has a t distribution with $n-1$ degrees of freedom.

7. Understand the interpretation of a level α test. If the null hypothesis is true, before the data are collected, the probability is α that the experiment will produce observations that lead to the rejection of the null hypothesis. Consequently, after many independent experiments, the proportion that lead to rejection of the null hypothesis will be nearly α.

Don'ts

1. Don't routinely apply the statistical procedures above if the sample is not random but collected from convenient units or the data show a trend in time.

Review Exercises

7.76 Specify the null hypothesis and the alternative hypothesis in each of the following cases.

(a) An engineer hopes to establish that an additive will increase the viscosity of an oil.

(b) An electrical engineer hopes to establish that a modified circuit board will give a computer a higher average operating speed.

7.77 With reference to Example 7 on page 19, find a 95% confidence interval for the mean strength of the aluminum alloy.

7.78 While performing a certain task under simulated weightlessness, the pulse rate of 32 astronaut trainees increased on the average by 26.4 beats per minute with a standard deviation of 4.28 beats per minute. What can one assert with 95% confidence about the maximum error if $\bar{x} = 26.4$ is used as a point estimate of the true average increase in the pulse rate of astronaut trainees performing the given task?

7.79 With reference to the preceding exercise, construct a 95% confidence interval for the true average increase in the pulse rate of astronaut trainees performing the given task.

7.80 It is desired to estimate the mean number of days of continuous use until a video surveillance system will first require repairs. If it can be assumed that $\sigma = 84$ days, how large a sample is needed so that one will be able to assert with 90% confidence that the sample mean is off by at most 10 days?

7.81 A sample of 12 camshafts intended for use in gasoline engines has an average eccentricity of 1.02 and a standard deviation of 0.044 inch. Assuming the data may be treated as a random sample from a normal population, determine a 95% confidence interval for the actual mean eccentricity of the camshaft.

7.82 In order to test the durability of a new paint, a highway department had test strips painted across heavily traveled roads in 15 different locations. If on the average the test strips disappeared after they had been crossed by 146,692 cars with standard deviation of 14,380 cars, construct a 99% confidence interval for the true average number of cars it takes to wear off the paint. Assume a normal population.

7.83 Referring to Exercise 7.82 and using 14,380 as an estimate of σ, find the sample size that would have been needed to be able to assert with 95% confidence that the sample mean is off by at most 10,000. [*Hint:* First estimate n_1 by using $z = 1.96$, then use $t_{0.025}$ for $n_1 - 1$ degrees of freedom to obtain a second estimate n_2, and repeat this procedure until the last two values of n thus obtained are equal.]

7.84 A laboratory technician is timed 20 times in the performance of a task, getting $\bar{x} = 7.9$ and $s = 1.2$ minutes. If the probability of a Type I error is to be at most 0.05, does this constitute evidence against the null hypothesis that the average time is less than or equal to 7.5 minutes?

7.85 Suppose that in the lithium car battery example on page 232, n is changed from 36 to 50 while the other quantities remain $\mu_0 = 1600$, $\sigma = 192$, and $\alpha = 0.03$. Find

(a) the new dividing line of the test criterion;

(b) the probability of Type II errors for the values of $\mu = 1620, 1640, 1660, 1680, 1700, 1720, 1740, 1760$ included in the table on page 249.

7.86 In an air-pollution study, ozone measurements were taken in a large California city at 5.00 P.M. The eight readings (in parts per million) were

   7.9   11.3   6.9   12.7   13.2   8.8   9.3   10.6

Assuming the population sampled is normal, construct a 95% confidence interval for the corresponding true mean.

7.87 An industrial engineer concerned with service at a large medical clinic recorded the duration of time from the time a patient called until a doctor or nurse returned the call. A sample of size 180 calls had a mean of 1.65 hours and a standard deviation of 0.82.

(a) Obtain a 95% confidence interval for the population mean of time to return a call.

(b) Does μ lie in your interval obtained in part (a)? Explain.

(c) In a long series or repeated experiments, with new random samples collected for each experiment, what proportion of the resulting confidence intervals will contain the true population mean? Explain your reasoning.

7.88 Refer to Exercise 7.87.

(a) Perform a test with the intention of establishing that the mean time to return a call is greater than 1.5 hours. Use $\alpha = 0.05$.

(b) In light of your conclusion in part (a), what error could you have made? Explain in the context of this problem.

(c) In a long series of repeated experiments, with new random samples collected for each experiment, what proportion of the resulting tests would reject the null hypothesis if it prevailed? Explain your reasoning.

7.89 The compressive strength of parts made from a composite material are known to be nearly normally distributed. A scientist, using the testing device for the first time, obtains the tensile strength (psi) of 20 specimens

   95   102   105   107   109   110   111   112   114   115
   134   135   136   138   139   141   142   144   150   155

shown in Figure 7.12. Should the scientist report the 95% confidence interval based on the t-distribution? Explain your reasoning.

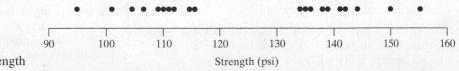

Figure 7.12
Dot diagram of tensile strength

Key Terms

CHAPTER

8

COMPARING TWO TREATMENTS

Advances occur in engineering when new ideas lead to better equipment, new materials, or revision of existing production processes. Any new procedure or device must be compared with the existing one and the amount of improvement assessed. As we describe in Section 8.1, statistical methods of comparison begin by deciding on one of two schemes for collecting data. Sections 8.2 and 8.3 concern the independent samples design and Section 8.4 the matched pairs sample design. These two designs are compared and contrasted in Section 8.5.

8.1 Experimental Designs for Comparing Two Treatments

Progress in science and engineering begins when new devices or materials are invented or when existing processes are revised. Advances occur whenever the new technique is shown to be better than the old. We perform experiments, collect data on performance, and then use statistical methods to make comparisons between the new and the old techniques.

It is common to use the statistical term **treatment** to refer to the procedures, machines, or processes that are being compared. The basic unit that is exposed to one treatment or the other is called the **experimental unit**, and the observed characteristic that forms the basis of the comparison is called the **response**.

EXAMPLE 1 **Randomly assigning treatment to units**

It is an extremely expensive event for a cell phone company when one of its relay towers breaks down. Because of a shortage of experts, sometimes a novice must be sent to fix the problem. A cell phone company wants to conduct an experiment to compare the average time for an expert to fix a problem and the average time for a novice to fix the problem.

The next 15 breakdowns can be used in the experiment. Describe the scheme for conducting a comparative experiment.

Solution A breakdown is the unit, and expert and novice will be Treatment 1 and Treatment 2, respectively. The response is the time for the tower to be fixed. One person will be assigned to each breakdown. The times for the group of breakdowns where novices are assigned should be independent of the times for the group of breakdowns where experts are assigned. This is the *independent samples design*.

Before we know anything about the breakdown, its location, or its possible severity, seven of the breakdowns should be chosen to receive Treatment 1, expert. This selection should be made using random numbers. Numbering the order of breakdowns from 1 to 15, seven different random numbers should be selected within this range. Randomization, without restriction, gives rise to the alternative name *completely randomized design*. ∎

EXAMPLE 2   **Pairing to eliminate a known source of variation**

A civil engineer needs to compare the durability of two different paints for marking lanes on a divided highway. One year after applying the paint, she will return and assign a number to the current quality of the marking. Ten widely separated 1-mile sections of divided highway are available. From experience, she expects that the sections would carry substantially different traffic volumes. Describe an experiment for making the comparison.

Solution   The two paints are the treatments which we call Treatment 1 and Treatment 2. The 1-mile sections, in either direction, will carry approximately the same traffic volume for the whole mile. A 1-mile section, with traffic going in one direction, is a unit. The response is quality of marking after one year. The civil engineer expects traffic volume to heavily influence the response. The traffic volume could be eliminated from the comparison if we pair the two 1-mile sections with traffic going in opposite directions. Then a comparison will only be made within the pair. This is called a *matched pairs design*.

Still the road for one side may be subject to more shade, higher temperature, or other conditions different from the other. For each 1-mile section, the engineer should flip a coin. If heads, the north or west direction receives Treatment 1. Randomization helps prevent these other uncontrolled variables from influencing the response in a systematic manner.

Note that by pairing like experimental units, we have eliminated traffic volume as an influencing variable.   ■

The term **experimental design** refers to the manner in which units are chosen and assigned to receive treatments. As introduced in the last two examples, there are two basic designs for comparing two treatments:

1. **Independent samples (complete randomization)**
2. **Matched pairs sample (randomization within each matched pair)**

We investigate the independent samples design in the next two sections and the matched pairs sample design in the following section.

8.2   Comparisons—Two Independent Large Samples

In this section, we consider the independent samples design when both sample sizes are large. To state the assumptions, we use X and Y for the observations and the subscript 1 or 2 for the mean and variance to distinguish the two populations.

Assumptions—Large Samples

1. $X_1, X_2, \ldots, X_{n_1}$ is a random sample of size n_1 from population 1 which has mean $= \mu_1$ and variance $= \sigma_1^2$.
2. $Y_1, Y_2, \ldots, Y_{n_2}$ is a random sample of size n_2 from population 2 which has mean $= \mu_2$ and variance $= \sigma_2^2$.
3. The two samples $X_1, X_2, \ldots, X_{n_1}$ and $Y_1, Y_2, \ldots, Y_{n_2}$ are independent.

Inferences will be made about the difference in means $\mu_1 - \mu_2 = \delta$. Since, by Theorem 6.1,

$$E(\overline{X}) = \mu_1 \qquad Var(\overline{X}) = \frac{\sigma_1^2}{n_1} \qquad E(\overline{Y}) = \mu_2 \qquad Var(\overline{Y}) = \frac{\sigma_2^2}{n_2}$$

the mean of $\overline{X} - \overline{Y}$ is $E(\overline{X} - \overline{Y}) = \mu_1 - \mu_2$ and, by independence (see Example 30, Chapter 5),

$$Var(\overline{X} - \overline{Y}) = \frac{\sigma_1^2}{n_1} + \frac{\sigma_2^2}{n_2}$$

When the sample sizes n_1 and n_2 are large, the central limit theorem implies that both $\overline{X}$ and $\overline{Y}$ are approximately normal. Because they are independent, their difference is also approximately normal and the **two sample Z statistic**

$$Z = \frac{\overline{X} - \overline{Y} - \delta}{\sqrt{\dfrac{\sigma_1^2}{n_1} + \dfrac{\sigma_2^2}{n_2}}}$$

is approximately standard normal. Because the sample sizes n_1 and n_2 are large— namely, both are greater than or equal to 30—the normal approximation remains valid when the sample variances replace the population variances.

<table>
<tr><td>

Statistic for large samples inference concerning difference between two means

</td><td>

When the sample sizes n_1 and n_2 are large—namely, $n_1, n_2 \geq 30$

$$Z = \frac{\overline{X} - \overline{Y} - \delta}{\sqrt{\dfrac{S_1^2}{n_1} + \dfrac{S_2^2}{n_2}}}$$

is approximately standard normal.

</td></tr>
</table>

Large Samples Confidence Intervals

Large samples confidence intervals, for the difference of means $\delta = \mu_1 - \mu_2$, are determined from the standard normal probability

$$1 - \alpha = P\left(-z_{\alpha/2} < \frac{\overline{X} - \overline{Y} - \delta}{\sqrt{\dfrac{S_1^2}{n_1} + \dfrac{S_2^2}{n_2}}} < z_{\alpha/2} \right)$$

$$= P\left(-z_{\alpha/2} \sqrt{\frac{S_1^2}{n_1} + \frac{S_2^2}{n_2}} < \overline{X} - \overline{Y} - \delta < z_{\alpha/2} \sqrt{\frac{S_1^2}{n_1} + \frac{S_2^2}{n_2}} \right)$$

or

$$1 - \alpha = P\left(\overline{X} - \overline{Y} - z_{\alpha/2} \sqrt{\frac{S_1^2}{n_1} + \frac{S_2^2}{n_2}} < \delta < \overline{X} - \overline{Y} + z_{\alpha/2} \sqrt{\frac{S_1^2}{n_1} + \frac{S_2^2}{n_2}} \right)$$

This last statement asserts that, before we obtain the data, the probability is $1 - \alpha$ that the random interval will cover the true unknown difference in the means $\delta = \mu_1 - \mu_2$.

<table>
<tr><td>

Confidence limits for large samples confidence interval for $\mu_1 - \mu_2$

</td><td>

Limits of $100(1 - d)\%$ confidence interval for $\mu_1 - \mu_2$

$$\overline{x} - \overline{y} \pm z_{\alpha/2} \sqrt{\frac{s_1^2}{n_1} + \frac{s_2^2}{n_2}}$$

</td></tr>
</table>

This confidence interval can also be obtained from the acceptance regions for the two-sided test on page 260.

EXAMPLE 3   **Large samples confidence interval for difference of means**

As a baseline for a study on the effects of changing electrical pricing for electricity during peak hours, July usage during peak hours was obtained for $n_1 = 45$ homes with air-conditioning and $n_2 = 55$ homes without.[1] The July on-peak usage (kWh) is summarized as

Population	Sample Size	Mean	Variance
With	45	204.4	13,825.3
Without	55	130.0	8,632.0

Obtain a 95% confidence interval for $\delta = \mu_1 - \mu_2$.

Solution   For a 95% confidence interval, $\alpha = 0.05$ and $z_{0.025} = 1.96$. We are given $n_1 = 45$, $n_2 - 55$, $\bar{x} = 204.4$, $s_1^2 = 13,825.3$, $\bar{y} = 130.0$, and $s_2^2 = 8,632.0$. Then the limits of the confidence interval are

$$\bar{x} - \bar{y} \pm z_{\alpha/2} \sqrt{\frac{s_1^2}{n_1} + \frac{s_2^2}{n_2}} = 204.4 - 130.0 \pm 1.96 \sqrt{\frac{13,825.3}{45} + \frac{8,632.0}{55}}$$

$$= 74.4 \pm 1.96 \sqrt{464.17}$$

so the 95% confidence interval is $(32.17, 116.63)$. The mean on-peak usage for homes with air-conditioning is higher than for homes without, from 32.17 to 116.63 kWh per month.

The confidence interval not only reveals that the two population means are statistically different, because the confidence interval does not cover 0, but also quantifies the amount of difference.   ∎

Large Samples Tests for Differences of Means

There are many statistical problems in which we are faced with decisions about the relative size of the means of two populations. For example, if two methods of welding are being considered, we may take samples and decide which is better by comparing their mean strengths.

Formulating the problem more generally, we shall consider two populations having the means μ_1 and μ_2 and the variances of σ_1^2 and σ_2^2. We want to test the null hypothesis

$$H_0: \mu_1 - \mu_2 = \delta_0$$

where δ_0 is a specified constant, on the basis of independent random samples of size n_1 and n_2. Analogous to the tests concerning one mean, we shall consider tests of this null hypothesis against each of the alternatives $\mu_1 - \mu_2 < \delta_0$, $\mu_1 - \mu_2 > \delta_0$, and $\mu_1 - \mu_2 \neq \delta_0$. The test itself will depend on the distance, measured in estimated standard deviation units, from the difference in sample means $\bar{X} - \bar{Y}$ to the hypothesized value, δ_0. When the sample sizes are large—namely, n_1 and n_2 are

---

[1] Richard Johnson and Dean Wichern (2007), *Applied Multivariate Statistical Analysis*, 6th ed., page 289, Prentice Hall: Upper Saddle River, NJ.

both greater than or equal to 30—we obtain this test statistic by specifying the null value δ_0 for the difference of means in the random Z defined above.

Test statistic for large samples concerning difference between two means

When $n_1, n_2 \geq 30$, test $H_0 : \mu_1 - \mu_2 = \delta_0$ using

$$Z = \frac{(\overline{X} - \overline{Y}) - \delta_0}{\sqrt{\dfrac{S_1^2}{n_1} + \dfrac{S_2^2}{n_2}}}$$

which has, approximately, a standard normal distribution.

Analogous to the table of level α rejection region on page 242, the rejection regions for testing the null hypothesis $\mu_1 - \mu_2 = \delta_0$, using the **two sample Z test** are;

Rejection Regions for Testing $\mu_1 - \mu_2 = \delta_0$ (normal populations and σ_1 and σ_2 known, or large samples $n_1, n_2 \geq 30$)	
Alternative hypothesis	**Reject null hypothesis if:**
$\mu_1 - \mu_2 < \delta_0$	$Z < -z_\alpha$
$\mu_1 - \mu_2 > \delta_0$	$Z > z_\alpha$
$\mu_1 - \mu_2 \neq \delta_0$	$Z < -z_{\alpha/2}$ or $Z > z_{\alpha/2}$

Although δ_0 can be any constant, it is worth noting that in the great majority of problems its value is zero and we test the null hypothesis of no difference, namely, the null hypothesis $\mu_1 = \mu_2$.

EXAMPLE 4   **A test for a the mean difference in driving performance**

With the goal of improving driving safety, engineers are quantifying the effects of such factors as drowsiness and alcohol on driver performance.[2] Volunteers drive a fixed course in a mid-sized car mounted in a sophisticated driving simulator. One performance measure is a standard deviation like score of the lateral deviation from center line.

We consider the experiment where the first treatment specifies that the driver has a blood alcohol reading of 0 percent and the second treatment specifies that the driver imbibe and then be carefully monitored to reach a blood alcohol reading of 0.10 percent. The summary statistics for one segment of the drive are (courtesy of John Lee)

Treatment 1 0 Blood Alcohol	Treatment 2 .1 % Blood Alcohol
$n_1 = 54$	$n_2 = 54$
$\bar{x} = 1.63$	$\bar{y} = 1.77$
$s_1 = 0.177$	$s_2 = 0.183$

[2]D. Das, S. Zhou, and John Lee, Differentiating alcohol-induced driving behavior using steering wheel signals, *IEEE Trans. Intell. Transp. Syst.* **13**, (2012), 1355–1368.

Conduct a test of hypotheses with the intent of establishing that the mean lateral deviation scores are different. Take $\alpha = 0.02$.

Solution   The test concerns $\delta = \mu_1 - \mu_2$ and the sample sizes $n_1 = 54$ and $n_2 = 54$ are large.

1. *Null hypothesis:* $\delta = 0$
   Alternative hypothesis: $\delta \neq 0$

2. *Level of significance:* $\alpha = 0.02$

3. *Criterion:* Reject the null hypothesis if $Z < -2.33$ or $Z > 2.33$ where Z is given by the large sample formula above.

4. *Calculations:* The observed value of the test statistic is

$$z = \frac{\bar{x} - \bar{y} - \delta_0}{\sqrt{\dfrac{s_1^2}{n_1} + \dfrac{s_2^2}{n_2}}} = \frac{1.63 - 1.77 - 0}{\sqrt{\dfrac{(0.177)^2}{54} + \dfrac{(0.183)^2}{54}}} = -4.04$$

5. *Decision:* Since $z = -4.04$ is less than -2.33, the null hypothesis must be rejected at level of significance 0.02. The small P-value 0.000053 (see Figure 8.1) provides very strong evidence that the mean lateral deviation scores for drinkers is different from that of non-drinkers.

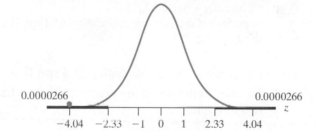

Figure 8.1
P-value for testing equality of mean lane deviation scores.

EXAMPLE 5   **Testing a difference in means with two large samples**

To test the claim that the resistance of electric wire can be reduced by more than 0.050 ohm by alloying, 32 values obtained for standard wire yielded $\bar{x} = 0.136$ ohm and $s_1 = 0.004$ ohm, and 32 values obtained for alloyed wire yielded $\bar{y} = 0.083$ ohm and $s_2 = 0.005$ ohm. At the 0.05 level of significance, does this support the claim?

Solution   1. *Null hypothesis:* $\mu_1 - \mu_2 = 0.050$
   Alternative hypothesis: $\mu_1 - \mu_2 > 0.050$

2. *Level of significance:* $\alpha = 0.05$

3. *Criterion:* Reject the null hypothesis if $Z > 1.645$, where Z is given by the large samples formula above.

4. *Calculations:*

$$z = \frac{0.136 - 0.083 - 0.050}{\sqrt{\dfrac{(0.004)^2}{32} + \dfrac{(0.005)^2}{32}}} = 2.65$$

5. *Decision:* Since $z = 2.65$ exceeds 1.645, the null hypothesis must be rejected; that is, the data substantiate the claim. From Table 3, the P-value is 0.004 (see Figure 8.2), so the evidence for alloying is very strong. Only 4 in 1,000 times would Z be at least 2.65, if the mean difference was 0.05.

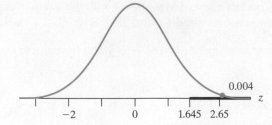

Figure 8.2
Large samples P-value =
0.004 for $Z = 2.65$

To judge the strength of support for the null hypothesis when it is not rejected, we consider Type II errors, for which the probabilities depend on the actual alternative differences $\delta' = \mu_1 - \mu_2$. Fortunately, these can be determined using the single sample results (as long as we are sampling from normal populations with known standard deviations or both samples are large).

The calculation of Type II error is based on the results for a single sample on page 250 and a fictitious single variance σ^2 and sample size n, where

$$\sigma^2 = \sigma_1^2 + \sigma_2^2 \qquad n = \frac{\sigma_1^2 + \sigma_2^2}{\dfrac{\sigma_1^2}{n_1} + \dfrac{\sigma_2^2}{n_2}}$$

Also, replace $\mu_0 - \mu_1$ by $\delta_0 - \delta'$.

The preferred method for calculation of Type II error is to use a computer software (see Exercise 7.72).

EXAMPLE 6   **Approximating the probability of Type II error**

With reference to the preceding example, what is the probability of a Type II error for $\delta' = 0.054$ ohms?

Solution   Refer to the formula on page 251. Since $\delta_0 = 0.050$, $\sigma^2 = 0.004^2 + 0.005^2 = 0.000041$, and $n = 32$

$$z_\alpha + \sqrt{n}\,\frac{(\delta_0 - \delta')}{\sigma} = 1.645 + \sqrt{32}\,\frac{(0.050 - 0.054)}{\sqrt{0.000041}} = 1.645 - 3.534 = -1.889$$

so the power is $\gamma(0.054) = P(Z > -1.889) = 0.971$. The Type II error probability $\beta = 1 - 0.971 = 0.029$. ■

8.3   Comparisons—Two Independent Small Samples

When n_1 and n_2 are both small and the population variances are unknown, we must impose additional assumptions, which we label 4 or 5 to emphasize that the three original assumptions still prevail.

Additional Assumptions for Small Samples

4. Both populations are normal.

5. The two standard deviations have a common value $\sigma_1 = \sigma_2 = \sigma$.

Because the populations are normal, $\overline{X}$ and $\overline{Y}$ are normal and, because they are independent, their difference is also normal (see Example 41, Chapter 5). Recall, $E(\overline{X} - \overline{Y}) = \mu_1 - \mu_2 = \delta$. Under the assumption of common standard deviation σ, the expression above for the variance of $\overline{X} - \overline{Y}$ becomes

$$Var(\overline{X} - \overline{Y}) = \frac{\sigma_1^2}{n_1} + \frac{\sigma_2^2}{n_2} = \sigma^2 \left(\frac{1}{n_1} + \frac{1}{n_2} \right)$$

and the standardized version of $\overline{X} - \overline{Y}$

$$Z = \frac{\overline{X} - \overline{Y} - \delta}{\sigma \sqrt{\frac{1}{n_1} + \frac{1}{n_2}}}$$

has a normal distribution.

The unknown σ must be estimated. Reasoning that each squared deviation $(X_i - \overline{X})^2$ is an estimate of σ^2 and so is each $(Y_i - \overline{Y})^2$ from the second sample, we estimate σ^2 by **pooling** the sums of squared deviations from the respective sample means. That is, we estimate σ^2 by the **pooled estimator**

$$S_p^2 = \frac{\sum_{i=1}^{n_1}(X_i - \overline{X})^2 + \sum_{i=1}^{n_2}(Y_i - \overline{Y})^2}{n_1 + n_2 - 2} = \frac{(n_1 - 1)S_1^2 + (n_2 - 1)S_2^2}{n_1 + n_2 - 2}$$

where $\sum(X_i - \overline{X})^2$ is the sum of the squared deviations from the mean for the first sample, while $\sum(Y_i - \overline{Y})^2$ is the sum of the squared deviations from the mean for the second sample. We divide by $n_1 + n_2 - 2$, since there are $n_1 - 1$ independent deviations from the mean in the first sample, $n_2 - 1$ in the second, and we have $n_1 + n_2 - 2$ independent deviations from their mean to estimate the population variance.

More specifically, from the single sample results we know that both S_1^2 and S_2^2 are estimates of σ^2 and that

$\frac{(n_1 - 1)S_1^2}{\sigma^2}$ has a chi square distribution with $n_1 - 1$ degrees of freedom

$\frac{(n_2 - 1)S_2^2}{\sigma^2}$ has a chi square distribution with $n_2 - 1$ degrees of freedom

and these two random quantities are independent since the samples on which they are based are independent. By either the result on page 201 or Example 14, Chapter 6, the sum of the two chi square variables has a chi square distribution with degrees of freedom equal to the sum of the two degrees of freedom $n_1 + n_2 - 2$. Further,

$$\frac{(n_1 - 1)S_1^2}{\sigma^2} + \frac{(n_2 - 1)S_2^2}{\sigma^2} = \frac{(n_1 + n_2 - 2)S_p^2}{\sigma^2}$$

so

$$\frac{\frac{(n_1 - 1)S_1^2}{\sigma^2} + \frac{(n_2 - 1)S_2^2}{\sigma^2}}{n_1 + n_2 - 2} = \frac{\frac{(n_1 + n_2 - 2)S_p^2}{\sigma^2}}{n_1 + n_2 - 2} = \frac{S_p^2}{\sigma^2}$$

and we conclude that

$$\frac{S_p}{\sigma} = \sqrt{\frac{\text{chi square variable}}{\text{degrees of freedom}}}$$

and this can be shown to be independent of the standard normal based on $\overline{X}_1 - \overline{X}_2$. Using the representation of t on page 201 as a standard normal over the square root of a chi square divided by its degrees of freedom, we obtain

Statistic for small sample test concerning difference between two means

$$t = \frac{\overline{X} - \overline{Y} - \delta}{S_p \sqrt{\dfrac{1}{n_1} + \dfrac{1}{n_2}}} \qquad \text{where } S_p^2 = \frac{(n_1 - 1)S_1^2 + (n_2 - 1)S_2^2}{n_1 + n_2 - 2}$$

has a t distribution with $n_1 + n_2 - 2$ degrees of freedom (d.f.).

Note that by substituting S_p for σ in the expression for Z on page 263, we arrive at the same statistic. With small sample sizes, the distribution is not standard normal but a t.

The criteria for the **two sample t test** based on this statistic are like those for Z for testing the null hypothesis: $H_0: \mu_1 - \mu_2 = \delta_0$.

Level α Rejection Regions for Testing $\mu_1 - \mu_2 = \delta_0$ (normal populations with $\sigma_1 = \sigma_2$) two sample t test	
Alternative hypothesis	**Reject null hypothesis if:**
$\mu_1 - \mu_2 < \delta_0$	$t < -t_\alpha$
$\mu_1 - \mu_2 > \delta_0$	$t > t_\alpha$
$\mu_1 - \mu_2 \neq \delta_0$	$t < -t_{\alpha/2}$ or $t > t_{\alpha/2}$

In the application of this test, n_1 and n_2 may be small, yet $n_1 + n_2 - 2$ may be 30 or more; in that case we use the normal critical value (also bottom line of Table 4.)

EXAMPLE 7   **A two sample t test to show a difference in strength**

To reduce the amount of recycled construction materials entering landfills it is crushed for use in the base of roadways. Green engineering practices require that their strength, resiliency modulus (MPa), be accessed. Measurements on $n_1 = n_2 = 6$ specimens of recycled materials from two different locations produce the data (Courtesy of Tuncer Edil)

Location 1 :	707	632	604	652	669	674
Location 2 :	552	554	484	630	648	610

Use the 0.05 level of significance to establish a difference in mean strength for materials from the two locations.

Solution   The test concerns $\delta = \mu_1 - \mu_2$ and the sample sizes $n_1 = n_2 = 6$ are small. There are no obvious departure from normality.

1. *Null hypothesis:* $\delta = 0$
   Alternative hypothesis: $\delta \neq 0$

2. *Level of significance:* $\alpha = 0.05$

3. *Criterion:* Reject the null hypothesis if $t < -t_{0.025}$ or $t > t_{0.025}$ where $t_{0.025} = 2.228$ for $6 + 6 - 2 = 10$ degrees of freedom.

4. *Calculations:* The means and variances of the two samples are

$$\bar{x} = 656.3 \qquad \bar{y} = 579.7$$

$$s_1^2 = \frac{6,389.3}{5} = 1,277.9 \qquad s_2^2 = \frac{18,699.3}{5} = 3,739.9$$

so that $s_P^2 = (6,389.3 + 18,699.3)/(5 + 5) = 2,508.5$, $s_P = 50.09$ and the observed value of the test statistic is

$$t = \frac{\bar{x} - \bar{y} - \delta_0}{s_P \sqrt{\frac{1}{n_1} + \frac{1}{n_2}}} = \frac{656.3 - 579.7 - 0}{50.09 \sqrt{\frac{1}{6} + \frac{1}{6}}} = 2.65$$

5. *Decision:* Since $t = 2.65$ is greater than 2.228, the null hypothesis must be rejected at the 0.05 level of significance. The P-value 0.0243 (see Figure 8.3) provides stronger evidence that the mean strength of recycled materials is different at the two locations.

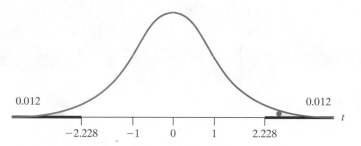

Figure 8.3
P-value for Example 7.

0.012   0.012

−2.228   −1   0   1   2.228 t

[Using **R**: data **x** and **y** **t.test(x,y,var.equal=T)**.] ∎

In the preceding example we went ahead and performed the two sample t test, tacitly assuming that the population variances are equal. Fortunately, the test is not overly sensitive to small differences between the population variances, and the procedure used in this instance is justifiable. As a rule of thumb, if one variance is four times the other, we should be concerned. A transformation will often improve the situation. As another alternative there is the Smith-Satterthwaite test discussed below.

Confidence intervals follow directly from the acceptance region for the tests. For two normal populations with equal variances,

Small sample confidence interval concerning difference between two means

> The $(1 - \alpha)100\%$ confidence interval for $\delta = \mu_1 - \mu_2$ has limits
>
> $$\bar{x} - \bar{y} \pm t_{\alpha/2} \sqrt{\frac{(n_1 - 1)s_1^2 + (n_2 - 1)s_2^2}{n_1 + n_2 - 2}} \sqrt{\frac{1}{n_1} + \frac{1}{n_2}}$$
>
> where $t_{\alpha/2}$ is based on $\nu = n_1 + n_2 - 2$ degrees of freedom.

EXAMPLE 8   **Graphics to accompany a two sample t test**

Example 7, Chapter 2, presents strength measurements on an aluminum alloy. A second alloy yielded measurements given in the following stem-and-leaf display. Find a 95% confidence interval for the difference in mean strength δ.

Solution   We first place the observations on the two alloys in stem-and-leaf displays. Note that the observations from the first alloy appear normal, but those on the second alloy may deviate from a normal distribution. Since the sample sizes are relatively large, this will not cause any difficulty.

Alloy 1, $N = 58$											Alloy 2, $N = 27$				
Leaf unit = 0.10											Leaf unit = 0.10				
66	4										66				
67	7										67				
68	0	0	3	4	6	8	9				68				
69	0	1	2	3	3	5	5	6	7	8 8 9	69				
70	0	0	1	2	3	3	4	5	6	6 8 9	70				
71	0	1	2	3	3	5	6	6	7	8 8 9	71	2	8		
72	1	2	3	4	6	7	9				72	6	8		
73	1	3	5								73	4	7	9	
74	2	5									74	4	9		
75	3										75	5	9		
76											76	3	5	7	9
77											77	1	3	6	7 8
78											78	1	2	4	6
79											79	0	3	8	

A computer calculation gives the sample means and standard deviations.

	N	MEAN	STDEV
ALLOY 1	58	70.70	1.80
ALLOY 2	27	76.13	2.42

From another computer calculation (or by interpolation in Table 4), we find $t_{0.025} = 1.99$ for 83 degrees of freedom, so the 95% confidence limits are

$$\bar{x} - \bar{y} \pm t_{\alpha/2} \sqrt{\frac{(n_1 - 1)s_1^2 + (n_2 - 1)s_2^2}{n_1 + n_2 - 2}} \sqrt{\frac{1}{n_1} + \frac{1}{n_1}}$$

$$= 70.70 - 76.13 \pm 1.99 \sqrt{\frac{57(1.80)^2 + 26(2.42)^2}{83}} \sqrt{\frac{1}{58} + \frac{1}{27}}$$

and

$$-6.4 < \mu_1 - \mu_2 < -4.5$$

We are 95% confident that the mean strength of alloy 2 is 4.5 to 6.4 thousand pounds per square inch higher than the mean strength of alloy 1.   ∎

It is good practice to show stem-and-leaf displays, boxplots, or histograms. Often they reveal more than a mean difference. For instance, in the last example, the first population is nearly symmetric but the second has a long tail to the left.

The large sample confidence interval is obtained from the acceptance regions for the test on page 260.

EXAMPLE 9   **Comparing the two confidence intervals**

Referring to the previous example, find the 95% large sample confidence interval.

Solution $$\bar{x} - \bar{y} \pm z_{\alpha/2} \sqrt{\frac{s_1^2}{n_1} + \frac{s_2^2}{n_2}} = 70.70 - 76.13 \pm 1.96 \sqrt{\frac{(1.80)^2}{58} + \frac{(2.42)^2}{27}}$$

so

$$-6.5 < \mu_1 - \mu_2 < -4.4$$

There is not much difference between this 95% confidence interval and the one in the previous example, where the variances were pooled.   ∎

Small Sample Sizes but Unequal Standard Deviations—Normal Populations

When we deal with independent random samples from normal populations whose variances seem to be unequal, we should not pool. As long as the populations are normal, an approximate t distribution is available for making inferences. The statistic t' is the same as the large samples statistic but, because sample sizes are small, its distribution is approximated as a t distribution.

Statistic for small samples inference, $\sigma_1 \neq \sigma_2$, normal populations

For normal populations, when the sample sizes n_1 and n_2 are not large and $\sigma_1 \neq \sigma_2$,

$$t' = \frac{(\overline{X} - \overline{Y}) - \delta}{\sqrt{\frac{S_1^2}{n_1} + \frac{S_2^2}{n_2}}}$$

is approximately distributed as a t with estimated degrees of freedom.

The estimated degrees of freedom for t' are calculated from the observed values of the sample variances s_1^2 and s_2^2.

$$\text{estimated degrees of freedom} = \frac{\left(\frac{s_1^2}{n_1} + \frac{s_2^2}{n_2} \right)^2}{\frac{(s_1^2/n_1)^2}{n_1 - 1} + \frac{(s_2^2/n_2)^2}{n_2 - 1}}$$

The estimated degrees of freedom are often rounded down to an integer so a t table can be consulted.

The test based on t' is called the **Smith-Satterthwaite test**.

EXAMPLE 10   **Testing equality of mean product volume**

One process of making green gasoline takes sucrose, which can be derived from biomass, and converts it into gasoline using catalytic reactions. This is not a process for making a gasoline additive but fuel itself, so research is still at the pilot plant stage. At one step in a pilot plant process, the product consists of carbon chains of length 3. Nine runs were made with each of two catalysts and the product volumes (gal) are

| Catalyst 1 | 0.63 | 2.64 | 1.85 | 1.68 | 1.09 | 1.67 | 0.73 | 1.04 | 0.68 |
| Catalyst 2 | 3.71 | 4.09 | 4.11 | 3.75 | 3.49 | 3.27 | 3.72 | 3.49 | 4.26 |

The sample sizes $n_1 = n_2 = 9$ and the summary statistics are

$$\bar{x} = 1.334, \qquad s_1^2 = 0.4548 \qquad \bar{y} = 3.766 \qquad s_2^2 = 0.1089$$

A chemical engineer wants to show that the mean product volumes are different. Test with $\alpha = 0.05$.

Solution   The test concerns $\delta = \mu_1 - \mu_2$ and the sample sizes $n_1 = n_2 = 9$ are small. We note that there are no outliers and no obvious departure from normality. However, $s_1^2 = 0.4548$ is more than four times $s_2^2 = 0.1089$. We should not pool.

1. *Null hypothesis*: $\delta = 0$
   Alternative hypothesis: $\delta \neq 0$

2. *Level of significance*: $\alpha = 0.05$

3. *Criterion*: We choose the Smith-Satterthwaite test statistic with $\delta_0 = 0$,

$$t' = \frac{\overline{X} - \overline{Y} - \delta_0}{\sqrt{\dfrac{S_1^2}{n_1} + \dfrac{S_2^2}{n_2}}}$$

The null hypothesis will be rejected if $t' < -t_{0.025}$ or $t' > t_{0.025}$, but the value of $t_{0.025}$ depends on the estimated degrees of freedom.

4. *Calculations*: As a first step, we estimate the degrees of freedom

$$\frac{\left(\dfrac{s_1^2}{n_1} + \dfrac{s_2^2}{n_2}\right)^2}{\dfrac{(s_1^2/n_1)^2}{n_1 - 1} + \dfrac{(s_2^2/n_2)^2}{n_2 - 1}} = \frac{\left(\dfrac{0.4548}{9} + \dfrac{0.1089}{9}\right)^2}{\dfrac{(0.4548/9)^2}{9 - 1} + \dfrac{(0.1089/9)^2}{9 - 1}} = 11.62$$

To use the t table, we round down to 11 and obtain $t_{0.025} = 2.201$. The observed value of the test statistic is

$$t' = \frac{\overline{X} - \overline{Y} - \delta_0}{\sqrt{\dfrac{S_1^2}{n_1} + \dfrac{S_2^2}{n_2}}} = \frac{1.334 - 3.766 - 0}{\sqrt{\dfrac{0.4548}{9} + \dfrac{0.1089}{9}}} = -9.71$$

5. *Decision*: Since $t' = -9.71$ is less than -2.201, the null hypothesis must be rejected at level of significance 0.05. The value of t' is so small that the *P*-value is 0.0000 when rounded. In other words, there is extremely strong evidence that the mean product volumes are different for the two catalysts.

[Using **R**: data **x** and **y**   **t.test(x,y)**   Use **alt="greater"**, for one-sided upper tail test.]   ∎

Confidence intervals can tell us what differences in means are plausible, not just that the means are different.

Confidence interval for
$\delta = \mu_1 - \mu_2$, **normal**
populations $\sigma_1 \neq \sigma_2$

A $100(1 - \alpha)\%$ confidence interval for $\delta = \mu_1 - \mu_2$

$$\left(\bar{x} - \bar{y} - t_{\alpha/2} \sqrt{\frac{s_1^2}{n_1} + \frac{s_2^2}{n_2}}, \; \bar{x} - \bar{y} + t_{\alpha/2} \sqrt{\frac{s_1^2}{n_1} + \frac{s_2^2}{n_2}} \right)$$

where $t_{\alpha/2}$ has the degrees of freedom estimated for t'.

EXAMPLE 11 **A confidence interval for the difference of mean yields when variances are unequal**

With reference to the previous example, obtain the 95% confidence interval for $\delta = \mu_1 - \mu_2$.

Solution From the previous example we have $\bar{x} = 1.334$, $s_1^2 = 0.4548$, $\bar{y} = 3.766$, $s_2^2 = 0.1089$, and $t_{0.025} = 2.201$ for 11 degrees of freedom. We get

$$\left(1.334 - 3.776 - 2.201\sqrt{\frac{0.45489}{9} + \frac{0.1089}{9}}, \right.$$

$$\left. 1.334 - 3.776 + 2.201\sqrt{\frac{0.45489}{9} + \frac{0.1089}{9}} \right)$$

or $(-2.982, -1.880)$ gallons. The mean product volume for the second catalyst is greater than that of the first catalyst by 1.880 to 2.982 gallons. ∎

Although the computations for unequal standard deviations seem tedious by hand, popular statistical software will allow this option in addition to pooling (see Exercise 8.35).

Exercises

8.1 Refer to Exercise 2.58, where $n_1 = 30$ specimens of 2×4 lumber have $\bar{x} = 1{,}908.8$ and $s_1 = 327.1$ psi. A second sample of size $n_2 = 40$ specimens of larger dimension, 2×6, lumber yielded $\bar{y} = 2{,}114.3$ and $s_2 = 472.3$. Test, with $\alpha = 0.05$, the null hypothesis of equality of mean tensile strengths versus the one-sided alternative that the mean tensile strength for the second population is greater than that of the first.

8.2 Refer to Exercise 8.1 and obtain a 95% confidence interval for the difference in mean tensile strength.

8.3 The dynamic modulus of concrete is obtained for two different concrete mixes. For the first mix, $n_1 = 33$, $\bar{x} = 115.1$, and $s_1 = 0.47$ psi. For the second mix, $n_2 = 31$, $\bar{y} = 114.6$, and $s_2 = 0.38$. Test, with $\alpha = 0.05$, the null hypothesis of equality of mean dynamic modulus versus the two-sided alternative.

8.4 Refer to Exercise 8.3 and obtain a 95% confidence interval for the difference in mean dynamic modulus.

8.5 An investigation of two kinds of photocopying equipment showed that 75 failures of the first kind of equipment took on the average 83.2 minutes to repair with a standard deviation of 19.3 minutes, while 75 failures of the second kind of equipment took on the average 90.8 minutes to repair with a standard deviation of 21.4 minutes.

(a) Test the null hypothesis $\mu_1 - \mu_2 = 0$ (namely, the hypothesis that on the average it takes an equal amount of time to repair either kind of equipment) against the alternative hypothesis $\mu_1 - \mu_2 \neq 0$ at the level of significance $\alpha = 0.05$.

(b) Using 19.3 and 21.4 as estimates of σ_1 and σ_2, find the probability of failing to reject the null hypothesis $\mu_1 - \mu_2 = 0$ with the criterion of part (a) when actually $\mu_1 - \mu_2 = -12$.

8.6 Studying the flow of traffic at two busy intersections between 4 P.M. and 6 P.M. (to determine the possible need for turn signals), it was found that on 40 weekdays there were on the average 247.3 cars approaching the first intersection from the south that made left turns while on 30 weekdays there were on the average 254.1 cars approaching the second intersection from the south that made left turns. The corresponding sample standard deviations are $s_1 = 15.2$ and $s_2 = 18.7$.

(a) Test the null hypothesis $\mu_1 - \mu_2 = 0$ against the alternative hypothesis $\mu_1 - \mu_2 \neq 0$ at the level of significance $\alpha = 0.01$.

(b) Using 15.2 and 18.7 as estimates of σ_1 and σ_2, find the probability of failing to reject (accepting) the null hypothesis $\mu_1 - \mu_2 = 0$ when actually $|\mu_1 - \mu_2| = 15.6$.

8.7 Given the $n_1 = 3$ and $n_2 = 2$ observations from Population 1 and Population 2, respectively,

Population 1	6	2	7
Population 2	14	10	

(a) Calculate the three deviations $x - \bar{x}$ and two deviations $y - \bar{y}$.

(b) Use your results from part (a) to obtain the pooled variance.

8.8 Two procedures for etching integrated circuits are to be compared. Given 10 units, five are prepared using etching procedure A and five are prepared using etching procedure B.

(a) How would you assign etching procedures to the 10 units?

(b) The response is the percent of area on the integrated circuit where the etching was inadequate. Suppose the results are

Procedure A	Procedure B
5	1
2	3
9	4
6	0
3	2

Find a 95% confidence interval for the difference in means.

(c) What assumptions did you make for your answer to part (b)?

8.9 Measuring specimens of nylon yarn taken from two spinning machines, it was found that 8 specimens from the first machine had a mean denier of 9.67 with a standard deviation of 1.81, while 10 specimens from the second machine had a mean denier of 7.43 with a standard deviation of 1.48. Assuming that the populations sampled are normal and have the same variance, test the null hypothesis $\mu_1 - \mu_2 = 1.5$ against the alternative hypothesis $\mu_1 - \mu_2 > 1.5$ at the 0.05 level of significance.

8.10 We know that silk fibers are very tough but in short supply. Breakthroughs by one research group result in the summary statistics for the stress (MPa) of synthetic silk fibers (Source: F. Teulé, et. al. (2012), Combining flagelliform and dragline spider silk motifs to produce tunable synthetic biopolymer fibers, *Biopolymers*, 97(6), 418–431.)

Small diameter	$n = 7$	$\bar{x} = 123.0$	$s_1 = 15.0$
Large diameter	$n = 6$	$\bar{y} = 92.0$	$s_2 = 21.0$

Use the 0.05 level of significance to test the claim that mean stress is largest for the small diameter fibers. Assume that both sampled populations have normal distributions with the same variance.

8.11 The following are the number of sales which a sample of 9 salespeople of industrial chemicals in California and a sample of 6 salespeople of industrial chemicals in Oregon made over a certain fixed period of time:

California: 59 68 44 71 63 46 69 54 48
Oregon: 50 36 62 52 70 41

Assuming that the populations sampled can be approximated closely with normal distributions having the same variance, test the null hypothesis $\mu_1 - \mu_2 = 0$ against the alternative hypothesis $\mu_1 - \mu_2 \neq 0$ at the 0.01 level of significance.

8.12 With reference to Example 5 construct a 95% confidence interval for the true difference between the average resistance of the two kinds of wire.

8.13 In each of the parts below, first decide whether or not to use the pooled estimator of variance. Assume that the populations are normal.

(a) The following are the Brinell hardness values obtained for samples of two magnesium alloys before testing:

Alloy 1: 66.3 63.5 64.9 61.8 64.3 64.7 65.1 64.5 68.4 63.2
Alloy 2: 71.3 60.4 62.6 63.9 68.8 70.1 64.8 68.9 65.8 66.2

Use the 0.05 level of significance to test the null hypothesis $\mu_1 - \mu_2 = 0$ against the alternative hypothesis $\mu_1 - \mu_2 < 0$.

(b) To compare two kinds of bumper guards, 6 of each kind, were mounted on a certain kind, of compact car. Then each car was run into a concrete wall at 5 miles per hour, and the following are the costs of the repairs (in dollars):

Bumper guard 1: 407 448 423 465 402 419
Bumper guard 2: 434 415 412 451 433 429

Use the 0.01 level of significance to test whether the difference between the two sample means is significant.

8.4 Matched Pairs Comparisons

In the application of the two sample t test we need to watch that the samples are really independent. For instance, the test cannot be used when we deal with "before and after" data, the IQs of husbands and wives, and numerous other kinds of situations where the data are naturally paired. Instead, comparisons are based on the **matched pairs**.

A manufacturer is concerned about the loss of weight of ceramic parts during a baking step. The readings before and after baking, on the same specimen, are naturally paired. It would make no sense to compare the before-baking weight of one specimen with the after-baking weight of another specimen. Let the pair of random variables (X_i, Y_i) denote the weight before and weight after baking for the ith specimen, for $i = 1, 2, \ldots, n$. A statistical analysis proceeds by considering the differences

$$D_i = X_i - Y_i \qquad \text{for } i = 1, 2, \ldots, n$$

This collection of (signed) differences is then treated as a random sample of size n from a population having mean μ_D. We interpret $\mu_D = 0$ as indicating that the means of the two responses are the same and $\mu_D > 0$ as indicating that the mean response of the first is higher than that of the second.

Tests of the null hypothesis $H_0: \mu_D = \mu_{D,0}$ are based on the ratio

$$\frac{\overline{D} - \mu_{D,0}}{S_D/\sqrt{n}} \quad \text{where} \quad \overline{D} = \frac{\sum\limits_{i=1}^{n} D_i}{n} \quad S_D^2 = \frac{\sum\limits_{i=1}^{n} (D_i - \overline{D})^2}{n-1}$$

If n is small, and the distribution of a difference is approximately normal, we treat this ratio as the one sample t statistic on page 243. Otherwise, we treat this ratio as the large sample statistic on page 242.

EXAMPLE 12   **Conducting a paired t test**

The following are the average weekly losses of worker-hours due to accidents in 10 industrial plants before and after a certain safety program was put into operation:

| Before: | 45 | 73 | 46 | 124 | 33 | 57 | 83 | 34 | 26 | 17 |
| After: | 36 | 60 | 44 | 119 | 35 | 51 | 77 | 29 | 24 | 11 |

Use the 0.05 level of significance to test whether the safety program is effective.

Solution   We cannot apply the independent samples test because the before and after weekly losses of worker-hours in the same industrial plant are correlated. Here there is the obvious pairing of these two observations.

1. *Null hypothesis*: $\mu_D = 0$
   Alternative hypothesis: $\mu_D > 0$

2. *Level of significance*: $\alpha = 0.05$

3. *Criterion*: Reject the null hypothesis if $t > 1.833$, the value of $t_{0.05}$ for $10 - 1 = 9$ degrees of freedom, where

$$t = \frac{\overline{D} - 0}{S_D/\sqrt{n}}$$

   and $\overline{D}$ and S_D are the mean and the standard deviation of the differences.

4. *Calculations*: The differences are

$$9 \quad 13 \quad 2 \quad 5 \quad -2 \quad 6 \quad 6 \quad 5 \quad 2 \quad 6$$

   their mean is $\overline{d} = 5.2$, their standard deviation is $s_D = 4.08$, so that

$$t = \frac{5.2 - 0}{4.08/\sqrt{10}} = 4.03$$

5. *Decision*: Since $t = 4.03$ exceeds 1.833, the null hypothesis must be rejected at level $\alpha = 0.05$. We conclude that the industrial safety program is effective. The evidence is very strong, since a computer calculation gives the P-value 0.0015 (see Figure 8.4). If $\mu_D = 0$, only in 15 out of 10,000 times would we observe t greater than or equal to 4.03.   ∎

Figure 8.4
The paired t test: P-value $=$ 0.0015 for $t = 4.03$ and degrees of freedom (d.f.) $= 9$

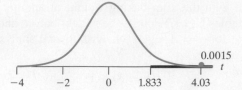

In connection with this kind of problem, the one sample t test is referred to as a **matched pairs t test** or just the **paired t test**.

EXAMPLE 13   **95% confidence interval for the mean of a paired difference**

Scientists are making a major breakthrough by creating devices that can smell toxic chemicals.[3] An array of sites, each coated with different nanoporous pigments, change colors when exposed to various chemicals. Computer software produces the numerical value of the change, or difference, by subtracting an initial scanned image from the image after exposure to the chemical. The red component of the difference of images, caused by exposure to a toxic level of formaldehyde, was measured seven times. (Courtesy of authors)

| 1.26 | 1.34 | 1.82 | 0.55 | 0.73 | 0.78 | 1.10 |

Construct a 95% confidence interval for the mean change of the red color component at this site when exposed to a toxic level of formaldehyde.

Solution   The sample size is $n = 7$ and $t_{0.025} = 2.447$ for $n - 1 = 6$ degrees of freedom. We first calculate

$$\bar{d} = 1.083 \quad \text{and} \quad s = 0.436$$

and the 95% confidence formula for μ_D becomes

$$1.083 - 2.447 \cdot \frac{0.436}{\sqrt{7}} < \mu_D < 1.083 + 2.447 \cdot \frac{0.436}{\sqrt{7}}$$

or $0.68 < \mu_D < 1.49$. We are 95% confident that the interval from 0.68 to 1.49 contains the mean change in the red color component. The mean change is different from zero.

This site by itself contributes a substantial information for detecting formaldehyde. By combining the measurements from all of the different sites in the array, scientists are actually able to identify many specific toxic chemicals. These arrays can actually smell.   ■

The next example illustrates some practical points when conducting a matched pairs experiment, including randomization.

EXAMPLE 14   **Comparing measurements made at two laboratories**

A state law requires municipal wastewater treatment plants to monitor their discharges into rivers and streams. A treatment plant could choose to send its samples to a commercial laboratory of its choosing. Concern over this self-monitoring led a civil engineer to design a matched pairs experiment.[4] Exactly the same bottle of

[3]Liang Feng, et. al., Colorimetric sensor array for determination and identification of toxic industrial chemicals, *Anal. Chem.*, 82 (2010), 9433–9440.

[4]R. Johnson and D. Wichern, (2007), *Applied Multivariate Statistical Analysis*, Prentice Hall, page 276.

effluent cannot be sent to two different laboratories. To match "identical" as closely as possible, she takes a sample of effluent in a large sample bottle and pours it back and forth over two open specimen bottles. When they are filled and capped, a coin is flipped to see if the one on the right was sent to Commercial Laboratory A or the Wisconsin State Laboratory of Hygiene. This process was repeated 11 times. The results, for the response suspended solids (SS) are

Sample	1	2	3	4	5	6	7	8	9	10	11
Commercial lab	27	23	64	44	30	75	26	124	54	30	14
State lab	15	13	22	29	31	64	30	64	56	20	21
Difference $x_i - y_i$	12	10	42	15	−1	11	−4	60	−2	10	−7

Obtain a 95% confidence interval and look for any unusual features in the data.

Solution   The sample size is relatively small so we assume normality and base the confidence interval on the t distribution. We have $n = 11$ and calculate $\bar{d} = 13.27$ and $s_D^2 = 418.61$. Then, with $n - 1 = 11 - 1 = 10$ degrees of freedom and $t_{0.025} = 2.228$, the 95% confidence interval is

$$\left(13.27 - 2.228 \sqrt{\frac{418.61}{11}}, \; 13.27 + 2.228 \sqrt{\frac{418.61}{11}} \right) \quad \text{or} \quad (-0.47, \; 27.01)$$

This 95% confidence interval just covers 0, so no difference is indicated with this small sample size. But wait, look at the dot diagram of the differences in Figure 8.5. There are two very large differences that would be unusual if the sample were taken from a normal population. The validity of the confidence interval is, at least, under suspicion. In Exercise 8.17 you are asked to try the square root transformation to see if it improves the situation.   ∎

Figure 8.5
Dot diagram of differences in suspended solids; outliers present

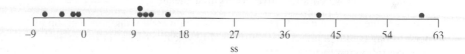

Exercises

8.14   A civil engineer wants to compare two instruments for measuring the amount of polychlorinated biphenyls (PCBs) in corn stalks. A sample of stalks is cut and crushed and two scoops of the material taken. One is measured with the first instrument and the other with the second instrument. This whole process is repeated five times. The results, in parts per billion, are as follows:

Sample No.	Instrument 1	Instrument 2
1	3	4
2	8	7
3	9	6
4	4	3
5	6	5

Find a 95% confidence interval for the mean difference in instrument readings assuming the differences have a normal distribution.

8.15 Refer to Exercise 8.14. Test, with $\alpha = 0.05$, that the mean difference is 0 versus a two-sided alternative.

8.16 The following data were obtained in an experiment designed to check whether there is a systematic difference in the weights obtained with two different scales:

	Weight in grams	
	Scale I	Scale II
Rock Specimen 1	11.23	11.27
Rock Specimen 2	14.36	14.41
Rock Specimen 3	8.33	8.35
Rock Specimen 4	10.50	10.52
Rock Specimen 5	23.42	23.41
Rock Specimen 6	9.15	9.17
Rock Specimen 7	13.47	13.52
Rock Specimen 8	6.47	6.46
Rock Specimen 9	12.40	12.45
Rock Specimen 10	19.38	19.35

Use the paired t test at the 0.05 level of significance to try to establish that the mean difference of the weights obtained with the two scales is nonzero.

8.17 Refer to Example 14 concerning suspended solids in effluent from a treatment plant. Take the square root of each of the measurements and then take the difference.

(a) Construct a 95% confidence interval for μ_D.

(b) Conduct a level $\alpha = 0.05$ level test of H_0: $\mu_D = 0$ against a two-sided alternative. Verify that the conclusion is the same as that obtained from the confidence interval.

(c) Make a dot diagram of these differences and decide if the transformation has essentially removed the outliers.

8.18 Refer to Example 14 concerning suspended solids in effluent from a treatment plant. Take the natural logarithm of each of the measurements and then take the difference.

(a) Construct a 95% confidence interval for μ_D.

(b) Conduct a level $\alpha = 0.05$ level test of H_0: $\mu_D = 0$ against a two-sided alternative. Verify that the conclusion is the same as that obtained from the confidence interval.

(c) Make a dot diagram of these differences and decide if the transformation has essentially removed the outliers.

8.19 A sunglass company wants potential customers to compare two types of lenses, the current lens A and one made of a new hi-tech material B. Glasses made with lens A and glasses made with lens B are available. Each person, in a sample of size 43, is asked to wear one of each type for a whole day. After a day in the sun, they are asked to score that day's pair on a scale of 1 to 7, with higher scores being best. The differences in scores

$$\text{(hi-tech lens } B) - \text{(current lens } A)$$

have mean 0.7 and variance 3.8. Construct a 95% confidence interval for the mean difference.

8.20 Referring to Example 13, conduct a test to show that the mean change μ_D is different from 0. Take $\alpha = 0.05$.

8.21 In a study of the effectiveness of physical exercise in weight reduction, a group of 16 persons engaged in a prescribed program of physical exercise for one month showed the following results:

Weight before (pounds)	Weight after (pounds)	Weight before (pounds)	Weight after (pounds)
209	196	170	164
178	171	153	152
169	170	183	179
212	207	165	162
180	177	201	199
192	190	179	173
158	159	243	231
180	180	144	140

Use the 0.01 level of significance to test whether the prescribed program of exercise is effective.

8.5   Design Issues—Randomization and Pairing

Often, the experimenter can make the choice of which of the two treatments is applied to an individual unit. Then it is possible to conduct either an independent samples comparison or a matched pairs comparison of means. The method of assignment of units to two groups or pairs can be crucial to the validity of the statistical procedures for comparing the means of two populations. We first emphasize the importance of randomization in the context of the independent samples design of a comparative experiment.

Independent Samples Design: Randomization

In many comparative studies the investigator applies one or the other treatment to an object we call an experimental unit. The method of assigning the treatments to the experimental units can be crucial to the validity of the statistical procedures. Suppose a chemist has a new formula for waterproofing that she applies to several pairs of shoes that are almost like new. She also applies the old formula to several pairs of scuffed shoes. At the end of a month, she will measure the ability of each pair of shoes to withstand water. It doesn't take a statistician to see that this is not a good experimental design. The persons with scuffed shoes probably walk a lot more and do so in all kinds of weather. These sources of variation could very well lead to systematic biases that make the new formula seem better than the old even when this is not the case. The pairs of shoes need to be assigned to the treatments with old and new waterproofing formula in a random manner.

When possible, the $n = n_1 + n_2$ experimental units should be assigned at random to the two treatments. This means that all $\binom{n}{n_1}$ possible selections of n_1 units to receive the first treatment are equally likely. Practically, the assignment is accomplished by selecting n_1 random integers between 1 and n. The corresponding experimental units are assigned to the first treatment. Generally, a test will have more power if the two sample sizes are equal.

In summary, we must actively assign treatments at random to experimental units. This process is called **randomization**.

Purpose of randomization

> **Randomization** of treatments helps prevent uncontrolled sources of variation from exerting a systematic influence on the responses.

Matched Pairs Design: Pairing and Randomization

The object of **pairing** experimental units, according to a characteristic that is likely to influence the response, is to eliminate this source of variation from the comparison. In the context of waterproofing for shoes, each person could have the old formula on one shoe and the new formula on the other. Since the paired t analysis only uses differences from the same pair, this experimental strategy should eliminate most of the variation in response due to different terrain, distance covered, and weather conditions.

> **Pairing** according to some variable(s) thought to influence the response will remove the effect of that variable from analysis.

Even after units are paired, there is a need for randomization. For each pair, a fair coin should be flipped to assign the treatments. In the context of the waterproofing example, the old formula could be applied to the right shoe if heads and left shoe if tails. The new formula is applied to the other shoe. This randomization, restricted to be within pairs, would prevent systematic influences such as those caused by the fact that a majority of persons would tend to kick things with their right shoe.

> **Randomizing** the assignment of treatments within a pair helps prevent any other uncontrolled variables from influencing the responses in a systematic manner.

Notice that in Example 12 the experimenter had no control over the before and after. Many uncontrolled variables may also have changed over the course of the experiment: fewer working hours due to strikes, phasing out of an old type of equipment, etc. One of these could have been the cause for the improvement rather than the safety program.

We pursue the ideas of randomization and blocking in Chapter 12. Our purpose here was to show what practical steps can be taken to meet the idealistic assumptions of random samples when comparing two treatments.

Exercises

8.22 An investigator wants to compare two busy network protocols by recording the number of messages that are successfully passed by the network in a day. Describe how to select 5 of the next 10 working days for trying Protocol 1. Protocol 2 would be tried on the other 5 days.

8.23 An electrical engineer has developed a modified circuit board for elevators. Suppose 3 modified circuit boards and 6 elevators are available for a comparative test of the old versus the modified circuit boards.

(a) Describe how you would select the 3 elevators in which to install the modified circuit boards. The old circuit boards will be installed in the other 3 elevators.

(b) Alternatively, describe how you would conduct a paired comparison and then randomize within the pair.

8.24 It takes an average of 10 weeks to train a typical employee to run a computer-aided machine. The instructor has a new approach that she feels will lead to faster learning. She intends to teach 5 persons by the new method and then compare their learning times with those of 5 randomly selected persons trained by the old method. In order to obtain 5 students from the 25 available candidates, she asks for volunteers. Why is this a bad idea?

8.25 How would you randomize, for a two sample test, if 50 cars are available for an emissions study and you want to compare a modified air pollution device with that used in current production?

Do's and Don'ts

Do's

1. When sample sizes are large, determine the limits of a $100(1 - \alpha)\%$ confidence interval for the difference of means $\mu_1 - \mu_2$ as

$$\bar{x} - \bar{y} \pm z_{\alpha/2} \sqrt{\frac{s_1^2}{n_1} + \frac{s_2^2}{n_2}}$$

2. When each of the two samples are from normal populations having the same variance, determine the limits of a $100(1 - \alpha)\%$ confidence interval for the difference of means $\mu_1 - \mu_2$ as

$$\bar{x} - \bar{y} \pm t_{\alpha/2}\, s_p \sqrt{\frac{1}{n_1} + \frac{1}{n_2}}$$

where the pooled estimate of variance

$$s_p^2 = \frac{(n_1 - 1)\,s_1^2 + (n_2 - 1)\,s_2^2}{(n_1 - 1) + (n_2 - 1)}$$

and $t_{\alpha/2}$ is based on $n_1 + n_2 - 2$ degrees of freedom.

3. When analyzing data from a matched pair design, use the results for one sample but applied to the differences from each matched pair. If the difference of paired measurements has a normal distribution, determine a $100(1 - \alpha)\%$ confidence interval for the mean difference μ_D as

$$\left(\bar{d} - t_{\alpha/2}\, \frac{s_D}{\sqrt{n}},\ \bar{d} + t_{\alpha/2}\, \frac{s_D}{\sqrt{n}} \right)$$

where $t_{\alpha/2}$ is based on $n - 1$ degrees of freedom.

4. When comparing two treatments using the independent samples design, randomly assign the treatments to groups whenever possible. With the matched pair design, randomly assign the treatments within each pair.

Don'ts

1. Don't pool the two sample variances s_1^2 and s_2^2 if they are very different. We suggest a factor of 4 as being too different.

Review Exercises

8.26 With reference to Exercise 2.64, test that the mean charge of the electron is the same for both tubes. Use $\alpha = 0.05$.

8.27 With reference to the previous exercise, find a 90% confidence interval for the difference of the two means.

8.28 Two chemical additives for drying paint are to be compared. Five spray cans are prepared using Additive A and six are prepared using Additive B. Then 11 different boards are sprayed, one can per board.

(a) The response is the time in minutes for the surface to dry, and the summary statistics are

	Sample size	Mean	Standard deviation
Additive A	5	16.3	2.7
Additive B	6	12.1	1.1

Should you pool or not pool the estimates of variance in order to conduct a test of hypotheses that is intended to show that there is a difference in means? Explain how you would proceed.

(b) Conduct the test for part (a) using $\alpha = 0.05$.

(c) Describe how you would randomize the assignment of paints when conducting this experiment.

8.29 With reference to Example 2, Chapter 2, test that the mean copper content is the same for both heats.

8.30 With reference to the previous exercise, find a 90% confidence interval for the difference of the two means.

8.31 Random samples are taken from two normal populations with $\sigma_1 = 10.8$ and $\sigma_2 = 14.4$ to test the null hypothesis $\mu_1 - \mu_2 = 53.2$ against the alternative hypothesis $\mu_1 - \mu_2 > 53.2$ at the level of significance

$\alpha = 0.01$. Determine the common sample size $n = n_1 = n_2$ that is required if the probability of not rejecting the null hypothesis is to be 0.09 when $\mu_1 - \mu_2 = 66.7$.

8.32 With reference to Example 8, find a 90% confidence interval for the difference of mean strengths of the alloys

   (a) using the pooled procedure;

   (b) using the large samples procedure.

8.33 How would you randomize, for a two sample test, in each of the following cases?

   (a) Twenty cars are available for a mileage study and you want to compare a modified spark plug with the regular.

   (b) A new oven will be compared with the old. Fifteen ceramic specimens are available for baking.

8.34 With reference to part (a) of Exercise 8.33, how would you pair and then randomize for a paired test?

8.35 Two samples in C1 and C2 can be analyzed using the *MINITAB* commands

---

Dialog box:

Stat > Basic Statistics > 2-Sample t
Pull down **Each sample in its own column**
Type *C*1 in **Sample 1** and *C*2 in **Sample 2**.
Click **Options** and then **Assume equal variances**.
Click **OK**. Click **OK**.

If you do not click **Assume equal variances**, the Smith-Satterthwaite test is performed.

The output relating to Example 8 is

TWO SAMPLE T FOR ALLOY 1 VS ALLOY 2

	N	MEAN	STDEV	SE MEAN
ALLOY 1	58	70.70	1.80	0.24
ALLOY 2	27	76.13	2.42	0.47

95 PCT C1 FOR MU ALLOY 1 − MU ALLOY 2: (−6.36, −4.50)

T TEST MU ALLOY 1 = MU ALLOY 2 (VS NE): T = −11.58 P = 0.000 DF = 83.0

Perform the test for the data in Exercise 8.11.

8.36 Refer to Example 13 concerning an array of sites that smell toxic chemicals. When exposed to the common manufacturing chemical Arsine, a product of arsenic and acid, the change in the red component is measured six times. (Courtesy of the authors)

   0.10 −0.33 −1.12 −1.95 −3.63 −1.48

   (a) Test, with $\alpha = 0.05$, that the mean change μ_D is different from 0.

   (b) Obtain a 95% confidence interval for the mean change μ_D.

8.37 Refer to Example 12 concerning the improvement in lost worker-hours. Obtain a 90% confidence interval for the mean of this paired difference.

Key Terms

Table 8.1 Summary of the formulas for inferences about a mean (μ), or a difference of two means ($\mu_1 - \mu_2$)

Confidence interval = Point estimator $\pm$ (Tabled value) (Estimated or true std. dev.)

$$\text{Test statistic} = \frac{\text{Point estimator} - \text{Parameter value at } H_0 \text{ (null hypothesis)}}{\text{(Estimated or true) std. dev. of point estimator}}$$

Population(s)	Single sample		Independent samples			Matched pairs
	General	Normal with unknown σ	Normal $\sigma_1 = \sigma_2 = \sigma$	Normal $\sigma_1 \neq \sigma_2$	General	Normal for the difference $D_i = X_i - Y_i$
Inference on	Mean μ	Mean μ	$\mu_1 - \mu_2 = \delta$	$\mu_1 - \mu_2 = \delta$	$\mu_1 - \mu_2 = \delta$	μ_D
Sample(s)	$X_1, \ldots, X_n$	$X_1, \ldots, X_n$	$X_1, \ldots, X_{n_1}$ $Y_1, \ldots, Y_{n_2}$	$X_1, \ldots, X_{n_1}$ $Y_1, \ldots, Y_{n_2}$	$X_1, \ldots, X_{n_1}$ $Y_1, \ldots, Y_{n_2}$	$D_1 = X_1 - Y_1$ $\ldots$ $D_n = X_n - Y_n$
Sample size n	Large $n \geq 30$	$n \geq 2$	$n_1 \geq 2$ $n_2 \geq 2$	$n_1 \geq 2$ $n_2 \geq 2$	$n_1 \geq 30$ $n_2 \geq 30$	$n \geq 2$
Point estimator	$\bar{X}$	$\bar{X}$	$\bar{X} - \bar{Y}$	$\bar{X} - \bar{Y}$	$\bar{X} - \bar{Y}$	$\bar{D} = \bar{X} - \bar{Y}$
Variance of point estimator	$\dfrac{\sigma^2}{n}$	$\dfrac{\sigma^2}{n}$	$\sigma^2\left(\dfrac{1}{n_1} + \dfrac{1}{n_2}\right)$	$\dfrac{\sigma_1^2}{n_1} + \dfrac{\sigma_2^2}{n_2}$	$\dfrac{\sigma_1^2}{n_1} + \dfrac{\sigma_2^2}{n_2}$	$\dfrac{\sigma_D^2}{n}$
Estimator std. dev.	$\dfrac{S}{\sqrt{n}}$	$\dfrac{S}{\sqrt{n}}$	$S_p\sqrt{\dfrac{1}{n_1} + \dfrac{1}{n_2}}$	$\sqrt{\dfrac{S_1^2}{n_1} + \dfrac{S_2^2}{n_2}}$	$\sqrt{\dfrac{S_1^2}{n_1} + \dfrac{S_2^2}{n_2}}$	$\dfrac{S_D}{\sqrt{n}}$
Distribution	Normal	t with d.f. $= n - 1$	t with d.f. $= n_1 + n_2 - 2$	t with d.f. estimated[†]	Normal	t with d.f. $= n - 1$
Test statistic	$\dfrac{\bar{X} - \mu_0}{S/\sqrt{n}}$	$\dfrac{\bar{X} - \mu_0}{S/\sqrt{n}}$	$\dfrac{(\bar{X} - \bar{Y}) - \delta_0}{S_p\sqrt{\dfrac{1}{n_1} + \dfrac{1}{n_2}}}$	$\dfrac{(\bar{X} - \bar{Y}) - \delta_0}{\sqrt{\dfrac{S_1^2}{n_1} + \dfrac{S_2^2}{n_2}}}$	$\dfrac{(\bar{X} - \bar{Y}) - \delta_0}{\sqrt{\dfrac{S_1^2}{n_1} + \dfrac{S_2^2}{n_2}}}$	$\dfrac{\bar{D} - \mu_{D,0}}{S_D/\sqrt{n}}$

$$S_p^2 = \frac{(n_1 - 1)S_1^2 + (n_2 - 1)S_2^2}{n_1 + n_2 - 2}$$

$S_D =$ sample std. dev. of the D_i's

†d.f. $= [(s_1^2/n_1) + (s_2^2/n_2)]^2 / [(n_1 - 1)^{-1}(s_1^2/n_1)^2 + (n_2 - 1)^{-1}(s_2^2/n_2)^2]$

CHAPTER

9

INFERENCES CONCERNING VARIANCES

In Chapters 7 and 8 we learned how to judge the size of the error in estimating a population mean, how to construct confidence intervals for means, and how to perform tests of hypotheses about the means of one and of two populations. Very similar methods apply to inferences about other population parameters.

In this chapter we shall concentrate on population variances, or standard deviations, which are not only important in their own right, but which must sometimes be estimated before inferences about other parameters can be made. Section 9.1 is devoted to the estimation of σ^2 and σ, and Sections 9.2 and 9.3 deal with tests of hypotheses about these parameters.

9.1 The Estimation of Variances

In the preceding chapters, there were several instances where we estimated a population standard deviation by means of a sample standard deviation—we substituted the sample standard deviation S for σ in the large sample confidence interval for μ on page 220, in the large sample test concerning μ on page 242, and in the large sample test concerning the difference between two means on page 260. There are many statistical procedures in which S is thus substituted for σ, or S^2 for σ^2.

Let

$$S^2 = \frac{\sum_{i=1}^{n}(X_i - \overline{X})^2}{n-1}$$

be the sample variance, based on a random sample from any population, discrete or continuous, having variance σ^2. It follows from Example 34 of Chapter 5 that the mean of the sampling distribution of S^2 is $E(S^2) = \sigma^2$ whatever the value if σ^2.

Unbiased estimation of a population variance

> The sample variance
>
> $$S^2 = \frac{\sum_{i=1}^{n}(X_i - \overline{X})^2}{n-1} \quad \text{is an unbiased estimator of } \sigma^2$$

Although the sample variance is an unbiased estimator of σ^2, it does not follow that the sample standard deviation is also an unbiased estimator of σ. In fact, it is not. However, for large samples the bias is small and it is common practice to estimate σ with s.

Besides s, population standard deviations are sometimes estimated in terms of the **sample range** R, which we defined in Section 2.6 as the largest value of a sample minus the smallest. Given a random sample of size n from a normal population, it can be shown that the sampling distribution of the range has the mean $d_2 \sigma$ and the standard deviation $d_3 \sigma$, where d_2 and d_3 are constants which depend on the size of the sample. For $n = 1, 2, \ldots$, and 10, their values are as shown in the following table:

n	2	3	4	5	6	7	8	9	10
d_2	1.128	1.693	2.059	2.326	2.534	2.704	2.847	2.970	3.078
d_3	0.853	0.888	0.880	0.864	0.848	0.833	0.820	0.808	0.797

Thus, R/d_2 is an unbiased estimate of σ, and for very small samples, $n \leq 5$, it provides nearly as good an estimate of σ as does s. As the sample size increases, it becomes more efficient to use s instead of R/d_2. Nowadays, the range is used to estimate σ primarily in problems of industrial quality control, where sample sizes are usually small and computational ease is of prime concern. This application will be discussed in Chapter 15, where we shall need the above values of the constant d_3.

EXAMPLE 1   **Using the sample range to estimate σ**

With reference to Example 7, Chapter 8, use the range of the first sample to estimate σ for the resiliency modulus of recycled materials from the first location.

Solution   Since the smallest value is 604, the largest value is 707, and $n = 6$ so that $d_2 = 2.534$, we get

$$\frac{R}{d_2} = \frac{707 - 604}{2.534} = 40.6$$

Note that this is moderately close to the sample standard deviation $s = 34.7$.   ∎

In most practical applications, interval estimates of σ or σ^2 are based on the sample standard deviation or the sample variance. For random samples from normal populations, we make use of Theorem 6.5, according to which

$$\frac{(n - 1)S^2}{\sigma^2}$$

is a random variable having the **chi square distribution** with $n - 1$ degrees of freedom.

As illustrated in Figure 9.1, with χ^2_α as defined on page 198, the two quantities $\chi^2_{1 - \alpha/2}$ and $\chi^2_{\alpha/2}$ cut off area $\alpha / 2$ in the left and right tail, respectively. We can then assert that, whatever the value of σ^2,

$$P\left(\chi^2_{1-\alpha/2} < \frac{(n - 1)S^2}{\sigma^2} < \chi^2_{\alpha/2} \right) = 1 - \alpha$$

Once the data have been obtained, we make the same assertion with $(1 - \alpha)100\%$ confidence. Solving each inequality for σ^2, we obtain the confidence interval:

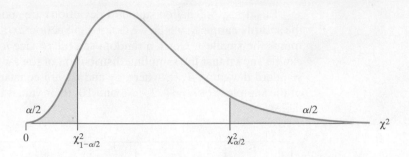

Figure 9.1
Two percentiles of the χ^2 distribution

Confidence interval for σ^2

$$\frac{(n-1)s^2}{\chi^2_{\alpha/2}} < \sigma^2 < \frac{(n-1)s^2}{\chi^2_{1-\alpha/2}}$$

If we take the square root of each member of this inequality, we obtain a corresponding $(1-\alpha)100\%$ confidence interval for σ.

Confidence interval for σ

$$\sqrt{\frac{(n-1)s^2}{\chi^2_{\alpha/2}}} < \sigma < \sqrt{\frac{(n-1)s^2}{\chi^2_{1-\alpha/2}}}$$

Note that confidence intervals for σ or σ^2 obtained by taking equal tails, as in the above formulae, do not actually give the narrowest confidence intervals, because the chi square distribution is not symmetrical. (See Exercise 7.26.) Nevertheless, they are used in most applications in order to avoid fairly complicated calculations. For moderate degrees of freedom, the choice of equal tails is inconsequential.

EXAMPLE 2   **A 95% confidence interval for the standard deviation σ of weight**

Referring to Example 8, Chapter 5, and the $n = 80$ measurements of the weight of cheese, construct a 95% confidence interval for the population standard deviation σ.

Solution   From Example 8, it is very reasonable to assume that the population is normal.
With $80 - 1 = 79$ degrees of freedom, we could interpolate in Table 5W but computer calculations first give $\chi^2_{.975} = 56.309$ and $\chi^2_{.025} = 105.473$. and then $s = 9.583$. Substituting into the formula for the confidence interval for σ^2 yields

$$\frac{(79)(9.583)^2}{105.473} < \sigma^2 < \frac{(79)(9.583)^2}{56.309}$$

or

$$68.8 < \sigma^2 < 128.8$$

and, taking the square root,

$$8.29 < \sigma < 11.35$$

This means we are 95% confident that the interval from 8.29 to 11.35 pounds contains σ, the population standard deviation of the weight of cheese. ∎

The method which we have discussed applies only to random samples from normal populations (or at least to random samples from populations which can be approximated closely with normal distributions).

Exercises

9.1 Use the data of Exercise 7.14 to estimate σ for the key performance indicator in terms of

(a) the sample standard deviation;

(b) the sample range.

Compare the two estimates by expressing their difference as a percentage of the first.

9.2 With reference to Example 7, Chapter 8, use the range of the second sample to estimate σ for the resiliency modulus of recycled materials from the second location. Compare the result with the standard deviation of the second sample.

9.3 Use the data of part (a) of Exercise 8.13 to estimate σ for the Brinell hardness of Alloy 1 in terms of

(a) the sample standard deviation;

(b) the sample range.

Compare the two estimates by expressing their difference as a percentage of the first.

9.4 With reference to Exercise 7.56, construct a 99% confidence interval for the variance of the yield.

9.5 With reference to Exercise 7.63, construct a 99% confidence interval for the variance of the population sampled.

9.6 Use the value s obtained in Exercise 9.3 to construct a 98% confidence interval for σ, measuring the actual variability in the hardness of Alloy 1.

9.2 Hypotheses Concerning One Variance

In this section we shall consider the problem of testing the null hypothesis that a population variance equals a specified constant against a suitable one-sided or two-sided alternative; that is, we shall test the null hypothesis $\sigma^2 = \sigma_0^2$ against one of the alternatives $\sigma^2 < \sigma_0^2, \sigma^2 > \sigma_0^2$, or $\sigma^2 \neq \sigma_0^2$. Tests like these are important whenever it is desired to control the uniformity of a product or an operation. For example, suppose that a silicon disk or wafer is to be cut into small squares, or dice, to be used in the manufacture of a semiconductor device. Since certain electrical characteristics of the finished device may depend on the thickness of the die, it is important that all dice cut from a wafer have approximately the same thickness. Thus, not only must the mean thickness of a wafer be kept within specifications, but also the variation in thickness from location to location on the wafer.

Using the same sampling theory as on page 281, we base such tests on the fact that for a random sample from a normal population with the variance σ_0^2

Statistic for test concerning variance (normal population)

$$\chi^2 = \frac{(n-1)S^2}{\sigma_0^2}$$

is a random variable having the chi square distribution with $n-1$ degrees of freedom. The critical regions for such tests are as shown in the following table:

Level α rejection regions for testing $\sigma^2 = \sigma_0^2$ (normal population)	
Alternative hypothesis	**Reject null hypothesis if:**
$\sigma^2 < \sigma_0^2$	$\chi^2 < \chi^2_{1-\alpha}$
$\sigma^2 > \sigma_0^2$	$\chi^2 > \chi^2_{\alpha}$
$\sigma^2 \neq \sigma_0^2$	$\chi^2 < \chi^2_{1-\alpha/2}$ or $\chi^2 > \chi^2_{\alpha/2}$

where χ^2_{α} is as defined on page 198. Note that equal tails are used for the two-sided alternative, although this is actually not the best procedure since the chi square distribution is not symmetrical. For moderate degrees of freedom, the two tests are nearly the same.

EXAMPLE 3   **Testing hypotheses concerning a standard deviation**

The lapping process which is used to grind certain silicon wafers to the proper thickness is acceptable only if σ, the population standard deviation of the thickness of dice cut from the wafers, is at most 0.50 mil. Use the 0.05 level of significance to test the null hypothesis $\sigma = 0.50$ against the alternative hypothesis $\sigma > 0.50$, if the thicknesses of 15 dice cut from such wafers have a standard deviation of 0.64 mil.

Solution   1. *Null hypothesis*: $\sigma = 0.50$
      Alternative hypothesis: $\sigma > 0.50$

2. *Level of significance*: $\alpha = 0.05$

3. *Criterion*: Reject the null hypothesis if $\chi^2 > 23.685$, the value of $\chi^2_{0.05}$ for 14 degrees of freedom, where

$$\chi^2 = \frac{(n-1)S^2}{\sigma_0^2}$$

4. *Calculations*:

$$\chi^2 = \frac{(15-1)(0.64)^2}{(0.50)^2} = 22.94$$

5. *Decision*: Since $\chi^2 = 22.94$ does not exceed 23.685, the null hypothesis cannot be rejected; even though the sample standard deviation exceeds 0.50, there is not sufficient evidence to conclude that the lapping process is unsatisfactory.

The rejection region and *P*-value are shown in Figure 9.2.

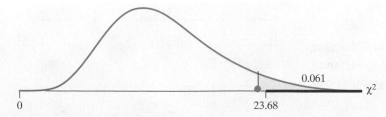

Figure 9.2
The rejection region and
P-value for Example 3

Statistical software is readily available to obtain other values of χ^2_{α}. See Exercise 9.23 for the *MINITAB* and Example 8, Chapter 6 for the *R* commands.

9.3   Hypotheses Concerning Two Variances

The two sample t test described in Section 8.2 requires that the variances of the two populations sampled are equal. In this section we describe a test of the null hypothesis $\sigma_1^2 = \sigma_2^2$, which applies to independent random samples from two normal populations. It must be used with some discretion as it is very sensitive to departures from this assumption.

If independent random samples of size n_1 and n_2 are taken from normal populations having the same variance, it follows from Theorem 6.6 that

Statistic for test of equality of two variances (normal populations)

$$F = \frac{S_1^2}{S_2^2}$$

is a random variable having the **F distribution** with $n_1 - 1$ and $n_2 - 1$ degrees of freedom. Thus, if the null hypothesis $\sigma_1^2 = \sigma_2^2$ is true, the ratio of the sample variances S_1^2 and S_2^2 provides a statistic on which tests of the null hypothesis can be based.

The critical region for testing the null hypothesis $\sigma_1^2 = \sigma_2^2$ against the alternative hypothesis $\sigma_1^2 > \sigma_2^2$ is $F > F_\alpha$, where F_α is as defined on page 199. Similarly, the critical region for testing the null hypothesis against the alternative hypothesis $\sigma_1^2 < \sigma_2^2$ is $F < F_{1-\alpha}$, and this causes some difficulties since Table 6W only contains values corresponding to right-hand tails of $\alpha = 0.05$ and $\alpha = 0.01$. However, we can use the reciprocal of the original test statistic and make use of the relation

$$F_{1-\alpha}(\nu_1, \nu_2) = \frac{1}{F_\alpha(\nu_2, \nu_1)}$$

first given on page 199. Thus, we base the test on the statistic $F = S_2^2/S_1^2$ and the critical region for testing the null hypothesis $\sigma_1^2 = \sigma_2^2$ against the alternative hypothesis $\sigma_1^2 < \sigma_2^2$ becomes $F > F_\alpha$, where F_α is the appropriate critical value of F for $n_2 - 1$ and $n_1 - 1$ degrees of freedom.

For the two-sided alternative $\sigma_1^2 \neq \sigma_2^2$, the critical region is $F < F_{1-\alpha/2}$ or $F > F_{\alpha/2}$, where $F = S_1^2/S_2^2$ and the degrees of freedom are $n_1 - 1$ and $n_2 - 1$. In practice, we modify this test as in the preceding paragraph, so that we can again use the table of F values corresponding to right-hand tails of $\alpha = 0.05$ and $\alpha = 0.01$. To this end we let S_M^2 represent the larger of the two sample variances, S_m^2 the smaller, and we write the corresponding sample sizes as n_M and n_m. Thus, the test statistic becomes $F = S_M^2/S_m^2$ and the critical region is as shown in the following table:

Level α rejection regions for testing $\sigma_1^2 = \sigma_2^2$ (normal populations)		
Alternative hypothesis	**Test statistic**	**Reject null hypothesis if:**
$\sigma_1^2 < \sigma_2^2$	$F = \dfrac{S_2^2}{S_1^2}$	$F > F_\alpha(n_2 - 1, n_1 - 1)$
$\sigma_1^2 > \sigma_2^2$	$F = \dfrac{S_1^2}{S_2^2}$	$F > F_\alpha(n_1 - 1, n_2 - 1)$
$\sigma_1^2 \neq \sigma_2^2$	$F = \dfrac{S_M^2}{S_m^2}$	$F > F_{\alpha/2}(n_M - 1, n_m - 1)$

The level of significance of these tests is α and the figures indicated in parentheses are the respective degrees of freedom. Note that, as in the chi square test, equal tails are used in the two-tailed test as a matter of mathematical convenience, even though the F distribution is not symmetrical.

EXAMPLE 4   **A one-sided F test of the equality of two variances**

It is desired to determine whether there is less variability in the silver plating done by Company 1 than in that done by Company 2. If independent random samples of size 12 of the two companies' work yield $s_1 = 0.035$ mil and $s_2 = 0.062$ mil, test the null hypothesis $\sigma_1^2 = \sigma_2^2$ against the alternative hypothesis $\sigma_1^2 < \sigma_2^2$ at the 0.05 level of significance.

Solution   1. *Null hypothesis*: $\sigma_1^2 = \sigma_2^2$
Alternative hypothesis: $\sigma_1^2 < \sigma_2^2$

2. *Level of significance*: $\alpha = 0.05$

3. *Criterion*: Reject the null hypothesis if $F > 2.82$, the value of $F_{0.05}$ for 11 and 11 degrees of freedom, where

$$F = \frac{S_2^2}{S_1^2}$$

4. *Calculations*:

$$F = \frac{(0.062)^2}{(0.035)^2} = 3.14$$

5. *Decision*: Since $F = 3.14$ exceeds 2.82, the null hypothesis must be rejected; at level $\alpha = 0.05$. The P-value 0.0352 is shown in Figure 9.3 along with the rejection region. The evidence against equality of variances, and in favor of Company 1's variance being smaller, is moderately strong. The data support the contention that the plating done by Company 1 is less variable than that done by Company 2.

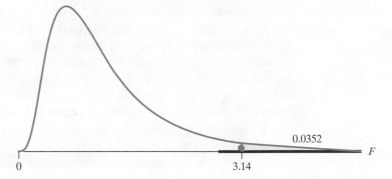

Figure 9.3
The rejection region and
P-value for Example 4

EXAMPLE 5   **A two-sided test for the equality of two variances**

Refer to Example 7, Chapter 8, dealing with the strength of recycled materials for use in pavements. Use the 0.02 level of significance to test for evidence that the variances are different.

Solution   1. *Null hypothesis*: $\sigma_1^2 = \sigma_2^2$
Alternative hypothesis: $\sigma_1^2 \neq \sigma_2^2$

2. *Level of significance*: $\alpha = 0.02$

3. *Criterion*: Reject the null hypothesis if $F > 10.97$, the value of $F_{0.01}$ for 5 and 5 degrees of freedom, where

$$F = \frac{S_2^2}{S_1^2}$$

since $s_1^2 = 1277.87$ is less than $s_2^2 = 3739.87$.

4. *Calculations*:

$$F = \frac{3739.87}{1277.87} = 2.93$$

5. *Decision*: Since $F = 2.93$ does not exceed 10.97, the null hypothesis cannot be rejected at level of significance 0.02. However, failure to reject the null hypothesis is not the same as showing it holds true. ∎

To obtain confidence intervals for the ratio of variances, we need a slightly more general sampling distribution. From the single sample result,

$$\frac{(n_1 - 1)S_1^2}{\sigma_1^2}$$ has a chi square distribution with $n_1 - 1$ degrees of freedom

$$\frac{(n_2 - 1)S_2^2}{\sigma_2^2}$$ has a chi square distribution with $n_2 - 1$ degrees of freedom

and these two random quantities are independent since the samples on which they are based are independent. Consequently,

$$\frac{S_1^2/\sigma_1^2}{S_2^2/\sigma_2^2} = \frac{\text{chi square/degrees of freedom}}{\text{chi square/degrees of freedom}} = F$$

where the right-hand side has an F distribution with $(n_1 - 1, n_2 - 1)$ degrees of freedom according to the representation on page 201. The $100(1 - \alpha)\%$ confidence intervals use the two percentiles illustrated in Figure 9.4.

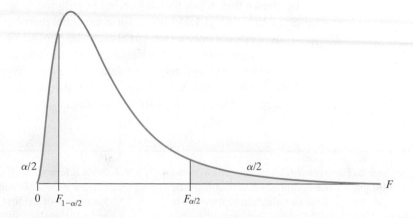

Figure 9.4
Two percentiles of the $F(n_1 - 1, n_2 - 1)$ distribution

Then, prior to sampling, we can assert that

$$1 - \alpha = P\left(F_{1-\alpha/2}(n_1 - 1, n_2 - 1) < \frac{S_1^2}{S_2^2}\frac{\sigma_2^2}{\sigma_1^2} < F_{\alpha/2}(n_1 - 1, n_2 - 1)\right)$$

Multiplying by S_2^2/S_1^2, we obtain a confidence interval in terms of the observed s_1^2 and s_2^2.

Confidence interval for σ_2^2/σ_1^2 normal populations

$$F_{1-\alpha/2}(n_1 - 1, n_2 - 1)\frac{s_2^2}{s_1^2} < \frac{\sigma_2^2}{\sigma_1^2} < F_{\alpha/2}(n_1 - 1, n_2 - 1)\frac{s_2^2}{s_1^2}$$

EXAMPLE 6   **Confidence interval for ratio of variances of yield**

Refer to Example 10 on page 267 of making green gasoline from sucrose. The equal sample sizes are $n_1 = n_2 = 9$, $s_1^2 = 0.4548$, and $s_2^2 = 0.1089$. Obtain a 98% confidence interval for σ_2^2/σ_1^2.

Solution   Since the degrees of freedom for the F are $(n_1 - 1, n_2 - 1) = (8, 8)$ and $\alpha/2 = 0.01$, we find $F_{0.01} = 6.03$ and $F_{0.99} = 1/F_{0.01} = 1/6.03$. The 98% confidence interval for σ_2^2/σ_1^2 becomes

$$\left(\frac{1}{6.03}\frac{0.1089}{0.4548}, \ 6.03\frac{0.1089}{0.4548}\right) \quad \text{or} \quad (\,0.04, 1.44\,)$$

The wideness of the interval illustrates the large amount of variability in variances when sample sizes are small. The second variance σ_2^2 could be as small as one-twenty-fifth of σ_1^2 or it could be larger than σ_1^2. ∎

Statistical software is readily available to obtain other values of $F_\alpha(\nu_1, \nu_2)$. See Exercise 9.23 for the *MINITAB* and Appendix C for the *R* commands.

Caution

In marked contrast to the procedures for making inferences about μ, the validity of the procedures in this chapter depends rather strongly on the assumption that the underlying population is normal. The sampling variance of S^2 can change when the population departs from normality by having, for instance, a single long tail. It can be shown that, when the underlying population is normal, the sampling variance of S^2 is $2\sigma^4/(n - 1)$. However, for nonnormal distributions, the sampling variance of S^2 depends not only on σ^2 but also on the population third and fourth moments, μ_3 and μ_4 (see page 104). Consequently, it could be much larger than $2\sigma^4/(n - 1)$. This behavior completely invalidates any tests of hypothesis or confidence intervals for σ^2. We say that these procedures for making inferences about σ^2 are not **robust** with respect to deviations from normality.

Exercises

9.7 With reference to Exercise 7.62, test the null hypothesis $\sigma = 600$ psi for the compressive strength of the given kind of steel against the alternative hypothesis $\sigma > 600$ psi. Use the 0.05 level of significance.

9.8 If 12 determinations of the specific heat of iron have a standard deviation of 0.0086, test the null hypothesis that $\sigma = 0.010$ for such determinations. Use the alternative hypothesis $\sigma \neq 0.010$ and the level of significance $\alpha = 0.01$.

9.9 With reference to Exercise 8.5, test the null hypothesis that $\sigma = 15.0$ minutes for the time that is required for repairs of the first kind of photocopying equipment against the alternative hypothesis that $\sigma > 15.0$ minutes. Use the 0.05 level of significance and assume normality.

9.10 Use the 0.01 level of significance to test the null hypothesis that $\sigma = 0.015$ inch for the diameters of certain bolts against the alternative hypothesis that

$\sigma \neq 0.015$ inch, given that a random sample of size 15 yielded $s^2 = 0.00011$.

9.11 Playing 10 rounds of golf on his home course, a golf professional averaged 71.3 with a standard deviation of 2.64.

  (a) Test the null hypothesis that the consistency of his game on his home course is actually measured by $\sigma = 2.40$, against the alternative hypothesis that he is less consistent. Use the level of significance 0.05. Assume that the distribution of his score, although discrete, is approximately normal.

  (b) If the distribution of his scores has a long right-hand tail, are your calculations in part (a) valid? Explain.

9.12 The security department of a large office building wants to test the null hypothesis that $\sigma = 2.0$ minutes for the time it takes a guard to walk his round against the alternative hypothesis that $\sigma \neq 2.0$ minutes. What can it conclude at the 0.01 level of significance if a random sample of size $n = 31$ yields $s = 1.8$ minutes?

9.13 Explore the use of the two sample t test in Exercise 8.9 by testing the null hypothesis that the two

populations have equal variances. Use the 0.02 level of significance.

9.14 With reference to Exercise 8.10, use the 0.10 level of significance to test the assumption that the two populations have equal variances.

9.15 Two different lighting techniques are compared by measuring the intensity of light at selected locations in areas lighted by the two methods. If 15 measurements in the first area had a standard deviation of 2.7 foot-candles and 21 measurements in the second area had a standard deviation of 4.2 foot-candles, can it be concluded that the lighting in the second area is less uniform? Use a 0.01 level of significance. What assumptions must be made as to how the two samples are obtained?

9.16 With reference to Exercise 8.6, where we had $n_1 = 40$, $n_2 = 30$, $s_1 = 15.2$, and $s_2 = 18.7$, use the 0.05 level of significance to test the claim that there is a greater variability in the number of cars which make left turns approaching from the south between 4 P.M. and 6 P.M. at the second intersection. Assume the distributions are normal.

Do's and Don'ts

Do's

1. Before applying the procedures in this chapter, always plot the data to look for outliers or presence of a long tail. Lack of normality can seriously affect tests of hypotheses and confidence intervals for variances.

Don'ts

1. Don't routinely calculate confidence intervals for variances or standard deviations using the formulas in this chapter. The confidence levels can deviate substantially from their specified level, say 95%, because of non-normality.

Review Exercises

9.17 With reference to Example 20, Chapter 7, construct a 95% confidence interval for the true standard deviation of the lead content.

9.18 If 31 measurements of the boiling point of sulfur have a standard deviation of 0.83 degree Celsius, construct a 98% confidence interval for the true standard deviation of such measurements. What assumption did you make about the population?

9.19 Past data indicate that the variance of measurements made on sheet metal stampings by experienced

quality-control inspectors is 0.18 (inch)^2. Such measurements made by an inexperienced inspector could have too large a variance (perhaps because of inability to read instruments properly) or too small a variance (perhaps because unusually high or low measurements are discarded). If a new inspector measures 101 stampings with a variance of 0.13 (inch)^2, test at the 0.05 level of significance whether the inspector is making satisfactory measurements. Assume normality.

9.20 Pull-strength tests on 10 soldered leads for a semiconductor device yield the following results in pounds-force required to rupture the bond:

15.8 12.7 13.2 16.9 10.6 18.8 11.1 14.3 17.0 12.5

Another set of 8 leads was tested after encapsulation to determine whether the pull strength has been increased by encapsulation of the device, with the following results:

24.9 23.6 19.8 22.1 20.4 21.6 21.8 22.5

9.23 *MINITAB calculation of t_α, χ^2_ν, and F_α*

The software finds percentiles, so to obtain F_α, we first convert from α to $1 - \alpha$. We illustrate with the calculation of $F_{0.025}(4, 7)$, where $1 - 0.025 = 0.975$.

Dialog box:

Calc> Probability distributions > F. Choose **Inverse cumulative probability**.
Type 4 in **Numerator degrees of freedom**, 7 in **Denominator degrees of freedom**.
Choose **Input constant** and *type* .975. Click **OK**.

Output:

F distribution with 4 DF in numerator and 7 DF in denominator

$$P(X <= x) \qquad x$$
$$0.975 \qquad 5.52259$$

In the first line, you may instead select **Chi square** or **t** and then there is only one **Degrees of Freedom** in the second line.

Obtain $F_{0.975}(7, 4)$ and check that it equals $1/F_{0.025}(4, 7) = 1/5.52259$.

9.24 A bioengineering company manufactures a device for externally measuring blood flow. Measurements of the electrical output (milliwatts) on a sample of 16 units yields the data

11	1	5	3	2	23	37	5
18	7	1	11	2	2	30	3

plotted in Figure 9.5.

(a) Should you report the 95% confidence interval for σ using the formula in this chapter? Explain.

(b) What is your answer to part (a) if you first take natural logarithms and then calculate the confidence interval for the variance of ln (output)?

(c) Does your conclusion in part (b) readily imply anything about variance on the original scale? Explain.

Use the 0.02 level of significance to test whether it is reasonable to assume that the two samples come from populations with equal variances.

9.21 With reference to the Example 8, Chapter 8, test the equality of the variances for the two aluminum alloys. Use the 0.02 level of significance.

9.22 With reference to the Example 8, Chapter 8, find a 98% confidence interval for the ratio of variances of the two aluminum alloys.

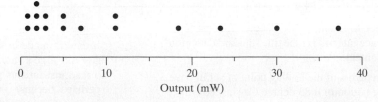

Figure 9.5
Output in milliwatts

Key Terms

Chi square distribution	281	Robust 288
F distribution 285		Sample range 281

INFERENCES CONCERNING PROPORTIONS

Many engineering problems deal with proportions, percentages, or probabilities. In acceptance sampling we are concerned with the proportion of defectives in a lot, and in life testing we are concerned with the percentage of certain components which will perform satisfactorily during a stated period of time, or the probability that a given component will last at least a given number of hours. It should be clear from these examples that problems concerning proportions, percentages, or probabilities are really equivalent; a percentage is merely a proportion multiplied by 100, and a probability may be interpreted as a proportion in a long series of trials.

Section 10.1 deals with the estimation of proportions; Section 10.2 deals with tests concerning proportions; Section 10.3 deals with tests concerning two or more proportions. In Section 10.4 we shall learn how to analyze data tallied into a two-way classification. In Section 10.5 we shall learn how to judge whether differences between an observed frequency distribution and corresponding expectations can be attributed to chance.

10.1 Estimation of Proportions

The information that is usually available for the estimation of a proportion is the number of times, X, that an appropriate event occurs in n trials, occasions, or observations. The point estimator of the population proportion, itself, is usually the **sample proportion** $\frac{X}{n}$, namely, the proportion of the time that the event actually occurs. If the n trials satisfy the assumptions underlying the binomial distribution listed on page 88, we know that the mean and the standard deviation of the number of successes are given by np and $\sqrt{np(1-p)}$. If we divide both of these quantities by n, we find that the mean and the standard deviation of the proportion of successes (namely, of the sample proportion) are given by

$$\frac{np}{n} = p \quad \text{and} \quad \frac{\sqrt{np(1-p)}}{n} = \sqrt{\frac{p(1-p)}{n}}$$

The first of these results shows that the sample proportion is an unbiased estimator of the binomial parameter p, namely, of the true proportion we are trying to estimate on the basis of a sample.

Parameter: Population proportion p

Data: X = number of times event occurs in n trials.

Estimator: $\widehat{p} = \frac{X}{n}$

Estimate of standard error: $\sqrt{\dfrac{\widehat{p}(1-\widehat{p})}{n}}$

EXAMPLE 1   **Point estimate of a binomial proportion p**

An engineering firm responsible for maintaining and improving the performance of thousands of wind turbines is asked to check on the sound levels. The purpose is to determine the proportion that currently would not meet proposed new sound level restrictions.

A random selection of $n = 55$ wind turbines reveals that 8 operate too loudly according to the proposed new restrictions. Obtain a point estimate of the proportion of their wind turbines that do not meet specifications.

Solution   The point estimate is

$$\widehat{p} = \frac{X}{n} = \frac{8}{55} = 0.1455$$

and the estimate of its standard error is

$$\sqrt{\frac{\widehat{p}(1 - \widehat{p})}{n}} = \sqrt{\frac{0.1455(1 - 0.1455)}{55}} = 0.0475$$

The point estimate is $\widehat{p} = 0.146$ with estimated standard error 0.048.   ∎

In the construction of confidence intervals for the binomial parameter p, we meet several obstacles. First, since x and $\frac{x}{n}$ are values of discrete random variables, it may be impossible to get an interval for which the degree of confidence is exactly $(1 - \alpha)100\%$. Second, the standard deviation of the sampling distribution of the number of successes, as well as that of the proportion of successes, involves the parameter p that we are trying to estimate.

A Conservative Confidence Interval of a Proportion

To construct a conservative confidence interval for p having approximately the degree of confidence $(1 - \alpha)100\%$, we first determine for a given set of values of p the corresponding quantities x_1 and x_2, where x_1 is the largest integer for which the binomial probabilities $b(k; n, p) = P[X = k]$ satisfy

$$\sum_{k=0}^{x_1} b(k; n, p) \leq \frac{\alpha}{2}$$

while x_2 is the smallest integer for which

$$\sum_{k=x_2}^{n} b(k; n, p) \leq \frac{\alpha}{2}$$

To emphasize the point that x_1 and x_2 depend on the value of p, we shall write these quantities as $x_1(p)$ and $x_2(p)$.

We can then assert, with a probability of at least $1 - \alpha$, but approximately $1 - \alpha$, that

$$P\left(x_1(p) \leq X \leq x_2(p)\right) \geq 1 - \alpha$$

where the value of p is the one that produces the binomial count X.

A confidence interval for p results from changing the inequalities to statements about a random interval that covers the true unknown p. To indicate what is involved in this step, we first give a graphical approach. However, we then eliminate the graphical step by recommending the use of computer software which works for almost any sample size. Suppose, for instance, that we want to find approximate 95% confidence intervals for p for samples of size $n = 20$. Using Table 1 at the end of the book, we first determine x_1 and x_2 for selected values of p such that x_1 is the largest integer for which

$$B(x_1; 20, p) \leq 0.025$$

while x_2 is the smallest integer for which

$$1 - B(x_2 - 1; 20, p) \leq 0.025$$

Letting p equal $0.1, 0.2, \ldots$, and 0.9, we thus obtain the values shown in the following table:

p	0.1	0.2	0.3	0.4	0.5	0.6	0.7	0.8	0.9
x_1	–	0	1	3	5	7	9	11	14
x_2	6	9	11	13	15	17	19	20	–

Plotting the points with coordinates p and $x(p)$ as in Figure 10.1, and drawing smooth curves, one through the x_1 points and one through the x_2 points, we can now "solve" for p. For any given value of x we can obtain approximate 95% confidence limits for p by going horizontally to the two curves and marking off the corresponding values of p. (See Figure 10.1.) Thus, for $x = 4$ we obtain the approximate 95% confidence interval

$$0.06 < p < 0.45$$

We again emphasize that this procedure is conservative. Before the count X is observed, the probability is at least $(1 - \alpha)$ that the interval will cover p. For instance, with $\alpha = 0.95, n = 20$, and $p = .3$, we find $B(1; 20.3) = 0.0076$ and $1 - B(10; 20.3) = 1 - 0.9829 = 0.0171$ so that

$$P(1 < X < 11) = B(10; 20.3) - B(1; 20.3) = 0.9829 - 0.0076 = 0.9753$$

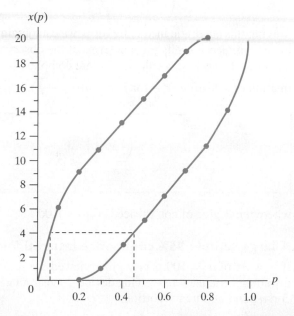

Figure 10.1
95% confidence intervals for proportions ($n = 20$)

which is somewhat larger than 0.95 With the aid of a computer software, this confidence interval is easily obtained for almost any sample size n.

EXAMPLE 2   **Conservative 95% confidence interval for binomial proportion p**

Refer to Example 1 where 8 out of 55 wind turbines were too noisy according to new restricted guidelines. Using computer software (see Exercise 10.13) obtain the 95% conservative confidence interval for the proportion of all wind turbines managed by the company that are too noisy.

Solution   A computer calculation (see Exercise 10.13) gives

```
CI for One Proportion

Sample  X   N   Sample p          95% CI
1       8   55  0.145455   (0.064951, 0.266632)
```

We are 95% confident that for proportion p of wind turbines that are too noisy is between 0.065 and 0.267. The larger values in this interval suggest their could be a major problem with noise.

[Using **R**: (a) **binom.confint(8, 55, conf.level=0.95)**]   ∎

A Large Sample Confidence Interval for a Proportion

On page 139 we gave the general rule of thumb that the normal distribution provides a good approximation to the binomial distribution when np and $n(1 - p)$ are both greater than 15. Thus, for $n = 50$ the normal curve approximation may be used if it can be assumed that p lies between 0.30 and 0.70; for $n = 200$ it may be used if it can be assumed that p lies between 0.075 and 0.925; and so forth. This is what we shall mean here, and later in this chapter, by "n being large."

When n is large, we can construct approximate confidence intervals for the binomial parameter p by using the normal approximation to the binomial distribution. Accordingly, we can assert that approximately

$$P\left(-z_{\alpha/2} < \frac{X - np}{\sqrt{np(1 - p)}} < z_{\alpha/2}\right) = 1 - \alpha$$

Solving this quadratic inequality for p, we can obtain a corresponding set of approximate confidence limits for p in terms of the observed value x (see Exercise 10.15), but since the necessary calculations are complex, we shall make the further approximation of substituting $\frac{x}{n}$ for p in $\sqrt{np(1 - p)}$. This yields

Large sample confidence interval for p

$$\frac{x}{n} - z_{\alpha/2}\sqrt{\frac{\frac{x}{n}\left(1 - \frac{x}{n}\right)}{n}} < p < \frac{x}{n} + z_{\alpha/2}\sqrt{\frac{\frac{x}{n}\left(1 - \frac{x}{n}\right)}{n}}$$

where the degree of confidence is $(1 - \alpha)100\%$.

EXAMPLE 3   **A large sample 95% confidence interval for p**

If $x = 36$ of $n = 100$ persons interviewed are familiar with the tax incentives for installing certain energy-saving devices, construct a 95% confidence interval for the corresponding true proportion.

Solution   Substituting

$$\frac{x}{n} = \frac{36}{100} = 0.36$$

and $z_{\alpha/2} = 1.96$ into the above formula, we get

$$0.36 - 1.96 \sqrt{\frac{(0.36)(0.64)}{100}} < p < 0.36 + 1.96 \sqrt{\frac{(0.36)(0.64)}{100}}$$

or

$$0.266 < p < 0.454$$

We are 95% confident that the population proportion of persons familiar with the tax incentives, p, is contained in the interval from 0.266 to 0.454. Note that if we had used the computer calculation in Exercise 10.13, we would have obtained

$$0.27 < p < 0.46$$ ■

The magnitude of the error we make when we use $\dfrac{X}{n}$ as an estimator of p is given by

$$\left| \frac{X}{n} - p \right|$$

Again using the normal approximation, we can assert that the

$$P\left(\left| \frac{X}{n} - p \right| \leq z_{\alpha/2} \sqrt{\frac{p(1-p)}{n}} \right) = 1 - \alpha$$

Namely, with probability $1 - \alpha$, the error will be at most

$$z_{\alpha/2} \sqrt{\frac{p(1-p)}{n}}$$

Maximum error of estimate

$$E = z_{\alpha/2} \sqrt{\frac{p(1-p)}{n}}$$

With the observed value $\dfrac{x}{n}$ substituted for p, we obtain an estimate of E.

EXAMPLE 4   **An estimate of the maximum error**

In a sample survey conducted in a large city, 136 of 400 persons answered yes to the question of whether their city's public transportation is adequate. With 99% confidence, what can we say about the maximum error, if

$$\frac{x}{n} = \frac{136}{400} = 0.34$$

is used as an estimate of the corresponding true proportion?

Solution   Substituting $\dfrac{x}{n} = 0.34$ and $z_{\alpha/2} = 2.575$ into the above formula, we estimate that the maximum error is at most

$$\hat{E} = 2.575 \sqrt{\frac{(0.34)(0.66)}{400}} = 0.061 \qquad \blacksquare$$

The preceding formula for E can also be used to determine the sample size that is needed to attain a desired degree of precision. Solving for n, we get

Sample size determination

$$n = p(1-p)\left[\frac{z_{\alpha/2}}{E}\right]^2$$

but this formula cannot be used as it stands unless we have some information about the possible size of p (on the basis of collateral data, say a pilot sample). If no such information is available, we can make use of the fact that $p(1-p)$ is at most $\frac{1}{4}$, corresponding to $p = \frac{1}{2}$, as can be shown by the methods of elementary calculus. If a range for p is known, the value closest to $\frac{1}{2}$ should be used.

Thus, if

Sample size (p unknown)

$$n = \frac{1}{4}\left[\frac{z_{\alpha/2}}{E}\right]^2$$

we can assert with a probability of at least $1 - \alpha$ that the error in using $\dfrac{X}{n}$ as an estimate of p will not exceed E. Once the data have been obtained, we will be able to assert with at least $(1 - \alpha)100\%$ confidence that the error does not exceed E.

EXAMPLE 5   **Selecting a sample size for estimating a proportion**

Suppose that we want to estimate the true proportion of defectives in a very large shipment of adobe bricks, and that we want to be at least 95% confident that the error is at most 0.04. How large a sample will we need if

(a) we have no idea what the true proportion might be;

(b) we know that the true proportion does not exceed 0.12?

Solution   **(a)** Using the second of the two formulas for the sample size, we get

$$n = \frac{1}{4}\left[\frac{1.96}{0.04}\right]^2 = 600.25$$

or $n = 601$ rounded up to the nearest integer.

(b) Using the first of the two formulas for the sample size with $p = 0.12$ (the possible value closest to $p = \frac{1}{2}$), we get

$$n = (0.12)(0.88)\left[\frac{1.96}{0.04}\right]^2 = 253.55$$

or $n = 254$ rounded up to the nearest integer. This serves to illustrate how some collateral information about the possible size of p can substantially reduce the size of the required sample. $\blacksquare$

Exercises

10.1 In a random sample of 200 claims filed against an insurance company writing collision insurance on cars, 84 exceeded $3,500. Construct a 95% confidence interval for the true proportion of claims filed against this insurance company that exceed $3,500, using the large sample confidence interval formula.

10.2 With reference to Exercise 10.1, what can we say with 99% confidence about the maximum error if we use the sample proportion as an estimate of the true proportion of claims filed against this insurance company that exceed $3,500?

10.3 In a random sample of 400 industrial accidents, it was found that 231 were due at least partially to unsafe working conditions. Construct a 99% confidence interval for the corresponding true proportion using the large sample confidence interval formula.

10.4 With reference to Exercise 10.3, what can we say with 95% confidence about the maximum error if we use the sample proportion to estimate the corresponding true proportion?

10.5 In a random sample of 90 sections of pipe in a chemical plant, 15 showed signs of serious corrosion. Construct a 95% confidence interval for the true proportion of pipe sections showing signs of serious corrosion, using the large sample confidence interval formula.

10.6 In a recent study, 69 of 120 meteorites were observed to enter the earth's atmosphere with a velocity of less than 26 miles per second. If we estimate the corresponding true proportion as $\frac{69}{120} = 0.575$, what can we say with 95% confidence about the maximum error?

10.7 Among 100 fish caught in a large lake, 18 were inedible due to the pollution of the environment. If we use $\frac{18}{100} = 0.18$ as an estimate of the corresponding true proportion, with what confidence can we assert that the error of this estimate is at most 0.065?

10.8 New findings suggest many persons possess symptoms of motion sickness after watching a 3D movie. One scientist administered a questionnaire to $n = 451$ adults after they watched a 3D movie of their choice. Based on these self-reported results, 247 are determined to have some motion sickness.

(Source: A. Solimini (2013) Are there side effects to watching 3D movies? A prospective crossover observational study on visually induced motion sickness, PLOS ONe, **8** (2), 1–8, e56160)

   Find a 98% confidence interval for the proportion of adults who would have some motion sickness after watching a 3D movie.

10.9 What is the size of the smallest sample required to estimate an unknown proportion of customers who would pay for an additional service, to within a maximum error of 0.06 with at least 95% confidence?

10.10 With reference to Exercise 10.9, how would the required sample size be affected if it is known that the proportion to be estimated is at least 0.75?

10.11 Suppose that we want to estimate what percentage of all drivers exceed the 55 mph speed limit on a certain stretch of road. How large a sample will we need to be at least 99% confident that the error of our estimate, the sample percentage, is at most 3.5%?

10.12 Refer to Example 1. How large a sample of wind turbines is needed to ensure that, with at least 95% confidence, the error in our estimate of the sample proportion is at most 0.06 if

   (a) nothing is known about the population proportion?

   (b) the population proportion is known not to exceed 0.20?

10.13 *MINITAB* **determination of confidence interval for** *p*

When the sample size is not large, the confidence interval for a proportion p can be obtained using the following commands. We illustrate the case $n = 20$ and $x = 4$.

Stat > Basic Statistics > 1-Proportion. Choose **Summarized outcomes**.
Type 4 in **Number of events** and 20 in **Number of trials**.
Click **Options** and then type 95.0 in **Confidence level**.
Choose *exact* in **Method**. Click **OK**. Click **OK**.

The partial output includes the 95% confidence interval (0.057334, 0.436614).

Obtain the 95% confidence interval when $n = 20$ and $x = 16$.

10.14 Use Exercise 10.13 or other software to obtain the interval requested in Exercise 10.3.

10.15 Show that the inequality on page 294 leads to the following $(1 - \alpha)100\%$ confidence limits:

$$\frac{x + \dfrac{1}{2} z_{\alpha/2}^2 \pm z_{\alpha/2} \sqrt{\dfrac{x(n-x)}{n} + \dfrac{1}{4} z_{\alpha/2}^2}}{n + z_{\alpha/2}^2}$$

10.16 Use the formula of Exercise 10.15 to rework Exercise 10.3.

10.17 A major automobile company was having trouble with the quality of paint on its gray cars. The first step was to collect data on the magnitude of the problem. Of 4,063 recently painted gray automobiles, 533 had paint problems that could easily be detected by visual inspection. Obtain a 95% confidence interval for the population proportion of defective gray paint jobs.

10.18 An international corporation needed several millions of words, from thousands of documents and manuals, translated. The work was contracted to a company that used computer-assisted translation, along with some human checks. The corporation conducted its own quality check by sampling the translation. Among the 2,037 mistakes found, 834 were an incorrect word. Obtain a 99% confidence interval for the population proportion of mistakes that are incorrect words.

10.2   Hypotheses Concerning One Proportion

Many of the methods used in sampling inspection, quality control, and reliability verification are based on tests of the null hypothesis that a proportion (percentage, or probability) equals some specified constant.

Exact Test with Conservative Significance Level

It is possible to construct tests of hypotheses that have level of significance no greater than a specified Type 1 error probability α_B. To test the null hypothesis $H_0 : p = p_0$ versus a two-sided alternative $H_1 : p \neq p_0$, choose the largest integer x_1 and smallest integer x_2 for which

$$\sum_{k=0}^{x_1} b(k; n, p_0) \leq \frac{\alpha_B}{2} \quad \text{and} \quad \sum_{k=x_2}^{n} b(k; n, p_0) \leq \frac{\alpha_B}{2}$$

are both satisfied. Alternatively, in terms of the cumulative distribution, both $B(x_1; n, p_0)$ and $1 - B(x_2 - 1, n, p_0)$ are less than $\alpha_B/2$. The test statistic is the binomial count X and the rejection region is then $X \leq x_1$ or $X \geq x_2$.

Suppose that the Type 1 error probability cannot exceed $\alpha_B = 0.05$ and the sample size is $n = 20$. Using Table 1, you may confirm that the rejection region for testing $H_0 : p = 0.4$ versus a two-sided alternative $H_1 : p \neq 0.4$, is $X \leq 3$ or $X \geq 13$. The level of significance is then

$$P(X \leq 3 \quad \text{or} \quad X \geq 13) = B(3\, 20, 0.4) + 1 - B(12\, 20, 0.4)$$
$$= 0.0160 + (1 - 0.9790) = 0.0370$$

The rejection regions, depending on the specified bound for the Type 1 error probability α_B, are given in the following table.

Level α rejection regions for testing $p = p_0$ when α_B = specified bound on Type 1 error probability	
Alternative hypothesis	**Rejection region**
$p \leq p_0$	$X \leq x_1$ where x_1 is the largest integer with $B(x_1; n, p_0) \leq \alpha_B$
$p \geq p_0$	$X \geq x_2$ where x_2 is the smallest integer with $1 - B(x_2; n, p_0) \leq \alpha_B$
$p \neq p_0$	$X \leq x_1$ or $X \geq x_2$ where x_1 is the largest and x_2 the smallest integer with $B(x_1; n, p_0) \leq \dfrac{\alpha_B}{2}$ and $1 - B(x_2; n, p_0) \leq \dfrac{\alpha_B}{2}$

Advances in computer software make it possible to perform this exact conservative test even for sample sizes in the thousands. When the sample size is moderate to large, the significance level is nearly equal to the specified bound.

EXAMPLE 6   **An exact test of a binomial proportion p**

Miniature drones are being programmed to posses swarm behavior. Engineers make 34 drones based on a new design. Each must be shown to fly for 2 hours before group activities can begin. Suppose 4 of the 34 drones fail this initial test.

(a) Using computer software (see Exercise 10.13), conduct a test that intends to establish that the probability of failing is less than 0.3 for any drone that can possibly be made using the new design. Control the Type 1 error probability to be below 0.05.

(b) Determine the level of significance for the test.

Solution   (a) A computer calculation gives

```
Test of p = 0.3 vs p < 0.3

                                    Exact
Sample   X    N   Sample p   P-Value
1        4    34   0.1176      0.012
```

The P-value $= 0.012$ provides quite strong evidence against the null hypothesis $H_0 : p = 0.3$ and in favor of the alternative that the probability is less than 0.3.

(b) Another computer calculation gives $B(5; 34, 0.3) = 0.0334$ and $B(6; 34, 0.3) = 0.0785$. The rejection region is then $X \leq 5$ and the level of significance for testing $H_0 : p = 0.3$ versus the one-sided alternative $H_1 : p < 0.3$ is $P(X \leq 5) = 0.0334$.

[Using **R**: (a) **binom.test(4, 34, conf.level=0.95, p=.3, alternative = "less")** (b) **pbinom(5,34,.3)**]   ∎

Large Sample Test of a Proportion

We now consider approximate large sample tests based on the normal approximation to the binomial distribution. In other words, we shall test the null hypothesis $p = p_0$ against one of the alternatives $p < p_0, p > p_0$, or $p \neq p_0$ with the use of the statistic

Statistic for large sample test concerning p

$$Z = \frac{X - n p_0}{\sqrt{n p_0 (1 - p_0)}} = \frac{\frac{X}{n} - p_0}{\sqrt{\dfrac{p_0(1 - p_0)}{n}}}$$

which is a random variable having approximately the standard normal distribution.[1] The second form emphasizes that Z is based on the difference between the sample proportion and the hypothesized probability p_0.

---

[1] Some authors write the numerator of this formula for Z as $X \pm \frac{1}{2} - np_0$, whichever is numerically smaller, but there is generally no need for this continuity correction so long as n is large.

The critical regions are like those shown in the table on page 240 with p and p_0 substituted for μ and μ_0.

Level α rejection Regions for Testing $p = p_0$ (large sample)	
Alternative hypothesis	**Reject null hypothesis if:**
$p < p_0$	$Z < -z_\alpha$
$p > p_0$	$Z > z_\alpha$
$p \neq p_0$	$Z < -z_{\alpha/2}$ or $Z > z_{\alpha/2}$

EXAMPLE 7   **A one-sided test of the proportion of transceivers**

Transceivers provide wireless communication among electronic components of consumer products. Responding to a need for a fast, low-cost test of Bluetooth-capable transceivers, engineers[2] developed a product test at the wafer level. In one set of trials with 60 devices selected from different wafer lots, 48 devices passed. Test the null hypothesis $p = 0.70$ against the alternative hypothesis $p > 0.70$ at the 0.05 level of significance.

Solution   1. *Null hypothesis*: $p = 0.70$
   Alternative hypothesis: $p > 0.70$

2. *Level of significance*: $\alpha = 0.05$

3. *Criterion*: Reject the null hypothesis if $Z > 1.645$, where

$$Z = \frac{X - np_0}{\sqrt{np_0(1 - p_0)}}$$

4. *Calculations*: Substituting $x = 48$, $n = 60$, and $p_0 = 0.70$ into the formula above, we get

$$z = \frac{48 - 60(0.70)}{\sqrt{60(0.70)(0.30)}} = 1.69$$

5. *Decision*: Since $z = 1.69$ is greater than 1.645, we reject the null hypothesis at level 0.05. In other words, there is sufficient evidence to conclude that the proportion of good transceivers that would be produced is greater than 0.70. The P-value, $P(Z > 1.69) = .0455$, somewhat strengthens this conclusion. ∎

10.3   Hypotheses Concerning Several Proportions

When we compare the consumer response (percentage favorable and percentage unfavorable) to two different products, when we decide whether the proportion of defectives of a given process remains constant from day to day, when we judge whether there is a difference in political persuasion among several nationality groups, and in many similar situations, we are interested in testing whether two or more binomial populations have the same parameter p. Referring to these parameters as $p_1, p_2, \ldots,$

---

[2]G. Srinivasan, F. Taenzler, and A. Chatterjee, Loopback DFT for low-cost test of single-VCO-based wireless Transceivers, *IEEE Design & Test of Computers* (2008), 150–159.

and p_k, we are, in fact, interested in testing the null hypothesis

$$p_1 = p_2 = \cdots = p_k = p$$

against the alternative hypothesis that these population proportions are not all equal. To perform a suitable large sample test of this hypothesis, we require independent random samples of size $n_1, n_2, \ldots,$ and n_k from the k populations; then, if the corresponding numbers of "successes" are $X_1, X_2, \ldots,$ and X_k, the test we shall use is based on the fact that

1. for large samples the sampling distribution of

$$Z_i = \frac{X_i - n_i p_i}{\sqrt{n_i p_i (1 - p_i)}}$$

is approximately the standard normal distribution,

2. the square of random variable having the standard normal distribution is a random variable having the chi square distribution with 1 degree of freedom, and

3. the sum of k independent random variables having chi square distributions with 1 degree of freedom is a random variable having the chi square distribution with k degrees of freedom. (See Examples 14 and Example 16, Chapter 6, for proofs of these last two results.) Thus,

$$\chi^2 = \sum_{i=1}^{k} \frac{(x_i - n_i p_i)^2}{n_i p_i (1 - p_i)}$$

is a value of a random variable having approximately the chi square distribution with k degrees of freedom. In practice we substitute for the p_i, which under the null hypothesis are all equal, the pooled estimate

$$\widehat{p} = \frac{x_1 + x_2 + \cdots + x_k}{n_1 + n_2 + \cdots + n_k}$$

Since the null hypothesis should be rejected if the differences between the x_i and the $n_i \widehat{p}$ are large, the critical region is $\chi^2 > \chi_\alpha^2$, where χ_α^2 is as defined on page 198 and the number of degrees of freedom is $k - 1$. The loss of one degree of freedom results from substituting for p the estimate $\widehat{p}$.

In actual practice, when we compare two or more sample proportions, it is convenient to determine the value of the χ^2 statistic by looking at the data as arranged in the following way:

	Sample 1	Sample 2	$\cdots$	Sample k	Total
Successes	x_1	x_2	$\cdots$	x_k	x
Failures	$n_1 - x_1$	$n_2 - x_2$	$\cdots$	$n_k - x_k$	$n - x$
Total	n_1	n_2	$\cdots$	n_k	n

The notation is the same as before, except for x and n, which represent, respectively, the total number of successes and the total number of trials for all samples combined. With reference to this table, the entry in the cell belonging to the ith row and jth column is called the **observed cell frequency** o_{ij} with $i = 1, 2$ and $j = 1, 2, \ldots, k$.

Under the null hypothesis $p_1 = p_2 = \cdots = p_k = p$, we estimate p, as before, as the total number of successes divided by the total number of trials, which we now write as $\widehat{p} = \dfrac{x}{n}$. Hence, the expected number of successes and failures for the jth sample are estimated by

$$e_{1j} = n_j \cdot \widehat{p} = \frac{n_j \cdot x}{n}$$

and

$$e_{2j} = n_j(1 - \widehat{p}) = \frac{n_j(n-x)}{n}$$

The quantities e_{1j} and e_{2j} are called the **expected cell frequencies** for $j = 1, 2, \ldots, k$. Note that **the expected frequency for any given cell may be obtained by multiplying the totals of the column and the row to which it belongs and then dividing by the grand total** n.

A **chi square test** is based on the χ^2 statistic on page 301, with $\widehat{p}$ substituted for the p_i. The χ^2 statistic can be written in the form

χ^2 statistic for test concerning difference among proportions

$$\chi^2 = \sum_{i=1}^{2} \sum_{j=1}^{k} \frac{(o_{ij} - e_{ij})^2}{e_{ij}}$$

as the reader will be asked to verify in Exercise 10.36. This formula has the advantage that it can easily be extended to the more general case, to be treated in Section 10.4, where each trial permits more than two possible outcomes. There are then more than two rows in the tabular presentation of the various frequencies.

EXAMPLE 8

Testing the equality of three proportions using the χ^2 statistic

Samples of three kinds of materials, subjected to extreme temperature changes, produced the results shown in the following table:

	Material A	Material B	Material C	Total
Crumbled	41	27	22	90
Remained intact	79	53	78	210
Total	120	80	100	300

Use the 0.05 level of significance to test whether, under the stated conditions, the probability of crumbling is the same for the three kinds of materials.

Solution

1. *Null hypothesis*: $p_1 = p_2 = p_3$
   Alternative hypothesis: $p_1, p_2,$ and p_3 are not all equal.

2. *Level of significance*: $\alpha = 0.05$

3. *Criterion*: Reject the null hypothesis if $\chi^2 > 5.991$, the value of $\chi^2_{0.05}$ for $3 - 1 = 2$ degrees of freedom, where χ^2 is given by the formula above.

4. *Calculations*: The expected frequencies for the first two cells of the first row are

$$e_{11} = \frac{90 \cdot 120}{300} = 36 \quad \text{and} \quad e_{12} = \frac{90 \cdot 80}{300} = 24$$

and, as it can be shown that **the sum of the expected frequencies for any row or column equals that of the corresponding observed frequencies** (see

Exercise 10.37), we find by subtraction that $e_{13} = 90 - (36 + 24) = 30$, and that the expected frequencies for the second row are $e_{21} = 120 - 36 = 84$, $e_{22} = 80 - 24 = 56$, and $e_{23} = 100 - 30 = 70$. Then, substituting these values together with the observed frequencies into the formula for χ^2, we get

$$\chi^2 = \frac{(41-36)^2}{36} + \frac{(27-24)^2}{24} + \frac{(22-30)^2}{30}$$
$$+ \frac{(79-84)^2}{84} + \frac{(53-56)^2}{56} + \frac{(78-70)^2}{70}$$
$$= 4.575$$

5. *Decision*: Since $\chi^2 = 4.575$ does not exceed 5.991, the null hypothesis cannot be rejected. In other words, the data do not refute the hypothesis that, under the stated conditions, the probability of crumbling is the same for all three materials.

[Using **R**: A box in Appendix C contains the commands for Example 8.] ∎

Most of the entries of Table 5W are given to three decimal places. But because its sampling distribution is only approximate, the final value of the χ^2 statistics is usually rounded to two decimals. We caution that it should not be used when one or more of the expected frequencies is less than 5. If this is the case, we can sometimes combine two or more of the samples in such a way that none of the e's is less than 5.

If the null hypothesis of equal proportions is rejected, it is a good practice to graph the confidence intervals (see page 294) for the individual proportions p_i. The graph helps illuminate differences between the proportions.

EXAMPLE 9 **Graphical display of confidence intervals**

Four methods are under development for turning metal disks into a superconducting material. Fifty disks are made by each method and they are checked for superconductivity when cooled with liquid nitrogen.

	Method 1	Method 2	Method 3	Method 4	Total
Superconductors	31	42	22	25	120
Failures	19	8	28	25	80
Total	50	50	50	50	200

Perform a chi square test with $\alpha = 0.05$. If there is a significant difference between the proportions of superconductors produced, plot the individual confidence intervals.

Solution 1. *Null hypothesis*: $p_1 = p_2 = p_3 = p_4$
Alternative hypothesis: p_1, p_2, p_3, and p_4 are not all equal.

2. *Level of significance*: $\alpha = 0.05$

3. *Criterion*: Reject the null hypothesis if $\chi^2 > 7.815$, the value of $\chi^2_{0.05}$ for $4 - 1 = 3$ degrees of freedom.

4. *Calculations*: Each cell in the first row has expected frequency

$$120 \cdot \frac{50}{200} = 30$$

and each cell in the second row has expected frequency

$$80 \cdot \frac{50}{200} = 20$$

The chi square statistic is

$$\chi^2 = \frac{1}{30} + \frac{144}{30} + \frac{64}{30} + \frac{25}{30} + \frac{1}{20} + \frac{144}{20} + \frac{64}{20} + \frac{25}{20}$$
$$= 19.50$$

5. *Decision*: Since 19.50 greatly exceeds 7.815, we reject the null hypothesis of equal proportions at the 5% level of significance.
The confidence intervals obtained from the large sample formula on page 294 have confidence limits

$$0.62 \pm 0.13, \quad 0.84 \pm 0.10, \quad 0.44 \pm 0.14, \quad 0.50 \pm 0.14.$$

They are plotted in Figure 10.2. Note how Method 2 stands out as being better.

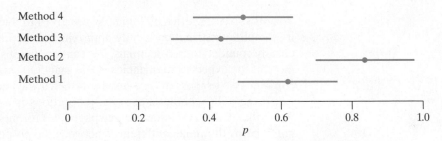

Figure 10.2
Confidence intervals for several proportions

Although there has been no mention of randomization in the development of the χ^2 statistic, wherever possible the experimental units should be randomly assigned to methods. In the example above, the disks could be numbered from 1 to 200 and random numbers selected from 1 to 200 without replacement. The disks corresponding to the first fifty numbers drawn would be assigned to method 1, and so on. This will prevent uncontrolled sources of variation from systematically influencing the test concerning the four methods.

So far, the alternative hypothesis has been that $p_1, p_2, \ldots,$ and p_k are not all equal, and for $k = 2$ this reduces to the alternative hypothesis $p_1 \neq p_2$. In problems where the alternative hypothesis may also be $p_1 < p_2$ or $p_1 > p_2$, we can base the test on the statistic

Statistic for test concerning difference between two proportions

$$Z = \frac{\dfrac{X_1}{n_1} - \dfrac{X_2}{n_2}}{\sqrt{\hat{p}(1 - \hat{p})\left(\dfrac{1}{n_1} + \dfrac{1}{n_2}\right)}} \quad \text{with} \quad \hat{p} = \frac{X_1 + X_2}{n_1 + n_2}$$

which, for large samples, is a random variable having approximately the standard normal distribution. The test based on this statistic is equivalent to the one based on the χ^2 statistic on page 302 with $k = 2$, in the sense that the square of this Z statistic actually equals the χ^2 statistic (see Exercise 10.38). The critical regions for this alternative test of the null hypothesis $p_1 = p_2$ are like those shown in the table on page 260 with p_1 and p_2 substituted for μ_1 and μ_2.

EXAMPLE 10   **A large sample test concerning two proportions**

A strategy called A/B testing is being implemented by many e-commerce companies to increase internet sales. An improvement project begins by selecting a web page to change. Maybe the product description is changed or maybe a picture changed. Next, a fraction of the incoming traffic to the product page is directed to the modified page. Daily counts of the number of visitors to the product page and the number who purchase are recorded both for the original page and for the modified page. The two different versions of the web page give rise to the terminology A and B.

A major online electronics retailer gets over a million hits each day on its web site. When a modified main page is compared with the existing main page, there will be millions of hits for each in a day. Further downstream, where traffic falls off, the object is to improve sales performance from a visitor who reach the selected product page.

For one product, a picture of someone using the product is added. Suppose the total weekly totals are

	Original page	Modified page
Number of visitors	2841	2297
Number that purchase	77	107

Is there strong evidence that the modified page increase sales?

Solution   Let p_1 be the probability a visitor to the original page purchases and item and let p_2 be the probability for the modified page.

1. We want to establish that $p_1 < p_2$ so
   Null hypothesis: $p_1 = p_2$
   Alternative hypothesis: $p_1 < p_2$

2. *Level of significance:* $\alpha = 0.01$

3. *Criterion:* Reject the null hypothesis if $Z < -2.33$ where Z is given by the formula on page 304.

4. *Calculations:* Substituting $x_1 = 77$, $n_1 = 2841$, $x_2 = 107$, $n_2 = 2297$, and

$$\widehat{p} = \frac{77 + 107}{2841 + 2297} = 0.0358$$

into the formula for Z, we obtain the observed value of the test statistic

$$\frac{\dfrac{77}{2841} - \dfrac{107}{2297}}{\sqrt{(0.0358)(0.9642)\left(\dfrac{1}{2841} + \dfrac{1}{2297}\right)}} = -3.74$$

5. *Decision:* Since $Z = -3.74$ is less than -2.33, the null hypothesis must be rejected at level of significance 0.01. We conclude that the proportion of purchasers is higher for the modified page than the original page. The significance probability

$$P\text{-value} = P[Z \leq -3.74] = .0001$$

shown in Figure 10.3, further strengthens the conclusion.

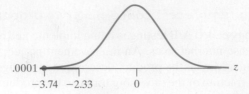

Figure 10.3
Rejection region and *P*-value
for Example 10

The statistic for testing $p_1 - p_2 = \delta_0$ leads to a confidence interval which provides the set of plausible values for $p_1 - p_2$. The confidence limits are

Large sample confidence limits for the difference of two proportions

$$\frac{x_1}{n_1} - \frac{x_2}{n_2} \pm z_{\alpha/2} \sqrt{\frac{\frac{x_1}{n_1}\left(1 - \frac{x_1}{n_1}\right)}{n_1} + \frac{\frac{x_2}{n_2}\left(1 - \frac{x_2}{n_2}\right)}{n_2}}$$

EXAMPLE 11   **A large sample confidence interval for the difference of two proportions**

With reference to the web site improvement in Example 9, find a 95% confidence interval for $p_1 - p_2$.

Solution   Since $\frac{x_1}{n_1} = \widehat{p}_1 = \frac{77}{2841} = 0.0271$ and $\frac{x_2}{n_2} = \widehat{p}_2 = \frac{107}{2297} = 0.0466$,

$$\frac{x_1}{n_1} - \frac{x_2}{n_2} \pm z_{\alpha/2} \sqrt{\frac{\frac{x_1}{n_1}\left(1 - \frac{x_1}{n_1}\right)}{n_1} + \frac{\frac{x_2}{n_2}\left(1 - \frac{x_2}{n_2}\right)}{n_2}}$$

$$= 0.0271 - 0.0466 \pm 1.96 \sqrt{\frac{(0.0271)(0.9729)}{2841} + \frac{(0.0466)(0.9534)}{2297}}$$

so the 95% confidence interval is $-0.030 < p_1 - p_2 < -0.009$

The modified web page leads to a higher proportion of purchasers. Although the increase in proportions is small, because of the extremely large number of visitors to the web page, there will be substantially more sales in a month.   ∎

Exercises

10.19 A manufacturer of submersible pumps claims that at most 30% of the pumps require repairs within the first 5 years of operation. If a random sample of 120 of these pumps includes 47 which required repairs within the first 5 years, test the null hypothesis $p = 0.30$ against the alternative hypothesis $p > 0.30$ at the 0.05 level of significance.

10.20 The performance of a computer is observed over a period of 2 years to check the claim that the probability is

0.20 that its downtime will exceed 5 hours in any given week. Testing the null hypothesis $p = 0.20$ against the alternative hypothesis $p \neq 0.20$, what can we conclude at the level of significance $\alpha = 0.05$, if there were only 11 weeks in which the downtime of the computer exceeded 5 hours?

10.21 To check on an ambulance service's claim that at least 40% of its calls are life-threatening emergencies, a random sample was taken from its files, and it was found that only 49 of 150 calls were life-threatening emergencies. Can the null hypothesis $p = 0.40$ be rejected against the alternative hypothesis $p < 0.40$ if the probability of a Type I error is to be at most 0.01?

10.22 In a random sample of 600 cars making a right turn at a certain intersection, 157 pulled into the wrong lane. Test the null hypothesis that actually 30% of all drivers make this mistake at the given intersection, using the alternative hypothesis $p \neq 0.30$ and the level of significance

(a) $\alpha = 0.05$;

(b) $\alpha = 0.01$.

10.23 An airline claims that only 6% of all lost luggage is never found. If, in a random sample, 17 of 200 pieces of lost luggage are not found, test the null hypothesis $p = 0.06$ against the alternative hypothesis $p > 0.06$ at the 0.05 level of significance.

10.24 Suppose that 4 of 13 undergraduate engineering students are going on to graduate school. Test the dean's claim that 60% of the undergraduate students will go on to graduate school, using the alternative hypothesis $p < 0.60$ and the level of significance $\alpha = 0.05$. [*Hint*: Use Table 1 to determine the probability of getting "at most 4 successes in 13 trials" when $p = 0.60$.]

10.25 A manufacturer of high-definition televisions (HDTVs) claims that 95% of a high-end model will not fail during the one-year warrantee period. You doubt this claim and want to refute it on the basis of a sample of 1,000 sets where 937 did not fail under warrantee. First,

(a) Conduct a test of hypotheses using $\alpha = 0.05$.

(b) In light of your decision in part (a), what error could you have made? Explain in the context of this exercise.

10.26 Refer to Exercise 10.25. Suppose a sample of 450 sets of an entry-level model HDTV yielded 380 sets that did not fail under warrantee. Obtain a 98% confidence interval for the difference in proportions.

10.27 Tests are made on the proportion of defective castings produced by 5 different molds. If there were 14 defectives among 100 castings made with Mold I, 33 defectives among 200 castings made with Mold II, 21 defectives among 180 castings made with Mold III, 17 defectives among 120 castings made with Mold IV, and 25 defectives among 150 castings made with Mold V, use the 0.01 level of significance to test whether the true proportion of defectives is the same for each mold.

10.28 A study showed that 64 of 180 persons who saw a photocopying machine advertised during the telecast of a baseball game and 75 of 180 other persons who saw it advertised on a variety show remembered the brand name 2 hours later. Use the χ^2 statistic to test

at the 0.05 level of significance whether the difference between the corresponding sample proportions is significant.

10.29 The following data come from a study in which random samples of the employees of three government agencies were asked questions about their pension plan:

	Agency 1	Agency 2	Agency 3
For the pension plan	67	84	109
Against the pension plan	33	66	41

Use the 0.01 level of significance to test the null hypothesis that the actual proportions of employees favoring the pension plan are the same.

10.30 The owner of a machine shop must decide which of two snack-vending machines to install in his shop. If each machine is tested 250 times and the first machine fails to work (neither delivers the snack nor returns the money) 13 times and the second machine fails to work 7 times, test at the 0.05 level of significance whether the difference between the corresponding sample proportions is significant, using

(a) the χ^2 statistic on page 302;

(b) the Z statistic on page 304.

10.31 With reference to the preceding exercise, verify that the square of the value obtained for Z in part (b) equals the value obtained for χ^2 in part (a).

10.32 Photolithography plays a central role in manufacturing integrated circuits made on thin disks of silicon. Prior to a quality-improvement program, too many rework operations were required. In a sample of 200 units, 26 required reworking of the photolithographic step. Following training in the use of Pareto charts and other approaches to identify significant problems, improvements were made. A new sample of size 200 had only 12 that needed rework.

Is this sufficient evidence to conclude at the 0.01 level of significance that the improvements have been effective in reducing the rework?

10.33 With reference to Exercise 10.32, find a large sample 99% confidence interval for the true difference of the proportions.

10.34 To test the null hypothesis that the difference between two population proportions equals some constant δ_0, not necessarily 0, we can use the statistic

$$Z = \frac{\dfrac{X_1}{n_1} - \dfrac{X_2}{n_2} - \delta_0}{\sqrt{\dfrac{\dfrac{X_1}{n_1}\left(1 - \dfrac{X_1}{n_1}\right)}{n_1} + \dfrac{\dfrac{X_2}{n_2}\left(1 - \dfrac{X_2}{n_2}\right)}{n_2}}}$$

which, for large samples, is a random variable having the standard normal distribution.

(a) With reference to Exercise 10.32, use this statistic to test at the 0.05 level of significance whether the true proportion of units requiring rework is now at least 4% less than before the improvements were made.

(b) In a true-false test, a test item is considered to be good if it discriminates between well-prepared students and poorly prepared students. If 205 of 250 well-prepared students and 137 of 250 poorly prepared students answer a certain item correctly, test at the 0.01 level of significance whether for the given item the proportion of correct answers can be expected to be at least 15% higher among well-prepared students than among poorly prepared students.

10.35 With reference to part (b) of Exercise 10.34, find a large sample 99% confidence interval for the true difference of the proportions.

10.36 Verify that the formulas for the χ^2 statistic on page 301 (with $\hat{p}$ substituted for the p_i) and on page 302 are equivalent.

10.37 Verify that if the expected frequencies are determined in accordance with the rule on page 302, the sum of the expected frequencies for each row and column equals the sum of the corresponding observed frequencies.

10.38 Verify that the square of the Z statistic on page 304 equals the χ^2 statistic on page 302 for $k = 2$.

10.4   Analysis of $r \times c$ Tables

As we suggested earlier, the method by which we analyzed the example on page 302 lends itself also to the analysis of **$r \times c$ tables**, or r-by-c tables; that is, tables in which data are tallied into a two-way classification having r rows and c columns. Such tables arise in essentially two kinds of problems. First, we might again have samples from several populations, with the distinction that now each trial permits more than two possible outcomes. This might happen, for example, if persons belonging to different income groups are asked whether they favor a certain political candidate, whether they are against him, or whether they are indifferent or undecided. The other situation giving rise to an $r \times c$ table is one in which we sample from one population but classify each item with respect to two (usually qualitative) categories. This might happen, for example, if a consumer testing service rates cars as excellent, superior, average, or poor with regard to performance and also with regard to appearance. Each car tested would then fall into one of the 16 cells of a 4×4 table, and it is mainly in connection with problems of this kind that $r \times c$ tables are referred to as **contingency tables**.

The essential difference between the two kinds of situations giving rise to $r \times c$ tables is that in the first case the column totals (the sample sizes) are fixed, while in the second case only the **grand total** (the total for the entire table) is fixed. As a result, there are also differences in the null hypotheses we shall want to test. In the first case we want to test whether the probability of obtaining an observation in the ith row is the same for each column; symbolically, we shall want to test the null hypothesis

$$p_{i1} = p_{i2} = \cdots = p_{ic} \qquad \text{for } i = 1, 2, \ldots, r$$

where p_{ij} is the probability of obtaining an observation belonging to the ith row and the jth column, and

$$\sum_{i=1}^{r} p_{ij} = 1$$

for each column, the alternative hypothesis is that the p's are not all equal for at least one row. In the second case we shall want to test the null hypothesis that the random variables represented by the two classifications are independent, so that p_{ij} is the product of the probability of getting a value belonging to the ith row and the probability of getting a value belonging to the jth column. The alternative hypothesis is that the two random variables are dependent.

In spite of the differences we have described, the analysis of an $r \times c$ table is the same for both cases. First we calculate the expected cell frequencies.

$$e_{ij} = \frac{(i\text{th row total}) \times (j\text{th column total})}{\text{grand total}}$$

When the column totals are fixed, the test is called a test of homogeneity.

Null hypothesis of homogeneity

$$H_0 : p_{i1} = p_{i2} = \cdots = p_{ic} \qquad \text{for all rows } i = 1, 2, \ldots, r$$

EXAMPLE 12 **Contingency table with column totals fixed**

Three different shops are used to repair electric motors. One hundred motors are sent to each shop. When a motor is returned, it is put in use and then the repair is classified as complete, requiring an adjustment, or an incomplete repair. The column totals are fixed at 100 each and the grand total at 300. Shop 1 produced 78 complete repairs, 15 minor adjustments, and 7 incomplete repairs. Shop 2 produced 56, 30, and 14, respectively; while Shop 3 produced 54, 31, and 15 complete, minor adjustments, and incomplete repairs, respectively.

		Shop 1	Shop 2	Shop 3	Total
	Complete	78	56	54	188
Repair	Adjustment	15	30	31	76
	Incomplete	7	14	15	36
	Total	100	100	100	300

Calculate the expected frequencies.

Solution For the Complete–Shop 2 cell of the table,

$$e_{12} = \frac{1\text{st row total} \times 2\text{nd column total}}{\text{grand total}} = \frac{188 \times 100}{300} = 62.67$$

Continuing, we obtain all of the expected frequencies, which are shown in bold below the frequencies.

		Shop 1	Shop 2	Shop 3	Total
	Complete	78	56	54	188
		62.7	**62.67**	**62.67**	
Repair	Adjustment	15	30	31	76
		25.33	**25.33**	**25.33**	
	Incomplete	7	14	15	36
		12.00	**12.00**	**12.00**	
	Total	100	100	100	300

Visually, Shop 1 has more than expected complete repairs and lower minor adjustments and incomplete repairs. A chi square test, described below, verifies that repair probabilities for the three shops are not homogeneous (see Exercise 10.39). ∎

The observed frequencies and the expected frequencies total the same for each row and column, so that only $(r-1)(c-1)$ of the e_{ij} have to be calculated directly, while the others can be obtained by subtraction from appropriate row or column totals. To perform a **chi square test** we then substitute into the formula

χ^2 statistic for analysis of $r \times c$ table

$$\chi^2 = \sum_{i=1}^{r} \sum_{j=1}^{c} \frac{(o_{ij} - e_{ij})^2}{e_{ij}}$$

and reject the null hypothesis if this statistic exceeds χ_α^2 for $(r-1)(c-1)$ degrees of freedom. The number of degrees of freedom is justified because after we determine $(r-1)(c-1)$ of the expected cell frequencies, the others are automatically determined. That is, they may be obtained by subtraction from appropriate row or column totals.

A **test of association** arises when each unit in a single sample is classified according to two characteristics. Only the total sample size is fixed but the chi square test remains the same. The null of hypothesis of independence specifies that each cell probability p_{ij} is the product of the marginal totals $p_{i\bullet} = \sum_{j=1}^{c} p_{ij}$ and $p_{\bullet j} = \sum_{i=1}^{r} p_{ij}$.

Null hypothesis of independence

$$H_0: p_{ij} = p_{i\bullet} \cdot p_{\bullet j} \qquad \text{for all } i, j$$

where $p_{i\bullet} = \sum_{j=1}^{c} p_{ij}$ and $p_{\bullet j} = \sum_{i=1}^{r} p_{ij}$

EXAMPLE 13   **The chi square test of independence**

To determine whether there really is a relationship between an employee's performance in the company's training program and his or her ultimate success in the job, the company takes a sample of 400 cases from its very extensive files and obtains the results shown in the following table:

		Performance in training program			
		Below average	*Average*	*Above average*	*Total*
	Poor	23	60	29	112
Success in job (*employer's rating*)	*Average*	28	79	60	167
	Very good	9	49	63	121
	Total	60	188	152	400

Use the 0.01 level of significance to test the null hypothesis that performance in the training program and success in the job are independent.

Solution   1. *Null hypothesis*: Performance in training program and success in job are independent.
Alternative hypothesis: Performance in training program and success in job are dependent.

2. *Level of significance*: $\alpha = 0.01$

3. *Criterion*: Reject the null hypothesis if $\chi^2 > 13.277$, the value of $\chi_{0.01}^2$ for $(3-1)(3-1) = 4$ degrees of freedom, where χ^2 is given by the formula above.

4. *Calculations*: Calculating first the expected cell frequencies for the first two cells of the first two rows, we get

$$e_{11} = \frac{112 \cdot 60}{400} = 16.80 \qquad e_{12} = \frac{112 \cdot 188}{400} = 52.64$$

$$e_{21} = \frac{167 \cdot 60}{400} = 25.05 \qquad e_{22} = \frac{167 \cdot 188}{400} = 78.49$$

Then, by subtraction, we find the expected frequencies $e_{13} = 42.56$ and $e_{23} = 63.46$. Those for the third row are 18.15, 56.87, and 45.98. Thus,

$$\chi^2 = \frac{(23 - 16.80)^2}{16.80} + \frac{(60 - 52.64)^2}{52.64} + \frac{(29 - 42.56)^2}{42.56}$$

$$+ \frac{(28 - 25.05)^2}{25.05} + \frac{(79 - 78.49)^2}{78.49} + \frac{(60 - 63.46)^2}{63.46}$$

$$+ \frac{(9 - 18.15)^2}{18.15} + \frac{(49 - 56.87)^2}{56.87} + \frac{(63 - 45.98)^2}{45.98}$$

$$= 20.179$$

5. *Decision*: Since $\chi^2 = 20.179$ exceeds 13.277, the null hypothesis must be rejected. Performance and success are dependent.

[Using **R**: Dat=as.table(rbind(c(23,60,29),c(28,79,60),c(9,49,63)) dim-names(Dat) = list(Success = c("Poor", "Average","Very Good"), Performance = c("Below avg", "Average", "Above avg")) Then (Xsq=chisq.test(Dat)) and Xsq$expected]   ■

EXAMPLE 14   **Exploring the form of dependence**

With reference Example 13, find a pattern in the departure from independence.

Solution   We display the contingency table, but this time we conclude the expected frequencies just below the observed frequencies.

		Performance in training program			
		Below average	Average	Above average	Total
	Poor	23 **16.80**	60 **52.64**	29 **42.56**	112
Success in job (employer's rating)	Average	28 **25.05**	79 **78.49**	60 **63.46**	167
	Very good	9 **18.15**	49 **56.87**	63 **45.98**	121
	Total	60	188	152	400

Also, we write the χ^2 statistic as the sum of the contributions.

$$\chi^2 = 2.288 + 1.029 + 4.320$$
$$+ 0.347 + 0.003 + 0.189$$
$$+ 4.613 + 1.089 + 6.300$$
$$= 20.179$$

From these two displays, it is clear that there is a positive dependence between performance in training and job success. For the three individual cells with the largest contributions to χ^2, the *above average–very good* cell frequency is high, whereas the *above average–poor* and *below average–very good* cell frequencies are low.   ■

10.5   Goodness of Fit

We speak of **goodness of fit** when we try to compare an observed frequency distribution with the corresponding values of an expected, or theoretical, distribution. To illustrate, suppose that during 400 five-minute intervals the air-traffic control of an airport received 0, 1, 2, ..., or 13 radio messages with respective frequencies of 3, 15, 47, 76, 68, 74, 46, 39, 15, 9, 5, 2, 0, and 1. Suppose, furthermore, that we want to check whether these data substantiate the claim that the number of radio messages which they receive during a 5-minute interval may be looked upon as a random variable having the Poisson distribution with $\lambda = 4.6$. Looking up the corresponding Poisson probabilities in Table 2W and multiplying them by 400 to get the expected frequencies, we arrive at the result shown in the following table, together with the original data:

Number of radio messages	Observed frequencies	Poisson probabilities	Expected frequencies
0	3 ⎫ 18	0.010	4.0 ⎫ 22.4
1	15 ⎭	0.046	18.4 ⎭
2	47	0.107	42.8
3	76	0.163	65.2
4	68	0.187	74.8
5	74	0.173	69.2
6	46	0.132	52.8
7	39	0.087	34.8
8	15	0.050	20.0
9	9	0.025	10.0
10	5 ⎫	0.012	4.8 ⎫
11	2 ⎪ 8	0.005	2.0 ⎪ 8.0
12	0 ⎪	0.002	0.8 ⎪
13	1 ⎭	0.001	0.4 ⎭
	400		400.0

Note that we combined some of the data so that none of the expected frequencies is less than 5.

To test whether the discrepancies between the observed and expected frequencies can be attributed to chance, we use the statistic

χ^2 statistic for test of goodness of fit

$$\chi^2 = \sum_{i=1}^{k} \frac{(o_i - e_i)^2}{e_i}$$

where the o_i and e_i are the observed and expected frequencies. The sampling distribution of this statistic is approximately the chi square distribution with $k - m$ degrees of freedom, where k is the number of terms in the formula for χ^2 and m is the number of quantities, obtained from the observed data, that are needed to calculate the expected frequencies.

EXAMPLE 15   **A chi square goodness of fit to the Poisson distribution**

With reference to the radio message data above, test at the 0.01 level of significance whether the data can be looked upon as values of a random variable having the Poisson distribution with $\lambda = 4.6$.

Solution   1. *Null hypothesis*: Random variable has a Poisson distribution with $\lambda = 4.6$.
   Alternative hypothesis: Random variable does not have the Poisson distribution with $\lambda = 4.6$.

2. *Level of significance*: $\alpha = 0.05$

3. *Criterion*: Reject the null hypothesis if $\chi^2 > 16.919$, the value of $\chi^2_{0.05}$ for $k - m = 10 - 1 = 9$ degrees of freedom, where χ^2 is given by the formula above. (The number of degrees of freedom is $10 - 1 = 9$, since only one quantity, the total frequency of 400, is needed from the observed data to calculate the expected frequencies.)

4. *Calculations*: Substitution into the formula for χ^2 yields

$$\chi^2 = \frac{(18 - 22.4)^2}{22.4} + \frac{(47 - 42.8)^2}{42.8} + \cdots + \frac{(9 - 10.0)^2}{10.0} + \frac{(8 - 8.0)^2}{8.0}$$

$$= 6.749$$

5. *Decision*: Since $\chi^2 = 6.749$ does not exceed 16.919, the null hypothesis cannot be rejected; we cannot reject that the Poisson distribution with $\lambda = 4.6$ provides a good fit at level $\alpha = 0.05$.   ∎

Exercises

10.39 Referring to Example 12 and the data on repair, use the 0.05 level of significance to test whether there is homogeneity among the shops' repair distributions.

10.40 A large electronics firm that hires many workers with disabilities wants to determine whether their disabilities affect such workers' performance. Use the level of significance $\alpha = 0.05$ to decide on the basis of the sample data shown in the following table whether it is reasonable to maintain that the disabilities have no effect on the workers' performance:

	Performance		
	Above average	*Average*	*Below average*
Disability	21	64	17
No disability	29	93	28

10.41 Tests of the fidelity and the selectivity of 190 digital radio receivers produced the results shown in the following table:

		Fidelity		
		Low	*Average*	*High*
Selectivity	*Low*	6	12	32
	Average	33	61	18
	High	13	15	0

Use the 0.01 level of significance to test whether there is a relationship (dependence) between fidelity and selectivity.

10.42 A quality-control engineer takes daily samples of $n = 4$ tractors coming off an assembly line and on 200 consecutive working days the data summarized in the following table are obtained:

Number requiring adjustments	Number of days
0	101
1	79
2	19
3	1

To test the claim that 10% of all the tractors coming off this assembly line require adjustments, look up the corresponding probabilities in Table 1, calculate the expected frequencies, and perform the chi square test at the 0.01 level of significance. (combine data so number of days ≥ 5.)

10.43 With reference to Exercise 10.42, verify that the mean of the observed distribution is 0.60, corresponding to 15% of the tractors requiring adjustments. Then look up the probabilities for $n = 4$ and $p = 0.15$ in Table 1, calculate the expected frequencies, and test at the 0.01 level of significance whether the binomial distribution with $n = 4$ and $p = 0.15$ provides a suitable model for this situation.

10.44 The following is the distribution of the hourly number of trucks arriving at a company's warehouse:

Trucks arriving per hour	Frequency
0	52
1	151
2	130
3	102
4	45
5	12
6	5
7	1
8	2

Find the mean of this distribution, and using it (rounded to one decimal place) as the parameter λ, fit a Poisson distribution. Test for goodness of fit at the 0.05 level of significance.

10.45 Among 100 purification filters used in an experiment, 46 had a service life of less than 20 hours, 19 had a service life of 20 or more but less than 40 hours, 17 had a service life of 40 or more but less than 60 hours, 12 had a service life of 60 or more but less than 80 hours, and 6 had a service life of 80 hours or more. Test at the 0.01 level of significance whether the lifetimes may be regarded as a sample from an exponential population with $\mu = 40$ hours.

Calculate $P[0 < X < 20]$ and multiply by 100 to obtain e_1, and so on.

10.46 A chi square test is easily implemented on a computer. With the counts

31	42	22	25
19	8	28	25

from Example 8 in columns 1–4, the *MINITAB* commands

Dialog box:
Stat > Tables > Chi-square Test for Association
Pull down **Summarized data in a two-way table.**
Type $C1 - C4$ in **columns containing the table.**
Click **Statistics** and then check **chi-square statistic**, **Display counts**, **Expected cell counts** and **Each cells contribution to chi-square.**
Click **OK**. Click **OK**.

produce the output
Expected counts are printed below observed counts

	Method 1	Method 2	Method 3	Method 4	Total
1	31	42	22	25	120
	30.00	30.00	30.00	30.00	
2	19	8	28	25	80
	20.00	20.00	20.00	20.00	
Total	50	50	50	50	200

Chi sq = $0.033 + 4.800 + 2.133$
$+ 0.050 + 7.200 + 3.200 + 1.250 = 19.550$
df = 3

Repeat the analysis using only the data from the first three methods.

10.47 The procedure in Exercise 10.46 also calculates the chi square test for independence. Do Exercise 10.40 using the computer.

Do's and Don'ts

Do's

1. Remember that it usually takes a sample size of a few hundred to get precise estimates of a proportion. When the sample size is large, calculate the approximate $100(1 - \alpha)\%$ confidence interval for the population proportion p

$$\frac{x}{n} - z_{\alpha/2} \sqrt{\frac{\frac{x}{n}\left(1 - \frac{x}{n}\right)}{n}} < p < \frac{x}{n} + z_{\alpha/2} \sqrt{\frac{\frac{x}{n}\left(1 - \frac{x}{n}\right)}{n}}$$

Base a test of the null hypothesis $H_0: p = p_0$ on the approximately standard normal test statistic

$$Z = \frac{X - n p_0}{\sqrt{n p_0 (1 - p_0)}}$$

2. When both sample sizes are large, compare two proportions by calculating the approximate $100(1 - \alpha)\%$ confidence interval for $p_1 - p_2$ with limits

$$\frac{x_1}{n_1} - \frac{x_2}{n_2} \pm z_{\alpha/2} \sqrt{\frac{\frac{x_1}{n_1}\left(1 - \frac{x_1}{n_1}\right)}{n_1} + \frac{\frac{x_2}{n_2}\left(1 - \frac{x_2}{n_2}\right)}{n_2}}$$

Base tests of the null hypothesis $H_0: p_1 = p_2$ on the approximately standard normal test statistic

$$Z = \frac{\dfrac{x_1}{n_1} - \dfrac{x_2}{n_2}}{\sqrt{\widehat{p}(1-\widehat{p})\left(\dfrac{1}{n_1} + \dfrac{1}{n_2}\right)}} \qquad \text{with} \qquad \widehat{p} = \frac{X_1 + X_2}{n_1 + n_2}$$

3. Use the chi square test to analyze $r \times c$ contingency tables. The test statistic

$$\chi^2 = \sum_{i=1}^{r} \sum_{j=1}^{c} \frac{(o_{ij} - e_{ij})^2}{e_{ij}}$$

with estimated expected values

$$e_{ij} = \frac{(i\text{th row total}) \times (j\text{th column total})}{\text{grand total}}$$

and observed values o_{ij} is approximately chi square distributed with $(r-1)(c-1)$ degrees of freedom. This statistic applies to comparing several proportions when samples of size n_j are selected from the jth population. The same statistic applies to testing independence of the two sets of categories when a single sample of size n is cross-tabulated to create the $r \times c$ table.

Don'ts

1. Don't routinely apply the inference procedures for a proportion p without confirming that the outcomes from different trials are independent.

2. When sampling without replacement from a finite population of size N, don't forget to account for dependence when the finite population correction factor is substantially different from 1.

Review Exercises

10.48 In a sample of 100 ceramic pistons made for an experimental diesel engine, 18 were cracked. Construct a 95% confidence interval for the true proportion of cracked pistons using the large sample confidence interval formula.

10.49 With reference to Exercise 10.48, test the null hypothesis $p = 0.20$ versus the alternative hypothesis $p < 0.20$ at the 0.05 level.

10.50 In a random sample of 160 workers exposed to a certain amount of radiation, 24 experienced some ill effects. Construct a 99% confidence interval for the corresponding true percentage using the large sample confidence interval formula.

10.51 With reference to Exercise 10.50, test the null hypothesis $p = 0.18$ versus the alternative hypothesis $p \neq 0.18$ at the 0.01 level.

10.52 In a random sample of 100 packages shipped by air freight, 16 had some damage. Construct a 95% confidence interval for the true proportion of damaged packages using the large sample confidence interval formula.

10.53 With reference to Exercise 10.52, test the hypothesis $p = 0.10$ versus the alternative hypothesis $p > 0.10$ at the 0.01 level.

10.54 Refer to Example 5 but suppose there are two additional design plans B and C for making miniature drones. Under B, 10 of 40 drones failed the initial test and under C 15 of 39 failed. Consider the results for all three design plans. Use the 0.05 level of significance to test the null hypothesis of no difference in the three probabilities of failing the test.

10.55 As a check on the quality of eye glasses purchased over the internet, glasses were individually ordered from several different online vendors. Among the 92 lenses with antireflection coating, 61 prescriptions required a thickness at the center greater than 1.9 mm and 31 were thinner. Of the 61 thicker lenses, 12 failed impact testing while 18 of the 31 thinner lens failed.

(Source: K. Citek, et al., Safety and compliance of prescription spectacles ordered by the public via the Internet,"*Optometry* **82** (2011), 549–555.)

Can we conclude, at the 0.05 level of significance, that a larger proportion of the thicker lenses will survive the impact test?

10.56 With reference to Exercise 10.55, find a large sample 95% confidence interval for the true difference of probabilities.

10.57 Two bonding agents, *A* and *B*, are available for making a laminated beam. Of 50 beams made with Agent *A*, 11 failed a stress test, whereas 19 of the 50 beams made with Agent *B* failed. At the 0.05 level, can we conclude that Agent *A* is better than Agent *B*?

10.58 With reference to Exercise 10.57, find a large sample 95% confidence interval for the true difference of the probabilities of failure.

10.59 Cooling pipes at three nuclear power plants are investigated for deposits that would inhibit the flow of water. From 30 randomly selected spots at each plant, 13 from the first plant, 8 from the second plant, and 19 from the third were clogged.

(a) Use the 0.05 level to test the null hypothesis of equality.

(b) Plot the confidence intervals for the three probabilities of being clogged.

10.60 Two hundred tires of each of four brands are individually placed in a testing apparatus and run until failure. The results are obtained the results shown in the following table:

	Brand A	Brand B	Brand C	Brand D
Failed to last 30,000 miles	26	23	15	32
Lasted from 30,000 to 40,000	118	93	116	121
Lasted more than 40,000 miles	56	84	69	47
Total	200	200	200	200

(a) Use the 0.01 level of significance to test the null hypothesis that there is no difference in the qual-

ity of the four kinds of tires with regard to their durability.

(b) Plot the four individual 99% confidence intervals for proportions that last more than 40,000 miles.

10.61 The following is the distribution of the daily number of power failures reported in a western city on 300 days:

Number of power failures	Number of days
0	9
1	43
2	64
3	62
4	42
5	36
6	22
7	14
8	6
9	2

Test at the 0.05 level of significance whether the daily number of power failures in this city is a random variable having the Poisson distribution with $\lambda = 3.2$.

10.62 With reference to Example 13, repeat the analysis after combining the categories *below average* and *average* in the training program and the categories *poor* and *average* in success. Comment on the form of the dependence.

10.63 Mechanical engineers, testing a new arc-welding technique, classified welds both with respect to appearance and an X-ray inspection.

		Appearance			
		Bad	Normal	Good	Total
X-ray	Bad	20	7	3	30
	Normal	13	51	16	80
	Good	7	12	21	40
	Total	40	70	40	150

Test for independence using $\alpha = 0.05$ and find the individual cell contributions to the χ^2 statistic.

Key Terms

Chi square test	302	Goodness of fit	312	Sample proportion	291
Contingency table	308	Observed cell frequency	301	Test of association	310
Expected cell frequency	302	$r \times c$ table	308		

CHAPTER

REGRESSION ANALYSIS

11

The main objective of many statistical investigations is to make predictions, preferably on the basis of mathematical equations. For instance, an engineer may wish to predict the amount of oxide that will form on the surface of a metal baked in an oven for one hour at 200 degrees Celsius, or the amount of deformation of a ring subjected to a compressive force of 1,000 pounds, or the number of miles to wear out a tire as a function of tread thickness and composition. Usually, such predictions require that a formula be found which relates the dependent variable (whose value one wants to predict) to one or more independent variables.

Problems relating to predictions based on the known value of one variable are treated in Sections 11.1 through 11.3 and 11.6; the case where predictions are based on the known values of several variables is treated in Section 11.4. The importance of checking the assumptions concerning the prediction model is considered in Section 11.5.

11.1 The Method of Least Squares

We introduce the ideas of regression analysis in the simple setting where the distribution of a random variable Y depends on the value x of one other variable. Calling the two variables x and y, the terminology is

$x =$ **independent variable**, also called **predictor variable**, or **input variable**.

$y =$ **dependent variable**, or **response variable**.

Typically, the independent variable is observed without error, or with an error which is negligible when compared with the error (chance variation) in the dependent variable. For example, in measuring the amount of oxide on the surface of a metal specimen, the baking temperature can usually be controlled with good precision, but the oxide-thickness measurement may be subject to considerable chance variation. Even though the independent variable may be fixed at x, repeated measurements of the dependent variable may lead to y values which differ considerably. Such differences among y values can be attributed to several causes, chiefly to errors of measurement and to the existence of other, uncontrolled variables which may influence the measured thickness, y, when x is fixed. Thus, measurements of the thickness of oxide layers may vary over several specimens baked for the same length of time at the same temperature because of the difficulty in measuring thickness as well as possible differences in the uncontrolled variables such as the composition of the oven atmosphere and surface conditions of the specimen.

It should be apparent from this discussion that, in this context, the measured thickness of oxide layers Y is a random variable whose distribution depends on x. In most situations of this sort we are interested mainly in the relationship between x and the mean $E[Y \mid x]$ of the corresponding distribution of Y. We refer to this relationship as the **regression of Y on x**. (For the time being we shall consider the case where x is fixed, that is, not random. In Section 11.6 we shall consider the case where x and y are both values of random variables.)

Let us first treat the case where the regression curve of Y on x is a **linear regression** curve. That is, for any given x, the mean of the distribution of the Y's is given by $\alpha + \beta x$. In general, Y will differ from this mean, and we shall denote this difference by ε, writing

$$Y = \alpha + \beta x + \varepsilon$$

Thus, ε is a random variable and under this linear regression model we can always choose α so that the mean of the distribution of this random variable is equal to zero. The value of ε for any given observation will depend on a possible error of measurement and on the values of variables other than x which might have an influence on Y.

An engineer conducts an experiment with the purpose of showing that adding a new component to the existing metal alloy increases the cooling rate. Faster cooling rates lead to stronger materials and improve other properties. Let

$x =$ percentage of the new component present in the metal.

$y =$ cooling rate, during a heat-treatment stage, in °F per hour .

The engineer decides to consider several different percentages of the new component. Suppose the observed data are

x	0	1	2	2	4	4	5	6
y	25	20	30	40	45	50	60	50

The first step in any analysis of the relationship between the two variables is to plot the data in a **scatter plot** or **scattergram**. The predictor variable x is located on the horizontal axes and the response variable y on the vertical axis.

First step in the analysis

> Creating a **scatter plot** is an important preliminary step preceding any statistical analysis of the two variables. The existence of any increasing, or decreasing, relationship becomes readily apparent.

The pattern of data in the scatter plot will suggest whether or not there is a straight line relationship. The scatter plot of the cooling data appears in Figure 11.1. The points cluster around a straight line but the linear relation is masked by moderately sized departures from a line.

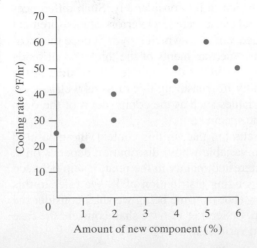

Figure 11.1

Scatter plot of cooling rate data suggests straight line model

Now we face the problem of using the observed data to estimate the parameters α and β of the regression line in a manner that somehow provides the best fit to the data. If different experimenters fit a line by eye, the lines would likely be different. Consequently, to handle problems of this kind, we must seek a nonsubjective method for fitting straight lines which reflects some desirable statistical properties.

To state the problem formally, we have n paired observations (x_i, y_i) for which it is reasonable to assume that the regression of Y on x is linear. We want to determine the line (that is, the equation of the line) which in some sense provides the best fit. There are several ways in which we interpret the word "best," and the meaning we shall give it here may be explained as follows. If we predict y by means of the equation

$$\widehat{y} = a + bx$$

where a and b are constants, then e_i, the error in predicting the value of y corresponding to the given x_i, is

$$e_i = y_i - \widehat{y}_i$$

and we shall want to determine a and b so that these errors are in some sense as small as possible.

Since we cannot simultaneously minimize each of the e_i individually, we might try to make their sum $\sum_{i=1}^{n} e_i$ as close as possible to zero. However, since this sum can be made equal to zero by many choices of totally unsuitable lines for which the positive and negative errors cancel, we shall minimize the sum of the squares of the e_i (for the same reason we worked with the squares of the deviations from the mean in the definition of the standard deviation). In other words, we apply the **principle of least squares** and choose a and b so that

$$\sum_{i=1}^{n} e_i^2 = \sum_{i=1}^{n} [y_i - (a + bx_i)]^2$$

is a minimum. This is equivalent to minimizing the sum of the squares of the vertical distances from the points to the line in any scatter plot (see Figure 11.2).

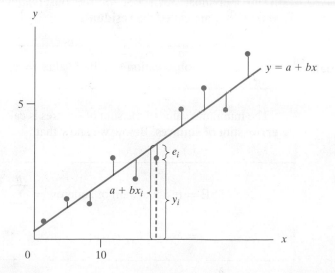

Figure 11.2
Diagram for least squares criterion showing the vertical deviations whose sum of squares is minimized

The procedure of finding the equation of the line which best fits a given set of paired data, called the **method of least squares**, yields values for a and b (estimates of α and β) that have many desirable properties; some of these are mentioned on page 326.

Before minimizing the sum of squared deviations to obtain the **least squares estimators**, it is convenient to introduce some notation for the sums of squares and sums of cross-products.

$$S_{xx} = \sum_{i=1}^{n} (x_i - \bar{x})^2 = \sum_{i=1}^{n} x_i^2 - \frac{\left(\sum_{i=1}^{n} x_i\right)^2}{n}$$

$$S_{yy} = \sum_{i=1}^{n} (y_i - \bar{y})^2 = \sum_{i=1}^{n} y_i^2 - \frac{\left(\sum_{i=1}^{n} y_i\right)^2}{n}$$

$$S_{xy} = \sum_{i=1}^{n} (x_i - \bar{x})(y_i - \bar{y}) = \sum_{i=1}^{n} x_i y_i - \frac{\left(\sum_{i=1}^{n} x_i\right)\left(\sum_{i=1}^{n} y_i\right)}{n}$$

The first expressions are preferred on conceptual grounds because they highlight deviations from the mean and on computing grounds because they are less susceptible to roundoff error. The second expressions are for handheld calculators.

Below, we show that the least squares estimates are

Least squares estimates

$$\widehat{\alpha} = \bar{y} - b \cdot \bar{x} \quad \text{and} \quad \widehat{\beta} = \frac{S_{xy}}{S_{xx}}$$

where $\bar{x}$ and $\bar{y}$ are, respectively, the means of the values of x and y.

The least squares estimates determine the best-fitting line

Fitted (or estimated) regression line

$$\widehat{y} = \widehat{\alpha} + \widehat{\beta} x$$

where the $\widehat{}$ on y, α, and β indicates the estimated value.

The individual deviations of the observations y_i from their fitted values $\widehat{y}_i = \widehat{\alpha} + \widehat{\beta} x_i$ are called the **residuals**.

Residuals

$$\text{observation} - \text{fitted value} = y_i - \widehat{\alpha} - \widehat{\beta} x_i$$

The minimum value of the sum of squares is called the **residual sum of squares** or **error sum of squares**. Below we show that

$$\text{SSE} = \text{residual sum of squares} = \sum_{i=1}^{n} (y_i - \widehat{\alpha} - \widehat{\beta} x_i)^2$$

$$= S_{yy} - S_{xy}^2 / S_{xx}$$

Example 1 illustrates the least squares calculations when the sample means are subtracted to center the data.

EXAMPLE 1

Least squares calculations for the cooling rate data

Calculate the least squares estimates and sum of squares error for the cooling rate data.

Solution   The structure of the table guides the calculations.

x	y	$x - \bar{x}$	$y - \bar{y}$	$(x - \bar{x})^2$	$(x - \bar{x})(y - \bar{y})$	$(y - \bar{y})^2$	residual
0	25	-3	-15	9	45	225	3
1	20	-2	-20	4	40	400	-8
2	30	-1	-10	1	10	100	-4
2	40	-1	0	1	0	0	6
4	45	1	5	1	5	25	-1
4	50	1	10	1	10	100	4
5	60	2	20	4	40	400	8
6	50	3	10	9	30	100	-8
$\bar{x} = 3$	$\bar{y} = 40$	0	0	$S_{xx} = 30$	$S_{xy} = 180$	$S_{yy} = 1350$	

so $\widehat{\beta} = S_{xy}/S_{xx} = 180/30 = 6$ and $\widehat{\alpha} = \bar{y} - \widehat{\beta}\bar{x} = 40 - 6(3) = 22$.

Since $\widehat{\beta} = 6$ and $\widehat{\alpha} = 22$, the *least squares line* is

$$\widehat{y} = \widehat{\alpha} + \widehat{\beta}x = 22 + 6x$$

The *residuals* are $y_i - \widehat{\alpha} - \widehat{\beta}x_i = y_i - 22 - 6x_i$, or, $25 - 22 - 6(0) = 3, -8, -4,$ $6, -1, 4, 8, -8$.

The sum of squares error is then

$$\text{SSE} = \sum_{i=1}^{n}(y_i - \hat{\alpha} - \hat{\beta}x_i)^2$$

$$= 3^2 + (-8)^2 + (-4)^2 + 6^2 + (-1)^2 + 4^2 + 8^2 + (-8)^2 = 270$$

Alternatively,

$$\text{SSE} = S_{yy} - S_{xy}^2/S_{xx} = 1350 - 180^2/30 = 270 \qquad ∎$$

When the observations themselves, or the means, have several digits, the approach in Example 1 proves tedious and cumbersome. Although the second expressions for S_{xx}, S_{yy}, and S_{xy} are somewhat easy to evaluate on hand held calculators, we strongly recommend performing regression analysis with computer software. The specifics for using *MINITAB* and R are given later in Examples 4 to 6 and Exercise 11.22.

EXAMPLE 2

A numerical example of fitting a straight line by least squares

The following are measurements of the air velocity and evaporation coefficient of burning fuel droplets in an impulse engine:

Air velocity (cm/s) x	Evaporation coefficient (mm²/s) y
20	0.18
60	0.37
100	0.35
140	0.78
180	0.56
220	0.75
260	1.18
300	1.36
340	1.17
380	1.65

Fit a straight line to these data by the method of least squares, and use it to estimate the evaporation coefficient of a droplet when the air velocity is 190 cm/s.

Solution   Typical output from statistical software includes the information

```
The regression equation is

y = 0.0692 + 0.003829x


Source          SS
Regression    1.93507
Error         0.20238
Total         2.13745
```

In our notation, $\widehat{\alpha} = 0.692$, $\widehat{\beta} = 0.003829$, and SSE $= 0.20238$. The least squares line predicts an increase of 0.003829 mm²/s in the evaporation coefficient for each increase of 1 cm/s in air velocity.

It is instructive, at least once, to confirm the computer software using a simple hand held calculator.

For these $n = 10$ pairs (x_i, y_i) we first calculate

$$\sum_{i=1}^{n} x_i = 2,000 \quad \sum_{i=1}^{n} x_i^2 = 532,000$$

$$\sum_{i=1}^{n} y_i = 8.35 \quad \sum_{i=1}^{n} x_i y_i = 2,175.40$$

$$\sum_{i=1}^{n} y_i^2 = 9.1097$$

and then we obtain

$$S_{xx} = 532,000 - (2,000)^2/10 = 132,000$$
$$S_{xy} = 2,175.40 - (2,000)(8.35)/10 = 505.40$$
$$S_{yy} = 9.1097 - (8.35)^2/10 = 2.13745$$

Consequently, the estimate of slope is

$$\widehat{\beta} = \frac{S_{xy}}{S_{xx}} = \frac{505.40}{132,000} = 0.003829$$

and then the estimate of intercept becomes

$$\widehat{\alpha} = \bar{y} - \widehat{\beta}\bar{x} = \frac{8.35}{10} - 0.003829 \frac{2,000}{10} = 0.0692$$

The equation of the straight line that best fits the given data in the sense of least squares,

$$\widehat{y} = 0.0692 + 0.003829\, x$$

confirms the computer software calculation.

For $x = 190$, we predict that the evaporation coefficient will be

$$\widehat{y} = 0.0692 + 0.003829(190) = 0.80 \text{ mm}^2/\text{s}$$

Finally, the residual sum of squares is

$$S_{yy} - \frac{S_{xy}^2}{S_{xx}} = 2.13745 - \frac{(505.40)^2}{132,000} = 0.20238 \qquad \blacksquare$$

To avoid confusion, we make it clear that there are two possible regression lines.

EXAMPLE 3

One scatter plot but two different fitted lines

Engineers fabricating a new transmission-type electron multiplier[1] created an array of silicon nanopillars (see Figure 2.5) on a flat silicon membrane. The precise structure can influence the electrical properties so, subsequently, the height and widths of 50 nanopillars (see Exercise 11.23) were measured in nanometers (nm) or $10^{-9} \times$ meters. The summary statistics, with $x =$ width and $y =$ height, are

$$n = 50 \qquad \bar{x} = 88.34 \qquad \bar{y} = 305.58$$
$$S_{xx} = 7,239.22 \qquad S_{xy} = 17,840.1 \qquad S_{yy} = 66,976.2$$

(a) Find the least squares line for predicting height from width.

(b) Find the least squares line for predicting width from height.

(c) Make a scatter plot and show both lines. Comment.

Solution (a) Here $y =$ height and the least squares estimates are

$$\text{slope} = \widehat{\beta} = \frac{S_{xy}}{S_{xx}} = \frac{17,840.1}{7,239.22} = 2.464 \qquad \text{and}$$

$$\widehat{\alpha} = \bar{y} - \widehat{\beta}\bar{x} = 305.58 - \frac{17,840.1}{7,239.22} \times 88.34 = 87.88$$

The fitted line is

$$\text{height} = 87.88 + 2.464 \text{ width}.$$

---

[1] H. Qin, H. Kim, and R. Blick, Nanopillar arrays on semiconductor membranes as electron emission amplifiers, *Nanotechnology* 19 (2008), 095504 (5pp).

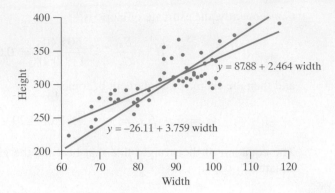

Figure 11.3
Scatter plot and two fitted lines

(b) Width is now the response variable and height the predictor, so x and y must be interchanged.

$$\text{slope} = \widehat{\beta} = \frac{17{,}840.1}{66{,}976.2} = 0.266 \qquad \text{and}$$

$$\widehat{\alpha} = 88.34 - 0.2664 \times 305.58 = 6.944$$

The fitted line is

$$\text{width} = 6.944 + 0.266\,\text{height}.$$

(c) Using the data in Exercise 11.23, we construct the scatter plot in Figure 11.3 and include the two lines. The line from part (b) is written as

$$\text{height} = -\frac{6.944}{0.266} + \frac{1}{0.266}\,\text{width} = -26.11 + 3.759\,\text{width}$$

Notice that both pass through the mean point $(\bar{x}, \bar{y}) = (88.34, 305.58)$. The choice of fitted line depends on which variable you wish to predict.   ■

Determining the Least Squares Estimators

We now show that the choice of estimates

$$\widehat{\beta} = \frac{S_{xy}}{S_{xx}} \qquad \widehat{\alpha} = \bar{y} - b\bar{x}$$

minimizes the sum of squares,

$$S(a, b) = \sum_{i=1}^{n} (y_i - a - bx_i)^2$$

over all choices of a and b. We first find an alternative expression for the sum of squares for any a and b. Adding and subtracting $b\bar{x} - \bar{y}$, we have

$$y_i - a - bx_i = (y_i - \bar{y}) - b(x_i - \bar{x}) + (\bar{y} - a - b\bar{x})$$

Squaring both sides, we obtain

$$\begin{aligned}
(y_i - a - bx_i)^2 = {} & (y_i - \bar{y})^2 + b^2(x_i - \bar{x})^2 + (\bar{y} - a - b\bar{x})^2 \\
& - 2b(y_i - \bar{y})(x_i - \bar{x}) \\
& - 2b(\bar{y} - a - b\bar{x})(x_i - \bar{x}) \\
& + 2(y_i - \bar{y})(\bar{y} - a - b\bar{x})
\end{aligned}$$

Now, summing over both sides, the last two terms vanish so we obtain

$$S(a, b) = S_{yy} + b^2 S_{xx} + n(\bar{y} - a - b\bar{x})^2 - 2b S_{xy}$$

$$= n(\bar{y} - a - b\bar{x})^2 + \left(b^2 S_{xx} - 2b S_{xy} + \frac{S_{xy}^2}{S_{xx}}\right) + S_{yy} - \frac{S_{xy}^2}{S_{xx}}$$

$$= n(\bar{y} - a - b\bar{x})^2 + \left(b\sqrt{S_{xx}} - \frac{S_{xy}}{\sqrt{S_{xx}}}\right)^2 + S_{yy} - \frac{S_{xy}^2}{S_{xx}}$$

According to the principle of least squares, we must select values for a and b which will minimize this sum of squares. All three terms on the right-hand side are non-negative and the third term does not depend on the choice of a and b. The second term can be made to equal zero, its minimum value, by taking $\widehat{\beta} = S_{xy}/S_{xx}$. With this choice for b, the first term can be made equal to zero by taking $\widehat{\alpha} = y - \widehat{\beta}\bar{x}$. This confirms the formula for the least squares estimators on page 320.

Further, we have shown that the minimized sum of squares equals the third term. That is, the sum of squares error

$$\text{SSE} = S_{yy} - S_{xy}^2/S_{xx}$$

as stated on page 320.

The estimate of σ^2 is

$$s_e^2 = \frac{\sum_{i=1}^{n}(y_i - \widehat{\alpha} - \widehat{\beta}x_i)^2}{n - 2} \quad \text{or} \quad s_e^2 = \frac{S_{yy} - S_{xy}^2/S_{xx}}{n - 2}$$

Normal Equations for the Least Squares Estimators

A necessary condition that the sum of squared deviations,

$$\sum_{i=1}^{n}(y_i - a - bx_i)^2$$

be a minimum is the vanishing of the partial derivatives with respect to a and b. We thus have

$$2\sum_{i=1}^{n}[y_i - (a + bx_i)](-1) = 0$$

$$2\sum_{i=1}^{n}[y_i - (a + bx_i)](-x_i) = 0$$

and we can rewrite these two equations as

Normal equations

$$\sum_{i=1}^{n} y_i = an + b\sum_{i=1}^{n} x_i$$

$$\sum_{i=1}^{n} x_i y_i = a\sum_{i=1}^{n} x_i + b\sum_{i=1}^{n} x_i^2$$

This set of two linear equations in the unknowns a and b, called the **normal equations**, gives the same values of $\widehat{\alpha}$ and $\widehat{\beta}$ for the line which provides the best fit to a given set of paired data in accordance with the criterion of least squares.

EXAMPLE 4   **The least squares estimates obtained from the normal equations**

Solve the normal equations for the data in Example 2 and confirm the values for the least squares estimates.

Solution   Using the calculations in Example 2, the normal equations are

$$8.35 = 10\,a + 2{,}000\,b$$
$$2{,}175.40 = 2{,}000\,a + 532{,}000\,b$$

Solving this system of equations by use of determinants or the method of elimination, we obtain $a = 0.069$ and $b = 0.00383$.

As they must be, these values are the same, up to rounding error, as those obtained in Example 2.   ■

It is impossible to make any exact statement about the "goodness" of an estimate like this unless we make some assumptions about the underlying distributions of the random variables with which we are concerned and about the true nature of the regression. Looking upon $\widehat{\alpha}$ and $\widehat{\beta}$ as estimators of the actual regression coefficients α and β, the reader will be asked to show in Exercise 11.20 that these estimators are linear in the observations Y_i and that they are unbiased estimators of α and β. With these properties, we can refer to the remarkable **Gauss-Markov theorem**, which states that among all unbiased estimators for α and β which are linear in the Y_i, the least squares estimators have the smallest variance. In other words, the least squares estimators are the most reliable in the sense that they are subject to the smallest chance variations. A proof of the Gauss-Markov theorem may be found in the book by Johnson and Wichern referred to in the bibliography.

11.2 Inferences Based on the Least Squares Estimators

The method of the preceding section is used when the relationship between x and the mean of Y is linear or close enough to a straight line so that the least squares line yields reasonably good predictions. In what follows we shall assume that the regression *is* linear in x and, furthermore, that the n random variables Y_i are independently normally distributed with the means $\alpha + \beta x_i$ and the common variance σ^2 for $i = 1, 2, \ldots, n$. Equivalently, we write the model as

Statistical model for straight-line regression

$$Y_i = \alpha + \beta x_i + \varepsilon_i \qquad \text{for } i = 1, 2, \ldots, n$$

where it is assumed that the ε_i are independent normally distributed random variables having zero means and the common variance σ^2.

The various assumptions we have made here are illustrated in Figure 11.4, showing the distributions of Y_i for several values of x_i. Note that these additional assumptions are required to discuss the goodness of predictions based on least squares equations, the properties of $\widehat{\alpha}$ and $\widehat{\beta}$ as estimators of α and β, and so on. They were not required to obtain the original estimates based on the method of least squares.

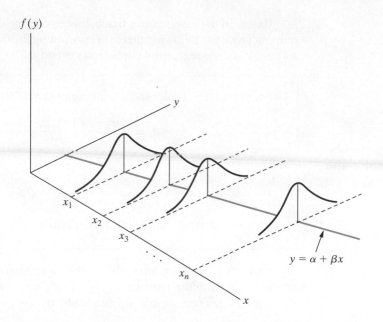

Figure 11.4
Diagram showing assumptions
underlying Theorem 11.1

Before we state a theorem concerning the distribution of the least squares estimators of α and β, we review the formulas for calculating the least squares estimators and then present an estimate of the error variance σ^2.

Recall from page 320 that the values for the least squares estimators of α and β are given by

$$\widehat{\alpha} = \bar{y} - \widehat{\beta} \cdot \bar{x} \qquad \widehat{\beta} = \frac{S_{xy}}{S_{xx}}$$

where S_{xx} and S_{xy} are defined also along with S_{yy}.

Note the close relationship between S_{xx} and S_{yy} and the respective sample variances of the x's and the y's; in fact, $s_x^2 = S_{xx}/(n-1)$ and $s_y^2 = S_{yy}/(n-1)$, and we shall sometimes use this alternative notation.

The variance σ^2 defined on page 326 is usually estimated in terms of the vertical deviations of the sample points from the least squares line. The ith such deviation is $y_i - \widehat{y_i} = y_i - (\widehat{\alpha} + \widehat{\beta} x_i)$ and the estimate of σ^2 is

$$s_e^2 = \frac{1}{n-2} \sum_{i=1}^{n} [y_i - (\widehat{\alpha} + \widehat{\beta} x_i)]^2$$

Traditionally, s_e is referred to as the **standard error of the estimate**. The s_e^2 estimate is the **residual sum of squares**, or the **error sum of squares**, divided by $n-2$. An equivalent formula for this estimate of σ^2, which is more convenient for handheld calculators, is given by

Estimate of σ^2

$$s_e^2 = \frac{S_{yy} - (S_{xy})^2 / S_{xx}}{n-2}$$

In these formulas the divisor $n-2$ is used to make the resulting estimator for σ^2 unbiased. It can be shown that under the given assumptions $(n-2)s_e^2/\sigma^2$ is a value of a random variable having the chi square distribution with $n-2$ degrees of freedom. The "loss" of two degrees of freedom is explained by the fact that the two regression coefficients α and β had to be replaced by their least squares estimates.

Based on the assumptions made concerning the distribution of the values of Y, one can prove the following theorem concerning the distributions of the least squares estimators of the regression coefficients α and β.

Statistics for inferences about α and β

Theorem 11.1   Under the assumptions given on page 326 the statistics

$$t = \frac{(\widehat{\alpha} - \alpha)}{s_e} \sqrt{\frac{nS_{xx}}{S_{xx} + n(\bar{x})^2}}$$

and

$$t = \frac{(\widehat{\beta} - \beta)}{s_e} \sqrt{S_{xx}}$$

are random variables having the t distribution with $n - 2$ degrees of freedom.

To construct confidence intervals for the regression coefficients α and β, we substitute for the middle term of $-t_{\alpha/2} < t < t_{\alpha/2}$ the appropriate t statistic of Theorem 11.1. Then, simple algebra leads to

---

Confidence limits for regression coefficients

and

$$\alpha : \widehat{\alpha} \pm t_{\alpha/2} \cdot s_e \sqrt{\frac{1}{n} + \frac{(\bar{x})^2}{S_{xx}}}$$

$$\beta : \widehat{\beta} \pm t_{\alpha/2} \cdot s_e \frac{1}{\sqrt{S_{xx}}}$$

EXAMPLE 5   **A confidence interval for the intercept**

With reference to Example 2, construct a 95% confidence interval for the regression coefficient α.

Solution   The relevant output of regression analysis software is

```
Coefficients:
              Estimate Std. Error t value Pr(>|t|)
(Intercept) 0.0692424  0.1009737   0.686    0.512
velocity    0.0038288  0.0004378   8.746 2.29e-05
```

Residual standard error: 0.1591 on 8 degrees of freedom

Since $t_{0.025} = 2.306$ for $10 - 2 = 8$, the 95% confidence limits are

$$\text{Estimate} \pm t_{0.025} \text{ Std. Error} = 0.0692424 \pm 2.306 \, (0.1009737)$$

The 95% confidence interval becomes

$$-0.164 < \alpha < 0.302$$

Note that 0 is a plausible value for the intercept α so the line could pass through the origin.

To use the formula above this example, recall from Example 2 that $n = 10$, $\bar{x} = 200$, and $S_{yy} - S_{xy}^2/S_{xx} = 0.20238$. Then,

$$s_e^2 = \frac{2.13745 - (505.40)^2/132,000}{8} = \frac{0.20238}{8} = 0.0253$$

$s_e = \sqrt{.0253} = 0.1591$. The 95% confidence limits are

$$0.0692 \pm (2.306)(0.1591) \sqrt{\frac{1}{10} + \frac{(200)^2}{132,000}}$$

and, consistent with the computer-based calculation, the 95% confidence interval is

$$-0.164 < \alpha < 0.302$$

[Using **R**: Read the file C11Ex12.txt into **Dat** and then use **summary(lm(evap ~ velocity), data = Dat)**]   ∎

In connection with tests of hypotheses concerning the regression coefficients α and β, those concerning β are of special importance because β is the **slope of the regression line**. That is, β is the change in the mean of Y corresponding to a unit increase in x. If $\beta = 0$, the regression line is horizontal and the mean of Y does not depend linearly on x. For tests of the null hypothesis $\beta = \beta_0$, we use the second statistic of Theorem 11.1 and the criteria are like those in the table on page 243 with μ replaced by β.

EXAMPLE 6   **A test of hypotheses concerning the slope parameter**

With reference to Example 2, test the null hypothesis $\beta = 0$ against the alternative hypothesis $\beta \neq 0$ at the 0.05 level of significance.

Solution   1. *Null hypothesis*: $\beta = 0$
Alternative hypothesis: $\beta \neq 0$

2. *Level of significance*: $\alpha = 0.05$

3. *Criterion*: Reject the null hypothesis if $t < -2.306$ or $t > 2.306$, where 2.306 is the value of $t_{0.025}$ for $10 - 2 = 8$ degrees of freedom, and t is given by the second formula of Theorem 11.1.

4. *Calculations*: Using the quantities obtained in Examples 2 and 5, we get

$$t = \frac{0.003829 - 0}{0.1591} \sqrt{132,000} = 8.744$$

5. *Decision*: Since $t = 8.744$ exceeds 2.306, the null hypothesis must be rejected; we conclude that there is a relationship between air velocity and the average evaporation coefficient. (According to the scatter plot in Figure 11.2, the assumption of a straight-line relationship appears to be valid.)

The evidence for nonzero slope β is extremely strong with P-value less than 0.00003. (See the computer output in Example 5 and Figure 11.5.)   ∎

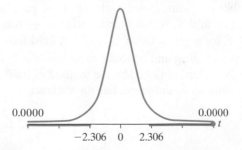

Figure 11.5
The extremely small P-value

Most statistical software programs include a least squares fit of a straight line. When the number of (x_i, y_i) pairs is moderate to large, a computer should be used. We illustrate with the cooling rate data on page 318. The first step is to plot the data; but this is already done on page 318. Typical output from a regression analysis program includes

```
    REGRESSION EQUATION
①   Y = 22.00 + 6.00x

    COEFFICIENTS

    TERM             COEF      SE COEF     T-VALUE      P-VALUE
    CONSTANT    ②   22.00      4.37    ③  5.03    ④   0.002
    X                6.00      1.22        4.90         0.003

⑤  S = 6.70820
```

The least squares line ① is $\widehat{y} = 22.00 + 6.00x$ and the estimate of σ^2 is $s_e^2 = (6.70820)^2 = 45.00$ using ⑤ $s_e = 6.70820$.

Since ③ $t = 4.90$, with $n-2 = 10$ degrees of freedom, is highly significant, the slope is different from zero. The small P-value ④ P-value $= 0.002$ for the constant term confirms that it is needed in the model data.

In Exercise 11.22, more MINITAB output is described.

Figure 11.6 gives the SAS output for a least squares fit with the cooling rate data. Notice that more decimal places are given. For instance the ④ P-value $0.0027 = Prob > |T|$ is given for testing $\beta = 0$ versus $\beta \neq 0$.

Dependent Variable: y

Analysis of Variance

Source	DF	Sum of Squares	Mean Square	F Value	Pr > F
Model	1	1080.00000	1080.00000	24.00	0.0027
Error	6	270.00000	45.00000		
Corrected Total	7	1350.00000			

Root MSE	⑤ 6.70820	
Dependent Mean	40.00000	

Parameter Estimates

Variable	DF	Parameter Estimate	Standard Error	t Value	Pr > |t|
Intercept	1	② 22.00000	4.37321	③ 5.03	④ 0.0024
x	1	6.00000	1.22474	4.90	0.0027

Figure 11.6
Selected SAS output for a regression analysis using the cooling rate data on page 318

We will return to this least squares fit in Section 11.5, where we investigate the assumptions of a straight-line model and the normal distribution for errors.

Another problem, closely related to the problem of estimating the regression coefficients α and β, is that of estimating $\alpha + \beta x$, namely, the mean of the distribution of Y for a given value of x. If x is held fixed at x_0, the quantity we want to estimate is $\alpha + \beta x_0$ and it would seem reasonable to use $\widehat{\alpha} + \widehat{\beta} x_0$, where $\widehat{\alpha}$ and $\widehat{\beta}$ are again the values obtained by the method of least squares. In fact, it can be shown that this estimator is unbiased, has the variance

$$\sigma^2 \left[\frac{1}{n} + \frac{(x_0 - \bar{x})^2}{S_{xx}} \right]$$

and that $(1 - \alpha)100\%$ confidence limits for $\alpha + \beta x_0$ are given by

<table>
<tr><td rowspan="2">**Confidence limits
for $\alpha + \beta x_0$**</td><td>$$(\widehat{\alpha} + \widehat{\beta}x_0) \pm t_{\alpha/2} \cdot s_e \sqrt{\frac{1}{n} + \frac{(x_0 - \bar{x})^2}{S_{xx}}}$$</td></tr>
<tr><td>where the number of degrees of freedom for $t_{\alpha/2}$ is $n - 2$.</td></tr>
</table>

The next example, illustrates a complete approach to fitting and interpreting a straight line. Typically, we rely on computer software to produce the regression analysis (see Exercise 11.22 for MINITAB commands).

EXAMPLE 7   **Modeling and making inferences concerning the effect of prestressing sheets of aluminum alloy**

Because of their strength and lightness, sheets of an aluminum alloy are an attractive option for structural members in automobiles. Engineers discovered that prestraining a sheet of one aluminum alloy may increase its strength. One aspect of their experiments concerns the effect of prestrain(%) on the peak load(kN) that corresponds to the critical buckling load.

Prestrain	Peak load	Peak load	Peak load
0	8.6	8.9	9.1
3	9.0	9.3	9.4
6	9.5	9.8	9.8
12	10.2	10.2	10.3

(Source: Data read from Figure 5, Wowk, E. and Pilkey, K (2013), An experimental and numerical study of prestrained AA5754 sheet in bending. *J Materials Processing Technology*, **213**(1), 1–10)

(a) Does prestraining increase the strength of the aluminum alloy?

(b) Obtain a 95% confidence interval for the mean increase in strength when strain is increased by 1 percent.

(c) Obtain a 95% confidence interval for the mean peak load when the prestrain is set at 9 percent.

(d) Comment on the fit of the straight line model.

Solution   (a) The scatter plot in Figure 11.7 suggests fitting a straight line model. Using computer software, we obtain

```
Predictor     Coef   SE Coef        T       P
Constant   8.90667   0.08078   110.26   0.000
C1         0.11460   0.01175     9.75   0.000
```

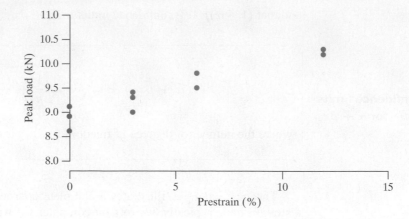

Figure 11.7
Scatter plot suggests straight
line model

Analysis of Variance

Source	DF	SS	MS	F	P
Regression	1	3.1029	3.1029	95.10	0.000
Residual Error	10	0.3263	0.0326		
Total	11	3.4292			

Summary of Model

S = 0.180634

The estimated slope of the regression line is $\widehat{\beta} = 0.1146$ which is positive.
Since the value of test statistic for testing $H_0: \beta = 0$

$$t = \frac{\widehat{\beta}}{s_e / \sqrt{S_{xx}}} = \frac{\text{Coef}}{\text{SE Coef}} = \frac{0.1146}{0.01175} = 9.75$$

is so large, the P-value for the one-sided test is less than 0.0000 when rounded
down. This is very strong evidence that prestressing results in stronger
material.

According to the definition of slope, we estimate that increasing x by one unit
(1%) will result in an increase of strength of 0.1146 kN.

(b) The estimated regression line is

$$\widehat{y} = \widehat{\alpha} + \widehat{\beta} x = 8.90667 + 0.1146 x$$

With $x = 9$ percent prestrain, we estimate $\widehat{y} = 8.90667 + 0.1146(9) = 9.938$ kN
Next, a further simple calculation gives $\bar{x} = 5.25$ and $S_{xx} = \sum_{i=1}^{12}(x_i - \bar{x})^2 =$
236.250. Since $t_{0.25} = 2.228$ for $n - 2 = 10$ d.f., the half length of the 95%
confidence interval is

$$t_{0.025}\, s_e \sqrt{\frac{1}{n} + \frac{(x_0 - \bar{x})^2}{S_{xx}}} = 2.228\,(0.180634)\sqrt{\frac{1}{12} + \frac{(9 - 5.25)^2}{236.250}} = 0.1521$$

The 95% confidence interval becomes

$$(\,9.938 - 0.152, 9.938 + 0.152\,)\quad\text{or}\quad(\,9.79, 10.09\,)\text{ kN}$$

We are 95% confident that mean strength is between 9.79 and 10.09 kN for all
alloy sheets that could undergo a prestrain of 9 percent.

(c) The scatter plot Figure 11.7 confirms the model is reasonable. Further, as you are asked to confirm to Exercise 11.44, there are no outliers in the residuals and the assumption of normal errors appears reasonable. We can rely upon the statistical conclusions above.

[Using **R**: (a) With the data in **x** and **y**, use **summary(lm(y~x))** (b) **new=data .frame(x=c(9))** then **predict (lm(y ~ x), new, level=.95, interval="confidence")**]

∎

To emphasize the danger inherent when extrapolating beyond the range of experimentation, consider the plot of conductivity versus temperature in Figure 11.8. (*Source*: K. Onnes (1912) *Communications of the Physical Laboratory at the University of Leiden*, no. 124, unnumbered figure.) If the line is extended to predict conductivity at 4.10° Kelvin, we would predict 0.10 ohms. This is much greater than the value of 0, which Onnes observed when he discovered superconductivity at 4.19° Kelvin. That is, the physical model changes drastically outside the experimental range shown.

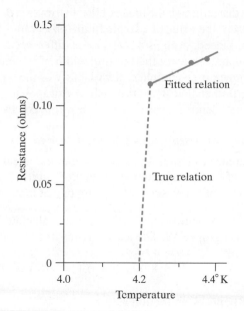

Figure 11.8
Resistance versus temperature for a specimen of mercury. A model change invalidates extrapolation below 4.2° K

Now let us indicate a method of constructing an interval in which a future observation, Y, can be expected to lie with a given probability (or confidence) when $x = x_0$. If α and β were known, we could use the fact that Y is a random variable having a normal distribution with the mean $\alpha + \beta x_0$ and the variance σ^2 (or that $Y - \alpha - \beta x_0$ is a random variable having a normal distribution with zero mean and the variance σ^2). However, if α and β are not known, we must consider the quantity $Y - \widehat{\alpha} - \widehat{\beta} x_0$, where $Y, \widehat{\alpha}$, and $\widehat{\beta}$ are all random variables, and the resulting theory leads to the following **limits of prediction** for a future observed value Y when $x = x_0$:

Limits of prediction for future Y at x_0

$$(\widehat{\alpha} + \widehat{\beta} x_0) \pm t_{\alpha/2} \cdot s_e \sqrt{1 + \frac{1}{n} + \frac{(x_0 - \bar{x})^2}{S_{xx}}}$$

where the number of degrees of freedom for $t_{\alpha/2}$ is $n - 2$.

EXAMPLE 8   **95% prediction limits for the strength of a future sheet of aluminum alloy**

With reference to Example 7, find 95% limits of prediction for an observation of the peak load for a new sheet of aluminum that was subject to a prestrain of 9 percent.

Solution   Adding a 1 under the radical in Example 7, we get

$$9.938 \pm 2.228\,(0.180634\,)\sqrt{1 + \frac{1}{12} + \frac{(9-5.25\,)^2}{236.250}}$$

and the limits of prediction are $9.94 - 0.43 = 9.51$ and $10.37 = 9.94 + 0.43$ We are 95% confident that the observed value of peak load for this new sheet of aluminum lies between 9.51 and 10.37 kN.

[Using **R**: **predict(lm(y~x), new, level=.95, interval = "prediction")** with **x**, **y** and **new** as in Example 7, part(b).]   ∎

Note that although the mean of the distribution of Y when $x = 9$ can be estimated fairly closely, the value of a single future observation cannot be predicted with much precision. Indeed, even as $n \to \infty$, the difference between the limits of prediction does not approach zero; the limiting width of the interval of prediction depends on s_e, which measures the inherent variability of the data.

Note further that if we do wish to extrapolate, the interval of prediction (and also the confidence interval for $\alpha + \beta x_0$) becomes increasingly wide.

EXAMPLE 9   **The limits of prediction become wider if x_0 is farther from $\bar{x}$**

With reference to Example 7, assume that the linear relationship continues beyond the range of experimentation and find 95% limits of prediction for an observation of the peak load of a new sheet of aluminum that is subject to a prestrain of 18 percent.

Solution   Substituting the various quantities already calculated in Example 7 and Example 8, but now using $x_0 = 18$. The least squares line predicts

$$8.90667 + 0.1146(18\,) = 10.9695$$

The 95% limits of prediction are

$$10.9695 \pm 2.228\,(0.180634\,)\sqrt{1 + \frac{1}{12} + \frac{(18-5.25\,)^2}{236.250}}$$

or, $10.9695 - 0.5356 = 10.4339$ and $11.5051 = 10.9695 + 0.5356$ kN. We are 95% confident that the observed value of peak load for this new sheet of aluminum lies between 10.43 and 11.51 kN. This confidence is misplaced if the linear relationship does not extend quite far beyond the range of values for x included in the experiment.

Even if the linear relationship holds, out to $z = 18$, the width of this interval of prediction is $2\,(0.54\,) = 1.08$ compared to the width of $2\,(0.43\,) = 0.86$ obtained in Example 8 for $x_0 = 9$ which is closer to $\bar{x}$. The band that results from calculating the 95% prediction limits for $0 \le x \le 20$ is shown as the solid line in Figure 11.9. The band for confidence intervals for the mean are the dotted lines. Notice that this band is substantially narrower except for extrapolations to the right of the data.

[Using **R**: **new = data.frame(x=c(18))** then **predict(lm(y~x), new, level=.95, interval = "prediction")** but with with **x** and **y** as in Example 7.]   ∎

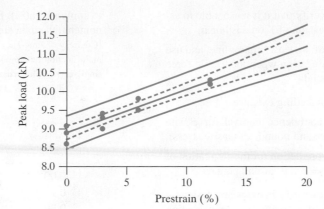

Figure 11.9
The 95% limits of prediction
(solid) and confidence intervals
(dotted) for aluminum sheets.

Exercises

11.1 A chemical engineer found that by adding different amounts of an additive to gasoline, she could reduce the amount of nitrous oxides (NOx) coming from an automobile engine. A specified amount will be added to a gallon of gas and the total amount of NOx in the exhaust collected. Initially, five runs with 1, 2, 3, 4, and 5 units of additive will be conducted.

(a) How would you randomize in this experiment?

(b) Suppose you properly calculate a point estimate of the mean value of NOx when the amount of additive is 8. What additional danger is there in using this estimate?

11.2 An engineer found that by including small amounts of a compound in rechargeable batteries for portable computers, she could extend their lifetimes. She experimented with different amounts of the additive and the data are

Amount of additive	Life (hours)
0	2
1	4
2	3
3	7
4	9

(a) Obtain the least squares fit of a straight line to the amount of additive.

(b) Test whether or not the slope $\beta = 0$. Take $\alpha = 0.10$ as your level of significance.

(c) Give a point estimate of the mean battery life when the amount of additive is 12.

(d) What additional danger is there when using your estimate in part (c)?

(e) How would you randomize in this experiment?

11.3 A chemical company, wishing to study the effect of extraction time on the efficiency of an extraction operation, obtained the data shown in the following table:

Extraction time (minutes) x	Extraction efficiency (%) y
27	57
45	64
41	80
19	46
35	62
39	72
19	52
49	77
15	57
31	68

(a) Draw a scatter plot to verify that a straight line will provide a good fit to the data, draw a straight line by eye, and use it to predict the extraction efficiency one can expect when the extraction time is 35 minutes.

(b) Fit a straight line to the given data by the method of least squares and use it to predict the extraction efficiency one can expect when the extraction time is 35 minutes.

11.4 In the accompanying table, x is the tensile force applied to a steel specimen in thousands of pounds, and y is the resulting elongation in thousandths of an inch:

x	1	2	3	4	5	6
y	14	33	40	63	76	85

(a) Graph the data to verify that it is reasonable to assume that the regression of Y on x is linear.

(b) Find the equation of the least squares line, and use it to predict the elongation when the tensile force is 3.5 thousand pounds.

11.5 With reference to the preceding exercise,

(a) construct a 95% confidence interval for β, the elongation per thousand pounds of tensile stress;

(b) find 95% limits of prediction for the elongation of a specimen with $x = 3.5$ thousand pounds.

11.6 The following table shows how many weeks a sample of 6 persons have worked at an automobile inspection station and the number of cars each one inspected between noon and 2 P.M. on a given day:

Number of weeks employed	Number of cars inspected
x	y
2	13
7	21
9	23
1	14
5	15
12	21

(a) Find the equation of the least squares line which will enable us to predict y in terms of x.

(b) Use the result of part (a) to estimate how many cars someone who has been working at the inspection station for 8 weeks can be expected to inspect during the given 2-hour period.

11.7 With reference to the preceding exercise, test the null hypothesis $\beta = 1.2$ against the alternative hypothesis $\beta < 1.2$ at the 0.05 level of significance.

11.8 With reference to Exercise 11.6, find

(a) a 95% confidence interval for the average number of cars inspected in the given period of time by a person who has been working at the inspection station for 8 weeks;

(b) 95% limits of prediction for the number of cars that will be inspected in the given period of time by a person who has worked at the inspection station for 8 weeks.

11.9 Scientists searching for higher performance flexible structures created a diode with organic and inorganic layers. It has excellent mechanical bending properties. Applying a bending strain to the diode actually leads to higher current density(mA/cm^2). Metal curvature molds, each having a different radius, were used

to apply strain(%). For one demonstration of the phenomena, the data are

(Courtesy of Jung-Hun Seo see Jung-Hun Seo et. al. (2013), A multifunction heterojunction formed between pentacene and a single-crystal silicon nanomembrane, *Advanced Functional Materials*, **23**(27), 3398–3403.)

Strain(%)	Current density(mA/cm^2)
x	y
0.00	3.47
0.25	3.57
0.49	3.68
0.64	3.73
0.80	3.86
1.08	3.99

(a) Obtain the least squares line.

(b) Predict the current density when the strain $x = 0.50$.

11.10 With reference to Exercise 11.9, construct a 95% confidence interval for α.

11.11 With reference to Exercise 11.9, test the null hypothesis $\beta = 0.40$ against the alternative hypothesis $\beta > 0.40$ at the 0.05 level of significance.

11.12 Raw material used in the production of a synthetic fiber is stored in a place which has no humidity control. Measurements of the relative humidity in the storage place and the moisture content of a sample of the raw material (both in percentages) on 12 days yielded the following results:

Humidity	Moisture content
x	y
42	12
35	8
50	14
43	9
48	11
62	16
31	7
36	9
44	12
39	10
55	13
48	11

(a) Make a scatter plot to verify that it is reasonable to assume that the regression of Y on x is linear.

(b) Fit a straight line by the method of least squares.

(c) Find a 99% confidence interval for the mean moisture content of the raw material when the humidity of the storage place is 40%.

11.13 With reference to the preceding exercise, find 95% limits of prediction for the moisture content of the raw material when the humidity of the storage place is 40%. Also indicate to what extent the width of the interval is affected by the size of the sample and to what extent it is affected by the inherent variability of the data.

11.14 With reference to Exercise 11.3, express 95% limits of prediction for the extraction efficiency in terms of the extraction time x_0. Choosing suitable values of x_0, sketch graphs of the loci of the upper and lower limits of prediction on the diagram of part (a) of Exercise 11.3. Note that since any two sets of limits of prediction obtained from these bands are dependent, they should be used only for a single extraction time x_0.

11.15 In Exercise 11.4 it would have been entirely reasonable to impose the condition $\alpha = 0$ before fitting a straight line by the method of least squares.

(a) Use the method of least squares to derive a formula for estimating β when the regression line has the form $y = \beta x$.

(b) With reference to Exercise 11.4, use the formula obtained in part (a) to estimate β and compare the result with the estimate previously obtained without the condition that the line must pass through the origin.

11.16 Recycling concrete aggregate is an important component of green engineering. The strength of any potential material, expressed in terms of its resilient modulus, must meet standards before it is incorporated in the base of new roadways. There are two methods of obtaining the resilient modulus. The exterior measurement M_{ext} (MPa) and interior measurement M_{int} (MPa) can be measured on each sample. The results for $n = 9$ samples are (Courtesy of Tuncer Edil)

M_{ext}	204.7	184.8	181.1	166.5	165.2	154.1	135.8	173.4	142.7
M_{int}	707.1	632.2	603.6	522.4	554.4	483.5	449.7	545.1	473.8

(a) Draw a scatter plot to verify the assumption that the relationship is linear, letting M_{int} be x and M_{ext} be y.

(b) Fit a straight line to these data by the method of least squares, and draw its graph on the diagram obtained in part (a).

11.17 With reference to Exercise 11.16, find a 90% confidence interval for α.

11.18 With reference to Exercise 11.16, fit a straight line to the data by the method of least squares, using M_{ext} as the independent variable, and draw its graph on the diagram obtained in part (a) of Exercise 11.16. Note that the two estimated regression lines do not coincide.

11.19 When the sum of the x values is equal to zero, the calculation of the coefficients of the regression line of Y on x is greatly simplified; in fact, their estimates are given by

$$\widehat{\alpha} = \frac{\sum y}{n} \quad \text{and} \quad \widehat{\beta} = \frac{\sum xy}{\sum x^2}$$

This simplification can also be attained when the values of x are equally spaced; that is, when they are in arithmetic progression. We then code the data by substituting for x the values $\ldots, -2, -1, 0, 1, 2, \ldots,$ when n is odd, or the values $\ldots, -3, -1, 1, 3, \ldots,$ when n is even. The preceding formulas are then used in connection with the coded data.

(a) Because of high lead residue, a faucet manufacturer cannot sell his product for home use unless each item is labeled as being hazardous to health. A consulting engineer suggests adding an acid bath at the end of the production line. An experiment is conducted with the bath having 0.2, 0.4, 0.6, 0.8, 1.0, 1.2 and 1.4 percent acid solutions. Suppose the corresponding values of lead residue are 4.6, 4.0, 3.3, 3.6, 3.0, 2.4, and 1.6 ppm.

Fit a least squares line and give the point prediction of the lead residue when using a 1.3 percent solution.

(b) Encouraged by the responses in Part (a), one further test was conducted with a 1.6 percent solution. Suppose the resulting lead residue is 1.1 ppm. Fit a least squares line using all eight runs. Again predict the lead residue for a 1.3 percent solution.

11.20 Using the formulas on page 320 for $\widehat{\alpha}$ and $\widehat{\beta}$, show that

(a) the expression for $\widehat{\alpha}$ is linear in the Y_i

(b) $\widehat{\alpha}$ is an unbiased estimate of α

(c) the expression for $\widehat{\beta}$ is linear in the Y_i

(d) $\widehat{\beta}$ is an unbiased estimate of β

11.21 The decomposition of the sums of squares into a contribution due to error and a contribution due to regression underlies the least squares analysis. Consider the identity

$$y_i - \bar{y} - (\widehat{y}_i - \bar{y}) = (y_i - \widehat{y}_i)$$

Note that

$$\widehat{y}_i = a + bx_i = \bar{y} - b\bar{x} + bx_i = \bar{y} + b(x_i - \bar{x}) \quad \text{so}$$

$$\widehat{y}_i - \bar{y} = b(x_i - \bar{x})$$

Using this last expression, then the definition of b and again the last expression, we see that

$$\sum(y_i - \bar{y})(\hat{y}_i - \bar{y}) = b \sum(y_i - \bar{y})(x_i - \bar{x})$$
$$= b^2 \sum(x_i - \bar{x})^2 = \sum(\hat{y}_i - \bar{y})^2$$

and the sum of squares about the mean can be decomposed as

$$\underset{total\ sum\ of\ squares}{\sum_{i=1}^{n}(y_i - \bar{y})^2}$$

$$= \underset{error\ sum\ of\ squares}{\sum_{i=1}^{n}(y_i - \hat{y}_i)^2} + \underset{regression\ sum\ of\ squares}{\sum_{i=1}^{n}(\hat{y}_i - \bar{y})^2}$$

Generally, we find the straight-line fit acceptable if the ratio

$$r^2 = \frac{regression\ sum\ of\ squares}{total\ sum\ of\ squares} = 1 - \frac{\sum_{i=1}^{n}(y_i - \hat{y}_i)^2}{\sum_{i=1}^{n}(y_i - \bar{y})^2}$$

is near 1.

Calculate the decomposition of the sum of squares and calculate r^2 using the observations in Exercise 11.9.

11.22 It is tedious to perform a least squares analysis without using a computer. We illustrate here a computer-based analysis using the *MINITAB* package. The observations on page 318 are entered in C1 and C2 of the worksheet.

DATA 11-22.DAT

x:	0	1	2	2	4	4	5	6
y:	25	20	30	40	45	50	60	50

We first obtain the scatter plot to see if a straight-line pattern is evident.

Dialog box:
Graph > Scatterplot. Click on **Simple**. Click **OK**.
Type $C2$ in **Y** column and Type $C1$ in **X** column. Click **OK**.

Then

Dialog box:
Stat > Regression > Regression > Fit Regression model
Type $C2$ in **Response**. Type $C1$ in **Continuous predictors**.
Click **OK**.

---

produces the output

Analysis of Variance

Source	DF	SS	MS	F-Value	P-Value
Regression	1	1080.00	1080.00	24.00	0.003
Error	6	270.00	45.00		
Total	7	1350.00			

Model Summary

S	R-sq
6.70820	80.00%

Coefficients

Term	Coef	SE Coef	T-Value	P-Value
Constant	22.00	4.37	5.03	0.002
x	6.00	1.22	4.90	0.003

Regression Equation
$y = 22.00 + 6.00\ x$

After the first two steps above and before you Click **OK**,
Click **Graphs**. Choose **Four in one**. Click **OK**
This will produce the three graphs that we introduce later for checking the assumptions.

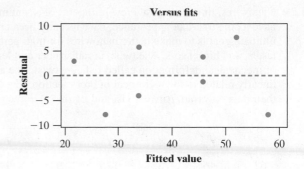

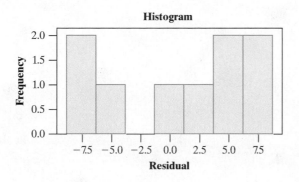

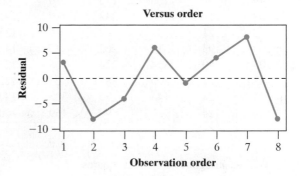

(a)   One further run with 3.5% of the new component produced the cooling rate 42. Obtain the regression equation using all 9 cases.

(b)   Referring to your computer output, identify the decomposition of sums of squares given as the analysis of variance.

11.23   Referring to Example 3, the nanopillar data on height (nm) and width (nm) are

Width	Height	Width	Height	Width	Height	Width	Height
62	221	77	290	102	298	93	323
68	234	80	292	95	312	92	343
69	245	83	289	90	297	98	330
80	266	73	284	98	314	101	333
68	265	79	271	86	305	97	346
79	253	100	292	93	296	102	364
83	274	93	308	91	304	91	366
70	278	92	303	90	310	87	355
74	290	101	308	95	315	110	390
73	276	87	315	97	311	106	373
74	272	96	309	87	337	118	391
75	276	99	300	89	338		
80	276	94	305	100	336		

(a)   Fit a straight line with y = height and x = width by least squares.

(b)   Test, with $\alpha = 0.05$, that the slope is different from zero.

(c)   Find a 95% confidence interval for the mean height when width = 100.

(d)   Plot the residuals versus the predicted values.

11.24 Nanowires, tiny wires just a few millionths of a centimeter thick, which spiral and have a pine tree-like appearance, have recently been created.[2] The investigators' ultimate goal is to make better nanowires for high-performance integrated circuits, lasers, and biosensors. The twist, in radians per unit length, should follow a theory of mechanical deformation called Eshelby twist. The amount of twist should be linearly related to the reciprocal of cross section. The authors provided some of their data on $y\,(\text{rad}/\mu\text{m}) = $ twist and $x\,(\mu\text{m}^{-2}) = $ reciprocal cross section.

x	y	x	y	x	y
62	46	286	126	92	94
57	46	72	162	138	112
49	90	168	172	250	306
161	113	286	248	189	291
180	121	337	288	96	133
103	193	315	262	45	143
103	89	354	224	168	120
43	122	509	381	25	122
144	124	144	76	326	137
182	124	127	90	169	100

(Courtesy of Song Jin)

(a) Fit a straight line by least squares.

(b) Test, with $\alpha = 0.05$, that the slope is different from zero.

(c) Find a 98% confidence interval for the mean twist angle when $x = 148$.

(d) Does this aspect of the Eshelby twist theory seem to apply?

11.3 Curvilinear Regression

So far we have studied only the case where the regression curve of Y on x is a straight line; that is, where for any given x, the mean of the distribution of Y is given by $\alpha + \beta x$. In this section we first investigate cases where the regression curve is nonlinear but where the methods of Section 11.1 can nevertheless be applied. Then we take up the problem of polynomial regression where for any given x the mean of the distribution of Y is given by

$$\beta_0 + \beta_1 x + \beta_2 x^2 + \cdots + \beta_p x^p$$

Polynomial curve fitting is also used to obtain approximations when the exact functional form of the regression curve is unknown.

It is common practice for engineers to plot paired data on various transformed scales such as square root or logarithm scale, in order to determine whether the transformed points will fall close to a straight line. If there is the transformation that suggests a straight-line regression equation, the necessary constants (parameters) can be estimated by applying the method of Section 11.1 to the transformed data. For instance, if a set of paired data consisting of n points (x_i, y_i) "straightens out" when $\log y_i$ is plotted versus x_i, this indicates that the regression curve of Y on x is

[2]M. Bierman, Y. Lau, A. Kvit, A. Schmitt, and S. Jin. Dislocation-driven nanowire growth and Eshelby Twist, *Science*, 23 May 2008, Vol. 320, 1060–1063.

exponential, namely, that for any given x, the mean of the distribution of values of Y is given by $\alpha \cdot \beta^x$. If we take logarithms to the base 10 (or any convenient base), the predicting equation $y = \alpha \cdot \beta^x$ becomes

$$\log y = \log \alpha + x \cdot \log \beta$$

and we can now get estimates of $\log \alpha$ and $\log \beta$, and hence of α and β, by applying the method of Section 11.1 to the n pairs of values $(x_i, \ \log y_i)$.

EXAMPLE 10   **A plot of ln y versus x leads to fitting a straight line**

Electric and hybrid cars require NI-MN batteries having a high capacity. Battery capacity decreases as the rate of discharge increases. Let y = battery capacity, measured in amp-hours, and x = rate of discharge in amps. Suppose tests of six NI-MN batteries, of the same model produce the results

Rate of discharge(A) x	Capacity(Ah) y
2	164.7
3	156.1
6	142.5
10	133.8
15	114.6
20	107.1

(a) Plot $\log y_i$ versus x_i to verify that it is reasonable to assume that the relationship is exponential.

(b) Fit an exponential curve by applying the method of least squares to the data points $(x_i, \ \log y_i)$.

(c) Use the result of part (b) to estimate the capacity when the discharge rate is 5 amps.

Solution   (a) Although there are too few points to firmly establish a linear pattern, the pattern in Figure 11.10 is approximately linear. This suggests we try fitting an exponential curve.

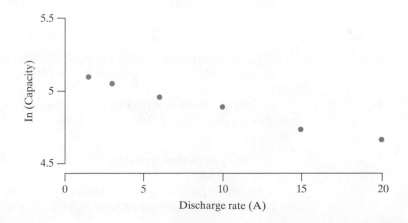

Figure 11.10
Plot of transformed data for
Example 10

(b) We prefer statistical software to fit ln y, the natural logarithm of y, to x. *MINITAB* output includes

```
Model Summary

        S      R-sq
0.0256849    98.17%

Coefficients

Term          Coef   SE Coef   T-Value   P-Value
Constant    5.1257    0.0184    278.55     0.000
x          -0.02372   0.00162   -14.64     0.000

Regression Equation

ln y = 5.1257 - 0.02372 x
```

The output tells us that the estimated regression curve is

$$\widehat{\ln y} = 5.1257 - 0.02372\,x$$

in **logarithmic form** which becomes

$$\text{predicted } y = 168.29\,e^{-0.02372\,x}$$

in **exponential form**.

A hand-held calculator solution could require the calculations

$$\bar{x} = 56/6 = 9.3333 \quad \text{and} \quad \overline{\ln y} = 29.42552/6 = 4.9043$$

$$S_{x\,\ln y} = 268.6762 - 29.42552(56)\,/\,6 = -5.961987$$

$$S_{xx} = 774 - (56)^2\,/\,6 = 251.3333$$

Then, the estimated slope $= -5.961987\,/\,251.3333 = -0.02372$ and the estimated intercept is $4.9043 + 9.3333(0.02372) = 5.1257$

(c) Using the logarithmic form which is more convenient, we predict

$$\widehat{\ln y} = 5.1257 - 0.02372\,(5) = 5.007$$

or, on the original scale, predicted $y = 149.5$ Ah.

[Using **R**: **summary(lm(log(y)~x))** when **x** contains the discharge rates and **y** contains the capacities.] ∎

The analysis of transformed relationships is easily implemented on a computer (see Exercise 11.41).

Two other relationships that frequently arise in engineering applications and can be fitted by the method of Section 11.1 after suitable transformations are the

reciprocal function

$$y = \frac{1}{\alpha + \beta x}$$

and the **power function** $y = \alpha \cdot x^\beta$. The first of these represents a linear relationship between x and $\frac{1}{y}$, namely,

$$\frac{1}{y} = \alpha + \beta x$$

and we obtain estimates of α and β by applying the method of Section 11.1 to the points $\left(x_i, \dfrac{1}{y_i} \right)$. The second represents a linear relationship between $\log x$ and $\log y$, namely,

$$\log y = \log \alpha + \beta \, \log x$$

and we obtain estimates of $\log \alpha$ and β, and hence of α and β, by applying the method of Section 11.1 to the points $(\log x_i, \, \log y_i)$. Another example of a curve that can be fitted by the method of least squares after a suitable transformation is given in Exercise 11.29.

If there is no clear indication about the functional form of the regression of Y on x, we often assume that the underlying relationship is at least "well-behaved" to the extent that it has a Taylor series expansion and that the first few terms of this expansion will yield a fairly good approximation. We thus fit to our data a **polynomial regression**, that is, the mean of Y at x has the form

$$\beta_0 + \beta_1 x + \beta_2 x^2 + \cdots + \beta_p x^p$$

where the degree is determined by inspection of the data or by a more rigorous method to be discussed below.

Given a set of data consisting of n points (x_i, y_i), we estimate the coefficients $\beta_0, \beta_1, \beta_2, \ldots, \beta_p$ of the pth-degree polynomial by minimizing

$$\sum_{i=1}^{n} [y_i - (\beta_0 + \beta_1 x_i + \beta_2 x_i^2 + \cdots + \beta_p x_i^p)]^2$$

In other words, we are now applying the least squares criterion by minimizing the sum of the squares of the vertical distances from the points to the curve (see Figure 11.11).

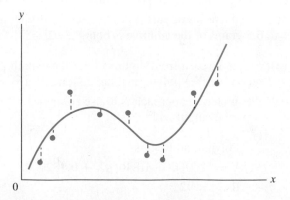

Figure 11.11

Least squares criterion for polynomial curve fitting

Taking the partial derivatives with respect to $\beta_0, \beta_1, \beta_2, \ldots, \beta_p$, equating these partial derivatives to zero, rearranging some of the terms, and letting b_i be the estimate of β_i, we obtain the $p + 1$ **normal equations**

Normal equations for polynomial regression

$$\sum y = n\,b_0 \qquad\qquad + b_1 \sum x \quad + \cdots + b_p \sum x^p$$
$$\sum xy = b_0 \sum x \; + b_1 \sum x^2 \quad + \cdots + b_p \sum x^{p+1}$$
$$\vdots$$
$$\sum x^p y = b_0 \sum x^p + b_1 \sum x^{p+1} + \cdots + b_p \sum x^{2p}$$

where the subscripts and limits of summation are omitted for simplicity. Note that this is a system of $p + 1$ linear equations in the $p + 1$ unknowns $b_0, b_1, b_2, \ldots,$ and b_p. If the x's include $p + 1$ distinct values, then the normal equations will have a unique solution.

EXAMPLE 11   **Fitting a quadratic function by the method of least squares**

The following are data on the drying time of a certain varnish and the amount of an additive that is intended to reduce the drying time:

Amount of varnish additive (grams) x	Drying time (hours) y
0	12.0
1	10.5
2	10.0
3	8.0
4	7.0
5	8.0
6	7.5
7	8.5
8	9.0

(a) Draw a scatter plot to verify that it is reasonable to assume that the relationship is parabolic.

(b) Fit a second-degree polynomial by the method of least squares.

(c) Use the result of part (b) to predict the drying time of the varnish when 6.5 grams of the additive is being used.

Solution   **(a)** As can be seen from Figure 11.12, the overall pattern suggests fitting a second-degree polynomial having one relative minimum.

(b) The preferred approach is to use computer software (see Exercise 11.42). A partial output includes

    Polynomial Regression

    $Y = 12.1848 - 1.84654X + 0.182900X ** 2$

    R-Sq $= 92.3\%$.

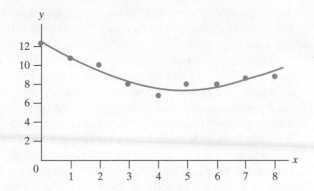

Figure 11.12
Parabola fitted to data of
Example 11

Alternatively, the summations required for substitution into the normal equations are

$$\sum x = 36 \qquad \sum x^2 = 204 \qquad \sum x^3 = 1{,}296 \qquad \sum x^4 = 8{,}772$$

$$\sum y = 80.5 \qquad \sum xy = 299.0 \qquad \sum x^2 y = 1{,}697.0$$

Thus we have to solve the following system of three linear equations in the unknowns b_0, b_1, and b_2:

$$80.5 = \quad 9\,b_0 + \quad 36\,b_1 + \quad 204\,b_2$$
$$299.0 = \quad 36\,b_0 + \quad 204\,b_1 + 1{,}296\,b_2$$
$$1{,}697.0 = 204\,b_0 + 1{,}296\,b_1 + 8{,}772\,b_2$$

Getting $\widehat{\beta}_0 = 12.2$, $\widehat{\beta}_1 = -1.85$, and $\widehat{\beta}_2 = 0.183$, the equation of the least squares polynomial is

$$\widehat{y} = 12.2 - 1.85x + 0.183x^2$$

(c) Substituting $x = 6.5$ into this equation, we get

$$\widehat{y} = 12.2 - 1.85(6.5) + 0.183(6.5)^2$$
$$= 7.9$$

that is, a predicted drying time of 7.9 hours. ∎

Note that it would have been rather dangerous in the preceding example to predict the drying time that corresponds to, say, 24.5 grams of the additive. The risks inherent in extrapolation, discussed on page 333 in connection with fitting straight lines, increase greatly when polynomials are used to approximate unknown regression functions.

In actual practice, it may be difficult to determine the degree of the polynomial to fit to a given set of paired data. As it is always possible to find a polynomial of degree at most $n - 1$ that will pass through each of n points corresponding to n distinct values of x, it should be clear that what we actually seek is a polynomial of lowest possible degree that "adequately" describes the data. As we did in our example, it is often possible to determine the degree of the polynomial by inspection of the data.

There also exists a more rigorous method for determining the degree of the polynomial to be fitted to a given set of data. Essentially, it consists of first fitting a straight line as well as a second-degree polynomial and testing the null hypothesis $\beta_2 = 0$, namely, that nothing is gained by including the quadratic term. If this null hypothesis can be rejected, we then fit a third-degree polynomial and test the hypothesis $\beta_3 = 0$, namely, that nothing is gained by including the cubic term. This procedure is continued until the null hypothesis $\beta_i = 0$ cannot be rejected in two successive steps and there is, thus, no apparent advantage to carrying the extra terms. Note that in order to perform these tests it is necessary to impose the assumptions of normality, independence, and equal variances introduced in Section 11.2. Also, these tests should never be used blindly; that is, without inspection of the overall pattern of the data.

The use of this technique is fairly tedious and we shall not illustrate it in the text. In Exercise 11.33 the reader will be given detailed instructions to apply it to the varnish-additive drying-time data in order to check whether it was really worthwhile to carry the quadratic term.

11.4   Multiple Regression

Before we extend the methods of the preceding sections to problems involving more than one independent variable, let us point out that the curves obtained (and the surfaces we will obtain) are not used only to make predictions. They are often used also for purposes of optimization—namely, to determine for what values of the independent variable (or variables) the dependent variable is a maximum or minimum. For instance, in Example 11 we might use the polynomial fitted to the data to conclude that the drying time is a minimum when the amount of varnish additive used is 5.1 grams (see Exercise 11.34).

Statistical methods of prediction and optimization are often referred to under the general heading of **response surface analysis**. Within the scope of this text, we shall be able to introduce two further methods of response surface analysis: **multiple regression** here and related problems of **factorial experimentation** in Chapter 13.

In multiple regression, we deal with data consisting of n $(r + 1)$-tuples $(x_{i1}, x_{i2}, \ldots, x_{ir}, y_i)$, where the x's are predictor variables whose values are assumed to be known without error while the y's are values of random variables. Data of this kind arise, for example, in studies designed to determine the effect of various climatic conditions on a metal's resistance to corrosion; or the effect of kiln temperature, humidity, and iron content on the strength of a ceramic coating.

As in the case of one independent variable, we shall first treat the problem where the regression equation is linear, namely, where for any given set of values $x_1, x_2, \ldots,$ and x_r, for the r independent variables, the mean of the distribution of Y is given by

$$\beta_0 + \beta_1 x_1 + \beta_2 x_2 + \cdots + \beta_r x_r$$

For two independent variables, this is the problem of fitting a plane to a set of n points with coordinates (x_{i1}, x_{i2}, y_i) as is illustrated in Figure 11.13.

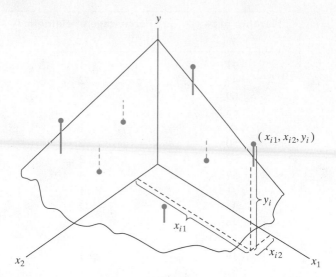

Figure 11.13
Regression plane

Applying the method of least squares to obtain estimates of the coefficients β_0, β_1, and β_2, we minimize the sum of the squares of the vertical distances from the observations y_i to the plane (see Figure 11.13); symbolically, we minimize

$$\sum_{i=1}^{n} [y_i - (b_0 + b_1 x_{i1} + b_2 x_{i2})]^2$$

and it will be left to the reader to verify in Exercise 11.35 that the resulting normal equations are

Normal equations for multiple regression with $r = 2$

$$\sum y = n b_0 \qquad + b_1 \sum x_1 \; + b_2 \sum x_2$$

$$\sum x_1 y = b_0 \sum x_1 + b_1 \sum x_1^2 \; + b_2 \sum x_1 x_2$$

$$\sum x_2 y = b_0 \sum x_2 + b_1 \sum x_1 x_2 + b_2 \sum x_2^2$$

As before, we write the least squares estimates of β_0, β_1, and β_2 as $\widehat{\beta}_0$, $\widehat{\beta}_1$, and $\widehat{\beta}_2$. Note that in the abbreviated notation $\sum x_1$ stands for $\sum_{i=1}^{n} x_{i1}$, $\sum x_1 x_2$ stands for $\sum_{i=1}^{n} x_{i1} x_{i2}$, $\sum x_1 y$ stands for $\sum_{i=1}^{n} x_{i1} y_i$, and so forth.

EXAMPLE 12

A multiple regression with two predictor variables

The following are data on the number of twists required to break a certain kind of forged alloy bar and the percentages of two alloying elements present in the metal:

Number of twists	Percentage of element A	Percentage of element B
y	x_1	x_2
41	1	5
49	2	5
69	3	5
65	4	5
40	1	10
50	2	10
58	3	10
57	4	10
31	1	15
36	2	15
44	3	15
57	4	15
19	1	20
31	2	20
33	3	20
43	4	20

Fit a least squares regression plane and use its equation to estimate the number of twists required to break one of the bars when $x_1 = 2.5$ and $x_2 = 12$.

Solution   Computers remove the drudgery of calculations from a multiple-regression analysis. (See Exercise 11.40.) Typical output includes

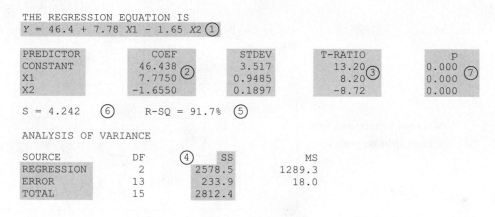

```
THE REGRESSION EQUATION IS
Y = 46.4 + 7.78 X1 - 1.65 X2  (1)

PREDICTOR       COEF        STDEV      T-RATIO         p
CONSTANT      46.438        3.517       13.20      0.000
X1             7.7750  (2)  0.9485  (3)  8.20  (3)  0.000  (7)
X2            -1.6550       0.1897       -8.72      0.000

S = 4.242  (6)     R-SQ = 91.7%  (5)

ANALYSIS OF VARIANCE

SOURCE        DF    (4)     SS          MS
REGRESSION     2          2578.5      1289.3
ERROR         13           233.9        18.0
TOTAL         15          2812.4
```

We now identify some important parts of the output.

1. The least squares regression plane is

$$\text{(1)} \quad \widehat{y} = 46.4 + 7.78\,x_1 - 1.65\,x_2$$

This equation estimates that the average number of twists required to break a bar increases by 7.78 if the percent of element A is increased by 1% and x_2 remains fixed.

2. The least squares estimates and their corresponding estimated standard error are

$$\widehat{\beta}_0 = 46.438 \quad \text{with estimated standard error } 3.517$$
$$\text{(2)} \quad \widehat{\beta}_1 = 7.7750 \quad \text{with estimated standard error } 0.9485$$
$$\widehat{\beta}_2 = -1.6550 \quad \text{with estimated standard error } 0.1897$$

3. The t ratios 13.20, 8.20, -8.72 are all highly significant, so all the terms are needed in the model ③.

4. In any regression analysis having a β_0 term, the decomposition

$$y_i - \bar{y} = (y_i - \widehat{y}_i) + (\widehat{y}_i - \bar{y})$$

produces the decomposition of the sum of squares

$$④\quad \underbrace{\sum_{i=1}^{n}(y_i - \bar{y})^2}_{\text{total sum of squares}} = \underbrace{\sum_{i=1}^{n}(y_i - \widehat{y}_i)^2}_{\text{error sum of squares}} + \underbrace{\sum_{i=1}^{n}(\widehat{y}_i - \bar{y})^2}_{\text{regression sum of squares}}$$

or

$$2{,}812.4 = 233.9 + 2{,}578.5$$

Thus, the proportion of variability explained by the regression is (see Exercise 11.61)

$$⑤\quad R^2 = \frac{2{,}578.5}{2{,}812.4} = 1 - \frac{233.9}{2{,}812.4} = 0.917$$

5. The estimate of σ^2 is $s_e^2 = 233.9/13 = 18.0$ so $s_e = 4.242$. ⑥

6. The ⑦ P-values confirm the significance of the t ratios and thus the fact that all the terms are required in the model.

Alternatively, a hand held calculator can be used to obtain the sums and then the normal equations.

$$723 = 16\,b_0 + 40\,b_1 + 200\,b_2$$
$$1{,}963 = 40\,b_0 + 120\,b_1 + 500\,b_2$$
$$8{,}210 = 200\,b_0 + 500\,b_1 + 3{,}000\,b_2$$

The unique solution of this system of equations is $\widehat{\beta}_0 = 46.4$, $\widehat{\beta}_1 = 7.78$, $\widehat{\beta}_2 = -1.65$, and the equation of the estimated regression plane is

$$\widehat{y} = 46.4 + 7.78\,x_1 - 1.65\,x_2$$

Finally, substituting $x_1 = 2.5$ and $x_2 = 12$ into this equation, we get

$$\widehat{y} = 46.4 + 7.78\,(2.5) - 1.65\,(12)$$
$$= 46.0 \quad\blacksquare$$

Note that $\widehat{\beta}_1$ and $\widehat{\beta}_2$ are estimates of the change in the mean of Y resulting from a unit increase in the corresponding independent variable when the other independent variable is held fixed.

Categorical variables can be included in any regression analysis. When there are only two categories, we create a **dummy variable** $x_1 = 1$ if the case corresponds to the second category and 0 otherwise. When a variable has 3 categories, two dummy variables need to be constructed. Let $x_1 = 1$ if the case corresponds to the second category and $x_2 = 1$ if it corresponds to the third. The rapidly increasing number of predictor variables places strong limits on the number of categorical variables that can be included in most regression problems. The next example illustrates the dummy variable technique.

EXAMPLE 13

Multiple regression to understand problems fixing relay towers

Wireless providers lose a great deal of income when relay towers do not function properly. Breakdowns must be assessed and fixed in a timely manner. To gain understanding of the problems involved, engineers collected the data (Source: Courtesy of Don Porter but extracted from Tables 6.20 and Exercise 7.27 of R. A. Johnson and D. W. Wichern (2006), *Applied multivariate statistical analysis*, Prentice Hall. A suitable number of years replaces the three categories of experience).

Table 11.1 already contains the dummy variable for difficulty.

Table 11.1 Time to assess problem when a relay tower breaks down			
Difficulty Level	**Difficulty**	**Experience(yr)**	**Assessment Time**
Simple	0	1.5	3.0
Simple	0	2.0	2.3
Simple	0	4.5	1.7
Simple	0	8.0	1.2
Complex	1	1.5	6.7
Complex	1	0.5	7.1
Complex	1	2.5	5.3
Complex	1	3.0	5.0
Complex	1	5.0	5.6
Complex	1	6.0	4.5
Simple	0	0.0	4.5
Simple	0	0.5	4.7
Simple	0	3.5	4.0
Simple	0	4.0	4.5
Simple	0	5.0	3.1
Simple	0	6.0	3.0
Complex	1	0.0	7.9
Complex	1	3.0	6.9
Complex	1	5.5	5.0
Complex	1	5.0	5.3
Complex	1	3.5	6.9

Fit a multiple regression of assessment time to difficulty and experience.

Solution   We use software to produce the statistical analysis.

```
The regression equation is
AssessTime = 4.47 + 2.72 Difficulty -0.364 Exper

Predictor          Coef        SE Coef           T           p
Constant         4.4743         0.3986       11.23       0.000
Difficulty       2.7189         0.3681        7.39       0.000
Exper           -0.36407        0.08485      -4.29       0.000

S = 0.840713       R-Sq = 81.1%

Analysis of Variance

Source           DF          SS           MS           F           P
Regression        2      54.616       27.308       38.64       0.000
Residual Error   18      12.722        0.707
Total            20      67.338
```

All of the parameter estimates are significantly different from 0. The estimated regression

$$\widehat{y} = \widehat{\alpha} + \widehat{\beta}_1 x_1 + \widehat{\beta}_2 x_2 = 4.474 + 2.719 x_1 - 0.3641 x_2$$

The value $\widehat{\beta}_1 = 2.719$ tells us that if x_1 is increased by one unit, while x_2 is held constant, the estimated mean assessment time will increase by 2.719 hours. This change in x_1 corresponds to changing from a simple to difficult problem. Similarly, $\widehat{\beta}_2 = -0.3641$ implies that if x_2 is increased by one unit, while x_1 is held constant, the estimated mean assessment time decreases by 0.3641 hours. That is, for either a simple or difficult problem, one more year experience decreases the estimated mean assessment time by 0.3641 hours.

Notice that $R^2 = 0.81$ is quite large. As you are asked to check in Exercise 11.43, there are no unusual values among the residuals and the assumption of normal errors is reasonable.

[Using **R** : Read the file C11Ex13.txt into **Dat** and then use **summary(lm(Time ~ Exper+Difficult), data = Dat)**]   ∎

11.5   Checking the Adequacy of the Model

Assuming that the regression model is adequate, we can use the fitted equation to make inferences. Before doing so, it is imperative that we check the assumptions underlying the analysis. In the context of the regression model with two predictors, we question whether Y_i is equal to $\beta_0 + \beta_1 x_{i1} + \beta_2 x_{i2} + \varepsilon_i$, where the errors ε_i are independent and have the same variation σ^2.

All of the information on lack of fit is contained in the residuals

$$e_1 = y_1 - \widehat{y}_1 = y_1 - \widehat{\beta}_0 - \widehat{\beta}_1 x_{11} - \widehat{\beta}_2 x_{12}$$
$$e_2 = y_2 - \widehat{y}_2 = y_2 - \widehat{\beta}_0 - \widehat{\beta}_1 x_{21} - \widehat{\beta}_2 x_{22}$$
$$\vdots$$
$$e_n = y_n - \widehat{y}_n = y_n - \widehat{\beta}_0 - \widehat{\beta}_1 x_{n1} - \widehat{\beta}_2 x_{n2}$$

The residuals should be plotted in various ways to detect systematic departures from the assumptions.

A plot of the residuals versus the predicted values is a major diagnostic tool. Figure 11.14 shows (a) the ideal constant band and two typical violations; (b) that variance increases with the response and a transformation is needed; and (c) that the model $\beta_0 + \beta_1 x_1 + \beta_2 x_2$ is not adequate. In the latter case, terms with x_1^2 and x_2^2 may be needed.

We also recommend plotting the residuals versus time in order to detect possible trends over time.

Figure 11.15 gives the plot of residual versus predicted value for the velocity-evaporation rate data in Example 2. The horizontal band pattern is somewhat pinched on the left-hand side. Nevertheless, 10 is too few residuals to confirm this pattern. There appears to be no serious violation of the constant variance assumption.

Although it is nearly impossible to assess normality with only 10 residuals, the normal scores plot in Figure 11.16 has a bit of an S shaped appearance. It would have a straighter line appearance if the smallest residual were smaller and the largest residual were larger. The tails of the distribution of a residual may be thinner than those of a normal distribution. Fortunately, the normal assumption is generally not critical for inference as long as serious outliers are not present.

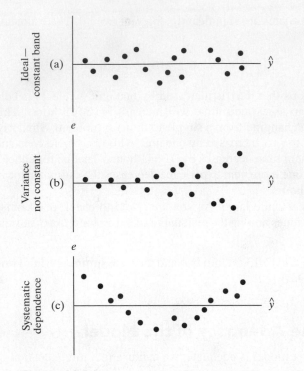

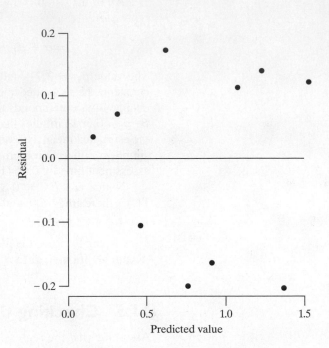

Figure 11.14

Residual plots versus $\widehat{y}$

Figure 11.15

A plot of the residuals versus predicted values $\widehat{y}$ for Example 2

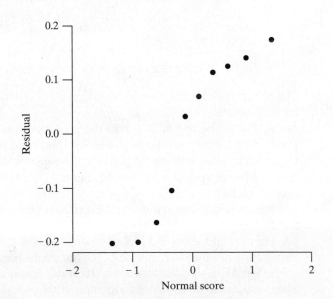

Figure 11.16

A normal-scores plot of the residuals for Example 2

Exercises

11.25 The following data pertain to the growth of a colony of bacteria in a culture medium:

Days since inoculation x	Count y
3	115,000
6	147,000
9	239,000
12	356,000
15	579,000
18	864,000

(a) Plot $\log y_i$ versus x_i to verify that it is reasonable to fit an exponential curve.

(b) Fit an exponential curve to the given data.

(c) Use the result obtained in part (b) to estimate the bacteria count at the end of 20 days.

11.26 The following data pertain to the cosmic ray doses measured at various altitudes:

Altitude (feet) x	Dose rate (mem/year) y
50	28
450	30
780	32
1,200	36
4,400	51
4,800	58
5,300	69

(a) Fit an exponential curve.

(b) Use the result obtained in part (a) to estimate the mean dose at an altitude of 3,000 feet.

11.27 With reference to the preceding exercise, change the equation obtained in part (a) to the form $\hat{y} = a \cdot e^{-cx}$, and use the result to rework part (b).

11.28 Refer to Example 10. Two new observations are available.

Rate of discharge(A)	Capacity(Ah)	ln(Capacity)
5	149.4	5.0066
18	108.2	6.5840

Add these observations to the data set in Example 10 and rework the example.

11.29 Fit a **Gompertz curve** of the form

$$y = e^{e^{\alpha x + \beta}}$$

to the data of Exercise 11.26.

11.30 Plot the curve obtained in the preceding exercise and the one obtained in Exercise 11.26 on one diagram and compare the fit of these two curves.

11.31 The number of inches which a newly built structure is settling into the ground is given by

$$y = 3 - 3\,e^{-\alpha x}$$

where x is its age in months.

x	2	4	6	12	18	24
y	1.07	1.88	2.26	2.78	2.97	2.99

Use the method of least squares to estimate α. [*Hint*: Note that the relationship between $\ln(3 - y)$ and x is linear.]

11.32 The following data pertain to the amount of hydrogen present, y, in parts per million in core drillings made at 1-foot intervals along the length of a vacuum-cast ingot, x, core location in feet from base:

x	1	2	3	4	5	6	7	8	9	10
y	1.28	1.53	1.03	0.81	0.74	0.65	0.87	0.81	1.10	1.03

(a) Draw a scatter plot to check whether it is reasonable to fit a parabola to the given data.

(b) Fit a parabola by the method of least squares.

(c) Use the equation obtained in part (b) to estimate the amount of hydrogen present at $x = 7.5$.

11.33 When fitting a polynomial to a set of paired data, we usually begin by fitting a straight line and using the method on page 329 to test the null hypothesis $\beta_1 = 0$. Then we fit a second-degree polynomial and test whether it is worthwhile to carry the quadratic term by comparing $\hat{\sigma}_1^2$, the **residual variance** after fitting the straight line, with $\hat{\sigma}_2^2$, the residual variance after fitting the second-degree polynomial. Each of these residual variances is given by the formula

$$\frac{\sum (y - \hat{y})^2}{\text{degrees of freedom}} = \frac{\text{SSE}}{\nu}$$

with $\hat{y}$ determined, respectively, from the equation of the line and the equation of the second-degree polynomial. The decision whether to carry the quadratic term is based on the statistic

$$F = \frac{\text{SSE}_1 - \text{SSE}_2}{\hat{\sigma}_2^2} = \frac{\nu_1\,\hat{\sigma}_1^2 - \nu_2\,\hat{\sigma}_2^2}{\hat{\sigma}_2^2}$$

which (under the assumptions of Section 11.2) is a value of a random variable having the F distribution with 1 and $n - 3$ degrees of freedom.

(a) Fit a straight line to the varnish-additive drying-time data in Example 11, test the null hypothesis $\beta_1 = 0$ at the 0.05 level of significance, and calculate $\widehat{\sigma}_1^2$.

(b) Using the result in the varnish-additive example, calculate $\widehat{\sigma}_2^2$ for the given data and test at the 0.05 level whether we should carry the quadratic term. (Note that we could continue this procedure and test whether to carry a cubic term by means of a corresponding comparison of residual variances. Then we could test whether to carry a fourth-degree term, and so on. It is customary to terminate this procedure after two successive steps have not produced significant results.)

11.34 With reference to Example 11, verify that the predicted drying time is minimum when the amount of additive used is 5.1 grams.

11.35 Verify that the system of normal equations on page 347 corresponds to the minimization of the sum of squares.

11.36 Twelve specimens of cold-reduced sheet steel, having different copper contents and annealing temperatures, are measured for hardness with the following results:

Hardness (Rockwell 30-T)	Copper content (%)	Annealing temperature (degrees F)
78.9	0.02	1,000
65.1	0.02	1,100
55.2	0.02	1,200
56.4	0.02	1,300
80.9	0.10	1,000
69.7	0.10	1,100
57.4	0.10	1,200
55.4	0.10	1,300
85.3	0.18	1,000
71.8	0.18	1,100
60.7	0.18	1,200
58.9	0.18	1,300

Fit an equation of the form $y = \beta_0 + \beta_1 x_1 + \beta_2 x_2$, where x_1 represents the copper content, x_2 represents the annealing temperature, and y represents the hardness.

11.37 With reference to Exercise 11.36, estimate the hardness of a sheet of steel with a copper content of 0.05% and an annealing temperature of 1,150 degrees Fahrenheit.

11.38 A compound is produced for a coating process. It is added to an otherwise fixed recipe and the coating process is completed. Adhesion is then measured. The following data concern the amount of adhesion and its relation to the amount of an additive and temperature of a reaction.

Additive x_1	Temperature x_2	Adhesion y
0	100	10
70	100	48
35	140	41
0	180	40
70	180	39
70	140	44
0	140	24
35	100	31
35	180	44

(Courtesy of Asit Banerjee)

Fit an equation of the form $y = \beta_0 + \beta_1 x_1 + \beta_2 x_2$ to the given data and use it to estimate the amount of adhesion when the amount of additive is 40 and the temperature is 130.

11.39 The following sample data were collected to determine the relationship between processing variables and the current gain of a transistor in the integrated circuit:

Diffusion time (hours) x_1	Sheet resistance (Ω-cm) x_2	Current gain y
1.5	66	5.3
2.5	87	7.8
0.5	69	7.4
1.2	141	9.8
2.6	93	10.8
0.3	105	9.1
2.4	111	8.1
2.0	78	7.2
0.7	66	6.5
1.6	123	12.6

Fit a regression plane and use its equation to estimate the expected current gain when the diffusion time is 2.2 hours and the sheet resistance is 90 Ω-cm.

11.40 Multiple regression is best implemented on a computer. The following *MINITAB* commands fits the y values in C1 to predictor values in C2 and C3.

Dialog box:
Stat > Regression > Regression > Fit Regression Model.
Type *C*1 in **Response**. Type *C*2 and *C*3 in **Continuous predictors**. Click **OK**.

It produces output like that on page 350. Use a computer to perform the multiple regression analysis in Exercise 11.36.

11.41 Using *MINITAB* we can transform the x values in C1 and/or the y values in C2. For instance, to obtain the logarithm to the base 10 of y, select

---
Dialog box:
Calc > Calculator
Type *C*3 in **Store**, *LOGT(C2)* in **Expression**. Click **OK**.

Use the computer to repeat the analysis of Exercise 11.27.

11.42 To fit the quadratic regression model using *MINITAB*, when the x values are in C1 and the y values in C2, you must select

---
Dialog box:
Stat >Regression > Fitted Line Plot
Enter *C*2 in **Response (Y)** and enter *C*1 in **Predictor (X)**.
Under **Type of Regression Model** choose **Quadratic**. Click **OK**.

Use the computer to repeat the analysis of Example 11.

11.43 With reference to Exercise 11.40, in order to plot residuals, before clicking **OK**, you must select

---
Dialog box:
Click **Storage**. Check **Residuals** and **Fits**. Click **OK** twice.

The additional steps, before clicking the second **OK**.

Dialog box:
Click **Graphs**. Check **Residuals versus fits**.
Click **OK** twice.

---

will produce a plot of the residuals versus $\hat{y}$. See Exercise 5.102, page 174, to obtain a normal-scores plot of the residuals.
Use a computer to analyze the residuals from the multiple-regression analysis in Example 13.

11.44 With reference to Exercise 11.39, analyze the residuals from the regression plane.

11.45 The following residuals and predicted values were obtained from an experiment that related yield of a chemical process (y) to the initial concentration (x) of a component (the time order of the experiments is given in parentheses):

Predicted	Residual	Predicted	Residual
4.1 (5)	-2	3.5 (3)	0
3.2 (9)	-1	4.0 (12)	3
3.5 (13)	3	4.2 (4)	-2
4.3 (1)	-3	3.9 (11)	2
3.3 (7)	-1	4.3 (2)	-5
4.6 (14)	5	3.7 (10)	0
3.6 (8)	0	3.2 (6)	1

Examine the residuals for evidence of a violation of the assumptions.

11.6 Correlation

So far in this chapter, we have studied problems where the independent variable (or variables) was assumed to be known without error. Although this applies to many experimental situations, there are also problems where the x's as well as the y's are values assumed by random variables. This would be the case, for instance, if we studied the relationship between input and output of a wastewater treatment plant, the relationship between the tensile strength and the hardness of aluminum, or the relationship between impurities in the air and the incidence of a certain disease. Problems like these are referred to as problems of **correlation analysis**, where it is assumed that the data points (x_i, y_i) for $i = 1, 2, \ldots, n$ are values of a pair of random variables whose joint density is given by $f(x, y)$.

The scatter plot provides a visual impression of the relation between the x and y values in a bivariate data set. Often the points appear to scatter about a straight line. The closeness of the scatter to a straight line can be expressed numerically in terms of the correlation coefficient. The best interpretation of the sample correlation coefficient is in terms of the standardized observations

$$\frac{\text{Observation} \ - \ \text{Sample mean}}{\text{Sample standard deviation}} = \frac{x_i - \bar{x}}{s_x}$$

where the subscript x on s distinguishes the sample variance of the x observations,

$$s_x^2 = \sum_{i=1}^{n} (x_i - \bar{x})^2 / (n - 1) = S_{xx} / (n - 1)$$

from the sample variance of the y observations.

The **sample correlation coefficient** r is the sum of products of the standardized variables divided by $n - 1$, the same divisor used for sample variance.

Sample correlation coefficient

$$r = \frac{1}{n - 1} \sum_{i=1}^{n} \left(\frac{x_i - \bar{x}}{s_x} \right) \left(\frac{y_i - \bar{y}}{s_y} \right)$$

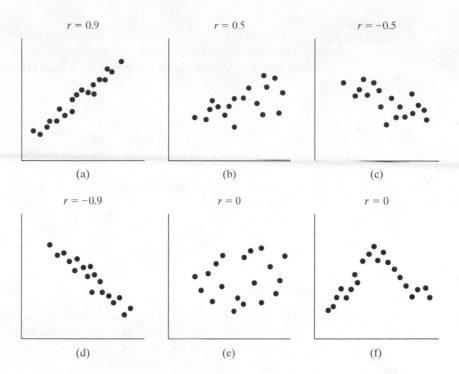

Figure 11.17
Correspondence between the values of r and the pattern of scatter

When most of the pairs of observations are such that either both components are simultaneously above their sample means or both are simultaneously below their sample means, the products of the standardized values will tend to be large and positive so r will be positive. This case corresponds to a southwest to northeast pattern in the scatter plot. [See Figure 11.17 (a)–(b).]

Alternatively, if one component of the pair tends to be large when the other is small, and vice versa, the correlation coefficient r is negative. This case corresponds to a northwest to southeast pattern in the scatter plot.

It can be shown that the value of r is always between -1 and 1, inclusive.

1. The magnitude of r describes the strength of a linear relation and its sign indicates the direction.
$r = +1$ if all pairs (x_i, y_i) lie exactly on a straight line having a positive slope.
$r > 0$ if the pattern in the scatter plot runs from lower left to upper right.
$r < 0$ if the pattern in the scatter plot runs from upper left to lower right.
$r = -1$ if all pairs (x_i, y_i) lie exactly on a straight line having a negative slope.
A value of r near -1 or $+1$ describes a strong linear relation.

2. A value of r close to zero implies that the *linear* association is weak. There may still be a strong association along a curve. [See Figure 11.17(f).]

From the definitions of S_{xx}, S_{xy}, and S_{yy} on page 320, we obtain a simpler calculation formula for r.

Alternative calculation for the sample correlation coefficient

$$r = \frac{S_{xy}}{\sqrt{S_{xx} \cdot S_{yy}}}$$

EXAMPLE 14   **Calculating the sample correlation coefficient**

The following are the numbers of minutes it took 10 mechanics to assemble a piece of machinery in the morning, x, and in the late afternoon, y:

x	y
11.1	10.9
10.3	14.2
12.0	13.8
15.1	21.5
13.7	13.2
18.5	21.1
17.3	16.4
14.2	19.3
14.8	17.4
15.3	19.0

Calculate r.

Solution   The first step is always to plot the data to make sure a linear pattern exists and that there are no outliers. A computer calculation provides the scatter plot in Figure 11.18 and the value $r = 0.732$ (see Exercise 11.64 for details).

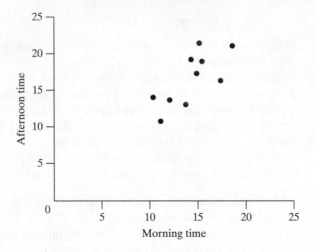

Figure 11.18
(Pearson) Correlation of x and y, $r = 0.732$

Alternatively, using a calculator, we determine the summations needed for the formulas on page 320. We get

$$S_{xx} = 2{,}085.31 - (142.3)^2/10 = 60.381$$
$$S_{xy} = 2{,}434.69 - (142.3)(166.8)/10 = 61.126$$
$$S_{yy} = 2{,}897.80 - (166.8)^2/10 = 115.576$$

so,

$$r = \frac{61.126}{\sqrt{(60.381)(115.576)}} = 0.732$$

The positive value for r confirms a positive association where long assembly times tend to pair together and so do short assembly times. Further, it captures the orientation of the pattern in Figure 11.18, which runs from lower left to upper right. Since $r = 0.732$ is moderately large, the pattern of scatter is moderately narrow.

[Using **R**: **cor(x , y)** with afternoon times in **y** and morning times in **x**.]   ∎

EXAMPLE 15   **Exploring the interpreting correlation**

Heavy metals can inhibit the biological treatment of waste in municipal treatment plants. Monthly measurements were made at a state-of-the-art treatment plant of the amount of chromium ($\mu g/l$) in both the influent and effluent. (Courtesy of Paul Berthouex)

influent	250	290	270	100	300	410	110	130	1100
effluent	19	10	17	11	70	60	18	30	180

(a) Make a scatter plot.

(b) Make a scatter plot after taking the natural logarithm of both variables.

(c) Calculate the correlation coefficient, r, in part (a) and part (b).

(d) Comment on the appropriateness of r in each case.

Solution   The scatter plots are shown in Figures 11.19 (a) and (b), respectively.

(c) A computer calculation gives $r = 0.942$ and $r = 0.747$, respectively.

(d) r is not really appropriate for the original data since the one large observation in the upper-right-hand corner has too much influence. If the pair (1100, 180) is dropped, r drops to 0.578. The situation in (b) is better.   ∎

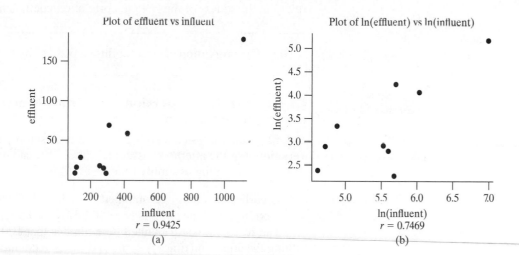

Figure 11.19
Scatter plots, original and transformed data

Correlation and Regression

There are two important relationships between r and the least squares fit of a straight line. First,

$$r = \frac{S_{xy}}{\sqrt{S_{xx}S_{yy}}} = \frac{\sqrt{S_{xx}}}{\sqrt{S_{yy}}}\frac{S_{xy}}{S_{xx}} = \frac{\sqrt{S_{xx}}}{\sqrt{S_{yy}}}\widehat{\beta}$$

so the sample correlation coefficient, r, and the least squares estimate of slope, $\widehat{\beta}$, have the same sign.

The second relationship concerns the proportion of variability in y explained by x in a least squares fit. The total variation in y is

$$S_{yy} = \sum_{i=1}^{n}(y_i - \bar{y})^2$$

while the unexplained part of the variation is the sum of squares residuals or $S_{yy} - S_{xy}^2 / S_{xx}$. This leaves the difference $S_{yy} - (S_{yy} - S_{xy}^2 / S_{xx}) = S_{xy}^2 / S_{xx}$ as the **regression sum of squares** due to fitting x. This decomposes the total variability in y into two components: one due to regression and the other due to error.

Decomposition of variability

$$
\underset{\substack{\text{Total variability} \\ \text{of } y}}{S_{yy}} = \underset{\substack{\text{Variability explained} \\ \text{by the linear relation}}}{S_{xy}^2 / S_{xx}} + \underset{\substack{\text{Residual or unexplained} \\ \text{variability}}}{S_{yy} - S_{xy}^2 / S_{xx}}
$$

For the straight line to provide a good fit to the data, the sum of squares due to regression, S_{xy}^2 / S_{xx}, should be a substantial proportion of the total sum of squares S_{yy}.

The **proportion of the y variability explained by the linear relation** is

$$
\frac{\text{Sum of squares due to regression}}{\text{Total sum of squares of } y} = \frac{S_{xy}^2 / S_{xx}}{S_{yy}} = \frac{S_{xy}^2}{S_{xx} S_{yy}} = r^2
$$

where r is the sample correlation coefficient. To summarize, the strength of the linear relationship is measured by the proportion of the y variability explained by the linear relation, the square of the sample correlation coefficient.

The proportion of y variability explained by the linear relation $= r^2$

EXAMPLE 16   **Calculating the proportion of y variation attributed to the linear relation**

Refer to Example 14 concerning the data on assembly times. Find the proportion of variation in y, the afternoon assembly times, that can be explained by a straight-line fit to x, the morning assembly times.

Solution   In the earlier example, we obtained $r = 0.732$. Consequently, the proportion of variation in y attributed to x is $r^2 = (0.732)^2 = 0.536$.

The result we have obtained here implies that $100 \, r^2 = 53.6\%$ of the variation among the afternoon times is explained by (is accounted for or may be attributed to) the corresponding differences among the morning times.   ∎

Correlation and Causation

Scientists have sometimes jumped to unjustified conclusions by mistaking a high observed correlation for a cause-and-effect relationship. The observation that two variables tend to vary simultaneously in the same direction does not imply a direct relationship between them. It would not be surprising, for example, to obtain a high positive correlation between the annual sales of chewing gum and the incidence of crime in cities of various sizes within the United States, but one cannot conclude that crime might be reduced by prohibiting the sale of chewing gum. Both variables depend upon the size of the population, and it is this mutual relationship with a third variable (population size) which produces the positive correlation. This third variable, called a **lurking variable**, is often overlooked when mistaken claims are made about x causing y.

When using the correlation coefficient as a measure of relationship, we must try to avoid the possibility that an important lurking variable is influencing the calculation.

A causal relationship may also exist that is opposite to the observed correlation. During the manufacture of high-quality printer paper, the whiteness and amount of metallic particles are measured. The whitest runs of paper have the lowest metal content and vice versa. That is, there is a strong but negative correlation. To someone unacquainted with the paper-making process, this would suggest eliminating the metallic particles. However, the causal relationship is just the opposite—there is a strong positive correlation. The metallic particles improve whiteness but they are only added to borderline papers to make sure they pass the standards test.

Inference about the Correlation Coefficient (Normal Populations)

To develop a population measure of association, or correlation, for two random variables X and Y, we begin with the two standardized variables

$$\frac{X - \mu_1}{\sigma_1} \text{ and } \frac{Y - \mu_2}{\sigma_2}$$

Each of these two standardized variables is free of its unit of measurement so their product is free of both units of measurement. The expected value of this product, which is also the covariance, is then the measure of association between X and Y called the **population correlation coefficient**. This measure of relationship or association is denoted by ρ (rho).

Population correlation coefficient

$$\rho = E\left[\left(\frac{X - \mu_1}{\sigma_1} \right) \left(\frac{Y - \mu_2}{\sigma_2} \right) \right]$$

The population correlation coefficient ρ is positive when both components (X, Y) are simultaneously large or simultaneously small with high probability. A negative value for ρ prevails when, with high probability, one member of the pair (X, Y) is large and the other is small. The value of ρ is always between -1 and 1, inclusive. The extreme values ± 1 arise only when probability 1 is assigned to pairs (x, y) where $(y - \mu_2)/\sigma_2 = (\pm 1)(x - \mu_1)/\sigma_1$, respectively. That is, probability 1 is assigned to a straight line and probability 0 to the rest of the plane.

In summary, when $\rho = \pm 1$, we say that there is a perfect linear correlation (relationship, or association) between the two random variables; when $\rho = 0$, we say there is no correlation (relationship, or association) between the two random variables.

Although the sample correlation coefficient is not an unbiased estimator of ρ, it is widely used as a point estimator whatever the form of the bivariate population.

Tests of hypotheses about ρ and confidence intervals require more restrictive assumptions. In the remainder of this section, we assume that the joint distribution of X and Y is, to a reasonable approximation, the **bivariate normal distribution**. This distribution has joint density function

$$f(x, y) = \frac{1}{2\pi \cdot \sigma_1 \sigma_2 \sqrt{1 - \rho^2}} \cdot$$

$$e^{-\frac{1}{2(1 - \rho^2)} \left[\left(\frac{x - \mu_1}{\sigma_1} \right)^2 - 2\rho \left(\frac{x - \mu_1}{\sigma_1} \right) \left(\frac{y - \mu_2}{\sigma_2} \right) + \left(\frac{y - \mu_2}{\sigma_2} \right)^2 \right]}$$

for $-\infty < x < \infty$ and $-\infty < y < \infty$.

For the bivariate normal distribution, we have a stronger property. When $\rho = 0$, the joint density factors (see Exercise 11.60) so zero correlation also implies that the two random variables are independent.

Inferences about ρ are based on the sample correlation coefficient. Whenever r is based on a random sample from a bivariate normal population, we can perform a test of significance (a test of the null hypothesis $\rho = \rho_0$) or construct a confidence interval for ρ on the basis of the **Fisher $\mathcal{Z}$ transformation**:

Fisher $\mathcal{Z}$ transformation

$$\mathcal{Z} = \frac{1}{2} \ln \frac{1+r}{1-r}$$

This statistic is a value of a random variable having approximately a normal distribution with

$$\text{mean} \quad \mu_{\mathcal{Z}} = \frac{1}{2} \ln \frac{1+\rho}{1-\rho} \quad \text{and} \quad \text{variance} \quad \frac{1}{n-3}$$

Thus, we can base inferences about ρ on

Statistic for inferences about ρ

$$Z = \frac{\mathcal{Z} - \mu_Z}{1/\sqrt{n-3}} = \frac{\sqrt{n-3}}{2} \cdot \ln \frac{(1+r)(1-\rho)}{(1-r)(1+\rho)}$$

which is a random variable having approximately the standard normal distribution.

In particular, we can test the null hypothesis of no correlation, namely, the null hypothesis $\rho = 0$, with the statistic

Statistic for test of null hypothesis $\rho = 0$

$$Z = \sqrt{n-3} \cdot \mathcal{Z} = \frac{\sqrt{n-3}}{2} \cdot \ln \frac{1+r}{1-r}$$

EXAMPLE 17

Testing for nonzero correlation in a normal population

With reference to Example 14, where $n = 10$ and $r = 0.732$, test the null hypothesis $\rho = 0$ against the null hypothesis $\rho \neq 0$ at the 0.05 level of significance.

Solution

1. *Null hypothesis*: $\rho = 0$
   Alternative hyptohesis: $\rho \neq 0$

2. *Level of significance*: $\alpha = 0.05$

3. *Criterion*: Reject the null hypothesis if $Z < -1.96$ or $Z > 1.96$, where $Z = \sqrt{n-3} \cdot \mathcal{Z}$.

4. *Calculations*: The value of $\mathcal{Z}$ corresponding to $r = 0.732$ is

$$\frac{1}{2} \ln \left(\frac{1+0.732}{1-0.732} \right) = 0.933$$

so that

$$Z = \sqrt{10-3} \cdot (0.933) = 2.47$$

5. *Decision*: Since $z = 2.47$ exceeds 1.96, the null hypothesis must be rejected; we conclude that there is a relationship between the morning and later afternoon times it takes a mechanic to assemble the given kind of machinery. ∎

To construct a confidence interval for ρ, we first construct a confidence interval for $\mu_{\mathcal{Z}}$, the mean of the sampling distribution of $\mathcal{Z}$, and convert to r and ρ using the inverse transformation. To obtain this transformation, we solve

$$\mathcal{Z} = \frac{1}{2} \ln \frac{1+r}{1-r}$$

for r to obtain

$$r = \frac{e^{\mathcal{Z}} - e^{-\mathcal{Z}}}{e^{\mathcal{Z}} + e^{-\mathcal{Z}}}$$

Making use of the theory above, we can write the first of these confidence intervals as

Confidence interval for $\mu_{\mathcal{Z}}$ (normal population)

$$\mathcal{Z} - \frac{z_{\alpha/2}}{\sqrt{n-3}} < \mu_{\mathcal{Z}} < \mathcal{Z} + \frac{z_{\alpha/2}}{\sqrt{n-3}}$$

Example 18 gives the steps for converting this interval into a confidence interval for ρ.

EXAMPLE 18   **Determining a confidence interval for ρ (normal population)**

If $r = 0.70$ for the mathematics and physics grades of 30 students, construct a 95% confidence interval for the population correlation coefficient.

Solution   The value of $\mathcal{Z}$ that corresponds to $r = 0.70$ is

$$\mathcal{Z} = \frac{1}{2} \ln \left(\frac{1+r}{1-r} \right) = \frac{1}{2} \ln \left(\frac{1+.7}{1-.7} \right) = 0.867$$

Substituting it together with $n = 30$ and $z_{0.025} = 1.96$ into the preceding confidence interval formula for $\mu_{\mathcal{Z}}$, we get

$$0.867 - \frac{1.96}{\sqrt{27}} < \mu_{\mathcal{Z}} < 0.867 + \frac{1.96}{\sqrt{27}}$$

or

$$0.490 < \mu_{\mathcal{Z}} < 1.244$$

Then, transforming the confidence limits back to the corresponding values of r,

$$r = \frac{e^{0.490} - e^{-0.490}}{e^{0.490} + e^{-0.490}} = 0.45 \quad \text{and} \quad \frac{e^{1.244} - e^{-1.244}}{e^{1.244} + e^{-1.244}} = 0.85$$

we get the 95% confidence interval

$$0.45 < \rho < 0.85$$

for the true strength of the linear relationship between grades of students in the two given subjects. ∎

Note that, in Example 18, the confidence interval for ρ is fairly wide. This illustrates the fact that correlation coefficients based on relatively small samples are generally not very informative.

Two serious pitfalls in the interpretation of the coefficient of correlation are worth repeating. First, it must be emphasized that r is an estimate of the strength of the linear relationship between the values of two random variables. Thus, as is shown in Figure 11.17(f), r may be close to 0 when there is actually a strong (but nonlinear) relationship. Second, and perhaps of greatest importance, a significant correlation does not necessarily imply a causal relationship between the two random variables.

(Optional) The Bivariate Normal Distribution and the Straight-Line Regression Model

Here we develop the bivariate normal joint density function in terms of the conditional density $f_2(y \mid x)$ and the marginal density $f_1(x)$, as defined in Section 5.10. So far as $f_2(y \mid x)$ is concerned, the conditions we shall impose are practically identical with the ones we used in connection with the sampling theory of Section 11.2. For any given x, it will be assumed that $f_2(y \mid x)$ is a normal distribution with the mean $\alpha + \beta x$ and the variance σ^2.

$$E(Y \mid x) \text{ is called the } \textbf{regression} \text{ of } Y \text{ on } x$$

ant it is linear. The variance of the conditional density does not depend on x. Interchanging x and y, we get the other regression which is estimated in Figure 11.3.

Furthermore, we shall assume that the marginal density $f_1(x)$ is normal with the mean μ_1 and the variance σ_1^2. Making use of the relationship $f(x, y) = f_1(x) \cdot f_2(y \mid x)$ given on page 157, we thus obtain

$$f(x, y) = \frac{1}{\sqrt{2\pi}\,\sigma_1} e^{-\frac{(x - \mu_1)^2}{2\sigma_1^2}} \cdot \frac{1}{\sqrt{2\pi}\,\sigma} e^{-\frac{[y - (\alpha + \beta x)]^2}{2\sigma^2}}$$

$$= \frac{1}{2\pi \cdot \sigma \cdot \sigma_1} e^{-\left\{\frac{[y - (\alpha + \beta x)]^2}{2\sigma^2} + \frac{(x - \mu_1)^2}{2\sigma_1^2}\right\}}$$

for $-\infty < x < \infty$ and $-\infty < y < \infty$. Note that this joint distribution involves the *five* parameters μ_1, σ_1, α, β, and σ.

For reasons of symmetry and other considerations to be explained later, it is customary to express the bivariate normal density in terms of the parameters μ_1, σ_1, μ_2, σ_2, and ρ. Here μ_2 and σ_2^2 are the mean and the variance of the marginal distribution $f_2(y)$, while ρ, in this notation, is given by

$$\rho^2 = 1 - \frac{\sigma^2}{\sigma_2^2}$$

with ρ taken to be positive when β is positive and negative when β is negative. Leaving it to the reader to show in Exercise 11.59 that

$$\mu_2 = \alpha + \beta \mu_1 \quad \text{and} \quad \sigma_2^2 = \sigma^2 + \beta^2 \sigma_1^2$$

we next substitute into the preceding expression for $f(x, y)$ and obtain the following form of the bivariate normal distribution:

$$f(x, y) = \frac{1}{2\pi \cdot \sigma_1 \sigma_2 \sqrt{1 - \rho^2}} \cdot$$

$$e^{-\frac{1}{2(1 - \rho^2)}\left[\left(\frac{x - \mu_1}{\sigma_1}\right)^2 - 2\rho\left(\frac{x - \mu_1}{\sigma_1}\right)\left(\frac{y - \mu_2}{\sigma_2}\right) + \left(\frac{y - \mu_2}{\sigma_2}\right)^2\right]}$$

for $-\infty < x < \infty$ and $-\infty < y < \infty$ (see Exercise 11.59).

Concerning the correlation coefficient ρ, note that $-1 \leq \rho \leq +1$ since $\sigma_2^2 = \sigma^2 + \beta^2\sigma_1^2$ and, hence, $\sigma_2^2 \geq \sigma^2$. Furthermore, ρ can equal -1 or $+1$ only when $\sigma^2 = 0$, which represents the degenerate case where all the probability is concentrated along the line $y = \alpha + \beta x$ and there is, thus, a perfect linear relationship between the two random variables. (That is, for a given value of x, Y must equal $\alpha + \beta x$.)

Exercises

11.46 Data, collected over seven years, reveals a positive correlation between the annual starting salary of engineers and the annual sales of diet soft drinks. Will buying more diet drinks increase starting salaries? Explain your answer and suggest a possible lurking variable.

11.47 Data, collected from cities of widely varying sizes, revealed a high positive correlation between the amount of beer consumed and the number of weddings in the past year. Will consuming lots of beer increase the number of weddings? Explain your answer.

11.48 Use the expressions on page 357, involving the deviations from the mean, to calculate r for the following data.

x	y
8	4
1	5
5	1
4	3
7	2

11.49 Calculate r for the air velocities and evaporation coefficients of Example 2. Also, assuming that the necessary assumptions can be met, test the null hypothesis $\rho = 0$ against the alternative hypothesis $\rho \neq 0$ at the 0.05 level of significance.

11.50 The following data pertain to the resistance (ohms) and the failure time (minutes) of certain overloaded resistors:

Resistance	Failure time
43	32
29	20
44	45
33	35
33	22
47	46
34	28
31	26
48	37
34	33
46	47
37	30
36	36
39	33
36	21
47	44
28	26
40	45
42	39
33	25
46	36
28	25
48	45
45	36

Calculate r.

11.51 With reference to Exercise 11.50, test $\rho = 0$ against $\rho \neq 0$ at $\alpha = 0.01$.

11.52 Calculate r for the extraction times and extraction efficiencies of Exercise 11.3. Assuming that the necessary assumptions can be met, test the null hypothesis $\rho = 0.75$ against the alternative hypothesis $\rho > 0.75$ at the 0.05 level of significance.

11.53 Calculate r for the humidities and moisture contents of Exercise 11.12. Assuming that the necessary assumptions can be met, construct a 95% confidence interval for the population correlation coefficient ρ.

11.54 The following are measurements of the carbon content and the permeability index of 22 sinter mixtures:

Carbon content (%)	Permeability index
4.4	12
5.5	14
4.2	18
3.0	35
4.5	23
4.1	13
4.9	19
4.7	22
5.0	20
4.6	16
4.9	29
4.6	16
5.0	12
4.7	18
5.1	21
4.4	27
3.6	27
4.9	21
5.1	13
4.8	18
5.2	17
5.2	11

(a) Calculate r.

(b) Find 99% confidence limits for ρ.

11.55 Referring to Example 3 concerning nanopillars, calculate the correlation coefficient between height and width.

11.56 If $r = 0.83$ for one set of paired data and $r = 0.60$ for another, compare the strengths of the two relationships.

11.57 If data on the ages and prices of 25 pieces of equipment yielded $r = -0.58$, test the null hypothesis $\rho = -0.40$ against the alternative hypothesis $\rho < -0.40$ at the 0.05 level of significance. Assume bivariate normality.

11.58 Assuming that the necessary assumptions are met, construct a 95% confidence interval for ρ when

(a) $r = 0.72$ and $n = 19$;

(b) $r = 0.35$ and $n = 25$;

(c) $r = 0.57$ and $n = 40$.

11.59 (a) Evaluating the necessary integrals, verify the identities

$$\mu_2 = \alpha + \beta\,\mu_1 \quad \text{and} \quad \sigma_2^2 = \sigma^2 + \beta^2\,\sigma_1^2$$

on page 364.

(b) Substitute $\mu_2 = \alpha + \beta\mu_1$ and $\sigma_2^2 = \sigma^2 + \beta^2\sigma_1^2$ into the formula for the bivariate density given on page 364, and show that this gives the final form shown on page 365.

11.60 Show that for the bivariate normal distribution

(a) independence implies zero correlation;

(b) zero correlation implies independence.

11.61 Instead of using the computing formula on page 357, we can obtain the correlation coefficient r with the formula

$$r = \pm\sqrt{1 - \frac{\sum(y - \widehat{y})^2}{\sum(y - \bar{y})^2}}$$

which is analogous to the formula used to define ρ. Although the computations required by the use of this formula are tedious, the formula has the advantage that it can be used also to measure the strength of nonlinear relationships or relationships in several variables. For instance, in the multiple linear regression in Example 12, one could calculate the predicted values by means of the equation

$$\widehat{y} = 46.4 + 7.78\,x_1 - 1.65\,x_2$$

and then determine r as a measure of how strongly y, the twists required to break one of the forged alloy bars, depends on both percentages of alloying elements present.

(a) Using the data in Example 12, find $\sum(y - \bar{y})^2$.

(b) Using the regression equation obtained in Example 12, calculate $\widehat{y}$ for the 16 points and then determine $\sum(y - \widehat{y})^2$.

(c) Substitute the results obtained in (a) and (b) into the above formula for r. The result is called the **multiple correlation coefficient**.

11.62 With reference to Exercise 11.39, use the theory of the preceding exercise to calculate the multiple correlation coefficient (which measures how strongly the current gain is related to the two independent variables).

11.63 Referring to the nano twisting data in Exercise 11.24, calculate the correlation coefficient.

11.64 To calculate r using *MINITAB* when the x values are in C1 and the y values are in C2, use

> **Dialog box:**
> **Stat>Basic Statistics>Correlation**
> Type C1 and C2 in **Variables**. Click **OK**.

Also, you can make a scatter plot using the plot procedure in Exercise 11.22.
Use the computer to do Exercise 11.50.

11.7 Multiple Linear Regression (Matrix Notation)

The model we are using in multiple linear regression lends itself uniquely to a unified treatment in matrix notation.[4] This notation makes it possible to state general results in compact form and to use to great advantage many of the results of matrix theory.

It is customary to denote matrices by capital letters in boldface type and vectors by lowercase boldface type. To express the normal equations on page 347 in matrix notation, let us define the following three matrices.

$$\mathbf{X} = \begin{bmatrix} 1 & x_{11} & x_{12} \\ 1 & x_{21} & x_{22} \\ \vdots & \vdots & \vdots \\ 1 & x_{n1} & x_{n2} \end{bmatrix}$$

$$\mathbf{y} = \begin{bmatrix} y_1 \\ y_2 \\ \vdots \\ y_n \end{bmatrix} \quad \text{and} \quad \mathbf{b} = \begin{bmatrix} b_0 \\ b_1 \\ b_2 \end{bmatrix}$$

The first one, $\mathbf{X}$, is an $n \times (1+2)$ matrix consisting essentially of the given values of the x's with the column of 1's appended to accommodate the constant term. $\mathbf{y}$ is an $n \times 1$ matrix (or column vector) consisting of observed values of the response variable and $\mathbf{b}$ is the $(1 + 2) \times 1$ matrix (or column vector) consisting of possible values of the regression coefficients.

Using these matrices, we can now write the following symbolic $\widehat{\boldsymbol{\beta}}$, the least squares estimates of the multiple regression coefficients are given by

$$\widehat{\boldsymbol{\beta}} = (\mathbf{X}'\mathbf{X})^{-1}\mathbf{X}'\mathbf{y}$$

where $\mathbf{X}'$ is the transpose of $\mathbf{X}$ and $(\mathbf{X}'\mathbf{X})^{-1}$ is the inverse of $\mathbf{X}'\mathbf{X}$.

---

[4]It is assumed for this section that the reader is familiar with the material ordinarily covered in a first course on matrix algebra. Since matrix notation is not used elsewhere in this book, this section may be omitted without loss of continuity.

To verify this relation, we first determine $\mathbf{X'X}$, $\mathbf{X'Xb}$, and $\mathbf{X'y}$.

$$\mathbf{X'X} = \begin{bmatrix} n & \sum x_1 & \sum x_2 \\ \sum x_1 & \sum x_1^2 & \sum x_1 x_2 \\ \sum x_2 & \sum x_2 x_1 & \sum x_2^2 \end{bmatrix}$$

$$\mathbf{X'Xb} = \begin{bmatrix} b_0 n + b_1 \sum x_1 + b_2 \sum x_2 \\ b_0 \sum x_1 + b_1 \sum x_1^2 + b_2 \sum x_1 x_2 \\ b_0 \sum x_2 + b_1 \sum x_2 x_1 + b_2 \sum x_2^2 \end{bmatrix}$$

$$\mathbf{X'y} = \begin{bmatrix} \sum y \\ \sum x_1 y \\ \sum x_2 y \end{bmatrix}$$

Identifying the elements $\mathbf{X'Xb}$ as the expressions on the right-hand side of the normal equations on page 347 and those of $\mathbf{X'y}$ as the expressions on the left-hand side, we can write

$$\mathbf{X'Xb} = \mathbf{X'y}$$

Multiplying on the left by $(\mathbf{X'X})^{-1}$, we get

$$(\mathbf{X'X})^{-1}\mathbf{X'Xb} = (\mathbf{X'X})^{-1}\mathbf{X'y}$$

and, finally, the solution $\widehat{\boldsymbol{\beta}}$ satisfies

$$\widehat{\boldsymbol{\beta}} = (\mathbf{X'X})^{-1}\mathbf{X'y}$$

since $(\mathbf{X'X})^{-1}\mathbf{X'X}$ equals the $(1+2) \times (1+2)$ identity matrix $\mathbf{I}$, and by definition $\mathbf{Ib} = \mathbf{b}$. We have assumed here that $\mathbf{X'X}$ is nonsingular, so that its inverse exists.

EXAMPLE 19   **Calculating the least squares estimates using $(\mathbf{X'X})^{-1}\mathbf{X'y}$**

With reference to the Example 12, use the matrix expressions to determine the least squares estimates of the multiple regression coefficients.

Solution   Substituting $\sum x_1 = 40$, $\sum x_2 = 200$, $\sum x_1^2 = 120$, $\sum x_1 x_2 = 500$, $\sum x_2^2 = 3{,}000$, and $n = 16$ into the expression for $\mathbf{X'X}$ above, we get

$$\mathbf{X'X} = \begin{bmatrix} 16 & 40 & 200 \\ 40 & 120 & 500 \\ 200 & 500 & 3{,}000 \end{bmatrix}$$

Then the inverse of this matrix can be obtained by any one of a number of different techniques; using the one based on cofactors, we find that

$$(\mathbf{X'X})^{-1} = \frac{1}{160{,}000} \begin{bmatrix} 110{,}000 & -20{,}000 & -4{,}000 \\ -20{,}000 & 8{,}000 & 0 \\ -4{,}000 & 0 & 320 \end{bmatrix}$$

where $160{,}000$ is the value of $|\mathbf{X'X}|$, the determinant of $\mathbf{X'X}$.

Substituting $\sum y = 723$, $\sum x_1 y = 1963$, and $\sum x_2 y = 8210$ into the expression for $\mathbf{X'y}$ on page 368, we then get

$$\mathbf{X'y} = \begin{bmatrix} 723 \\ 1{,}963 \\ 8{,}210 \end{bmatrix}$$

and, finally,

$$\widehat{\boldsymbol{\beta}} = (\mathbf{X'X})^{-1}\mathbf{X'y} = \frac{1}{160{,}000} \begin{bmatrix} 110{,}000 & -20{,}000 & -4{,}000 \\ -20{,}000 & 8{,}000 & 0 \\ -4{,}000 & 0 & 320 \end{bmatrix} \begin{bmatrix} 723 \\ 1{,}963 \\ 8{,}210 \end{bmatrix}$$

$$= \frac{1}{160{,}000} \begin{bmatrix} 7{,}430{,}000 \\ 1{,}244{,}000 \\ -264{,}800 \end{bmatrix}$$

$$= \begin{bmatrix} 46.4375 \\ 7.7750 \\ -1.6550 \end{bmatrix}$$

Note that the results obtained here are identical with those shown in the computer printout on page 348. ∎

The residual sum of squares also has a convenient matrix expression. The predicted values $\widehat{y}_i = \widehat{\beta}_0 + \widehat{\beta}_1 x_{i1} + \widehat{\beta}_2 x_{i2}$ can be collected as a matrix (column vector).

$$\widehat{\mathbf{y}} = \begin{bmatrix} \widehat{y}_1 \\ \widehat{y}_2 \\ \vdots \\ \widehat{y}_n \end{bmatrix} = \begin{bmatrix} 1 & x_{11} & x_{12} \\ 1 & x_{21} & x_{22} \\ \vdots & \vdots & \vdots \\ 1 & x_{n1} & x_{n2} \end{bmatrix} \begin{bmatrix} \widehat{\beta}_0 \\ \widehat{\beta}_1 \\ \widehat{\beta}_2 \end{bmatrix} = X\widehat{\boldsymbol{\beta}}$$

Then the residual sum of squares

$$\sum_{i=1}^{n} (y_i - \widehat{y}_i)^2 = (\mathbf{y} - \widehat{\mathbf{y}})'(\mathbf{y} - \widehat{\mathbf{y}}) = (\mathbf{y} - \mathbf{X}\widehat{\boldsymbol{\beta}})'(\mathbf{y} - \mathbf{X}\widehat{\boldsymbol{\beta}})$$

Consequently, the estimate s_e^2 of σ^2 can be expressed as

$$s_e^2 = \frac{1}{n-3}(\mathbf{y} - \mathbf{X}\widehat{\boldsymbol{\beta}})'(\mathbf{y} - \mathbf{X}\widehat{\boldsymbol{\beta}})$$

The same matrix expressions for $\mathbf{b}$ and the residual sum of squares hold for any number of predictor variables. If the mean of Y has the form $\beta_0 + \beta_1 x_1 + \beta_2 x_2 + \cdots + \beta_k x_k$, we define the matrices

$$\mathbf{X} = \begin{bmatrix} 1 & x_{11} & x_{12} & \cdots & x_{1k} \\ 1 & x_{21} & x_{22} & \cdots & x_{2k} \\ \vdots & \vdots & \vdots & \ddots & \vdots \\ 1 & x_{n1} & x_{n2} & \cdots & x_{nk} \end{bmatrix} \qquad \widehat{\boldsymbol{\beta}} = \begin{bmatrix} \widehat{\beta}_0 \\ \widehat{\beta}_1 \\ \widehat{\beta}_2 \\ \vdots \\ \widehat{\beta}_k \end{bmatrix} \qquad \mathbf{y} = \begin{bmatrix} y_1 \\ y_2 \\ \vdots \\ y_n \end{bmatrix}$$

Then

$$\widehat{\boldsymbol{\beta}} = (\mathbf{X'X})^{-1}\mathbf{X'y} \quad \text{and} \quad s_e^2 = \frac{1}{n-k-1}(\mathbf{y} - \mathbf{X}\widehat{\boldsymbol{\beta}})'(\mathbf{y} - \mathbf{X}\widehat{\boldsymbol{\beta}})$$

Generally, the sum of squares error, *SSE*, has degrees of freedom (d.f.)

$$\text{d.f.} = n - \text{number of } \beta's \text{ in model} = n - (k+1)$$

EXAMPLE 20 **Fitting a straight line using the matrix formulas**
Use the matrix relations to fit a straight line to the data

x	0	1	2	3	4
y	8	9	4	3	1

Solution Here $k = 1$ and, dropping the subscript 1, we have

$\mathbf{X}'$	$\mathbf{y}$	$\mathbf{X}'\mathbf{X}$	$(\mathbf{X}'\mathbf{X})^{-1}$	$\mathbf{X}'\mathbf{y}$

$$\begin{bmatrix} 1 & 1 & 1 & 1 & 1 \\ 0 & 1 & 2 & 3 & 4 \end{bmatrix} \quad \begin{bmatrix} 8 \\ 9 \\ 4 \\ 3 \\ 1 \end{bmatrix} \quad \begin{bmatrix} 5 & 10 \\ 10 & 30 \end{bmatrix} \quad \begin{bmatrix} 0.6 & -0.2 \\ -0.2 & 0.1 \end{bmatrix} \quad \begin{bmatrix} 25 \\ 30 \end{bmatrix}$$

Consequently,

$$\widehat{\boldsymbol{\beta}} = (\mathbf{X}'\mathbf{X})^{-1}\mathbf{X}'\mathbf{y} = \begin{bmatrix} 0.6 & -0.2 \\ -0.2 & 0.1 \end{bmatrix} \begin{bmatrix} 25 \\ 30 \end{bmatrix} = \begin{bmatrix} 9 \\ -2 \end{bmatrix}$$

and the fitted equation is

$$\widehat{y} = 9 - 2x$$

The vector of fitted values is

$$\widehat{\mathbf{y}} = \mathbf{X}\widehat{\boldsymbol{\beta}} = \begin{bmatrix} 1 & 0 \\ 1 & 1 \\ 1 & 2 \\ 1 & 3 \\ 1 & 4 \end{bmatrix} \begin{bmatrix} 9 \\ -2 \end{bmatrix} = \begin{bmatrix} 9 \\ 7 \\ 5 \\ 3 \\ 1 \end{bmatrix}$$

so the vector of residuals

$$\mathbf{y} - \widehat{\mathbf{y}} = \begin{bmatrix} 8 \\ 9 \\ 4 \\ 3 \\ 1 \end{bmatrix} - \begin{bmatrix} 9 \\ 7 \\ 5 \\ 3 \\ 1 \end{bmatrix} = \begin{bmatrix} -1 \\ 2 \\ -1 \\ 0 \\ 0 \end{bmatrix}$$

and the residual sum of squares is

$$\begin{bmatrix} -1 & 2 & -1 & 0 & 0 \end{bmatrix} \begin{bmatrix} -1 \\ 2 \\ -1 \\ 0 \\ 0 \end{bmatrix} = 6$$

Finally,

$$s_e^2 = \frac{1}{n-k-1}(\mathbf{y}-\widehat{\mathbf{y}})'(\mathbf{y}-\widehat{\mathbf{y}}) = \frac{1}{5-2}(6) = 2.00$$

The elegance of the expressions using matrices goes one step further. We can express the estimated variances and covariances of the least squares estimators as

$$
\begin{bmatrix}
\widehat{Var}(\widehat{\beta}_0) & \widehat{Cov}(\widehat{\beta}_0, \widehat{\beta}_1) & \cdots & \widehat{Cov}(\widehat{\beta}_0, \widehat{\beta}_k) \\
\widehat{Cov}(\widehat{\beta}_1, \widehat{\beta}_0) & \widehat{Var}(\widehat{\beta}_1) & \cdots & \widehat{Cov}(\widehat{\beta}_1, \widehat{\beta}_k) \\
\vdots & \vdots & \ddots & \vdots \\
\widehat{Cov}(\widehat{\beta}_k, \widehat{\beta}_0) & \widehat{Cov}(\widehat{\beta}_k, \widehat{\beta}_1) & \cdots & \widehat{Var}(\widehat{\beta}_k)
\end{bmatrix}
= s_e^2 (\mathbf{X'X})^{-1}
$$

That is, to obtain the estimated variance, $\widehat{Var}(\widehat{\beta}_i)$, of $\widehat{\beta}_i$, we multiply the corresponding diagonal entry of $(\mathbf{X'X})^{-1}$ by s_e^2, which is the estimate of σ^2.

EXAMPLE 21 **Estimating the variance of the least squares estimators**

With reference to the preceding example, use the matrix relations to obtain the estimated variances $\widehat{Var}(\widehat{\beta}_0)$ and $\widehat{Var}(\widehat{\beta}_1)$.

Solution We have

$$
\begin{bmatrix}
\widehat{Var}(\widehat{\beta}_0) & \widehat{Cov}(\widehat{\beta}_0, \widehat{\beta}_1) \\
\widehat{Cov}(\widehat{\beta}_1, \widehat{\beta}_0) & \widehat{Var}(\widehat{\beta}_1)
\end{bmatrix}
= s_e^2 (\mathbf{X'X})^{-1}
$$

$$
= (2.00)
\begin{bmatrix}
0.6 & -0.2 \\
-0.2 & 0.1
\end{bmatrix}
=
\begin{bmatrix}
1.2 & -0.4 \\
-0.4 & 0.2
\end{bmatrix}
$$

where the values for $(\mathbf{X'X})^{-1}$ and s_e^2 are those obtained in Example 20. Therefore, the estimates are $\widehat{Var}(\widehat{\beta}_0) = 1.2$ and $\widehat{Var}(\widehat{\beta}_1) = 0.2$. Note also that the estimated covariance of $\widehat{\beta}_0$ and $\widehat{\beta}_1$ is $\widehat{Cov}(\widehat{\beta}_0, \widehat{\beta}_1) = -0.4$. ∎

Do's and Don'ts

Do's

1. As a first step, plot the response variable versus the predictor variable. If there is more than one predictor variable, make separate plots for each. Examine the plot to see if a linear or other relationship exists.

2. Apply the principle of least squares to obtain estimates of the coefficients when fitting a straight line or a multiple regression model.

3. Estimate the least squares line $\widehat{y} = a + bx$ with the least squares estimates

$$
\widehat{\beta} = \frac{S_{xy}}{S_{xx}} = \frac{\sum_{i=1}^{n} (x_i - \bar{x})(y_i - \bar{y})}{\sum_{i=1}^{n} (x_i - \bar{x})^2} \qquad \widehat{\alpha} = \bar{y} - b\bar{x}
$$

and the variance σ^2 of the error term by

$$
s_e^2 = \frac{\sum_{i=1}^{n} (y_i - (\widehat{\alpha} + \widehat{\beta} x_i))^2}{n - 2} = \frac{S_{yy} - S_{xy}^2/S_{xx}}{n - 2}
$$

where $S_{xx} = \sum_{i=1}^{n} (x_i - \bar{x})^2$, $S_{xy} = \sum_{i=1}^{n} (x_i - \bar{x})(y_i - \bar{y})$, and

$S_{yy} = \sum_{i=1}^{n} (y_i - \bar{y})^2$.

4. Determine the $100(1-\alpha)\%$ confidence intervals using the confidence limits

$$\text{intercept } \alpha: \quad \hat{\alpha} \pm t_{\alpha/2} \cdot s_e \sqrt{\frac{1}{n} + \frac{\bar{x}^2}{S_{xx}}}$$

$$\text{slope } \beta: \quad \hat{\beta} \pm t_{\alpha/2} \cdot s_e \frac{1}{\sqrt{S_{xx}}}$$

5. Remember that the sample correlation coefficient

$$r = \frac{S_{xy}}{\sqrt{S_{xx} \cdot S_{yy}}}$$

is a unit free measure of the linear association between the two variables.

Don'ts

1. Don't routinely accept the regression analysis presented in computer output. Instead, examine the model by inspecting the residuals for outliers and moderate to severe lack of normality. A normal-scores plot is useful if there are more than 20 or so residuals. It may suggest a transformation.

2. Don't confuse a high correlation with a causal relationship.

Review Exercises

11.65 The data below pertains to the number of hours jet aircraft engines have been used and the number of hours required for repair.

(a) Use the first set of expressions on page 320, involving the deviations from the mean, to fit a least squares line to the observations.

(b) Use the equation of the least squares line to estimate mean repair time at $x = 4.5$.

Number of hours (hundreds)	Repair time (hours)
1	10
2	40
3	30
4	80
5	90

(c) What difficulty might you encounter if you use the least squares line to predict the mean repair time for a jet aircraft engine with 700 hours?

11.66 With reference to Exercise 11.65, construct a 95% confidence interval for α.

11.67 With reference to Exercise 11.65, test the null hypothesis $\beta = 5$ against the alternative hypothesis $\beta > 5$ at the 0.05 level of significance.

11.68 With reference to Exercise 11.65,

(a) find a 95% confidence interval for the mean repair time at $x = 4.5$;

(b) find 95% limits of prediction for the time to repair an engine that will be run for $x = 4.5$ hundred hours.

11.69 A chemical engineer found that by adding different amounts of an additive to gasoline, she could reduce the amount of nitrous oxides (NOx) coming from an automobile engine. A specified amount was added to a gallon of gas and the total amount of NOx in the exhaust collected. Suppose, in suitable units, the data are

Amount of additive	NOx
1	19
2	17
3	14
4	13
5	12

(a) Obtain the least squares fit, of a straight line, to the amount of NOx.

(b) Test whether or not the slope $\beta = 0$. Take $\alpha = 0.10$ as your level of significance.

(c) Give a 99% confidence interval for the mean value of NOx when the amount of additive is 9.

(d) What additional danger is there when using your estimate in part (c)?

11.70 With reference to Exercise 11.69, find the 95% limits of prediction when the amount of additive is 4.5.

11.71 With reference to Exercise 11.69, find the proportion of variance in the amount of NOx explained by the amount of additive.

11.72 To determine how well existing chemical analyses can detect lead in test specimens in water, a civil engineer submits specimens spiked with known concentrations of lead to a laboratory. The chemists are told only that all samples are from a study about measurements on "low" concentrations, but they are not told the range of values to expect. This is sometimes called a calibration problem because the goal is to relate the measured concentration (y) to the known concentration (x). Given the data (Courtesy of Paul Berthouex)

x	0.00	0.00	1.25	1.25	2.50	2.50	2.50	5.00	10.00	10.00
y	0.7	0.5	1.1	2.0	2.8	3.5	2.3	5.3	9.1	9.4

(a) plot measured concentration versus known concentration; comment on the pattern;

(b) fit a straight line by least squares;

(c) if the chemical test is correct, on average, we would expect a straight line that has slope 1. Obtain a 95% confidence interval for β;

(d) test $H_0: \beta = 1$ versus $H_1: \beta \neq 1$ at level $\alpha = 0.05$.

11.73 With reference to the preceding exercise, construct a 95% confidence interval for α.

11.74 With reference to Example 15,

(a) find the least squares line for predicting the chromium in the effluent from that in the influent after taking natural logarithms of each variable;

(b) predict the mean ln (effluent) when the influent has 500 $\mu g/l$ chromium.

11.75 In an experiment designed to determine the specific heat ratio γ for a certain gas, measurements of the volume and corresponding pressure p produced the data:

p (lb/in.^{2})	16.6	39.7	78.5	115.5	195.3	546.1
V (in.^{3})	50	30	20	15	10	5

Assuming the ideal gas law $p \cdot V^\gamma = C$, use these data to estimate γ for this gas.

11.76 With reference to Exercise 11.75, use the method of Section 11.2 to construct a 95% confidence interval for γ. State what assumptions will have to be made.

11.77 The rise of current in an inductive circuit having the time constant τ is given by

$$I = 1 - e^{-t/\tau}$$

where t is the time measured from the instant the switch is closed, and I is the ratio of the current at time t to the full value of the current given by Ohm's law. Given the measurements

I	0.073	0.220	0.301	0.370	0.418	0.467	0.517	0.578
t (sec)	0.1	0.2	0.3	0.4	0.5	0.6	0.7	0.8

estimate the time constant of this circuit from the experimental results given. [*Hint*: Note that the relationship between ln $(1 - I)$ and t is linear.]

11.78 The following are sample data provided by a moving company on the weights of six shipments, the distances they are moved, and the damage that was incurred:

Weight (1,000 pounds) x_1	Distance (1,000 miles) x_2	Damage (dollars) y
4.0	1.5	160
3.0	2.2	112
1.6	1.0	69
1.2	2.0	90
3.4	0.8	123
4.8	1.6	186

(a) Fit an equation of the form $y = \beta_0 + \beta_1 x_1 + \beta_2 x_2$.

(b) Use the equation obtained in part (a) to estimate the damage when a shipment weighing 2,400 pounds is moved 1,200 miles.

11.79 With reference to Exercise 11.9,

(a) find a 95% confidence interval for the mean current density when the strain is $x = 0.50$;

(b) find 95% limits of prediction for the current density when a new diode has stress $x = 0.50$.

11.80 Use the expression on page 357, involving the deviations from the mean, to calculate r for the following data.

x	y
8	4
5	5
6	1
2	3
9	2

11.81 If $r = 0.41$ for one set of paired data and $r = 0.29$ for another, compare the strengths of the two relationships.

11.82 If for certain paired data $n = 18$ and $r = 0.44$, test the null hypothesis $\rho = 0.30$ against the alternative hypothesis $\rho > 0.30$ at the 0.01 level of significance.

11.83 Assuming that the necessary assumptions are met, construct a 95% confidence interval for ρ when

(a) $r = 0.78$ and $n = 15$;

(b) $r = -0.62$ and $n = 32$;

(c) $r = 0.17$ and $n = 35$.

11.84 With reference to Exercise 11.78, use the theory of Exercise 11.61 to calculate the multiple correlation coefficient (which measures how strongly the damage is related to both weight and distance).

11.85 Robert A. Millikan (1865–1953) produced the first accurate measurements on the charge e of an electron. He devised a method to observe a single drop of water or oil under the influence of both electric and gravitational fields. Usually, a droplet carried multiple electrons, and direct calculations based on voltage, time of fall, etc., provided an estimate of the total charge. [*Source: Philosophical Magazine* **19** (1910); 209–228.]

x (No. of e's)	Observations (10^9 × charge)									
3	1.392	1.392	1.398	1.368	1.368	1.368	1.345			
4	1.768	1.768	1.910	1.768	1.746	1.746	1.886	1.768	1.768	1.768
5	2.471	2.471	2.256	2.256	2.471					
2	0.944	0.992								
6	2.981	2.688								

(a) Find the equation of the least squares line for Millikan's data.

(b) Find a 95% confidence interval for the slope β, the charge e on a single electron.

(c) Test the null hypothesis $\alpha = 0$ against the alternative hypothesis $\alpha \neq 0$.

(d) Examine the residuals.

11.86 Robert Boyle (1627–1691) established the law that (pressure × volume) = constant for a gas at a constant temperature. By pouring mercury into the open top of the long side of a J-shaped tube, he increased the pressure on the air trapped in the short leg. The volume of trapped air = h × cross section, where h is the height of the air in the short leg. If y = height of mercury, adjusted for the pressure of the atmosphere on the open end, then y and $x = 1/h$ should obey a straight-line relationship. [*Source: The Laws of Gases*, edited by Carl Barus (1899), New York: Harper and Brothers Publishers.]

h	48	46	44	42	40	38	36	34	32	30	28	26	24
y	$29\frac{2}{16}$	$30\frac{9}{16}$	$31\frac{15}{16}$	$33\frac{8}{16}$	$35\frac{5}{16}$	37	$39\frac{5}{16}$	$41\frac{10}{16}$	$44\frac{3}{16}$	$47\frac{1}{16}$	$50\frac{5}{16}$	$54\frac{5}{16}$	$58\frac{13}{16}$

h	23	22	21	20	19	18	17	16	15	14	13	12
y	$61\frac{5}{16}$	$64\frac{1}{16}$	$67\frac{1}{16}$	$70\frac{11}{16}$	$74\frac{2}{16}$	$77\frac{14}{16}$	$82\frac{12}{16}$	$87\frac{14}{16}$	$93\frac{1}{16}$	$100\frac{7}{16}$	$107\frac{13}{16}$	$117\frac{9}{16}$

(a) Fit a straight line by least squares to Boyle's data.

(b) Check the residuals for a possible violation of the assumptions.

Key Terms

CHAPTER

12

ANALYSIS OF VARIANCE

Some examples in Chapter 11 show us that considerable economies in calculation result from planning an experiment in advance. More importantly, proper experimental planning can give a reasonable assurance that the experiment will provide clear-cut answers to questions under investigation. Two experimental designs, (*i*) independent samples design and (*ii*) matched pairs design, appear in Chapter 8.

We begin by presenting some general principles of experimental design. Sections 12.2 and 12.3 discuss the often used one-way design and randomized block design. The first generalizes the independent samples design and the second the matched pairs design. In the remainder of the chapter, we introduce tests for comparing several means. Besides developing tests for one-way classifications and randomized block experiments, we also treat a balanced experiment in the presence of covariate. The presence of a covariate allows us to test the equality of means after adjusting for its values on each experimental unit.

12.1 Some General Principles

Many of the most important aspects of **experimental design** can be illustrated by means of an example drawn from the important field of engineering measurement.

Suppose that a steel mill supplies tin plate to 3 can manufacturers, the major specification being that the tin-coating weight should be at least 0.25 pound per base box. The mill and each can manufacturer has a laboratory where measurements are made of the tin-coating weights of samples taken from each shipment. Because some disagreement has arisen about the actual tin-coating weights of the tin plate being shipped, it is decided to plan an experiment to determine whether the 4 laboratories are making consistent measurements. A complicating factor is that part of the measuring process consists of the chemical removal of the tin from the surface of the base metal; thus, it is impossible to have the identical sample measured by each laboratory.

One possibility is to send several samples (in the form of circular disks having equal areas) to each of the laboratories. Although these disks may not actually have identical tin-coating weights, it is hoped that such differences will be small and that they will more or less average out. In other words, it will be assumed that whatever differences there will be among the means of the 4 samples can be attributed to no other causes but systematic differences in measuring techniques and chance variability. This would make it possible to determine whether the results produced by the laboratories are consistent by comparing the variability of the 4 sample means with an appropriate measure of chance variation.

Now there remains the problem of deciding how many disks are to be sent to each laboratory and how the disks are actually to be selected. The question of sample size can be answered in many different ways, one of which is to use the formula on page 258 to obtain the standard deviation of the difference between two means.

Substituting known values of σ_1 and σ_2 and specifying what differences between the true means of any 2 of the laboratories would be detected with a probability of at least 0.95 (or 0.98, or 0.99), it is possible to determine $n_1 = n_2 = n$ (see Exercise 12.13). Suppose that this method and, perhaps, also considerations of cost and availability of the necessary specimens lead to the decision to send a sample of 12 disks to each laboratory.

The problem of selecting the required 48 disks and allocating 12 to each laboratory is not as straightforward as it may seem at first. To begin with, suppose that a sheet of tin plate, sufficiently long and sufficiently wide, is selected and that the 48 disks are cut as shown in Figure 12.1. The 12 disks cut from strip 1 are sent to the first laboratory, the 12 disks from strip 2 are sent to the second laboratory, and so forth.

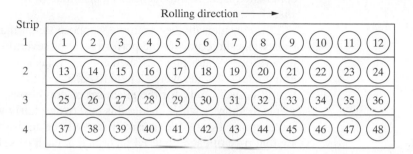

Figure 12.1
Numbering of tin-plate samples

If the 4 mean coating weights are subsequently found to differ significantly, would this allow us to conclude that these differences can be attributed to lack of consistency in the measuring techniques? Suppose, for instance, that additional investigation shows that the amount of tin deposited electrolytically on a long sheet of steel has a distinct and repeated pattern of variation perpendicular to the direction in which it is rolled. (Such a pattern might be caused by the placement of electrodes, edge effects, and so forth.) Thus, even if all 4 laboratories measured the amount of tin consistently and without error, there could be cause for differences in the tin-coating weight determinations. The allocation of an entire strip of disks to each laboratory is such that the inconsistencies among the laboratories' measuring techniques are inseparable from (or **confounded** with) whatever differences may exist in the actual amount of tin deposited perpendicular to the direction in which the sheet of steel is rolled.

One way to avoid this kind of confounding is to number the disks and allocate them to the four laboratories at random. With the aid of random numbers, we obtain

Laboratory A: 3   38   17   32   24   30   48   19   11   31   22   41
Laboratory B: 44   20   15   25   45   4   14   5   39   7   40   34
Laboratory C: 12   21   42   8   27   16   47   46   18   43   35   26
Laboratory D: 9   2   28   23   37   1   10   6   29   36   33   13

If there is any actual pattern of tin-coating thickness on the sheet of tin plate, it will be broken up by the **randomization**.

Although we have identified and counteracted one possible systematic pattern of variation, there is no assurance that there will be no others. For instance, there may be systematic differences in the areas of the disks caused by progressive wear of the cutting instrument, or there may be scratches or other imperfections on one part of the sheet that could affect the measurements. There is always the possibility that differences in means attributed to inconsistencies among the laboratories are actually caused by some other uncontrolled variable. **It is the purpose of randomization to avoid confounding the variable under investigation with other uncontrolled variables**.

By distributing the 48 disks among the 4 laboratories entirely at random, we have no choice but to include whatever variation may be attributable to extraneous causes under the heading of *chance variation*. This may give us an excessively large estimate of chance variation, which, in turn, may make it difficult to detect differences between the true laboratory means. In order to avoid this, we could, perhaps, use only disks cut from the same strip (or from an otherwise homogeneous region). Unfortunately, this kind of **controlled experimentation** presents us with new complications. Of what use would it be, for example, to perform an experiment that allows us to conclude that the laboratories are consistent (or inconsistent), *if such a conclusion is limited to measurements made at a fixed distance from one edge of a sheet?*

To consider another example, suppose that a manufacturer of plumbing materials wishes to compare the performance of several kinds of material to be used in underground water pipes. If such conditions as soil acidity, depth of pipe, and mineral content of water were all held fixed, any conclusions as to which material is best would be valid only for the given set of conditions. What the manufacturer really wants to know is which material is best over a fairly wide variety of conditions, and in designing a suitable experiment it would be advisable (indeed, necessary) to specify that pipe of each material be buried at each of several depths in each of several kinds of soil, and in locations where the water varies in hardness.

It is seldom desirable to hold all or most extraneous factors fixed throughout an experiment in order to obtain an estimate of chance variation that is not inflated by variations due to other causes. (In fact, it is rarely, if ever, possible to exercise such strict control; that is, to hold *all* extraneous variables fixed.) In actual practice, experiments should be planned so that known sources of variability are deliberately varied over as wide a range as necessary. Furthermore, they should be varied in such a way that their variability can be eliminated from the estimate of chance variation. One way to accomplish this is to repeat the experiment in several **blocks**, where known sources of variability (that is, extraneous variables) are held fixed in each block, but vary from block to block.

Recall the reasoning behind the matched pairs design. In the tin-plating problem we might account for variations across the sheet of steel by randomly allocating 3 disks from each strip to each of the laboratories as in the arrangement:

	Strip 1	*Strip 2*	*Strip 3*	*Strip 4*
Laboratory A	8, 4, 10	23, 24, 19	26, 29, 35	37, 44, 48
Laboratory B	2, 6, 12	21, 15, 22	34, 33, 32	45, 43, 46
Laboratory C	1, 5, 11	16, 20, 13	36, 27, 30	41, 38, 47
Laboratory D	7, 3, 9	17, 18, 14	28, 31, 25	39, 40, 42

In this experimental layout, the strips form the blocks, and if we base our estimate of chance variation on the variability *within* each of the 16 sets of 3 disks, this estimate will not be inflated by the extraneous variable; that is, differences among the strips. (Note also that, with this arrangement, differences among the means obtained from the 4 laboratories cannot be attributed to differences among the strips. The arrangement on page 377 does not have this property, since, for instance, 5 disks from strip 1 are allocated to Laboratory D.)

The analysis of experiments in which **blocking** is used to eliminate one source of variability is discussed in Section 12.3.

12.2   Completely Randomized Designs

In this section we consider the statistical analysis of the **completely randomized design**, or **one-way classification**. We shall suppose that the experimenter has available the results of k independent random samples, from k different populations (that is, data concerning k treatments, k groups, k methods of production, etc.). A primary goal is to test the hypothesis that the means of these k populations are all equal.

In general, we denote the jth observation in the ith sample by y_{ij}, and the scheme for a one-way classification is as follows:

	Observations	*Means*	*Sum of Squares*
Sample 1 :	$y_{11}, y_{12}, \ldots, y_{1j}, \ldots, y_{1 n_1}$	$\bar{y}_1$	$\sum_{j=1}^{n_1} (y_{1j} - \bar{y}_1)^2$
Sample 2 :	$y_{21}, y_{22}, \ldots, y_{2j}, \ldots, y_{2 n_2}$	$\bar{y}_2$	$\sum_{j=1}^{n_2} (y_{2j} - \bar{y}_2)^2$
$\vdots$	$\vdots$	$\vdots$	$\vdots$
Sample i :	$y_{i1}, y_{i2}, \ldots, y_{ij}, \ldots, y_{i n_i}$	$\bar{y}_i$	$\sum_{j=1}^{n_i} (y_{ij} - \bar{y}_i)^2$
$\vdots$	$\vdots$	$\vdots$	$\vdots$
Sample k :	$y_{k1}, y_{k2}, \ldots, y_{kj}, \ldots, y_{k n_k}$	$\bar{y}_k$	$\sum_{j=1}^{n_k} (y_{kj} - \bar{y}_k)^2$

To simplify the calculations below, we use the notation $T.$ for the sum of all the observations and N for the total sample size.

$$T. = \sum_{i=1}^{k} \sum_{j=1}^{n_i} y_{ij} \qquad N = \sum_{i=1}^{k} n_i$$

The overall sample mean $\bar{y}$ is

$$\bar{y} = \frac{\sum_{i=1}^{k} \sum_{j=1}^{n_i} y_{ij}}{\sum_{i=1}^{k} n_i} = \frac{\sum_{i=1}^{k} n_i \bar{y}_i}{\sum_{i=1}^{k} n_i} = \frac{T.}{N}$$

An example of such an experiment with $k = 4$ treatments and equal sample sizes $n_i = 12$ is given on page 377, where y_{ij} is the jth tin-coating weight measured by the ith laboratory, $\bar{y}_i$ is the mean of the measurements obtained by the ith laboratory, and $\bar{y}$ is the overall mean (or **grand mean**) of all $n = 48$ observations.

The statistical analysis leading to a comparison of the k different population means consists essentially of splitting the sum of squares about the overall grand mean $\bar{y}$ into a component due to treatment differences and a component due to error or variation within a sample. It is instructive to see how this analysis emanates from a decomposition of the individual observations.

Suppose 3 drying formulas for curing a glue are studied and the following times observed.

Formula A:	13	10	8	11	8
Formula B:	13	11	14	14	
Formula C:	4	1	3	4	2	4

There are $N = 5 + 4 + 6 = 15$ observations in all, and these total $T. = 120$.

The grand mean is

$$\bar{y} = T_{..}/N = \sum_{i=1}^{k} \sum_{j=1}^{n_i} y_{ij}/N = 120/15 = 8$$

Each observation y_{ij} will be decomposed as

$$\underset{\text{observation}}{y_{ij}} \quad = \quad \underset{\substack{\text{grand} \\ \text{mean}}}{\bar{y}} \quad + \quad \underset{\substack{\text{deviation due} \\ \text{to treatment}}}{(\bar{y}_i - \bar{y})} \quad + \underset{\text{error}}{(y_{ij} - \bar{y}_i)}$$

For instance, $13 = 8 + (10 - 8) + (13 - 10) = 8 + 2 + 3$. Repeating the decomposition for each observation, we obtain the arrays

$$
\begin{matrix}
\text{observation} \\
y_{ij}
\end{matrix}
\qquad\qquad
\begin{matrix}
\text{grand mean} \\
\bar{y}
\end{matrix}
$$

$$
\begin{bmatrix}
13 & 10 & 8 & 11 & 8 \\
13 & 11 & 14 & 14 & \\
4 & 1 & 3 & 4 & 2 & 4
\end{bmatrix}
=
\begin{bmatrix}
8 & 8 & 8 & 8 & 8 \\
8 & 8 & 8 & 8 & \\
8 & 8 & 8 & 8 & 8 & 8
\end{bmatrix}
$$

$$
\begin{matrix}
\text{treatment effects} \\
\bar{y}_i - \bar{y}
\end{matrix}
\qquad\qquad
\begin{matrix}
\text{error} \\
y_{ij} - \bar{y}_i
\end{matrix}
$$

$$
+
\begin{bmatrix}
2 & 2 & 2 & 2 & 2 \\
5 & 5 & 5 & 5 & \\
-5 & -5 & -5 & -5 & -5 & -5
\end{bmatrix}
+
\begin{bmatrix}
3 & 0 & -2 & 1 & -2 \\
0 & -2 & 1 & 1 & \\
1 & -2 & 0 & 1 & -1 & 1
\end{bmatrix}
$$

Taking the sum of squares as a measure of variation for the whole array,

$$
\text{treatment sum of squares} = \sum_{i=1}^{k} n_i (\bar{y}_i - \bar{y})^2
$$

$$
= 5(2)^2 + 4(5)^2 + 6(-5)^2 = 270
$$

$$
\text{error sum of squares} = \sum_{i=1}^{k} \sum_{j=1}^{n_i} (y_{ij} - \bar{y}_i)^2
$$

$$
= 3^2 + 0^2 + (-2)^2 + \cdots + (-1)^2 + 1^2 = 32
$$

Their sum, $302 = 270 + 32$, the total sum of squares, also equals the sum of squared entries in the observation array minus the sum of squares of the entries in the grand mean array. That is, the array for total sum of squares has entries $y_{ij} - \bar{y}$ whose sum of squares is 302.

The decomposition also provides us with an interpretation of the degrees of freedom associated with each sum of squares. In this example, the treatment effects array has only 3 possibly distinct entries: $\bar{y}_1 - \bar{y}$, $\bar{y}_2 - \bar{y}$, and $\bar{y}_3 - \bar{y}$. Further, the sum of entries in the treatment effects array,

$$
\sum_{i=1}^{k} n_i (\bar{y}_i - \bar{y})
$$

is always zero. So, for instance, the third value is determined by the first two. Consequently, there are $3 - 1 = 2$ degrees of freedom associated with treatments. In the general case, there are $k - 1$ degrees of freedom.

Among the entries of the error array, each row sums to zero, so 1 degree of freedom is lost for each row. The array has $n_1 + n_2 + n_3 - 3 = 5 + 4 + 6 - 3 = 12$ degrees of freedom. In the general case, there are $n_1 + n_2 + \cdots + n_k - k$ degrees of freedom.

The grand mean array has a single value $\bar{y}$ and hence has 1 degree of freedom, whereas the observation array has $n_1 + n_2 + \cdots + n_k$ possible distinct entries and hence that number of degrees of freedom. The total sum of squares, based on the difference of these last two arrays, has

$$n_1 + n_2 + \cdots + n_k - 1 = 5 + 4 + 6 - 1 = 14$$

degrees of freedom.

To summarize our calculations for the curing times, we enter the degrees of freedom and sums of squares in a table called an *analysis of variance* table.

Analysis of Variance Table for Cure Times		
Source of variation	Degrees of freedom	Sum of squares
Treatment	2	270
Error	12	32
Total	14	302

For further reference, we also summarize the decomposition of the degrees of freedom associated with total, treatment, and error sum of squares, for the general one-way analysis of variance.

Decomposition of the degrees of freedom

$$\text{d.f. Total} = \text{d.f. Treatment} + \text{d.f. Error}$$
$$\sum_{i=1}^{k} n_i - 1 = k - 1 \qquad + \sum_{i=1}^{k} n_i - k$$

With reference to the **total sum of squares**

$$\sum_{i=1}^{k} \sum_{j=1}^{n_i} (y_{ij} - \bar{y})^2$$

we shall now prove the following theorem.

Theorem 12.1

Identity for one-way analysis of variance

$$\sum_{i=1}^{k} \sum_{j=1}^{n_i} (y_{ij} - \bar{y})^2 = \sum_{i=1}^{k} \sum_{j=1}^{n_i} (y_{ij} - \bar{y}_i)^2 + \sum_{i=1}^{k} n_i (\bar{y}_i - \bar{y})^2$$

The proof of this theorem is based on the identity

$$y_{ij} - \bar{y} = (y_{ij} - \bar{y}_i) + (\bar{y}_i - \bar{y})$$

Squaring both sides and summing on i and j, we obtain

$$\sum_{i=1}^{k} \sum_{j=1}^{n_i} (y_{ij} - \bar{y})^2 = \sum_{i=1}^{k} \sum_{j=1}^{n_i} (\bar{y}_{ij} - \bar{y}_i)^2 + \sum_{i=1}^{k} \sum_{j=1}^{n_i} (\bar{y}_i - \bar{y})^2$$
$$+ 2 \sum_{i=1}^{k} \sum_{j=1}^{n_i} (y_{ij} - \bar{y}_i)(\bar{y}_i - \bar{y})$$

Next, we observe that

$$\sum_{i=1}^{k} \sum_{j=1}^{n_i} (y_{ij} - \bar{y}_i)(\bar{y}_i - \bar{y}) = \sum_{i=1}^{k} (\bar{y}_i - \bar{y}) \sum_{j=1}^{n_i} (y_{ij} - \bar{y}_i) = 0$$

since $\bar{y}_i$ is the mean of the ith sample and, hence,

$$\sum_{j=1}^{n_i} (y_{ij} - \bar{y}_i) = 0 \text{ for all } i$$

To complete the proof of Theorem 12.1, we have only to observe that the summand of the second sum of the right-hand side of the above identity does not involve the subscript j and that, consequently,

$$\sum_{i=1}^{k} \sum_{j=1}^{n_i} (\bar{y}_i - \bar{y})^2 = \sum_{i=1}^{k} n_i (\bar{y}_i - \bar{y})^2$$

It is customary to denote the total sum of squares, the left-hand member of the identity of Theorem 12.1, by *SST*. We refer to the first term on the right-hand side as the **error sum of squares**, *SSE*. The term *error sum of squares* expresses the idea that the quantity estimates random (or chance) error. The second term on the right-hand side of the identity of Theorem 12.1 we refer to as the **between-samples sum of squares** or the **treatment sum of squares**, *SS(Tr)*. (Most of the early applications of this kind of analysis were in the field of agriculture, where the k populations represented different **treatments**, such as fertilizers, applied to agricultural plots.)

To be able to test the hypothesis that the samples were obtained from k populations with equal means, we shall make several assumptions. Specifically, it will be assumed that we are dealing with *normal populations* having *equal variances*. However, the methods we develop in this chapter are fairly **robust**; that is, they are relatively insensitive to violations of the assumption of normality as well as the assumption of equal variances.

If μ_i denotes the mean of the ith population and σ^2 denotes the common variance of the k populations, we can express each observation Y_{ij} as μ_i, plus the value of a random component; that is, we can write[1]

$$Y_{ij} = \mu_i + \varepsilon_{ij} \quad \text{for} \quad i = 1, 2, \ldots, k \qquad j = 1, 2, \ldots, n_i$$

In accordance with the preceding assumptions, the ε_{ij} are independent, normally distributed random variables with zero means and the common variance σ^2.

---

[1] Note that this equation, or model, can be regarded as a multiple regression equation; by introducing the variables x_{il} that equal 0 or 1, depending on whether the two subscripts are unequal or equal, we can write

$$Y_{ij} = \mu_1 x_{i1} + \mu_2 x_{i2} + \cdots + \mu_k x_{ik} + \varepsilon_{ij}$$

The parameters μ_i can be interpreted as regression coefficients, and they can be estimated by the least squares methods of Chapter 11.

To attain uniformity with corresponding equations for more complicated kinds of designs, it is customary to replace μ_i by $\mu + \alpha_i$, where μ is the mean

$$\sum_{i=1}^{k} n_i \mu_i / N$$

of the μ_i in the experiment and α_i is the **effect** of the ith treatment; hence,

$$\sum_{i=1}^{k} n_i \alpha_i = 0$$

(See Exercise 12.14.)

Using these new parameters, we can write the **model equation** for the one-way classification as

Model equation for one-way classification

$$Y_{ij} = \mu + \alpha_i + \varepsilon_{ij} \quad \text{for} \quad i = 1, 2, \ldots, k; \qquad j = 1, 2, \ldots, n_i$$

and the null hypothesis that the k population means are all equal is replaced by the null hypothesis that $\alpha_1 = \alpha_2 = \cdots = \alpha_k = 0$. The alternative hypothesis that at least two of the population means are unequal is equivalent to the alternative hypothesis that $\alpha_i \neq 0$ for some i.

To test the null hypothesis that the k population means are all equal, we shall compare two estimates of σ^2—one based on the variation, or differences between, the sample means, and one based on the variation within the samples.

Each sum of squares is first converted to a **mean square** so a test for the equality of treatment means can be performed.

Mean square

$$\text{mean square} = \frac{\text{sum of squares}}{\text{degrees of freedom}}$$

When the population means are equal, both the

$$\text{treatment mean square: } MS(Tr) = \frac{\sum_{i=1}^{k} n_i (\bar{y}_i - \bar{y})^2}{k - 1}$$

and the

$$\text{error mean square: } MSE = \frac{\sum_{i=1}^{k} \sum_{j=1}^{n_i} (y_{ij} - \bar{y}_i)^2}{N - k}$$

are estimates of σ^2. However, when the null hypothesis is *false*, the **treatment** or **between-sample mean square** can be expected to exceed the **error** or **within-sample mean square**. If the null hypothesis is true, it can be shown that the two mean squares are independent and that their ratio

F ratio for treatments

$$F = \frac{\displaystyle\sum_{i=1}^{k} n_i (\overline{Y}_i - \overline{Y})^2 / (k-1)}{\displaystyle\sum_{i=1}^{k} \sum_{j=1}^{n_i} (Y_{ij} - \overline{Y}_i)^2 / (N-k)} = \frac{SS(Tr)/(k-1)}{SSE/(N-k)}$$

has an F distribution with $k-1$ and $N-k$ degrees of freedom.

A large value for F indicates large differences between the sample means. Therefore, the null hypothesis will be rejected, at level α, if the value of F exceeds F_α where F_α is obtained from Table 6W with $k-1$ and $N-k$ degrees of freedom.

To test for the equality of the mean curing times, we complete the **Analysis of Variance (ANOVA) table** by including the mean square errors and value of F.

Analysis of Variance Table for Cure Times				
Source of variation	Degrees of freedom	Sum of squares	Mean square	F
Treatment	2	270	135	50.6
Error	12	32	2.667	
Total	14	302		

The value of $F_{0.05}$ with 2 and 12 degrees of freedom is 3.89 so we reject the null hypothesis of equal means.

In general, the results obtained in analyzing the total sum of squares into its components are conveniently summarized by means of the **analysis of variance table**:

Source of variation	Degrees of freedom	Sum of squares	Mean square	F
Treatments	$k-1$	$SS(Tr)$	$MS(Tr) = SS(Tr)/(k-1)$	$\dfrac{MS(Tr)}{MSE}$
Error	$N-k$	SSE	$MSE = SSE/(N-k)$	
Total	$N-1$	SST		

where

$$N = \sum_{i=1}^{k} n_i$$

Note that each **mean square** is obtained by dividing the corresponding sum of squares by its degrees of freedom.

The calculations can become quite cumbersome and we recommend the use of a statistical software program. (See Exercises 12.18 for using *MINITAB*.)

EXAMPLE 1   **Conducting a one-way analysis of variance**

To illustrate the **analysis of variance** (as this technique is appropriately called) for a one-way classification, suppose that in accordance with the layout on page 377 each laboratory measures the tin-coating weights of 12 disks and that the results are as follows:

Laboratory A	Laboratory B	Laboratory C	Laboratory D
0.25	0.18	0.19	0.23
0.27	0.28	0.25	0.30
0.22	0.21	0.27	0.28
0.30	0.23	0.24	0.28
0.27	0.25	0.18	0.24
0.28	0.20	0.26	0.34
0.32	0.27	0.28	0.20
0.24	0.19	0.24	0.18
0.31	0.24	0.25	0.24
0.26	0.22	0.20	0.28
0.22	0.29	0.21	0.22
0.28	0.16	0.19	0.21

Construct an analysis of variance table, and test the equality of mean weights with $\alpha = 0.05$.

Solution   1. *Null hypothesis:* $\mu_1 = \mu_2 = \mu_3 = \mu_4$
   Alternative hypothesis: The μ_i's are not all equal

2. *Level of significance:* $\alpha = 0.05$

3. *Criterion:* Reject the null hypothesis if $F > 2.82$, the value of $F_{0.05}$ with 3 and 44 degrees of freedom.

4. *Calculations:* Statistical software produces an ANOVA table that includes the P-value.

Source of variation	Degrees of freedom	Sum of squares	Mean square	F	P-value
Laboratory	3	0.01349	0.00450	2.96	0.042
Error	44	0.06683	0.00152		
Total	47	0.08033			

5. *Decision:* Since the observed value of F exceeds $2.82 = F_{0.05}$, the null hypothesis of equal mean weights is rejected at the 0.05 level of significance. We conclude that the laboratories are *not* obtaining consistent results.

   The P-value 0.042, shown in Figure 12.2, only provides minimal additional support for the conclusion.

   [Using **R** : Read the file C12Ex1.TXT into **Dat**. Then, use **anova(lm(Weight~Laboratory,data=Dat))**]   ∎

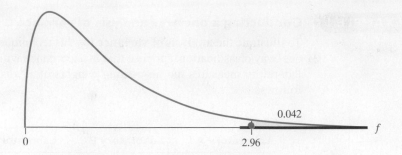

Figure 12.2
P-value and rejection region
for Example 1

To estimate the parameters $\mu, \alpha_1, \alpha_2, \alpha_3, \ldots, \alpha_k$ (or $\mu_1, \mu_2, \mu_3, \ldots, \mu_k$), we can use the method of least squares, minimizing

$$\sum_{i=1}^{k} \sum_{j=1}^{n_i} (y_{ij} - \mu - \alpha_i)^2$$

with respect to μ and α_i subject to the restriction that

$$\sum_{i=1}^{k} n_i \alpha_i = 0$$

This may be done by eliminating one of the α's or, better, by using the method of Lagrange multipliers, which is treated in most texts on advanced calculus. In either case we obtain the intuitively obvious estimate $\widehat{\mu} = \bar{y}$ and $\widehat{\alpha}_i = \bar{y}_i - \bar{y}$ for $i = 1, 2, \ldots, k$, and the corresponding estimates for the μ_i are given by $\widehat{\mu}_i = \bar{y}_i$.

EXAMPLE 2

Estimating the treatment effects

Estimate the parameters of the one-way classification model for the tin-coating weights given in the preceding example.

Solution   For the data from the 4 laboratories we get

$$\widehat{\mu} = \frac{11.70}{48} = 0.244, \quad \widehat{\alpha}_1 = \frac{3.22}{12} - 0.244 = 0.024$$

$$\widehat{\alpha}_2 = \frac{2.72}{12} - 0.244 = -0.017, \quad \widehat{\alpha}_3 = \frac{2.76}{12} - 0.244 = -0.014$$

and

$$\widehat{\alpha}_4 = \frac{3.00}{12} - 0.244 = 0.006 \qquad \blacksquare$$

When the null hypothesis of equal treatment effects is rejected, the magnitudes of the differences should be estimated using confidence intervals. The difference of sample means for treatment i and treatment l, $\bar{y}_i - \bar{y}_l$ estimates the difference in mean response $\mu + \alpha_i - \mu - \alpha_l$. Using the mean square error, s^2, from the ANOVA table, the $(1 - \alpha)100\%$ confidence interval for the true difference in mean response is

$$\bar{y}_i - \bar{y}_l \pm t_{\alpha/2} \sqrt{s^2 \left(\frac{1}{n_i} + \frac{1}{n_l} \right)}$$

When several means need to be compared, α can be modified according to Bonferroni's procedure, discussed in Section 12.4.

EXAMPLE 3   **Confidence intervals for different resins**

The internal bonding strengths of 3 different resins, ED, MD, and PF, need to be compared. Five specimens were prepared with each of the resins.

Resin	Strength					Mean
ED	0.99	1.19	0.79	0.95	0.90	0.964
MD	1.11	1.53	1.37	1.24	1.42	1.334
PF	0.83	0.68	0.94	0.86	0.57	0.776

The analysis of variance table shows that there is a statistically significant difference at level $\alpha = 0.05$. (See Exercise 12.44.)

Source of variation	Degrees of freedom	Sum of squares	Mean square	F	P value
Resin	2	0.8060	0.4030	17.2	0.000
Error	12	0.2810	0.0234		
Total	14	1.0870			

Determine the individual 95% confidence intervals for the 3 differences of means.

Solution   The confidence intervals use the MSE = 0.0234 as the estimate s^2 and the degrees of freedom = 12, so $t_{0.025} = 2.179$. The three confidence intervals become

$$MD - ED: \ 1.334 - 0.964 \pm 2.179 \sqrt{0.0234 \left(\frac{1}{5} + \frac{1}{5}\right)} \text{ or } (0.159, 0.581)$$

$$MD - PF: \ 1.334 - 0.776 \pm 2.179 \sqrt{0.0234 \left(\frac{1}{5} + \frac{1}{5}\right)} \text{ or } (0.347, 0.769)$$

$$ED - PF: \ 0.964 - 0.776 \pm 2.179 \sqrt{0.0234 \left(\frac{1}{5} + \frac{1}{5}\right)} \text{ or } (-0.023, 0.399)$$

The resin MD has a higher internal bound strength than the other two, which cannot be distinguished.   ∎

EXAMPLE 4   **Confidence intervals quantify the amount of difference**

The old way of testing the strength of paper with a special tearing machine is by testing a single sheet (ply). It has been suggested that measuring 5 sheets together (5 plys) and then adjusting to the single-thickness strength would be a better procedure. The first question is whether or not the two procedures give essentially the same value for strength. There is a strong element of experimental design involved here. Five pieces of paper are cut in half. One-half of each piece is randomly selected and its strength obtained. Next, the 5 remaining halves are tested together as the

5-ply specimen. The scientist then calculates the average of the 5 individual readings minus the adjusted 5-ply reading. This procedure is repeated 3 times for each of 4 different but representative types of paper. The observations of the differences are given in the following table. (Courtesy of Steve Verrill)

Paper type	Rep. 1	Rep. 2	Rep. 3	Mean
No. 1	2.80	0.75	3.70	2.417
No. 2	0.00	−0.10	3.45	1.117
No. 3	1.15	1.75	4.20	2.367
No. 4	1.88	2.65	2.70	2.410

Test, at level 0.05, the null hypothesis that there is no difference between the two methods of testing paper strength. Also summarize the results using confidence intervals.

Solution   Formally, we test whether the mean difference changes with paper type. Omitting the details, a computer calculation gives the ANOVA table:

Source of variation	Degrees of freedom	Sum of squares	Mean square	F
Type	3	3.70	1.23	0.54
Error	8	18.39	2.30	
Total	11	22.08		

Because $F = 0.54$ is less than 1, we cannot reject $H_0: \alpha_1 = \alpha_2 = \alpha_3 = \alpha_4 = 0$, even at the 50% level. Because of this evidence, it is reasonable to treat all 12 of the differences as coming from the same population. We calculate $\bar{y} = 2.078$ and the standard deviation $= 1.417$. Since, with the degrees of freedom $12 - 1 = 11$, $t_{0.025} = 2.201$ and the 95% confidence interval for the mean of the differences is

$$2.078 \pm 2.201 \frac{1.417}{\sqrt{12}} \quad \text{or} \quad (1.18, 2.98)$$

The tearing strength based on the 5-ply reading is about 1 to 3 units lower than that of the individual readings. It is up to the scientist to decide if this discrepancy is of importance in the engineering application.   ∎

Alternative Calculation of Sums of Squares

We conclude our discussion of the completely randomized design by presenting alternative formulas that help simplify the calculations of the sums of squares when statistical software is unavailable or when students are restricted to simple handheld calculators. The reader is asked to very the formulas in Exercise 12.15.

First calculate SST and $SS(Tr)$ using the formulas

Alternative formulas —
Sums of squares

$$SST = \sum_{i=1}^{k} \sum_{j=1}^{n_i} y_{ij}^2 - C$$

$$SS(Tr) = \sum_{i=1}^{k} \frac{T_i^2}{n_i} - C$$

where C, called the **correction term for the mean**, is given by

$$C = \frac{T_{\boldsymbol{\cdot}}^2}{N}$$

with

$$N = \sum_{i=1}^{k} n_i, \quad T_{\boldsymbol{\cdot}} = \sum_{i=1}^{k} T_i, \quad \text{and} \quad T_i = \sum_{j=1}^{n_i} y_{ij}$$

That is, in these formulas, T_i is the total of the n_i observations in the ith sample, whereas $T_{\boldsymbol{\cdot}}$ is the grand total of all N observations. The error sum of squares, SSE, is then obtained by subtraction; according to Theorem 12.1, we can write

Error sum of squares

$$SSE = SST - SS(Tr)$$

EXAMPLE 5   **Referring to the weights of disks in Example 1, use the alternative formulas to verify the analysis of variance table.**

Solution   The totals for the $k = 4$ samples all of sample size $n_i = 12$, are 3.22, 2.72, 2.76, and 3.00, respectively. The grand total is $T_{\boldsymbol{\cdot}} = 11.70$. The alternative calculations become

$$N = 12 + 12 + 12 + 12 = 48$$

$$C = \frac{(11.70)^2}{48} = 2.8519$$

$$SST = (0.25)^2 + (0.27)^2 + \cdots + (0.21)^2 - 2.8519 = 0.0803$$

$$SS(Tr) = \frac{(3.22)^2}{12} + \frac{(2.72)^2}{12} + \frac{(2.76)^2}{12} + \frac{(3.00)^2}{12} - 2.8519 = 0.0135$$

$$SSE = 0.0809 - 0.0130 = 0.0679$$

The resulting analysis of variance table

Source of variation	Degrees of freedom	Sum of squares	Mean square	F
Laboratory	3	0.0135	0.0045	2.96
Error	44	0.0668	0.0015	
Total	47	0.0803		

verifies, up to the number of decimal places retained here, the table in Example 1. ∎

Exercises

12.1 An experiment is performed to compare the cleansing action of two detergents, Detergent A and Detergent B. Twenty swatches of cloth are soiled with dirt and grease, each is washed with one of the detergents in an agitator-type machine and then measured for whiteness. Criticize the following aspects of the experiment:

(a) The entire experiment is performed with soft water.

(b) Fifteen of the swatches are washed with Detergent A and five with Detergent B.

(c) To accelerate the testing procedure, very hot water and 30-second washing times are used in the experiment.

(d) The whiteness readings of all swatches with Detergent A are taken first.

12.2 A certain *bon vivant*, wishing to ascertain the cause of his frequent hangovers, conducted the following experiment. On the first night, he drank nothing but whiskey and water; on the second night, he drank vodka and water; on the third night, he drank gin and water; and on the fourth night, he drank rum and water. On each of the following mornings he had a hangover, and he concluded that it was the common factor, the water, that made him ill.

(a) This conclusion is obviously unwarranted, but can you state what principles of sound experimental design are violated?

(b) Give a less obvious example of an experiment having the same shortcoming.

(c) Suppose that our friend has modified his experiment so that each of the 4 alcoholic beverages was used both with and without water, so that the experiment lasted 8 nights. Could the results of this enlarged experiment serve to support or refute the hypothesis that water was the cause of the hangovers? Explain.

12.3 Using field theory, three settings are suggested for the magnetic field that helps particles bombard a target disk and guides released atoms to coat a computer hard disk. In an experiment conducted to compare the manufacturing yields under the three settings, experimenters record the number of hard disks coated before the target is worn out. Suppose the data are

Setting 1	Setting 2	Setting 3
573	591	572
577	581	577
567	587	576
571	585	579

Without using the alternative formulas, calculate

$$\sum_{i=1}^{k}\sum_{j=1}^{n_i}(y_{ij}-\bar{y})^2 \quad \sum_{i=1}^{k}\sum_{j=1}^{n_i}(y_{ij}-\bar{y}_i)^2$$

$$\text{and} \quad \sum_{i=1}^{k} n_i(\bar{y}_i-\bar{y})^2$$

and verify the identity of Theorem 12.1.

12.4 Using the sums of squares obtained in Exercise 12.3, test at the level of significance $\alpha = 0.05$ whether the differences among the means obtained for the 3 samples are significant.

12.5 The following are the numbers of mistakes made in 5 successive days for 4 technicians working for a photographic laboratory:

Technician I	Technician II	Technician III	Technician IV
5	17	9	9
12	12	11	13
9	15	6	7
8	14	14	10
11	17	10	11

Test at the level of significance $\alpha = 0.01$ whether the differences among the 4 sample means can be attributed to chance.

12.6 With reference to the example on page 379, suppose one additional observation $y_{25} = 8$ is available using formula B. Construct the analysis of variance table and test the equality of the mean curing times using $\alpha = 0.05$.

12.7 Given the following observations collected according to the one-way analysis of variance design,

Treatment 1:	6	4	5	
Treatment 2:	13	10	13	12
Treatment 3:	7	9	11	
Treatment 4:	3	6	1	4	1

(a) decompose each observation y_{ij} as

$$y_{ij} = \bar{y} + (\bar{y}_i - \bar{y}) + (y_{ij} - \bar{y}_i)$$

and obtain the sum of squares and degrees of freedom for each component;

(b) construct the analysis of variance table and test the equality of treatments using $\alpha = 0.05$.

12.8 The one-way analysis of variance is conveniently implemented using *MINITAB*. With reference to the example on page 379, we first set the observations in columns:

```
DATA:
  C1:   13    10     8    11     8
  C2:   13    11    14    14
  C3:    4     1     3     4     2     4
```

Dialog Box:
Stat > ANOVA > One-way. Pull down **Response data are in separate ···**
Enter $C1 - C3$ in **Responses**. Click **OK**.

One-Way Analysis of Variance

```
ANALYSIS  OF  VARIANCE
SOURCE    DF        SS       MS   F-Value  P-Value
FACTOR     2    270.00   135.00     50.62    0.000
ERROR     12     32.00    2.667
TOTAL     14    302.00

       S    R-sq
 1.63299  89.40%

Factor  N       MEAN   STDEV          95% CI
C1      5     10.000   2.121   ( 8.409, 11.591)
C2      4     13.000   1.414   (11.221, 14.779)
C3      6      3.000   1.265   ( 1.547,  4.453 )

POOLED STDEV = 1.63299
```

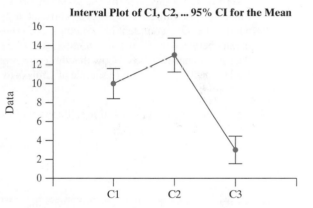

Use the computer to perform the analysis of variance suggested in Exercise 12.5.

12.9 To find the best arrangement of instruments on a control panel of an airplane, 3 different arrangements were tested by simulating an emergency condition and observing the reaction time required to correct the condition. The reaction times (in tenths of a second) of 28 pilots (randomly assigned to the different arrangements) were as follows:

Arrangement 1: 14 13 9 15 11 13 14 11
Arrangement 2: 10 12 9   7 11   8 12   9 10 13 9 10
Arrangement 3: 11   5 9 10   6   8   8   7

Test at the level of significance $\alpha = 0.01$ whether we can reject the null hypothesis that the differences among the arrangements have no effect.

12.10 Several different aluminum alloys are under consideration for use in heavy-duty circuit-wiring applications. Among the desired properties is low electrical resistance, and specimens of each wire are tested by applying a fixed voltage to a given length of wire and measuring the current passing through the wire. Given the following results, would you conclude that these alloys differ in resistance? (Use the 0.01 level of significance.)

Alloy	Current (amperes)			
1	1.085	1.016	1.009	1.034
2	1.051	0.993	1.022	
3	0.985	1.001	0.990	0.988	1.011
4	1.101	1.015		

12.11 Two tests are made of the compressive strength of each of 6 samples of poured concrete. The force required to crumble each of 12 cylindrical specimens, measured in kilograms, is as follows:

	Sample					
	A	B	C	D	E	F
Test 1	110	125	98	95	104	115
Test 2	105	130	107	91	96	121

Test at the 0.05 level of significance whether these samples differ in compressive strength.

12.12 Corrosion rates (percent) were measured for 4 different metals that were immersed in a highly corrosive solution:

Aluminum:	75	77	76	79	74	77	75	
Stainless Steel:	74	76	75	78	74	77	75	77
Alloy I:	73	74	72	74	70	73	74	71
Alloy II:	71	74	74	73	74	73	71	

(a) Perform an the analysis of variance and test for differences due to metals using $\alpha = 0.05$.

(b) Give the estimates of corrosion rates for each metal.

(c) Find 95% confidence intervals for the differences of mean corrosion rates.

12.13 Referring to the discussion on page 376, assume that the standard deviations of the tin-coating weights determined by any one of the 4 laboratories have the common value $\sigma = 0.012$, and that it is desired to be 95% confident of detecting a difference in means between any 2 of the laboratories in excess of 0.01 pound per base box. Show that these assumptions lead to the decision to send a sample of 12 disks to each laboratory.

12.14 Show that if $\mu_i = \mu + \alpha_i$ and μ is the mean of the μ_i, it follows that

$$\sum_{i=1}^{k} n_i \alpha_i = 0$$

12.15 Verify the alternative formulas for computing SST and $SS(Tr)$ given on page 389.

12.16 With reference to Exercise 12.9, determine individual 99% confidence intervals for the differences of mean reaction times.

12.17 Samples of peanut butter produced by 2 different manufacturers are tested for aflatoxin content, with the following results:

Aflatoxin Content (ppb)	
Brand A	**Brand B**
0.5	4.7
0.0	6.2
3.2	0.0
1.4	10.5
0.0	2.1
1.0	0.8
8.6	2.9

(a) Use analysis of variance to test whether the two brands differ in aflatoxin content.

(b) Test the same hypothesis using a two sample t test.

(c) We have shown on page 202 that the t statistic with ν degrees of freedom and the F statistic with 1 and ν degrees of freedom are related by the formula

$$F(1, \nu) = t^2(\nu)$$

Using this result, prove that the analysis of variance and two sample t test methods are equivalent in this case.

12.3 Randomized-Block Designs

As we observed in Section 12.1, the estimate of chance variation (the experimental error) can often be reduced—that is, freed of variability due to extraneous causes—by dividing the observations in each classification into blocks. This is accomplished when known sources of variability (that is, extraneous variables) are fixed in each block but vary from block to block.

In this section we shall suppose that the experimenter has available measurements pertaining to a treatments distributed over b blocks. First, we shall consider the case where there is exactly one observation from each treatment in each block. With reference to Figure 12.1 on page 377, this case would arise if each laboratory tested one disk from each strip. Letting y_{ij} denote the observation pertaining to the ith treatment and the jth block, $\bar{y}_{i.}$ the mean of the b observations for the ith treatment, $\bar{y}_{.j}$ the mean of the a observations in the jth block, and $\bar{y}_{..}$ the grand mean of all the ab observations, we shall use the following layout for this kind of **two-way classification**:

Blocks

	B_1	B_2	$\cdots$	B_j	$\cdots$	B_b	*Means*
Treatment 1	y_{11}	y_{12}	$\cdots$	y_{1j}	$\cdots$	y_{1b}	$\bar{y}_{1\cdot}$
Treatment 2	y_{21}	y_{22}	$\cdots$	y_{2j}	$\cdots$	y_{2b}	$\bar{y}_{2\cdot}$
				$\vdots$			
Treatment i	y_{i1}	y_{i2}	$\cdots$	y_{ij}	$\cdots$	y_{ib}	$\bar{y}_{i\cdot}$
				$\vdots$			
Treatment a	y_{a1}	y_{a2}	$\cdots$	y_{aj}	$\cdots$	y_{ab}	$\bar{y}_{a\cdot}$
Means	$\bar{y}_{\cdot 1}$	$\bar{y}_{\cdot 2}$	$\cdots$	$\bar{y}_{\cdot j}$	$\cdots$	$\bar{y}_{\cdot b}$	$\bar{y}_{\cdot\cdot}$

This kind of arrangement is also called a **randomized-block design**, provided the treatments are allocated at random *within* each block. Note that when a dot is used in place of a subscript, this means that the mean is obtained by summing over that subscript.

The statistical analysis of a randomized block experiment tests on the decomposition of the total sum of squares into three components.

Theorem 12.2

Identity for analysis of two-way classification

$$\sum_{i=1}^{a}\sum_{j=1}^{b}(y_{ij}-\bar{y}_{\cdot\cdot})^2 = \sum_{i=1}^{a}\sum_{j=1}^{b}(y_{ij}-\bar{y}_{i\cdot}-\bar{y}_{\cdot j}+\bar{y}_{\cdot\cdot})^2$$

$$+ b\sum_{i=1}^{a}(\bar{y}_{i\cdot}-\bar{y}_{\cdot\cdot})^2 + a\sum_{j=1}^{b}(\bar{y}_{\cdot j}-\bar{y}_{\cdot\cdot})^2$$

The left-hand side of this identity represents the total sum of squares, *SST*, and the terms of the right-hand side are, respectively, the error sum of squares, *SSE*; the treatment sum of squares, *SS(Tr)*; and the **block sum of squares**, *SS(Bl)*. To prove this theorem, we would make use of the identity

$$y_{ij}-\bar{y}_{\cdot\cdot} = (y_{ij}-\bar{y}_{i\cdot}-\bar{y}_{\cdot j}+\bar{y}_{\cdot\cdot}) + (\bar{y}_{i\cdot}-\bar{y}_{\cdot\cdot}) + (\bar{y}_{\cdot j}-\bar{y}_{\cdot\cdot})$$

and follow essentially the same argument as in the proof of Theorem 12.1.

Engineers are considering three different air filters for a clean room. Because production activities in the room vary from day to day, they will block according to the nuisance variable day. They run all three filters for two hours each day and measure the amount of particulate matter captured. Randomization is an important aspect of this design. Each day, there must be a random choice of which filter to run first, second, and third. This randomization ensures that any time of day differences in the amount of particulate matter cannot systematically influence the experimental results.

Suppose the results for four days are

Blocks

	1	2	3	4
Treatment 1	13	8	9	6
Treatment 2	7	3	6	4
Treatment 3	13	7	12	8

each observation y_{ij} will be decomposed as

$$y_{ij} = \bar{y}_{..} + (\bar{y}_{i.} - \bar{y}_{..}) + (\bar{y}_{.j} - \bar{y}_{..})$$

| observation | grand mean | deviation due to treatment | deviation due to block |

$$+ (y_{ij} - \bar{y}_{i.} - \bar{y}_{.j} + \bar{y}_{..})$$

error

For instance, $13 = 8 + (9 - 8) + (11 - 8) + (13 - 9 - 11 + 8) = 8 + 1 + 3 + 1$. Repeating this decomposition for each observation, we obtain the arrays

$$
\begin{array}{ccc}
observation & mean & treatment \\
y_{ij} & \bar{y}_{..} & \bar{y}_{i.} - \bar{y}_{..}
\end{array}
$$

$$
\begin{bmatrix} 13 & 8 & 9 & 6 \\ 7 & 3 & 6 & 4 \\ 13 & 7 & 12 & 8 \end{bmatrix} = \begin{bmatrix} 8 & 8 & 8 & 8 \\ 8 & 8 & 8 & 8 \\ 8 & 8 & 8 & 8 \end{bmatrix} + \begin{bmatrix} 1 & 1 & 1 & 1 \\ -3 & -3 & -3 & -3 \\ 2 & 2 & 2 & 2 \end{bmatrix}
$$

$$
\begin{array}{cc}
block & error \\
\bar{y}_{.j} - \bar{y}_{..} & y_{ij} - \bar{y}_{i.} - \bar{y}_{.j} + \bar{y}_{..}
\end{array}
$$

$$
+ \begin{bmatrix} 3 & -2 & 1 & -2 \\ 3 & -2 & 1 & -2 \\ 3 & -2 & 1 & -2 \end{bmatrix} + \begin{bmatrix} 1 & 1 & -1 & -1 \\ -1 & 0 & 0 & 1 \\ 0 & -1 & 1 & 0 \end{bmatrix}
$$

Taking the sum of squares for each array,

$$treatment\ sum\ of\ squares = b \sum_{i=1}^{a} (\bar{y}_{i.} - \bar{y}_{..})^2$$

$$= 4(1)^2 + 4(-3)^2 + 4(2)^2 = 56$$

$$block\ sum\ of\ squares = a \sum_{j=1}^{b} (\bar{y}_{.j} - \bar{y}_{..})^2$$

$$= 3(3)^2 + 3(-2)^2 + 3(1)^2 + 3(-2)^2 = 54$$

$$error\ sum\ of\ squares = \sum_{i=1}^{a} \sum_{j=1}^{b} (y_{ij} - \bar{y}_{i.} - \bar{y}_{.j} + \bar{y}_{..})^2$$

$$= 1^2 + 1^2 + \cdots + 1^2 + 0^2 = 8$$

we obtain the entries in the body of the analysis of variance table. Their sum, $56 + 54 + 8 = 118$, the total sum of squares, is also equal to the sum of squares of the observations, 886, minus the sum of squares $12 \times 8^2 = 768$ for the grand mean array.

The mean array, which has a single entry $\bar{y}_{..}$, has 1 degree of freedom. The 3 distinct values in the treatment array always sum to zero, so it has $3 - 1 = 2$ degrees of freedom. In general, it has $a - 1$ degrees of freedom when there are a treatments. Similarly, the block array has $4 - 1 = 3$ degrees of freedom in this example and $b - 1$ when there are b blocks.

The number of degrees of freedom associated with the error array is $(a - 1) \cdot (b - 1) = (2)(3) = 6$. Because every row sum is zero, the last column is always determined from the first $b - 1$. Similarly, the last row is always determined by the first $a - 1$. Thus there are $(a - 1)(b - 1)$ unconstrained entries. In summary, the degrees of freedom can be decomposed as

$$ab - 1 = \underset{total}{} \;\; \underset{treatment}{(a-1)} + \underset{blocks}{(b-1)} + \underset{error}{(a-1)(b-1)}$$

An analysis of variance table presents the breakdowns, or decompositions, for the sums of squares and degrees of freedom.

Source variation	Degrees of freedom	Sum of squares
Treatments	2	56
Blocks	3	54
Error	6	8
Total	11	118

We summarize the expressions for the sums of squares and their degrees of freedom, for the general case.

Sums of squares for two-way analysis of variance

Treatment sum of squares: $SS(TR) = b \sum_{i=1}^{a} (\bar{y}_{i\bullet} - \bar{y}_{\bullet\bullet})^2$

Block sum of squares: $SS(Bl) = a \sum_{j=1}^{b} (\bar{y}_{\bullet j} - \bar{y}_{\bullet\bullet})^2$

Error sum of squares: $SSE = \sum_{i=1}^{a} \sum_{j=1}^{b} (y_{ij} - \bar{y}_{i\bullet} - \bar{y}_{\bullet j} + \bar{y}_{\bullet\bullet})^2$

Total sum of squares: $SST = \sum_{i=1}^{a} \sum_{j=1}^{b} (y_{ij} - \bar{y}_{\bullet\bullet})^2$

Degrees of freedom for two-way ANOVA

$$ab - 1 = \underset{total}{} \;\; \underset{treatment}{(a-1)} + \underset{blocks}{(b-1)} + \underset{error}{(a-1)(b-1)}$$

The underlying model which we shall assume for the analysis of this kind of experiment with one observation per cell (that is, there is one observation corresponding to each treatment within each block) is given by

Model equation for randomized-block design

$$Y_{ij} = \mu + \alpha_i + \beta_j + \varepsilon_{ij} \qquad \text{for } i = 1, 2, \ldots, a; \quad j = 1, 2, \ldots, b$$

Here μ is the grand mean, α_i is the effect of the ith treatment, β_j is the effect of the jth block, and the ε_{ij} are *independent, normally distributed* random variables having *zero means* and the *common variance* σ^2. Analogous to the model for the one-way classification, we restrict the parameters by imposing the conditions that $\sum_{i=1}^{a} \alpha_i = 0$ and $\sum_{j=1}^{b} \beta_j = 0$. (See Exercise 12.29.)

In the analysis of a two-way classification where each treatment is represented once in each block, the major objective is to test the significance of the differences among the $\bar{y}_{i\cdot}$. That is, the null hypothesis becomes

$$\alpha_1 = \alpha_2 = \cdots = \alpha_a = 0$$

In addition, it may also be desirable to test whether the blocking has been effective. That is, we test the null hypothesis

$$\beta_1 = \beta_2 = \cdots = \beta_b = 0$$

can be rejected. In either case, the alternative hypothesis is that at least one of the effects is different from zero.

As in the one-way analysis of variance, we shall base these significance tests on comparisons of estimates of σ^2—one based on the variation among treatments, one based on the variation among blocks, and one measuring the experimental error.

Using the sums of squares, we can reject the null hypothesis that the α_i are all equal to zero at the level of significance α if

F ratio for treatments

$$F_{Tr} = \frac{MS(Tr)}{MSE} = \frac{SS(Tr)/(a-1)}{SSE/(a-1)(b-1)}$$

exceeds F_α with $a-1$ and $(a-1)(b-1)$ degrees of freedom. The null hypothesis that the β_j are all equal to zero can be rejected at the level of significance α if

F ratio for blocks

$$F_{Bl} = \frac{MS(Bl)}{MSE} = \frac{SS(Bl)/(b-1)}{SSE/(a-1)(b-1)}$$

exceeds F_α with $b-1$ and $(a-1)(b-1)$ degrees of freedom. Note that the mean squares, $MS(Tr)$, $MS(Bl)$, and MSE, are again defined as the corresponding sums of squares divided by their degrees of freedom.

The results obtained in this analysis are summarized in the following **analysis of variance table:**

Source of variation	Degrees of freedom	Sum of squares	Mean square	F
Treatments	$a-1$	$SS(Tr)$	$MS(Tr) = \dfrac{SS(Tr)}{(a-1)}$	$F_{Tr} = \dfrac{MS(Tr)}{MSE}$
Blocks	$b-1$	$SS(Bl)$	$MS(Bl) = \dfrac{SS(Bl)}{(b-1)}$	$F_{Bl} = \dfrac{MS(Bl)}{MSE}$
Error	$(a-1)(b-1)$	SSE	$MSE = \dfrac{SSE}{(a-1)(b-1)}$	
Total	$ab-1$	SST		

EXAMPLE 6 **Constructing the randomized-block analysis of variance table**

Construct the analysis of variance from the decomposition of the observations given on page 394.

Solution Using the sums of squares and their associated degrees of freedom, we have

Source of variation	Degrees of freedom	Sum of squares	Mean square	F
Treatments	2	56.0	28.000	21.0
Blocks	3	54.0	18.000	13.50
Error	6	8.0	1.333	
Total	11	118.0		

The value of $F_{0.05}$ with 2 and 6 degrees of freedom is 5.14, so we reject the null hypothesis of equal mean particulate material removal. Blocking was important because we also reject the null hypothesis of equal block means. ∎

In practice, computer calculations are preferrable.

EXAMPLE 7 **Comparing four detergents using an F test**

An experiment was designed to study the performance of 4 different detergents for cleaning fuel injectors. The following "cleanness" readings were obtained with specially designed equipment for 12 tanks of gas distributed over 3 different models of engines:

	Engine 1	Engine 2	Engine 3	Totals
Detergent A	45	43	51	139
Detergent B	47	46	52	145
Detergent C	48	50	55	153
Detergent D	42	37	49	128
Totals	182	176	207	565

Looking at the detergents as treatments and the engines as blocks, obtain the appropriate analysis of variance table and test at the 0.01 level of significance whether there are differences in the detergents or in the engines.

Solution 1. *Null hypotheses*: $\alpha_1 = \alpha_2 = \alpha_3 = \alpha_4 = 0$; $\beta_1 = \beta_2 = \beta_3 = 0$
   Alternative hypotheses: The α's are not all equal to zero; the β's are not all equal to zero.

2. *Level of significance*: $\alpha = 0.01$

3. *Criteria*: For treatments, reject the null hypothesis if $F > 9.78$, the value of $F_{0.01}$ with $a - 1 = 4 - 1 = 3$, and $(a - 1)(b - 1) = (4 - 1)(3 - 1) = 6$ degrees of freedom; for blocks, reject the null hypothesis if $F > 10.92$, the value of $F_{0.01}$ for $b - 1 = 3 - 1 = 2$, and $(a - 1)(b - 1) = (4 - 1)(3 - 1) = 6$ degrees of freedom.

4. *Calculations*: A statistical software program produces the analysis of variance table

Source of variation	Degrees of freedom	Sum of squares	Mean square	F	P
Detergents	3	110.917	36.972	11.78	0.006
Engines	2	135.167	67.583	21.53	0.002
Error	6	18.833	3.139		
Total	11	264.917			

5. *Decisions*: Since $F_{Tr} = 11.6$ exceeds 9.78, the value of $F_{0.01}$ with 3 and 6 degrees of freedom, we conclude that there are differences in the effectiveness of the 4 detergents. Also, since $F_{Bl} = 21.2$ exceeds 10.92, the value of $F_{0.01}$ with 2 and 6 degrees of freedom, we conclude that the differences among the results obtained for the 3 engines are significant. There is an effect due to the engines, so blocking was important. To make the effect of this blocking even more evident, the reader will be asked to verify in Exercise 12.25 that the test for differences among the detergents would *not* yield significant results if we looked at the data as a one-way classification.

[Using **R** : Read data file with **Dat = read.table("C12Ex7.TXT", header=T, colClasses=c("factor", "factor", "numeric"))**. Then use **anova(lm(Cleanness~Detergent + Engine, data=Dat))**]   ■

The effect of the ith detergent can be estimated by means of the formula $\widehat{\alpha}_i = \bar{y}_{i\cdot} - \bar{y}_{\cdot\cdot}$, which may be obtained by the method of least squares. The resulting estimates are

$$\widehat{\alpha}_1 = 46.3 - 47.1 = -0.8 \qquad \widehat{\alpha}_2 = 48.3 - 47.1 = 1.2$$
$$\widehat{\alpha}_3 = 51.0 - 47.1 = 3.9 \qquad \widehat{\alpha}_4 = 42.7 - 47.1 = -4.4$$

Similar calculations lead to $\widehat{\beta}_1 = -1.6$, $\widehat{\beta}_2 = -3.1$, and $\widehat{\beta}_3 = 4.7$ for the estimated effects of the engines.

It should be observed that a two-way classification automatically allows for repetitions of the experimental conditions; for example, in the preceding experiment each detergent was tested 3 times. Further repetitions may be handled in several ways, and care must be taken that the model used appropriately describes the situation. One way to provide further repetition in a two-way classification is to include additional blocks—for example, to test each detergent using several additional engines, randomizing the order of testing for each engine. Note that the model remains essentially the same as before, the only change being an increase in b and a corresponding increase in the degrees of freedom for blocks and for error. The latter is important because an increase in the degrees of freedom for error makes the test of the null hypothesis $\alpha_i = 0$ for all i *more sensitive* to small differences among the treatment means. In fact, the real purpose of this kind of repetition is to increase the degrees of freedom for error, thereby increasing the sensitivity of the F tests (see Exercise 12.28).

A second method is to repeat the entire experiment, using a new pattern of randomization to obtain $a \cdot b$ additional observations. This is possible only if the blocks are strips across the rolling direction of a sheet of tin plate, and, given a new sheet, it is possible to identify which is strip 1, which is strip 2, and so forth. In the example of this section, the kind of repetition (usually called **replication**) would require that the conditions of the engines be exactly duplicated; see also Exercises 12.26 and 12.27.

A third method of repetition is to include n observations for each treatment in each block. When an experiment is designed in this way, the n observations in each "cell" are regarded as duplicates, and it is to be expected that their variability will be somewhat less than experimental error. To illustrate this point, suppose that the tin-coating weights of 3 disks from adjacent positions in a strip are measured in sequence by one of the laboratories, using the same chemical solutions. The variability of these measurements will probably be considerably less than that of 3 disks from the same strip measured in that laboratory at different times, using different chemical solutions, and perhaps different technicians. The analysis of variance appropriate for this kind of repetition reduces essentially to a two-way analysis of variance applied to the *means* of the n duplicates in the $a \cdot b$ cells; thus, *there would be no gain in degrees of freedom for error*, and, consequently, *no gain in sensitivity of the F tests*. It can be expected, however, that there will be some reduction in the error mean square, since it now measures the residual variance of the *means* of several observations.

Alternative Calculation of Sums of Squares

We conclude our discussion of the randomized block design by presenting alternative formulas that help simplify the calculations of the sums of squares when statistical software is unavailable or when students are restricted to simple handheld calculators. The reader is asked to very the formulas in Exercise 12.30.

Convenient formulas are available to calculate SST, $SS(Tr)$, and $SS(Bl)$ using handheld calculators.

Sums of squares for two-way analysis of variance

$$SST = \sum_{i=1}^{a} \sum_{j=1}^{b} y_{ij}^2 - C$$

$$SS(Tr) = \frac{\sum_{i=1}^{a} T_{i\bullet}^2}{b} - C$$

$$SS(Bl) = \frac{\sum_{j=1}^{b} T_{\bullet j}^2}{a} - C$$

where C, the correction term, is given by

$$C = \frac{T_{\bullet\bullet}^2}{ab}$$

In these formulas, $T_{i\bullet}$ is the sum of the b observations for the ith treatment, $T_{\bullet j}$ is the sum of the a observations in the jth block, and $T_{\bullet\bullet}$ is the grand total of all the observations. Note that the divisors for $SS(Tr)$ and $SS(Bl)$ are the number of observations in the respective totals, $T_{i\bullet}$ and $T_{\bullet j}$. The error sum of squares is then obtained by subtraction; according to Theorem 12.2 we can write

Error sum of squares

$$SSE = SST - SS(Tr) - SS(Bl)$$

In Exercise 12.30, the reader will be asked to verify that all these computing formulas are, indeed, equivalent to the corresponding terms of the identity of Theorem 12.2.

EXAMPLE 8   **Referring to the detergent data in Example 7, use the alternative formulas to verify the analysis of variance table**

Solution   Substituting $a = 4$, $b = 3$, $T_{1.} = 139$, $T_{2.} = 145$, $T_{3.} = 153$, $T_{4.} = 128$, $T_{.1} = 182$, $T_{.2} = 176$, $T_{.3} = 207$, $T_{..} = 565$, and

$$\sum \sum y_{ij}^2 = 26{,}867$$

into the formulas for the sums of squares, we get

$$C = \frac{(565)^2}{12} = 26{,}602$$

$$SST = 45^2 + 43^2 + \cdots + 49^2 - 26{,}602 = 26{,}867 - 26{,}602 = 265$$

$$SS(Tr) = \frac{139^2 + 145^2 + 153^2 + 128^2}{3} - 26{,}602 = 111$$

$$SS(Bl) = \frac{182^2 + 176^2 + 207^2}{4} - 26{,}602 = 135$$

$$SSE = 265 - 111 - 135 = 19$$

Dividing the sums of squares by their respective degrees of freedom we obtain the appropriate mean squares, and then an analysis variance table that agrees up to the number of decimal places retained here, with the table in Example 7.   ∎

12.4 Multiple Comparisons

The F tests used so far in this chapter showed whether differences among several means are significant, but they did not tells us whether a given mean (or groups of means) differs significantly from another given mean (or group of means). In actual practice, the latter is the kind of information an investigator really wants. For instance, having determined in Example 1 that the means of the tin-coating weights obtained by the 4 laboratories differ significantly, it may be important to find out which laboratory (or laboratories) differs from which others.

If an experimenter is confronted with k means, it may seem reasonable at first to test for significant differences between all possible pairs, that is, to perform

$$\binom{k}{2} = \frac{k(k-1)}{2}$$

two sample t tests as described on page 264. Aside from the fact that this would require a large number of tests even if k is relatively small, these tests would not be independent, and it would be virtually impossible to assign an overall level of significance to this procedure.

Several **multiple comparisons** procedures have been proposed to overcome these difficulties. The goal of a multiple confidence interval method is to guarantee, with a specified probability, that all of the intervals will cover their respective differences in means. One method of multiple comparisons, called the **Bonferroni method**, takes a conservative approach by guaranteeing that all of the confidence intervals will cover their true differences of means with at least probability $1 - \alpha$.

To do so, it simply modifies the level from α to $2\alpha/k(k-1)$ for each of $k(k-1)/2$ differences. This requires entering the t table with $\alpha/k(k-1)$ rather than $\alpha/2$. The confidence level α then pertains to all of the $k(k-1)/2$ confidence intervals covering their respective differences of population means. (See Exercise 12.36.) The confidence intervals help to judge if statistically significant differences are large enough to be of practical importance.

The Bonferroni approach uses the MSE as the estimate s^2 of the common variance.

Bonferroni simultaenous confidence intervals

With probability at least $1 - \alpha$, simultaneously, each confidence interval

$$\bar{y}_i - \bar{y}_\ell \pm t_{\alpha/k(k-1)} \sqrt{MSE \left(\frac{1}{n_i} + \frac{1}{n_\ell} \right)}$$

will cover $\mu_i - \mu_\ell$ for all $i < \ell$,

EXAMPLE 9   **Calculating Bonferroni simultaneous confidence intervals for mean resin strength**

With reference to the resin strength data in Example 3, obtain the 94% Bonferroni simultaneous confidence intervals for the three differences of means.

Solution   The confidence intervals use the MSE = 0.0234 as the estimate s^2. Since $\alpha = 0.06$ and $k = 3$, the Bonferroni procedure uses the $\alpha/k(k-1) = 0.01$ point of the t distribution with 12 degrees of freedom so $t_{0.01} = 2.681$. The three confidence intervals become

$$\text{MD} - \text{ED}: \ 1.334 - 0.964 \pm 2.681 \sqrt{0.0234 \left(\frac{1}{5} + \frac{1}{5} \right)} \text{ or } (\,0.111, 0.629\,)$$

$$\text{MD} - \text{PF}: \ 1.334 - 0.776 \pm 2.681 \sqrt{0.0234 \left(\frac{1}{5} + \frac{1}{5} \right)} \text{ or } (\,0.299, 0.817\,)$$

$$\text{ED} - \text{PF}: \ 0.964 - 0.776 \pm 2.681 \sqrt{0.0234 \left(\frac{1}{5} + \frac{1}{5} \right)} \text{ or } (\,-0.071, 0.447\,)$$

Resin MD is higher internal bonding strength than the other resins. It is between 0.111 to 0.629 units stronger than resin ED and 0.299 to 0.817 stronger than resin PF. Notice that these are longer than the individual 95% intervals calculated in Example 3.

Even though 94% confidence is less than 95% confidence, all 3 of these Bonferroni intervals hold simultaneously.   ∎

The Bonferroni approach is conservative in the sense that it maintains a probability of at least $1 - \alpha$ that all pairwise mean differences $\mu_i - \mu_\ell$ with $i < \ell$ are covered by their confidence intervals. The method is, however, very general and the sample sizes can be very different.

A better simultaneous method is available for the equal sample size case when the sample means are independent. Called the **Tukey honest significant difference** method (**Tukey HSD**), it maintains probability exactly $1 - \alpha$ that all pairwise mean differences $\mu_i - \mu_\ell$ for $i < \ell$ will be covered by their confidence intervals. The

Tukey HSD method is based on the **Standarized range distribution** for the random variable

$$Q = \frac{max_i \overline{Y}_i - min_i \overline{Y}_i}{\sqrt{MSE / n}}$$

All of the random variables have the same normal distribution and are independent. The common sample size is n. Here q_α cuts off probability α to the right so $\alpha = P(Q \geq q_\alpha)$.

Notice that q_α depends on the number of means being compared, k, and the degrees of freedom for the mean square error. The value for q_α is available in many statistical software programs including SAS and R. Tukey defines the

$$\text{honest significant difference} = q_\alpha \sqrt{\frac{MSE}{n}} = \frac{q_\alpha}{\sqrt{2}} \sqrt{\frac{2\,MSE}{n}}$$

where the right hand side is the usual notation. We see that the Tukey HSD approach replaces the two sample value $t_{\alpha/2}$ with a larger value $q_\alpha/\sqrt{2}$.

Tukey HSD simultaneous confidence intervals

Let the sample size for each mean equal n.
With probability $1 - \alpha$, simultaneously, each confidence interval

$$\overline{y}_i - \overline{y}_\ell \pm \frac{q_\alpha}{\sqrt{2}} \sqrt{\frac{2\,MSE}{n}}$$

will cover $\mu_i - \mu_\ell$ for all $i < \ell$,

EXAMPLE 10 **Calculating the Tukey HSD confidence intervals for differences in mean resin strength**

With reference to the internal bonding strengths in Example 3, obtain the Tukey HSD 94% simultaneous confidence intervals for the differences in mean strength. Here $q_{0.04} = 3.6256$.

Solution The Tukey HSD method uses MSE = 0.0234 with 12 degrees of freedom and the means from Example 3. The three confidence intervals become

$$\text{MD} - \text{ED:} \quad 1.334 - 0.964 \pm \frac{3.6256}{\sqrt{2}} \sqrt{\frac{2 \times 0.0234}{5}} \quad \text{or} \quad (0.122, 0.618)$$

$$\text{MD} - \text{PF:} \quad 1.334 - 0.776 \pm \frac{3.6256}{\sqrt{2}} \sqrt{\frac{2 \times 0.0234}{5}} \quad \text{or} \quad (0.310, 0.806)$$

$$\text{ED} - \text{PF:} \quad 0.964 - 0.776 \pm \frac{3.6256}{\sqrt{2}} \sqrt{\frac{2 \times 0.0234}{5}} \quad \text{or} \quad (-0.060, 0.436)$$

Each of these intervals is contained in the corresponding 94% Bonferroni interval in Example 9.

[Using **R** : **qtukey(0.94,3,12)** produces $q_{0.06}$
With **Dat = read.table("C12Ex9.TXT", header=T)**,
use **summary(fm1 <−aov(lm(Strength~Resin, data=Dat)))**
TukeyHSD(fm1, "Resin", conf.level=0.94)]

In the equal sample size case, the Tukey HSD intervals are always shorter than the Bonferroni intervals. The probability is exactly $1-\alpha$ that all of the intervals cover their respective difference in means. Although the sample sizes must be equal for the exact probability to hold, see Exercise 12.34 for an approximation when sample sizes are not quite equal.

Exercises

12.18 A randomized-block experiment is run with three treatments and four blocks. The three treatment means are $\bar{y}_{1\bullet} = 6$, $\bar{y}_{2\bullet} = 7$, and $\bar{y}_{3\bullet} = 11$.

The total (corrected) sum of squares is

$$220 = \sum_{i=1}^{3} \sum_{j=1}^{b} (y_{ij} - \bar{y}_{\bullet\bullet})^2$$

The analysis of variance (ANOVA) table takes the form

Source of variation	Degrees of freedom	Sum of squares	Mean square	F
Treatments				
Blocks		132		
Error				
Total	11	220		

(a) Fill in all of the missing entries in the analysis table.

(b) Conduct the F test for treatments and the F test for blocks. Use $\alpha = 0.05$.

12.20 The analysis of variance for a randomized-block design is conveniently implemented using *MINITAB*.

With reference to Example 7, first open C12Ex7.MTW in the MINITAB data bank.

Dialog Box:
Stat > ANOVA > Balanced ANOVA.
Enter *Cleanness* in **Responses**. Enter *Engine* and *Detergent* in **Model**.
Click **OK**.

```
ANOVA: Cleanness versus Engine, Detergent

Factor        Type      Levels      Values
Engine        fixed         3       1, 2, 3
Detergent     fixed         4       A, B, C, D


Analysis of variance for Cleanness

Source       DF          SS          MS          F          P
Engine        2     135.167      67.583      21.53      0.002
Detergent     3     110.917      36.972      11.78      0.006
Error         6      18.833       3.139
Total        11     264.917
```

Use computer software to re-work Example 6.

12.19 Concern about the running temperature of a computer chip prompted the investigation of 4 different types of cooling fans. Five different computers were available and each type of cooling fan was tried in each computer. Given the observations on temperature, coded by subtracting the smallest value,

	Blocks				
	1	2	3	4	5
Treatment 1	14	6	11	0	9
Treatment 2	14	10	16	9	16
Treatment 3	12	7	10	9	12
Treatment 4	12	9	11	6	7

(a) decompose each observation y_{ij} as

$$y_{ij} = \bar{y}_{\bullet\bullet} + (\bar{y}_{i\bullet} - \bar{y}_{\bullet\bullet}) + (\bar{y}_{\bullet j} - \bar{y}_{\bullet\bullet})$$
$$+ (y_{ij} - \bar{y}_{i\bullet} - \bar{y}_{\bullet j} + \bar{y}_{\bullet\bullet})$$

(b) obtain the sums of squares and degrees of freedom for each component;

(c) construct the analysis of variance table and test for differences among the treatments using $\alpha = 0.05$.

12.21 Looking at the days (rows) as blocks, rework Exercise 12.5 by the method of Section 12.3.

12.22 Four different, though supposedly equivalent, forms of a standardized reading achievement test were given to each of 5 students, and the following are the scores which they obtained:

	Student 1	Student 2	Student 3	Student 4	Student 5
Form A	75	73	59	69	84
Form B	83	72	56	70	92
Form C	86	61	53	72	88
Form D	73	67	62	79	95

Treating students as blocks, perform an analysis of variance to test at the level of significance $\alpha = 0.01$ whether it is reasonable to treat the 4 forms as equivalent.

12.23 A laboratory technician measures the breaking strength of each of 5 kinds of linen thread by means of 4 different instruments and obtains the following results (in ounces):

Measuring Instrument

	I_1	I_2	I_3	I_4
Thread 1	20.6	20.7	20.0	21.4
Thread 2	24.7	26.5	27.1	24.3
Thread 3	25.2	23.4	21.6	23.9
Thread 4	24.5	21.5	23.6	25.2
Thread 5	19.3	21.5	22.2	20.6

Looking at the threads as treatments and the instruments as blocks, perform an analysis of variance at the level of significance $\alpha = 0.01$.

12.24 An industrial engineer tests 4 different shop-floor layouts by having each of 6 work crews construct a subassembly and measuring the construction times (minutes) as follows:

	Layout 1	Layout 2	Layout 3	Layout 4
Crew A	48.2	53.1	51.2	58.6
Crew B	49.5	52.9	50.0	60.1
Crew C	50.7	56.8	49.9	62.4
Crew D	48.6	50.6	47.5	57.5
Crew E	47.1	51.8	49.1	55.3
Crew F	52.4	57.2	53.5	61.7

Test at the 0.01 level of significance whether the 4 floor layouts produce different assembly times and whether some of the work crews are consistently faster in constructing this subassembly than the others.

12.25 To emphasize the importance of blocking, reanalyze the cleanness data Example 7 as a one-way classification with the 4 detergents being the different treatments.

12.26 If, in a two-way classification, the entire experiment is repeated r times, the model becomes

$$Y_{ijk} = \mu + \alpha_i + \beta_j + \varepsilon_{ijk}$$

for $i = 1, 2, \ldots, a, j = 1, 2, \ldots, b$, and $k = 1, 2, \ldots, r$, where the sum of the α's the sum and the β's are equal to zero. The ε_{ijk} are independent normally distributed random variables with zero means and the common variance σ^2.

(a) Write down (but do not prove) an identity analogous to the one of Theorem 12.2, subdividing the total sum of squares into components attributable to treatments, blocks, and error.

(b) Give the corresponding degrees of freedom.

12.27 The following are the number of defectives produced by the 4 workers operating, in turn, 3 different machines. In each case, the first figure represents the number of defectives produced on a Friday and the second figure represents the number of defectives produced on the following Monday:

Worker

	B_1	B_2	B_3	B_4
Machine A_1	37, 43	38, 44	38, 40	32, 36
Machine A_2	31, 36	40, 44	43, 41	31, 38
Machine A_3	36, 40	33, 37	41, 39	38, 45

Use the theory developed in Exercise 12.26 to analyze the combined figures for the 2 days as a two-way classification with replication. Use the level of significance $\alpha = 0.05$.

12.28 As was pointed out on page 298, two ways of increasing the size of a two-way classification experiment are (a) to double the number of blocks, and (b) to replicate the entire experiment. Discuss and compare the gain in degrees of freedom for the error sum of squares by the two methods.

12.29 Show that if $\mu_{ij} = \mu + \alpha_i + \beta_j$, the mean of the μ_{ij} (summed on j) is equal to $\mu + \alpha_i$, and the mean of μ_{ij} (summed on i and j) is equal to μ, it follows that

$$\sum_{i=1}^{a} \alpha_i = \sum_{j=1}^{b} \beta_j = 0$$

12.30 Verify that the computing formulas for SST, $SS(Tr)$, $SS(Bl)$, and SSE, given on page 399, are equivalent to the corresponding terms of the identity of Theorem 12.2.

12.31 Using $q_{0.05} = 4.199$ for the Tukey HSD method, examine differences among the treatments in Exercise 12.19.

12.32 Referring to Example 1, use $q_{0.05} = 3.776$ for the Tukey HSD method to examine differences in mean coating weights.

12.33 Referring to Exercise 12.3, use Bonferroni simultaneous confidence intervals with $\alpha = 0.03$ to compare the mean number of hard disks produced by equipment under the 3 different settings.

12.34 **An approximation to Tukey HSD confidence intervals for mildly unequal sample sizes** When the there are only small differences in the sample sizes, an approximation is available. The MSE and its degrees of freedom still come from the ANOVA table but $\sqrt{2\,MSE\,/\,n}$ is replaced by

$$\sqrt{MSE\left(\frac{1}{n_i} + \frac{1}{n_\ell}\right)}$$

Referring to the example concerning the drying time of glue on page 379, use $q_{0.05} = 3.773$ with this approximate Tukey HSD approach to examine the three differences in means.

12.35 (a) Using $q_{0.10} = 3.921$ for the Tukey HSD method, compare the strength of the 5 linen threads in Exercise 12.23.

(b) Use the Bonferroni confidence interval approach on page 401, with $\alpha = 0.10$, to compare the mean linen thread strengths in Exercise 12.20.

12.36 The **Bonferroni** inequality states that

$$P\left(\cap_i C_i\right) \geq 1 - \sum_i P\left(\overline{C_i}\right)$$

(a) Show that this holds for 3 events.

(b) Let C_i be the event that the ith confidence interval will cover the true value of the parameter for $i = 1, \ldots, m$. If $P(\overline{C_i}) \leq \alpha/m$, so the probability of not covering the ith parameter is at most α/m, show that the probability that all of the confidence intervals cover their respective parameters is at least $1 - \alpha$.

12.5   Analysis of Covariance

The purpose of the method of Section 12.3 was to free the experimental error from variability due to an identifiable and controllable extraneous causes. In this section, we shall introduce a method called the **analysis of covariance**. It applies when such extraneous, or **concomitant**, variables cannot be held fixed but can be measured. This would be the case, for example, if we wanted to compare the effectiveness of several industrial training programs and the results depended on the trainees' IQs; if we wanted to compare the durability of several kinds of leather soles and the results depended on the weight of the persons wearing the shoes; or if we wanted to compare the merits of several cleaning agents and the results depended on the original condition of the surfaces cleaned.

The analysis of covariance for comparing treatments, when a single **covariate** x is present, blends the linear regression method of Section 11.1 with the analysis of variance of Section 12.2. The underlying model has terms for treatment effects and a regression term.

The underlying model is given by

Model equation for analysis of covariance

$$Y_{ij} = \mu + \alpha_i + \beta x_{ij} + \varepsilon_{ij}$$

for $i = 1, 2, \ldots, k$; $j = 1, 2, \ldots, n$. As in the model on page 383, μ is the grand mean, treatment effects satisfy $\sum_{i=1}^{k} \alpha_i = 0$, and the ε_{ij} are independent, normally distributed variables with zero means and the common variance σ^2. As in the model on page 326, β is the slope of the linear regression equation.

Under the ith treatment, the expected value for an observation having covariate value x, is

$$E(Y|\text{ given }i\text{th treatment and }x) = \mu + \alpha_i + \beta x \quad \text{for} \quad i = 1, 2 \ldots, k$$

As a function of x, the expected value for each treatment is a straight line. There is only one slope parameter, common to all treatments, so all of lines are parallel. Figure 12.3 depicts the three lines of expected values for $k = 3$ treatments.

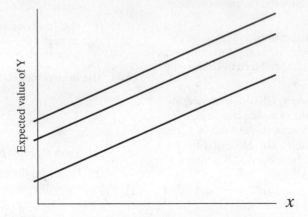

Figure 12.3
Parallel lines of expected values for each treatment

A manufacturer of activity tracking wristbands is developing an App to motivate users to exercise. Two versions are currently operational. Eight subjects are available and four are randomly chosen to use App 1 while the other four use App 2. The response Y is an index that combines the subjects walking, jogging, and running during a one-week trial period. The covariate x is the average index for the previous four weeks when no App was available. Suppose the data are

App 1		App 2	
x	y	x	y
1	7	2	2
2	8	3	3
2	10	5	6
3	11	6	9

We tentatively assume the analysis of covariance model above with $k = 2$ and $n = 4$.

Correlations between the observed treatment means and the covariate complicate the statistical analysis. We cannot decompose the observations as we did for the one-way analysis of variance. Our alternative approach is to fit the full model and then fit the model without the α_i treatment terms. Let

$SSE_{tr,x}$ = sum of squares error for the full model.

SSE_x = sum of squares error for the reduced model having μ and the β term.

$SSE_{tr.}$ = sum of squares error for the other reduced model having μ and the α_i terms.

Section 11.1 describes the calculation of the sum of squares error SSE_x for the reduced model with slope and Section 12.2 describes the calculation of $SSE_{tr.}$ for the one-way analysis of variance. For the full model, with both the treatment effects and a common slope term, we obtain $SSE_{tr.,x}$ by adjusting the sum of squares error $SSE_{tr.}$ from the one-way analysis of variance.

Sums of squares error for full model

$$SSE_{tr.,x} = SSE_{tr.}$$
$$- \frac{\left(\sum_{i=1}^{k} \sum_{j=1}^{n} (x_{ij} - \bar{x})(y_{ij} - \bar{y}) - \sum_{i=1}^{k} n(\bar{x}_i - \bar{x})(\bar{y}_i - \bar{y}) \right)}{\sum_{i=1}^{k} \sum_{j=1}^{n} (x_{ij} - \bar{x}_i)^2}$$

where $\bar{x}$ and $\bar{y}$ are the grand mean for the covariate and response, respectively.

The appropriate **sum of squares for treatments, adjusted** for the covariate, is the change in sum of squares error when the treatment terms are dropped. The sum of squares for the regression variable x is the change in sum of squares error when the x term is dropped.

Sums of squares for the analysis of covariance

$$SS(Tr) = SSE_x - SSE_{tr,x} \quad \text{with } k-1 \text{ degrees of freedom}$$
$$SS(x) = SSE_{tr.} - SSE_{tr.,x} \quad \text{with } 1 \text{ degree of freedom}$$
$$SSE_{tr.,x} \quad \text{with } nk - k - 1 \text{ degrees of freedom}$$

Note that the sum of squares error for the full model has 1 fewer degrees of freedom than in the one-way analysis of variance. This is due to the estimate of β.

Tests of hypotheses concerning treatment effects or the slope parameter involve the mean square errors

$$MS(Tr) = \frac{SS(Tr)}{k-1} \qquad MS(x) = \frac{SS(x)}{1}$$

and the mean square error from the full model

$$MSE = \frac{SSE_{tr.,x}}{nk - k - 1}$$

F ratios for the analysis of variance

Reject the null hypothesis $\alpha_1 = \alpha_2 = \cdots = \alpha_k = 0$ if

$$F = \frac{MS(Tr)}{MSE}$$

exceeds the tabled value F_α having $k-1$ and $nk - k - 1$ degrees of freedom
Reject the null hypothesis $\beta = 0$ if

$$F = \frac{MS(x)}{MSE}$$

exceeds the tabled value F_α having 1 and $nk - k - 1$ degrees of freedom

Concerning the fitness wristband data, we obtain the residuals from fitting the line as in Example 1, Chapter 11. The sum of squares error is then

$$SSE_x = (-0.25)^2 + (-1.00)^2 + (1.00)^2 + (0.25)^2 + (0.50)^2$$
$$+ (-0.25)^2 + (-0.75)^2 + (0.50)^2 = 70.75$$

Obtaining the decomposition of observations as in Section 12.2, the sum of squares error when only the treatment terms are included is

$$SSE_{tr.} = (-2)^2 + (-2)^2 + 1^2 + 2^2 + (-3)^2 + (-2)^2 + 1^2 + 4^2 = 40$$

The full model with both the treatments and common slope, we obtain $SSE_{tr., x}$ by adjusting the the smaller model sum of squares error $SSE_{tr.}$ according to

$$SSE_{tr.} - \frac{\left(\sum_{i=1}^{k} \sum_{j=1}^{n} (x_{ij} - \bar{x})(y_{ij} - \bar{y}) - \sum_{i=1}^{k} n(\bar{x}_i - \bar{x})(\bar{y}_i - \bar{y}) \right)^2}{\sum_{i=1}^{k} \sum_{j=1}^{n} (x_{ij} - \bar{x}_i)^2}$$

In our fitness wristband example,

$$\sum_{i=1}^{k} \sum_{j=1}^{n} (x_{ij} - \bar{x})(y_{ij} - \bar{y}) = -2 \cdot 0 - 1 \cdot 1 - 1 \cdot 3 + 0 \cdot 4 - 1(-5)$$
$$+ 0(-4) + 2(-1) + 3 \cdot 4 = 5$$

$$\sum_{i=1}^{k} n(\bar{x}_i - \bar{x})(\bar{y}_i - \bar{y}) = 4(-1)2 + 4 \cdot 1(-2) = -16$$

$$\sum_{i=1}^{k} \sum_{j=1}^{n} (x_{ij} - \bar{x}_i)^2 = (-1)^2 + 0^2 + 0^2 + 1^2 + (-2)^2 + (-1)^2 + 1^2 + 2^2 = 12$$

It follows that sum of squares error for the full model is

$$SSE_{tr., x} = SSE_{tr.} - \frac{(5 - (-16))^2}{12} = 40 - \frac{21^2}{12} = 3.25$$

Then

$$SS(Tr) = SSE_x - SSE_{tr., x} = 70.75 - 3.25 = 67.50 \text{ with } 2 - 1 \text{ degrees of freedom}$$

$$SS(x) = SSE_{tr.} - SSE_{tr., x} = 40 - 3.25 = 36.75 \quad \text{with } 1 \text{ degree of freedom}$$

The corresponding values of the F statistic are

$$F = \frac{67.5 / 1}{3.25 / 5} = 103.85 \quad \text{and} \quad F = \frac{36.75 / 1}{3.25 / 5} = 56.54$$

Both are highly significant with P-values 0.000 and 0.001, respectively. APP 1 better motivates users to exercise by walking.

An **analysis of covariance table** summarizes this analysis.

EXAMPLE 11   **Creating an analysis of covariance table with one covariate**

Suppose that a research worker has three different cleaning agents, A_1, A_2, and A_3, and he wishes to select the most efficient agent for cleaning a metallic surface. The cleanliness of a surface is measured by its reflectivity, expressed in arbitrary units as the ratio of the reflectivity observed to that of a standard mirror surface. Analysis of covariance must be used because the effect of a cleaning agent on reflectivity will depend on the original cleanliness, namely, the original reflectivity of the surface. The research worker obtained the following results:

A_1	Original reflectivity, x	0.90	0.95	0.80	0.50
	Final reflectivity, y	1.05	0.95	1.15	0.85
A_2	Original reflectivity, x	0.50	0.40	0.15	0.25
	Final reflectivity, y	1.10	1.00	0.90	0.80
A_3	Original reflectivity, x	0.20	0.55	0.30	0.40
	Final reflectivity, y	0.75	1.05	0.95	0.90

Perform an analysis of covariance to determine (at the 0.05 level of significance) whether there are differences in the reflectivity improvements by the 3 cleaning agents.

Solution
1. *Null hypothesis*: $\alpha_1 = \alpha_2 = \alpha_3 = 0$
   Alternative hypothesis: The α's are not all equal to zero.
2. *Level of significance*: $\alpha = 0.05$
3. *Criterion*: Reject the null hypothesis if $F > 4.46$, the value of $F_{0.05}$ for $k - 1 = 3 - 1 = 2$ and $nk - k - 1 = 4 \times 3 - 3 - 1 = 8$ degrees of freedom.
4. *Calculations*: The calculations associated with an analysis of covariance can become quite cumbersome and we recommend the use of a statistical software program. *MINITAB* (see Exercise 12.38) calculates the adjusted sum of squares, mean squares and F statistics. The corresponding *SAS* output is shown in Figure 12.4.

Sources of Variation	Degrees of Freedom	Adjusted Sum of Squares	Adjusted Mean Square	F-Value	P-Value
x	1	0.10506	0.105057	20.10	0.002
Treatment	2	0.06776	0.033878	6.48	0.021
Error	8	0.04182	0.005227		
Total	11	0.16229			

The sums of squares, mean squares and F-statistics are those based on the calculation of the sum of squares error under the full model and then under the reduced model. The interplay of the treatments and covariate then results in the sum of the three sums of squares in the table being greater than the sum of squares total, $SST = 0.16229$.

5. *Decision*: Since the F- ratio for treatments

$$F = \frac{0.033878}{0.005227} = 6.48$$

we reject the null hypothesis of equal α_i's in favor of the alternative that not all are equal. The P-value 0.021 further strengthens this conclusion.

The very small P-value of 0.002 confirms that it is very important to include the original reflection as a covariate when comparing the treatments.

Additional calculations give the P-value for each individual term in the model.

Coefficients Term	Coef	SE Coef	T-Value	P-Value
Constant	0.5339	0.0960	5.56	0.001
x	0.782	0.174	4.48	0.002
Treatment				
A1	−0.2571	0.0737	−3.49	0.008
A2	0.1620	0.0474	3.42	0.009

According to the constraint that the sum of treatment effects is zero, the estimated coefficient for A3 is $-(-0.2571 + 0.1620) = 0.0951$. The resulting estimates of the regression lines, all having the same slope, are

Regression Equation Treatment	
A1	$y = 0.2768 + 0.782\, x$
A2	$y = 0.6959 + 0.782\, x$
A3	$y = 0.6291 + 0.782\, x$

Cleaning agent A1 does not clean as well as the other two cleaning agents.

[Using **R** : Read data file with **Dat = read.table("C12Ex8.TXT", header=T)**. Then use **res =lm($y \sim x$ + treatment, data=Dat)** followed by **anova(res)** for the ANOVA table and **summary(res)** for the fit of the lines.

To obtain the correct F-statistic for x you could use **res2 =lm($y \sim$ treatment + x,data=Dat)** and then **anova(res2)**]   ■

Although the calculations may at first appear to be formidable, they are routine with many computer statistical programs. The output from the *SAS* program for this example is presented in Figure 12.4. Because only two decimal points are retained in the preceding example, the computer calculations for the *F statistic* are more accurate. However, the conclusions are the same. You must request Type III sum of squares to get the correct SAS output. See Exercise 2.38 for MINITAB commands.

Dependent Variable: y

Source	DF	Sum of Squares	Mean Square	F Value	Pr > F
Model	3	0.12047348	0.04015783	7.68	0.0097
Error	8	0.04181818	0.00522727		
Corrected Total	11	0.16229167			

R-Square	Root MSE
0.742327	0.072300

Source	DF	Type III SS	Mean Square	F Value	Pr > F
Treatment	2	0.06775668	0.03387834	6.48	0.0212
x	1	0.10505682	0.10505682	20.10	0.0020

Figure 12.4

Selected SAS output for the analysis of covariance using the data from Example 11

Analysis of covariance methods have not been widely used until recent years, due mainly to the rather extensive calculations that are required. Of course, with the widespread availability of computers and appropriate programs, this is no longer a problem. There are several ways in which the analysis of covariance method presented here can be generalized. First, there can be more than one concomitant variable; then the method can be applied to more complicated kinds of designs, say, to a randomized-block design, where the regression coefficient could even assume a different value for each block.

Exercises

12.37 An experimenter wants to compare the time to failure y after rebuilding a robotic welder by three different methods but adjusting for the covariate $x =$ age of robotic welder.

Suppose the data, in thousands of hours, are

Method 1		Method 2		Method 3	
x	y	x	y	x	y
7	2	6	4	5	6
11	5	8	5	3	5
6	2	4	3	4	4

The resulting sum of squares error are $SSE_x = 15.917$, $SSE_{tr.} = 10.000$, and $SSE_{tr., x} = 1.833$. Assuming that the analysis of covariance model is reasonable, conduct the F-test for treatment differences and the F-test for the covariate x. Take $\alpha = 0.05$.

12.38 *MINITAB* **calculation of balanced analysis of covariance** We illustrate the *MINTAB* commands for Example 11 concerning surface reflectivity.

Data:

C1(Tr) :	1	1	1	1	2	2	2	2	3	3	3	3
C2(x):	0.90	0.95	1.05	0.80	0.50	0.40	0.15	0.25	0.20	0.55	0.30	0.40
C3(y):	1.05	0.95	1.15	0.85	1.10	1.00	0.90	0.80	0.75	1.05	0.95	0.90

Dialog box:

Stat> ANOVA > General Linear Model > Fit General Linear Model.
Type y in **Responses**, *Treatment* in **Factors** and x in **Covariates**. Click **OK**.

The partial output includes

```
Analysis of variance

Source        DF      Adj SS      Adj MS    F-Value    P-Value
    x          1     0.10506    0.105057      20.10      0.002
 Treatment     2     0.06776    0.033878       6.48      0.021
Error          8     0.04182    0.005227
Total         11     0.16229

         S      R-sq
 0.0722999    74.23%
```

Use computer software to perform an analysis of covariance for the fitness wristband example on page 406.

12.39 Four different railroad-track cross-section configurations were tested to determine which is most resistant to breakage under use conditions. Ten miles of each kind of track were laid in each of 5 locations, and the number of cracks and other fracture-related conditions (y) was measured over a two-year usage period. To compare these track designs adequately, however, it is necessary to correct for extent of usage (x), measured in terms of the average number of trains per day that ran over each section of track. Use the following experimental results to test (0.01 level of significance) whether the track designs were equally resistant to breakage and to estimate the effect of usage on breakage resistance.

Track design A		Track design B		Track design C		Track design D	
x	y	x	y	x	y	x	y
10.4	3	16.9	8	17.8	5	19.6	9
19.3	7	23.6	11	24.4	9	25.4	8
13.7	4	14.4	7	13.5	5	35.5	16
7.2	0	17.2	10	20.1	6	16.8	7
16.3	5	9.1	4	11.0	4	31.2	11

12.40 To compare the life expectancy of a transistor under 3 storage conditions and account at the same time for a leakage current (collector to base), a laboratory technician obtained the following results, where the leakage current, x, is in microamperes, and the life times, y, are in hours:

Storage condition 1		Storage condition 2		Storage condition 3	
x	y	x	y	x	y
4.8	9,912	6.4	9,952	8.8	9,596
7.2	9,383	8.7	9,482	6.2	9,697
5.5	9,734	7.1	9,435	7.5	9,700
6.0	9,551	5.3	9,915	4.9	9,610
8.3	8,959	4.6	9,492	5.4	10,145
7.6	9,474	6.0	9,565	5.8	10,191
5.9	9,179	7.2	9,704	7.3	9,855
8.0	9,359	8.8	9,636	8.6	9,682
4.3	9,580	5.4	9,608	8.8	10,160
5.1	9,245	7.8	9,548	6.0	9,982

Perform an analysis of covariance, using the level of significance $\alpha = 0.05$. Also, estimate the value of the regression coefficient.

12.41 Use computer software to work Exercise 12.37.

Do's and Don'ts

Do's

1. Whenever possible, randomize the assignment of treatments in the completely randomized design. In other designs, randomize the assignments of treatment within the restraints of the design.

2. When numerous comparisons must be made, use a multiple comparisons method to be able to make a confidence statement about the whole set of confidence intervals.

3. Consider applying the analysis of covariance when an important extraneous variable cannot be held constant.

Don'ts

1. Don't routinely accept the analysis of variance presented in a computer output. Instead, inspect the residuals for outliers and moderate to severe lack of normality. A normal-scores plot is useful if there are 20 or more residuals. It may suggest a transformation.

Review Exercises

12.42 Assume the following data obey the one-way analysis of variance model.

Treatment 1:	13	8	10	11	8	
Treatment 2:	1	0	3	0		
Treatment 3:	8	5	10	3	7	3

(a) Decompose each observation y_{ij} as

$$y_{ij} = \bar{y} + (\bar{y}_i - \bar{y}) + (y_{ij} - \bar{y}_i)$$

(b) Obtain the sums of squares and degrees of freedom for each array.

(c) Construct the analysis of variance table and test for differences among the treatments with $\alpha = 0.05$.

12.43 To determine the effect on exit dust loading in a precipitator, the following measurements were made:

Total flow (ft^3/hr)	Exit dust loading (grains per cubic yard in flue gas)				
200	1.5	1.7	1.6	1.9	1.9
300	1.5	1.8	2.2	1.9	2.2
400	1.4	1.6	1.7	1.5	1.8
500	1.1	1.5	1.4	1.4	2.0

Use the level of significance $\alpha = 0.05$ to test whether the flow through the precipitator has an effect on the exit dust loading.

12.44 Refer to Example 11. Ignore the covariate original reflectivity. Perform an analysis of variance take $\alpha = 0.05$

12.45 Given the following data from a randomized block design,

	Blocks			
	1	2	3	4
Treatment 1	9	10	2	7
Treatment 2	6	13	1	12
Treatment 3	9	16	9	14

(a) Decompose each observation y_{ij} as

$$y_{ij} = \bar{y}_{..} + (\bar{y}_{i.} - \bar{y}_{..}) + (\bar{y}_{.j} - \bar{y}_{..})$$
$$+ (y_{ij} - \bar{y}_{i.} - \bar{y}_{.j} + \bar{y}_{..})$$

(b) Obtain the sums of squares and degrees of freedom for each array.

(c) Construct the analysis of variance table and test for differences among the treatments with $\alpha = 0.05$.

12.46 Using $q_{0.05} = 4.339$ for the Tukey HSD method, compare the treatments in Exercise 12.45.

12.47 Samples of groundwater were taken from 5 different toxic-waste dump sites by each of 3 different agencies: the EPA, the company that owned each site, and an independent consulting engineer. Each sample was analyzed for the presence of a certain contaminant by whatever laboratory method was customarily used by the agency collecting the sample, with the following results:

Concentration (parts per million)

	Site A	Site B	Site C	Site D	Site E
Agency 1	23.8	7.6	15.4	30.6	4.2
Agency 2	19.2	6.8	13.2	22.5	3.9
Agency 3	20.9	5.9	14.0	27.1	3.0

Use the $\alpha = 0.05$ level of significance to decide:

(a) Is there reason to believe that the agencies are not consistent with one another in their measurements?

(b) Do the dump sites differ from one another in their level of contamination?

12.48 Using $q_{0.05} = 4.041$ for the Tukey HSD method, compare the pollution levels of the three agencies in Exercise 12.47.

12.49 An experiment is conducted with $k = 4$ treatments, one covariate x, and $n = 7$. Calculations result in the sums of square error $SSE_x = 10.8$, $SSE_{tr.} = 4.38$, and $SSE_{tr., x} = 4.19$ Assuming that the analysis of covariance model is reasonable, conduct the F-test for treatment differences and the F-test for the covariate x. Take $\alpha = 0.05$.

12.50 The state highway department does an experiment to compare three types of surfacing treatments and the response y is road roughness. The following table also gives average daily traffic volume x.
Suppose the data are, in suitable units,

Treatment 1		Treatment 2		Treatment 3	
x	y	x	y	x	y
5	12	3	3	4	9
5	10	2	2	2	5
2	5	1	1	3	7

(a) Perform an analysis of variance on the response variable road roughness. Take $\alpha = 0.05$.

(b) Is there one surface treatment that is better than the others? Answer by finding the Bonferroni 95% simultaneous confidence intervals for the differences of the means.

12.51 Refer to Exercise 12.50.

(a) Perform an analysis of covariance. Test for a difference in treatments using level of significance 0.05.

(b) Compare your analysis in part (a) with the analysis of variance. Is the covariate important?

12.52 Three different instrument panel configurations were tested by placing airline pilots in flight simulators and testing their reaction time to simulated flight emergencies. Eight pilots were assigned to each instrument panel configuration. Each pilot was faced with 10 emergency conditions in a randomized sequence, and the total time required to take corrective action for all 10 conditions was measured, with the following results:

Instrument panel 1		Instrument panel 2		Instrument panel 3	
x	y	x	y	x	y
8.1	6.55	12.1	5.74	15.2	6.37
19.4	6.40	2.1	5.93	8.7	6.97
11.6	5.93	3.9	6.16	7.2	7.38
24.9	6.79	5.2	5.68	6.1	6.43
6.2	7.16	4.6	5.41	11.8	7.59
3.8	5.64	14.4	6.29	12.1	7.16
18.4	5.87	16.1	5.55	9.5	7.02
9.4	6.31	8.5	4.82	2.6	6.85

In this table, x is the number of years of experience of the pilot, and y is the total reaction time in seconds. Perform an analysis of covariance to test whether the instrument-panel configurations yield significantly different results ($\alpha = 0.05$). Also, perform a one-way analysis of variance (ignoring the covariate, x) and determine in that way what effect experience has on the results.

12.53 Using the alternative calculation formula verify analysis of variance table for the paper-strength in Example 4.

12.54 Benjamin Franklin (1706–1790) conducted an experiment to study the effect of water depth on the amount of drag on a boat being pulled up a canal. He made a 14-foot trough and a model boat 6 inches long. A thread was attached to the bow, put through a pulley, and then a weight was attached. Not having a second hand on his watch, he counted as fast as he could to 10 repeatedly. These times, for the model boat to traverse the trough at the different water depths, are

Water 1.5 inches:	100	104	104	106	100	99	100	100
Water 2.0 inches:	94	93	91	87	88	86	90	88
Water 4.5 inches:	79	78	77	79	79	80	79	81

(*Source*: Letter to John Pringle, May 10, 1768.)

(a) Perform an analysis of variance and test for differences due to water depth using $\alpha = 0.05$.

(b) Using $\alpha = 0.06$ and $q_{0.06} = 3.438$ for the Tukey HSD method, investigate differences.

(c) Use the Bonferroni 94% confidence interval approach on page 401 to compare the mean times.

Key Terms

Analysis of covariance 405
Analysis of covariance table 408
Analysis of variance 385
Analysis of variance table 384, 396
Between-sample mean square 383
Blocks 378
Bonferroni method 400
Completely randomized design 379
Concomitant variable 405
Confounded 377
Correction term (for mean) 389
Covariate 405

Effect 383
Experimental design 376
Grand mean 380
Mean square 383
Model equation 383
Multiple comparisons 400
One-way classification 379
Randomization 377
Randomized-block design 393
Replication 399
Robust 382
Standardized range distribution 402

Sum of squares:
  between-samples 382
  blocks 393
  error 382, 395
  total 381, 395
  treatment 382, 395
  treatment, adjusted 407
Two-way classification 392
Tukey HSD method 401
Within-sample mean square 383

FACTORIAL EXPERIMENTATION

C hapter 12 concentrates mainly on the effects of one variable whose values are referred to as "treatments". This chapter expands the discussion to explore the effects of several variables. The combinations of their values, or levels, now play the roles of treatments.

Sections 13.1 and 13.2 deal with the analysis of experiments whose treatments can be regarded as combinations of the levels of two or more factors. Section 13.3 studies the special case of factors having two levels. Section 13.4 illustrates a design for obtaining the levels of variables where the estimated response is a maximum. The remainder of the chapter takes up the analysis of experiments where there are too many combinations of all the factors to be included in the same block or experimental program.

13.1   Two-Factor Experiments

We introduce the idea of a **two-factor (two variable) experiment** in the context of recycling material for roadways. Recycling of construction and demolition waste can greatly extend the life of existing landfills. But, any recycled materials used in roadways must perform as well as the typical natural aggregates.

To establish specifications for the strength of recycled materials, the experimenter involves 3 locations, subject to different environmental conditions, with 2 types of recycled materials, recycled concrete aggregate(RCA) and recycled asphalt pavement aggregate (RAP). The locations are labeled by the state of origin MN, CO, and TX.

Factor A Location	Factor B Type of material
MN	Recycled Concrete Aggregate (RCA)
MN	Recycled Pavement Aggregate (RPA)
CO	Recycled Concrete Aggregate (RCA)
CO	Recycled Pavement Aggregate (RPA)
TX	Recycled Concrete Aggregate (RCA)
TX	Recycled Pavement Aggregate (RPA)

One key physical property is the resiliency modulus which is obtained by dynamic loading. Generally, higher values imply a stiffer base which increases pavement life.

Do changes in location or type of recycled material affect the resiliency modulus? If changes in the modulus are attributable to changes in location, is this change the same for both types of materials? It is possible to answer questions of this kind if the experimental conditions, the treatments, consist of appropriate combinations of the **levels** (or values) of the various **factors**. The factors in the preceding example are location and type of material; location has 3 levels, MN, CO, and TX while the type of material has 2 levels RCA and BPA. Note that the 6 treatments were chosen in such a way that each level of location is used once in conjunction with each level of type of material. In general, if 2 factors A and B are to be investigated at a and b levels, respectively, then there are $a \cdot b$ experimental conditions (treatments) corresponding to all possible combinations of the levels of the 2 factors. The resulting experiment is referred to as a **complete $a \times b$ factorial experiment**. Note that if one or more of the $a \cdot b$ experimental conditions is omitted, the experiment can still be analyzed as a two-way classification, but it cannot readily be analyzed as a factorial experiment. It is customary to omit the word *complete* so that an **$a \times b$ factorial experiment** is understood to contain experimental conditions corresponding to all possible combinations of the levels of the two factors.

In order to obtain an estimate of the experimental error in a two-factor experiment, it is necessary to replicate, that is, to repeat the entire set of $a \cdot b$ experimental conditions, say, a total of r times, randomizing the order of applying the condition in each replicate. If y_{ijk} is the observation in the kth replicate, taken at the ith level of factor A and the jth level of factor B, the model assumed for the analysis of this kind of experiment is usually written as

Model equation for two-factor experiment

$$Y_{ijk} = \mu + \alpha_i + \beta_j + (\alpha\beta)_{ij} + \varepsilon_{ijk}$$

for $i = 1, 2, \ldots, a$, $j = 1, 2, \ldots, b$, and $k = 1, 2, \ldots, r$. Here μ is the grand mean, α_i is the effect of the ith level of factor A, β_j is the effect of the jth level of factor B, $(\alpha\beta)_{ij}$ is the **interaction**, or joint effect, of the ith level of factor A and the jth level of factor B. As in the models used in Chapter 12 we shall assume that the ε_{ijk} are independent random variables having normal distributions with zero means and the common variance σ^2. Also, analogous to the restrictions imposed on the models on pages 383 and 396, we shall assume that

$$\sum_{i=1}^{a} \alpha_i = \sum_{j=1}^{b} \beta_j = \sum_{i=1}^{a} (\alpha\beta)_{ij} = \sum_{j=1}^{b} (\alpha\beta)_{ij} = 0$$

It can be shown that these restrictions will assure unique definitions of the parameters μ, α_i, β_j, and $(\alpha\beta)_{ij}$.

To illustrate the model underlying a two-factor experiment, let us consider an experiment with two replicates in which factor A occurs at two levels and factor B occurs at two levels. In view of the restrictions on the parameters, we also have

$$\alpha_2 = -\alpha_1 \quad \beta_2 = -\beta_1 \quad (\alpha\beta)_{21} = (\alpha\beta)_{12} = -(\alpha\beta)_{11} = -(\alpha\beta)_{22}$$

and the population means corresponding to the four experimental conditions defined by the 2 levels of factor A and the 2 levels of factor B can be written as

$$\mu_{111} = \mu_{112} = \mu + \alpha_1 + \beta_1 + (\alpha\beta)_{11}$$
$$\mu_{121} = \mu_{122} = \mu + \alpha_1 - \beta_1 - (\alpha\beta)_{11}$$
$$\mu_{211} = \mu_{212} = \mu - \alpha_1 + \beta_1 - (\alpha\beta)_{11}$$
$$\mu_{221} = \mu_{222} = \mu - \alpha_1 - \beta_1 + (\alpha\beta)_{11}$$

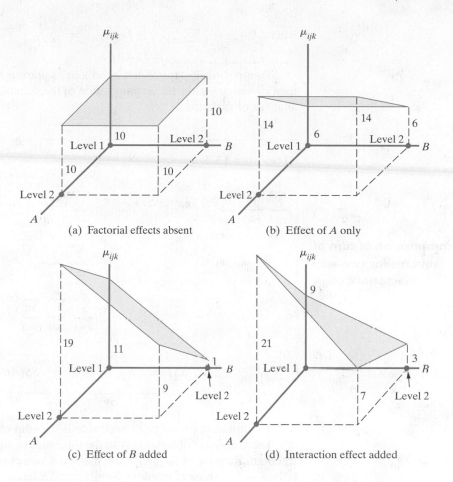

Figure 13.1
Factorial effects

(a) Factorial effects absent

(b) Effect of A only

(c) Effect of B added

(d) Interaction effect added

Substituting for $\mu_{ij1} = \mu_{ij2}$ the mean of all observations obtained for the ith level of factor A and the jth level of factor B, we get 4 simultaneous linear equations which can be solved for the parameters μ, α_1, β_1, and $(\alpha\beta)_{11}$. (See Exercise 13.9.)

To continue our illustration, let us now suppose that $\mu = 10$. If all the other effects equaled zero, each of the μ_{ijk} would equal 10, and the response surface would be the horizontal plane shown in Figure 13.1(a). If we now add an effect of factor A, with $\alpha_1 = -4$, the response surface becomes the tilted plane shown in Figure 13.1(b), and if we add to this an effect of factor B, with $\beta_1 = 5$, we get the plane shown in Figure 13.1(c). Note that, so far, the effects of factors A and B are *additive*; that is, the change in the mean for either factor in going from level 1 to level 2 does not depend on the level of the other factor, and the response surface is a plane. If we now include an interaction, with $(\alpha\beta)_{11} = -2$, the plane becomes twisted as shown in Figure 13.1(d), the effects are no longer additive, and the response surface is no longer a plane.

Generalizing these ideas from a 3×2 factorial experiment, our statistical analysis of an $a \times b$ factorial experiment is based on the decomposition of the observations according to the model.

$$y_{ijk} = \bar{y}_{...} + (\bar{y}_{i..} - \bar{y}_{...}) + (\bar{y}_{.j.} - \bar{y}_{...})$$

$$\text{observation} \quad \text{grand} \quad \text{factor } A \quad \text{factor } B$$
$$\text{mean} \quad \text{effect} \quad \text{effect}$$

$$+ (\bar{y}_{ij.} - \bar{y}_{i..} - \bar{y}_{.j.} + \bar{y}_{...}) + (y_{ijk} - \bar{y}_{ij.})$$
$$\text{AB interaction} \quad \text{error}$$

where $\bar{y}_{ij\cdot} = \sum_{k=1}^{r} y_{ijk}/r$, $\bar{y}_{i\cdot\cdot} = \sum_{j=1}^{b}\sum_{k=1}^{r} y_{ijk}/(br)$, and $\bar{y}_{\cdots} = \sum_{i=1}^{a}\sum_{j=1}^{b}\sum_{k=1}^{r} y_{ijk}/(abr)$.

Transposing $\bar{y}_{\cdots}$ to the left hand side, squaring both sides of the identity, and then summing yields the decomposition of the sum of squares that is the basis of the analysis of variance.

Theorem 13.1

Decomposition of sum of squares for two-way factorial design

$$\sum_{i=1}^{a}\sum_{j=1}^{b}\sum_{k=1}^{r}(y_{ijk} - \bar{y}_{\cdots})^2 = br\sum_{i=1}^{a}(\bar{y}_{i\cdot\cdot} - \bar{y}_{\cdots})^2 + ar\sum_{j=1}^{b}(\bar{y}_{\cdot j\cdot} - \bar{y}_{\cdots})^2$$

$$+ r\sum_{i=1}^{a}\sum_{j=1}^{b}(\bar{y}_{ij\cdot} - \bar{y}_{i\cdot\cdot} - \bar{y}_{\cdot j\cdot} + \bar{y}_{\cdots})^2$$

$$+ \sum_{i=1}^{a}\sum_{j=1}^{b}\sum_{k=1}^{r}(y_{ijk} - \bar{y}_{ij\cdot})^2$$

or

$$SST = SSA + SSB + SS(AB) + SSE$$

The last term on the right-hand side is the sum of squares due to error.

There are abr observations so the total sum of squares has $abr - 1$ degrees of freedom. Because of the constraint that the a values of $\bar{y}_{i\cdot\cdot} - \bar{y}_{\cdots}$ sum to 0, SSA has only $a - 1$ degrees of freedom. Similarly, SSB has $b - 1$ degrees of freedom. Next, for each fixed pair i, j, the error terms $y_{ijk} - \bar{y}_{ij\cdot}$ sum to 0. Consequently, they each contribute $r - 1$ degrees of freedom and SSE has $ab(r - 1)$ degrees of freedom. The degrees of freedom for the AB interaction can be obtained by subtraction

$$abr - 1 - (a-1) - (b-1) - ab(r-1) = (a-1)(b-1)$$

Decomposition of degrees of freedom for two-way factorial design

$$abr - 1 = \underset{Total}{a-1} + \underset{Factor\,A}{b-1} + \underset{Factor\,B}{(a-1)(b-1)} + \underset{AB\ Interaction}{ab(r-1)}$$
Total   Factor A   Factor B   AB Interaction   error

The mean square error of any term is obtained by dividing the sum of squares by its degrees of freedom. The corresponding F-statistic is then obtained by dividing the mean square error of a factor by the mean square error.

The next example illustrates the successive breakdown of the sum of squares. For the amount of calculation involved, you can appreciate the widespread availability of computer programs for creating an analysis of variance table. (See Exercise 13.10.)

EXAMPLE 1   **Conducting statistical tests for a 3 × 2 factorial experimental design**

Referring to the recycling example on page 415, the experimenter obtains values of the resiliency modulus (MPa) from 3 replications of the experiment. (Courtesy of Tuncer Edil)

Factor A Location	Factor B Type of Mat.	Rep. I	Rep.2	Rep. 3
MN	RCA	707	632	604
MN	RPA	652	669	674
CO	RCA	522	554	484
CO	RPA	630	648	610
TX	RCA	450	545	474
TX	RPA	845	810	682

Perform an analysis of variance based on this two-factor experiment and test for the significance of the factorial effects, using the 0.01 level of significance.

Solution Following steps analogous to those used in the analysis of a two-way classification, we get

1. *Null hypotheses*:

$$\alpha_1 = \alpha_2 = \alpha_3 = 0, \quad \beta_1 = \beta_2 = 0$$
$$(\alpha\beta)_{11} = (\alpha\beta)_{12} = (\alpha\beta)_{21} = (\alpha\beta)_{22} = (\alpha\beta)_{31} = (\alpha\beta)_{32} = 0$$

Alternative hypotheses: The α's are not all equal to zero; the β's are not all equal to zero; the $(\alpha\beta)$ terms are not all equal to zero.

2. *Level of significance*: $\alpha = 0.01$ for all tests.

3. *Criteria*: For replications, reject the null hypotheses if $F > 6.93$, the value of $F_{0.01}$ for $r - 1 = 3 - 1 = 2$ and $ab(r - 1) = (3 \times 2) \times (3 - 1) = 12$ degrees of freedom; for the main effect of factor A, reject the null hypothesis if $F > 9.33$, the value of $F_{0.01}$ for $a - 1 = 3 - 1 = 2$ and $ab(r - 1) = (3 \times 2)(3 - 1) = 12$ degrees of freedom; for the main effect of factor B, reject if $F > 10.04$, the value of $F_{0.01}$ for $b - 1 = 2 - 1 = 1$ and $ab(r - 1) = (3 \times 2)(3 - 1) = 12$ degrees of freedom; for the interaction effect, reject if $F > 6.93$, the value of $F_{0.01}$ for $(a - 1)(b - 1) = (3 - 1)(2 - 1) = 2$ and $ab(r - 1) = (3 \times 2)(3 - 1) = 12$ degrees of freedom.

4. *Calculations*: It is the best practice to use computer software to obtain the analysis of variance table. (See Exercise 13.2 for the *MINITAB* and the end of this example for the *R* commands.)

Source of Variation	Degrees of Freedom	Sum of Squares	Mean Square	F	P
Main Effects:					
A	2	21427	10714	4.48	0.035
B	1	86528	86528	36.1	0.000
Interaction	2	57424	28712	12.0	0.001
Error	12	28724	2394		
Total	17	194103			

5. *Decisions*: *P*-values are given in the last column of the table. The $F = 12.0$ for Factor B exceeds $F_{0.01} = 9.33$ for 1 and 12 degrees of freedom and that $F = 36.1$ for the AB interaction term exceeds $F_{0.01} = 6.93 =$ for 2 and 12 degrees of freedom. In fact, both *P*-values are extremely small.

Because the interaction effect is significant, at level 0.01, we *cannot* conclude that Factor A is unimportant. The effect of changing materials on the estimated response does depend on the location. Location cannot be ignored.

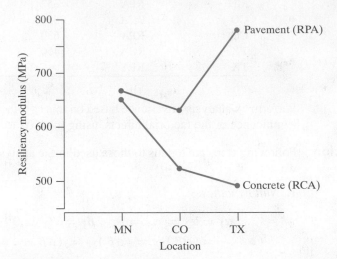

Figure 13.2
Results of recycled materials

The interaction plot in Figure 13.2 gives a visual representation of any interaction between the effects of the two factors.

It is apparent that the increase in resiliency modulus when changing from RCA to RPA is greater at the location TX than at MN. In view of this interaction, great care must be exercised in stating the results of the experiment. For instance, it would be very misleading to state that the effect of changing from RCA to RPA is to increase the resiliency modulus by

$$\frac{6220}{9} - \frac{4972}{9} = 138.7 \text{ MPa}$$

In fact, the resiliency modulus is increased, on average, by 17.3 when the location is MN and it is increased by 289.3 for TX.

Whenever you have two factors, always visually inspect the interaction plot for information about the nature of the interaction or lack of interaction. When there is no interaction, the two profiles will be nearly parallel.

Because of the presence of interaction, the summary must take the form of the two-way table of cell means:

Summary Table of Cell Means $\bar{y}_{ij\bullet}$				
		Factor B **Type of Material**		
		Concrete (RCA)	**Pavement (RPA)**	
Factor A	*MN*	647.7	665.0	656.3
Location	*CO*	520.0	629.3	574.7
	TX	489.7	779.0	634.3
		552.4	691.1	621.8

Statistical software easily calculates the ANOVA table and allows us to focus on the analysis. The **R** software commands

dat=read.table("C13Ex1.TXT", header=T)
model=lm(resilmod~A + B +A:B, data=dat)
anova(model)

produce the ANOVA table. Then

with(dat, tapply(resilmod, list(A,B), mean))

will produce the two-way table of means.

Repeating the analysis, using the *SAS* statistical package ANOVA program, we obtained the output presented in Figure 13.3. The two P values less than 0.0021 confirm our previous analysis and their small values strengthen the conclusions. Note that the F values in Figure 13.3 differ slightly from those above because more decimals are retained.

```
Dependent Variable: ResiliencyMod (MPa)
```

Source	DF	Sum of Squares	Mean Square	F Value	Pr>F
Model	5	165379.1111	33075.8222	13.82	0.0001
Error	12	28724.0000	2393.6667		
Corrected Total	17	194103.1111		Root MSE	
				48.92511	

Figure 13.3
Selected SAS output for ANOVA using the data in Example 1

Source	DF	Anova SS	Mean Square	F Value	Pr > F
Location	2	21427.11111	10713.55556	4.48	0.0353
Type	1	86528.00000	86528.00000	36.15	<.0001
Location*Type	2	57424.00000	28712.00000	11.99	0.0014

When replications and interactions are not significant, the influence of factor A and the influence of factor B can be interpreted separately. Then, when a factor is significant, many statisticians recommend comparing the levels by calculating confidence intervals using the two sample approach but using the mean square error. The confidence intervals for the difference in mean response at levels i_1 and i_2 of factor A have limits

$$\bar{y}_{i_1\cdot\cdot} - \bar{y}_{i_2\cdot\cdot} + t_{\alpha/2}\sqrt{s^2 \frac{2}{b\cdot r}}$$

where $s^2 = SSE/(ab(r-1))$ is the mean square error and the $t_{\alpha/2}$ value is based on $(ab(r-1))$ degrees of freedom.

Similarly, for levels j_1 and j_2 of factor B, the confidence interval for the difference in mean response has limits

$$\bar{y}_{\cdot j_1 \cdot} - \bar{y}_{\cdot j_2 \cdot} \pm t_{\alpha/2}\sqrt{s^2 \frac{2}{a\cdot r}}$$

EXAMPLE 2 **Using confidence intervals to compare means at different factor levels**

Illustrate the calculation of the confidence intervals for the difference in mean response using the means and s^2 from the previous example.

Solution From the analysis of variance table, $s^2 = 2394$ is the mean square error based on 12 degrees of freedom. For these degrees of freedom, we find $t_{0.025} = 2.179$.

Therefore, the confidence intervals for differences in mean due to the $a = 3$ levels of location, factor A, are

$$\bar{y}_{1..} - \bar{y}_{2..} \pm t_{0.025} \sqrt{s^2 \frac{2}{b \cdot r}} = 656.3 - 574.7 \pm 2.179 \sqrt{2394 \frac{2}{2 \cdot 3}}$$

$$\text{or} \quad 20.0 \quad \text{to} \quad 143.2 \quad \text{MPa}$$

$$\bar{y}_{1..} - \bar{y}_{3..} \pm t_{0.025} \sqrt{s^2 \frac{2}{b \cdot r}} = 656.3 - 634.3 \pm 2.179 \sqrt{2394 \frac{2}{2 \cdot 3}}$$

$$\text{or} \quad -39.6 \quad \text{to} \quad 83.6 \quad \text{MPa}$$

$$\bar{y}_{2..} - \bar{y}_{3..} \pm t_{0.025} \sqrt{s^2 \frac{2}{b \cdot r}} = 574.7 - 634.3 \pm 2.179 \sqrt{2394 \frac{2}{2 \cdot 3}}$$

$$\text{or} \quad -121.9 \quad \text{to} \quad 1.95 \quad \text{MPa}$$

Because the interaction was significant, we cannot interpret these intervals on differences of mean resiliency modulus as due to changing location alone.

Similarly, the confidence interval for the single difference in mean due to the $b = 2$ types of material is

$$\bar{y}_{.1.} - \bar{y}_{.2.} \pm t_{0.025} \sqrt{s^2 \frac{2}{a \cdot r}}$$

$$= 552.4 - 691.1 \pm 2.179 \sqrt{2394 \frac{2}{3 \cdot 3}}$$

$$\text{or} \quad -189.0 \quad \text{to} \quad -88.4 \quad \text{MPa} \quad \blacksquare$$

13.2 Multifactor Experiments

Much industrial research and experimentation is conducted to discover the individual and joint effects of several factors on variables thought to be most relevant to response variable under investigation. In the preceeding section, we analyze the $a \times b$ factorial experiment where the experimental conditions represent all possible combinations of the levels of two or factors A and B. In this section, we extend the discussion to factorial experiments involving more than 2 factors, that is, to experiments where the experimental conditions represent all possible combinations of the levels of 3 or more factors.

To illustrate the analysis of a **multifactor experiment**, let us consider the following situation. A warm sulfuric pickling bath is used to remove oxides from the surface of a metal prior to plating, and it is desired to determine what factors in addition to the concentration of the sulfuric acid might affect the electrical conductivity of the bath. As it is felt that the salt concentration as well as the bath temperature might also affect the electrical conductivity, an experiment is planned to determine the individual and joint effects of these 3 variables on the electrical conductivity of the bath. In order to cover the ranges of concentrations and temperatures normally encountered, it is decided to use the following levels of the 3 factors:

Factor	Level 1	Level 2	Level 3	Level 4
A. Acid concentration (%)	0	6	12	18
B. Salt concentration (%)	0	10	20	
C. Bath temperature (°F)	80	100		

The resulting factorial experiment requires $4 \cdot 3 \cdot 2 = 24$ experimental conditions in each replicate, where each experimental condition is a pickling bath made up according to specifications. The order in which these pickling baths are made up should be random. Let us suppose that 2 replicates of the experiment have actually been completed—that is, the electrical conductivities of the various pickling baths have been measured—and that the results are as shown in the following table:

Results of Acid-Bath Experiment					
Level of Factor			Conductivity (mhos/cm^2)		
A	B	C	Rep. 1	Rep. 2	Total
1	1	1	0.99	0.93	1.92
1	1	2	1.15	0.99	2.14
1	2	1	0.97	0.91	1.88
1	2	2	0.87	0.86	1.73
1	3	1	0.95	0.86	1.81
1	3	2	0.91	0.85	1.76
2	1	1	1.00	1.17	2.17
2	1	2	1.12	1.13	2.25
2	2	1	0.99	1.04	2.03
2	2	2	0.96	0.98	1.94
2	3	1	0.97	0.95	1.92
2	3	2	0.94	0.99	1.93
3	1	1	1.24	1.22	2.46
3	1	2	1.12	1.15	2.27
3	2	1	1.15	0.95	2.10
3	2	2	1.11	0.95	2.06
3	3	1	1.03	1.01	2.04
3	3	2	1.12	0.96	2.08
4	1	1	1.24	1.20	2.44
4	1	2	1.32	1.24	2.56
4	2	1	1.14	1.10	2.24
4	2	2	1.20	1.19	2.39
4	3	1	1.02	1.01	2.03
4	3	2	1.02	1.00	2.02
		Total	25.53	24.64	50.17

The model we shall assume for the analysis of this experiment (or any similar three-factor experiment) is an immediate extension of the one used in Section 13.1. If y_{ijkl} is the conductivity measurement obtained at the ith level of acid concentration, the jth level of salt concentration, the kth level of bath temperature, in the lth replicate, we write

Model equation for three-factor experiment

$$Y_{ijkl} = \mu + \alpha_i + \beta_j + \gamma_k + (\alpha\beta)_{ij} + (\alpha\gamma)_{ik} + (\beta\gamma)_{jk} + (\alpha\beta\gamma)_{ijk} + \varepsilon_{ijkl}$$

for $i = 1, 2, \ldots, a$, $j = 1, 2, \ldots, b$, $k = 1, 2, \ldots, c$, and $l = 1, 2, \ldots, r$. We also assume that the sums of the **main effects** (α's, β's, and γ's), that the sums of the **two-way interaction effects** summed on either subscript equal zero for any value of the other subscript, and that the sum of the **three-way interaction effects** summed on any one of the subscripts is zero for any values of the other two subscripts. As before, the ε_{ijkl} are assumed to be independent normal random variables having zero means and the common variance σ^2.

The analysis of variance consists of decomposing the total sum of squares

$$SST = \sum_{i=1}^{a} \sum_{j=1}^{b} \sum_{k=1}^{c} \sum_{\ell=1}^{r} (y_{ijk\ell} - \bar{y}_{\ldots})^2$$

into contributions from each of the model components.

$$SST = SSA + SSB + SSC + SS(AB) + SS(AC) + SS(BC) + SS(ABC) + SSE$$

What is new here is the three factor interaction term

$$SS(ABC) = r \sum_{i=1}^{a} \sum_{j=1}^{b} \sum_{k=1}^{c} (\bar{y}_{ijk\cdot} - \bar{y}_{ij\cdot\cdot} - \bar{y}_{i\cdot k\cdot} - \bar{y}_{\cdot jk\cdot} + \bar{y}_{i\cdot\cdot\cdot} + \bar{y}_{\cdot j\cdot\cdot} + \bar{y}_{\cdot\cdot k\cdot} - \bar{y}_{\ldots})^2$$

This term quantifies the changes in one, or more, of the two-way interactions as the third variable changes levels.

The sums of squares parallel those for SSA, SSB, and $SS(AB)$ from a two-factor experiment but with an extra dot in the subscript. The first of three main effects is

$$SSA = bcr \sum_{i=1}^{a} (\bar{y}_{i\cdots} - \bar{y}_{\ldots})^2$$

and the first of three two-way interaction terms is

$$SS(AB) = cr \sum_{i=1}^{a} \sum_{j=1}^{b} (\bar{y}_{ij\cdot\cdot} - \bar{y}_{i\cdots} - \bar{y}_{\cdot j\cdot\cdot} + \bar{y}_{\ldots})^2$$

and the sum of squares error is

$$SSE = \sum_{i=1}^{a} \sum_{j=1}^{b} \sum_{k=1}^{c} \sum_{\ell=1}^{r} (y_{ijk\ell} - \bar{y}_{ijk\cdot})^2$$

Using a computer software program removes the drudgery of calculation and produces the ANOVA table.

Note that the degrees of freedom for each main effect is one less than the number of levels of the corresponding factor. The degrees of freedom for each interaction is the *product* of the degrees of freedom for those factors appearing in the interaction. Thus, the degrees of freedom for the three main effects are 3, 2, and 1 in this example, while the degrees of freedom for the two-way interactions are 6, 3, and 2, and the degrees of freedom for the three-way interaction are 6.

Source of Variation	Degrees of Freedom	Sum of Squares	Mean Square	F	P
Main effects:					
A	3	0.27504	0.09168	24.0	0.000
B	2	0.22622	0.11311	29.6	0.000
C	1	0.00017	0.00017	<1	
Two factor interactions:					
AB	6	0.02882	0.00480	1.26	0.311
AC	3	0.00851	0.00284	<1	
BC	2	0.00420	0.00210	<1	
Three-factor interaction:					
ABC	6	0.02820	0.00470	1.23	0.323
Error	24	0.09125	0.00382		
Total	47	0.66241			

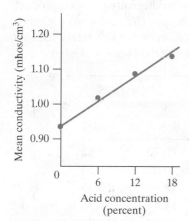

Figure 13.4

Effect of acid concentration

Obtaining the appropriate values of $F_{0.05}$ and $F_{0.01}$ from Table 6, we find that the test for the factor A and factor B main effects are significant at the 0.01 level. None of the other F's is significant at either level. We conclude from this analysis that variation in acid concentration and salt concentration affect the electrical conductivity, variations in bath temperature do not, and that there are no interactions. The extremely small P-values, less than 0.001, further strengthen this conclusion.

To go one step farther, we might investigate the *magnitudes* of the effects by studying graphs of means like those shown in Figure 13.4 and 13.5. Here we find that the conductivity increases as acid is added and decreases as salt is added; using the methods of Chapter 11, we might even fit lines, curves, or surfaces to describe the response surface relating conductivity to the variables under consideration.

[Using **R**:

```
dat=read.table("C13acidb.TXT","factor",
"factor","factor","numeric","factor", header=T)
model=lm(conduct~A + B + A:B + A:C+B:C+A:B:C, data=dat)
anova(model)
```

produce the ANOVA table. Then

```
with(dat, tapply(conduct, list(A), mean))
with(dat, tapply(conduct, list(B), mean))
```

produce the relevant means.]

Figure 13.5

Effect of salt concentration

For a factorial design having three factors, we can obtain confidence intervals for the difference of means corresponding to two different levels of any main effect that does not interact with the other factors. These intervals are based on the estimate of the variance s^2 that is the mean square error in the ANOVA table. This estimate of the variance of one response is based on $a \cdot b \cdot c (r - 1)$ degrees of freedom.

The $100 (1 - \alpha)$ confidence interval for the difference between the two levels i_1 and i_2 of Factor A has limits

$$\bar{y}_{i_1 \cdots} - \bar{y}_{i_2 \cdots} \pm t_{\alpha / 2} \sqrt{s^2 \frac{2}{b \cdot c \cdot r}}$$

The intervals for the difference between the two levels j_1 and j_2 of Factor B has limits

$$\bar{y}_{\cdot j_1 \cdot \cdot} - \bar{y}_{\cdot j_2 \cdot \cdot} \pm t_{\alpha/2} \sqrt{s^2 \frac{2}{a \cdot c \cdot r}}$$

The intervals for the difference between the two levels k_1 and k_2 of Factor C has limits

$$\bar{y}_{\cdot \cdot k_1 \cdot} - \bar{y}_{\cdot \cdot k_2 \cdot} \pm t_{\alpha/2} \sqrt{s^2 \frac{2}{a \cdot b \cdot r}}$$

Example 3 illustrates the calculation of these confidence intervals.

EXAMPLE 3 **Conducting an analysis of variance to improve the safety of an ignitor**

A customer requested improvements in the safety of an ignitor. Statistically designed experiments, run at the development stage, can help build quality into this product. It was decided to study 3 initiators (A), 2 booster charges (B), and 4 main charges (C). One response measured was the delay time (milliseconds). For safety reasons, this should remain under 30. Two replicates were run of the factorial design.

A	B	C	Delay Time (milliseconds) Rep. 1	Rep. 2
Initiator 1	Powder	Mc 1	10.70	9.82
Initiator 1	Pellet	Mc 1	10.02	13.66
Initiator 1	Powder	Mc 2	14.46	20.86
Initiator 1	Pellet	Mc 2	11.44	13.76
Initiator 1	Powder	Mc 3	15.04	16.02
Initiator 1	Pellet	Mc 3	27.26	21.42
Initiator 1	Powder	Mc 4	20.82	14.46
Initiator 1	Pellet	Mc 4	24.56	36.48
Initiator 2	Powder	Mc 1	18.42	18.62
Initiator 2	Pellet	Mc 1	22.80	25.14
Initiator 2	Powder	Mc 2	33.40	20.62
Initiator 2	Pellet	Mc 2	31.86	19.78
Initiator 2	Powder	Mc 3	22.94	31.12
Initiator 2	Pellet	Mc 3	32.92	21.38
Initiator 2	Powder	Mc 4	27.92	59.86
Initiator 2	Pellet	Mc 4	31.94	28.32
Initiator 3	Powder	Mc 1	7.14	7.98
Initiator 3	Pellet	Mc 1	24.32	10.26
Initiator 3	Powder	Mc 2	8.30	7.86
Initiator 3	Pellet	Mc 2	7.00	8.40
Initiator 3	Powder	Mc 3	8.40	10.94
Initiator 3	Pellet	Mc 3	17.82	15.28
Initiator 3	Powder	Mc 4	9.56	19.04
Initiator 3	Pellet	Mc 4	19.98	18.46

Analyze this experiment.

Solution   A computer program produced the ANOVA table.

SOURCE	DF	SS	MS	F	P
A	2	1973.18	986.59	22.22	0.000
B	1	74.90	74.90	1.69	0.206
C	3	864.39	288.13	6.49	0.002
AB	2	141.83	70.91	1.60	0.223
AC	6	200.97	33.50	0.75	0.612
BC	3	122.14	40.71	0.92	0.448
ABC	6	319.57	53.26	1.20	0.340
ERROR	24	1065.86	44.41		
TOTAL	47	4762.84	101.34		

Only the initiators and main charges are significant. They are significant even at $\alpha = 0.01$. The experiment can therefore be summarized by the two sets of sample means, those for initiators and those for main charges:

initiator 1	$\bar{y}_{1\cdots} = 17.55$	Mc 1	$\bar{y}_{\cdot\cdot1\cdot} = 14.91$
initiator 2	$\bar{y}_{2\cdots} = 27.94$	Mc 2	$\bar{y}_{\cdot\cdot2\cdot} = 16.48$
initiator 3	$\bar{y}_{3\cdots} = 12.55$	Mc 3	$\bar{y}_{\cdot\cdot3\cdot} = 20.05$
		Mc 4	$\bar{y}_{\cdot\cdot4\cdot} = 25.95$

According to the estimated model, where the initiator and main charge effects are additive,

$$\widehat{y}_{i\cdot k\cdot} = \bar{y}_{\cdots\cdot} + (\,\bar{y}_{i\cdots} - \bar{y}_{\cdots\cdot}\,) + (\,\bar{y}_{\cdot\cdot k\cdot} - \bar{y}_{\cdots\cdot}\,)$$

the lowest estimated delay time would come from using initiator 3 with main charge 1 (Mc 1). This experiment was successful in identifying better components for the ignitor.

The standard deviation of the delay times is estimated by s, which is the square root of the mean square error. Since $s = 6.66$ milliseconds, there is considerable variation in the individual delay times. The confidence intervals for the difference in mean delay times for two initiators are

$$\bar{y}_{i\cdots} - \bar{y}_{m\cdots} \pm t_{0.025} \sqrt{\frac{s^2}{2\cdot4}} = \bar{y}_{i\cdots} - \bar{y}_{m\cdots} \pm 2.064 \sqrt{\frac{44.41}{8}}$$

or $\bar{y}_{i\cdots} - \bar{y}_{m\cdots} \pm 4.86$, so differences of 4.86 are significant. All three initiator means are different. The confidence intervals for main charges

$$\bar{y}_{\cdot\cdot k\cdot} - \bar{y}_{\cdot\cdot m\cdot} \pm t_{0.025} \sqrt{\frac{s^2}{3\cdot2}}$$

lead to differences of 5.62 being significant. Consequently, these data do not establish any difference between the first three main charges Mc 1, Mc 2, and Mc 3.

[Using **R** : With short hand notation $A * B * C = A : B + A : C + B : C + A : B : C$, we write

```
dat=read.table("C13Ex3.TXT", header=T)
model=lm(delay~A + B +C+A*B*C*, data=dat)
anova(model)
```

produce the ANOVA table. Then

$$\text{with(dat, tapply(delay, list(A), mean))}$$
$$\text{with(dat, tapply(delay, list(C), mean))}$$

produce the relevant means.]   ∎

It is strongly recommended that you use a standard statistical software program to perform the statistical analysis of any multifactor experiment and to produce various plots of the residuals.

Exercises

13.1 Given the two replications of a 2 × 3 factorial experiment, calculate the analysis of variance table using the formulas on pages *** and ***.

Factor A	Factor B	Rep. 1	Rep. 2
1	1	15	21
1	2	1	3
1	3	10	8
2	1	1	1
2	2	16	14
2	3	5	13

13.2 MINITAB can create the analysis of variance table for Example 1 concerning recycled road materials. The three levels of A are coded 1, 2, and 3, and the two levels of B are coded 1 and 2. The third column contains the values of the resiliency modulus.

1	1	707
1	1	632
1	1	604
1	2	652
...		

> **Dialog box:**
> **Stat > ANOVA > Balanced ANOVA**
> Type Y in **Responses**. In **Model** type $A\ B\ A * B$. Click **OK**.

Source	DF	SS	MS	F	P
A	2	21427	10714	4.48	0.035
B	1	86528	86528	36.15	0.000
A*B	2	57424	28712	11.99	0.001
Error	12	28724	2394		
Total	17	194103			

To obtain the two-way table of means, click on **Results** and type $A * B$ in the **Display means** box. Then, click **OK**.
Repeat Exercise 13.1 using computer calculations.

13.3 To determine optimum conditions for a plating bath, the effects of sulfone concentration and bath temperature on the reflectivity of the plated metal are studied in a 2 × 5 factorial experiment. The results of three replicates are as follows:

Concentration	Temperature	Reflectivity		
(grams/liter)	(°F)	Rep. I	Rep. 2	Rep. 3
5	75	35	39	36
5	100	31	37	36
5	125	30	31	33
5	150	28	20	23
5	175	19	18	22
10	75	38	46	41
10	100	36	44	39
10	125	39	32	38
10	150	35	47	40
10	175	30	38	31

Analyze these results and determine the bath condition or conditions that produce the highest reflectivity. Also construct a 95% confidence interval for the reflectivity of the plating bath corresponding to these optimum conditions.

13.4   A spoilage-retarding ingredient is added in brewing beer. To determine the extent to which the taste of the beer is affected by the amount of this ingredient added to each batch, and how such taste changes might depend on the age of the beer, a 3×4 factorial experiment in two replications was designed. The taste of the beer was rated on a scale of 0, 1, 2, or 3 (3 being the most desirable) by a panel of trained experts, who reported the following mean ratings:

Amount of Ingredient	Aging Period	Mean Ratings	
(grams per batch)	(weeks)	Rep. I	Rep. 2
2	2	2.1	1.6
2	4	2.6	1.9
2	6	2.9	2.4
3	2	1.4	1.7
3	4	1.9	2.2
3	6	2.3	2.7
4	2	0.5	0.9
4	4	1.2	0.8
4	6	1.7	1.4
5	2	1.0	1.6
5	4	2.2	1.3
5	6	2.3	2.1

Interpret the results of this experiment.

13.5   Suppose that in the experiment described in Example 7, Chapter 12, it is desired to determine also whether there is an interaction between the detergents and the engines; that is, whether one detergent might perform better in Engine 1, another might perform better in Engine 2, and so on. Combining the data in Example 7, Chapter 12, with the following replicate of the experiment, test for a significant interaction and discuss the results.

	Engine 1	Engine 2	Engine 3
Detergent A	39	42	58
Detergent B	44	46	48
Detergent C	34	47	45
Detergent D	47	45	57

13.6 An experiment was designed to study the effects of melt flow index (rate) and filler content (% of weight) on the strength of resin as measured by the secant modulus.

Melt Flow Index	Filler Content	Rep. 1	Rep. 2
3	32	1.95	2.14
3	37	2.36	2.28
3	42	2.61	2.56
12	32	2.32	2.31
12	37	2.33	2.28
12	42	2.60	2.53
30	32	2.29	2.36
30	37	2.61	2.50
30	42	2.68	2.52

Analyze this experiment.

13.7 The commercial value of softwood species would be increased if the wood could be treated to meet preserver's standards. The response, y, is the amount of retention (lb/ft^3) of the preservative. Two treatments (preservatives) were considered and the samples were either incised or unincised. Further, the conditions of spruce trees were not defoliated, partially defoliated, or totally defoliated.

		Replicate 1		Replicate 2	
		Treatment 1	Treatment 2	Treatment 1	Treatment 2
Not	Incised	1.28	1.22	0.77	1.09
Defoliated	Unincised	1.28	1.22	1.22	1.17

		Replicate 1		Replicate 2	
		Treatment 1	Treatment 2	Treatment 1	Treatment 2
Partially	Incised	0.63	0.78	0.64	0.74
Defoliated	Unincised	0.39	0.49	0.39	0.53

		Replicate 1		Replicate 2	
		Treatment 1	Treatment 2	Treatment 1	Treatment 2
Totally	Incised	0.65	1.09	0.38	0.63
Defoliated	Unincised	0.40	0.82	0.40	0.60

Perform an appropriate analysis of variance and interpret the results.

13.8 A market test was performed to evaluate the impact of shelf position, and label color of a canned food product on sales.

Shelf Position	Label Color	Sales (dollars)		
		Day 1	Day 2	Day 3
Low	Red	70.10	68.00	69.50
Low	Red	72.25	71.90	74.70
Low	Red	78.05	74.85	82.60
Low	Red	61.50	62.10	59.15
Low	Green	65.75	62.35	68.60
Low	Green	69.45	71.05	75.45
Low	Green	75.15	70.70	71.25
Low	Green	64.80	60.85	59.90
Medium	Red	94.10	90.20	88.05
Medium	Red	104.85	99.55	96.80
Medium	Red	109.10	105.80	112.60
Medium	Red	59.90	62.50	54.75
Medium	Green	88.95	91.10	90.15
Medium	Green	100.60	94.05	101.35
Medium	Green	98.70	99.90	96.75
Medium	Green	62.50	53.85	59.40
High	Red	92.60	88.80	85.50
High	Red	100.55	102.15	99.10
High	Red	111.95	108.25	109.45
High	Red	61.40	65.20	59.70
High	Green	97.35	98.70	92.60
High	Green	120.65	115.45	108.65
High	Green	118.10	116.35	121.90
High	Green	70.30	65.05	71.40

Analyze this experiment.

13.9 Solve the 4 equations on page 416 for μ, α_1, β_1, and $(\alpha\beta)_{11}$ in terms of the population means μ_{ijl} corresponding to the 4 experimental conditions in the first replicate. Note that these equations serve as a guide for estimating the parameters in terms of the *sample means* corresponding to the various experimental conditions.

13.3   The Graphic Presentation of 2^2 and 2^3 Experiments

Too often in the past, engineers and scientists have taken the *change one variable at a time* approach to designing experiments. After first determining possible causal variables, one variable is changed at a time while the others are held fixed. This approach may initially appear reasonable, but it is both inefficient and can produce seriously misleading conclusions.

To see how important it is to change more than one variable at a time, suppose two input variables, x_1 and x_2, are varied in an attempt to locate the maximum response. In the situation illustrated in Figure 13.6, moving (x_1, x_2) toward the upper right-hand corner will increase the response from 6 to 10 and even higher. However, if the experimenter fixes x_1 at 1.6 and varies x_2, it will look like a maximum occurs at $x_2 = 1.7$. If x_2 is fixed at 1.7 and x_1 varied, it will then appear as if a maximum is confirmed. That is, the classical method of varying one variable at a time can lead to a false location for maximum response. Factorial designs are well suited for studying the influence of several factors on a response.

Experimental designs based on only two or three input variables, each having two possible values, can greatly aid our understanding of complex phenomena. The key ingredient is the systematic variation of all the input variables. Each input variable is called a **factor** and its values are called **levels**. All combinations of the **levels**

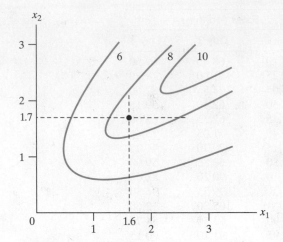

Figure 13.6
False location of a maximum response

of the factors are considered as treatments. When there are two factors and only 2 levels of each factor, there are $2 \times 2 = 2^2$, or four treatments or experimental conditions. This design is called a 2^2 **factorial design** and a 2^3 **factorial design** is created when three factors are included.

Both the 2^2 and 2^3 factorial designs have the added advantage that dramatic graphical displays are available to convey the results of the experiment. In spite of its simplicity, even the 2^2 design is a powerful tool to improve products and processes. The key is the decision to perform any designed experiment at all.

2^2 Design

Every 2^2 factorial design has two factors and each has two levels. To begin, we code the two levels each factor as -1 for the low level and $+1$ for the high level. If the two levels of a factor are non-quantitative, say two different brands of detergent, simply designate one brand as level -1 and the other as $+1$.

We record the four possible experimental conditions in the **standard order** for a 2^2 factorial experiment.

	Design	
x_1		x_2
-1		-1
1		-1
-1		1
1		1

The first column alternates signs between -1 and $+1$ starting with -1, while the second column alternates doubles starting with two -1's.

Geometrically, the four conditions comprising the 2^2 design are the corners of the square in Figure 13.7. Attached at these corners are the four values of the corresponding mean responses $\bar{y}_1$, $\bar{y}_2$, $\bar{y}_3$, and $\bar{y}_4$. Their subscripts match the row number in the design.

Suppose the yield of a new chemical process for growing crystals needs improvement. Because temperature and pH are thought to influence yield, a 2^2 design is attempted with two replicates. The following results are obtained for yield.

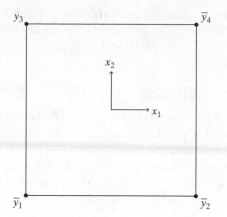

Figure 13.7
The graphical presentation of
the results of a 2^2 factorial
experiment

Factor A Temperature	Factor B pH	Rep. 1	Rep. 2	Mean
300	2	10	14	$\bar{y}_1 = 12$
350	2	21	19	$\bar{y}_2 = 20$
300	3	17	15	$\bar{y}_3 = 16$
350	3	20	24	$\bar{y}_4 = 22$

Coding the low levels as -1 and the high levels as $+1$ confirms that the design is in the standard order. Because both factors are quantitative, the coding can be expressed by formula.

$$x_1 = \frac{\text{Temperature} - (300 + 350)/2}{(350 - 300)/2} \quad \text{and} \quad x_2 = \frac{\text{pH} - (2 + 3)/2}{(3 - 2)/2}$$

Figure 13.8 shows the square representing the design with the sample means written at the corners. For instance $\bar{y}_1 = 12$ is attached to the lower-left hand corner.

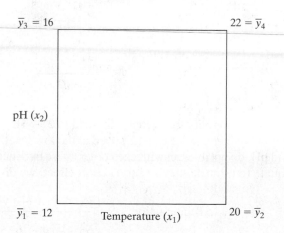

Figure 13.8
Visual presentation of yield
from a 2^2 factorial experiment

It is clear from Figure 13.8 that increasing the temperature from low ($300°$) to high ($350°$) increases the yield substantially at both levels of pH. Below we develop confidence intervals that enable us to confirm the size of the improvement. This figure very effectively presents the information contained in the experiment.

We continue our analysis by finding point estimates and confidence intervals for the magnitude of the difference in mean response that results when a factor is changed from its low to high level.

Estimates of effects

To estimate the main effect of temperature, we see that runs 1 and 2 both have x_2 at the low level. Their difference $(20 - 12) = 8$, along the bottom of the box in Figure 13.8, is one measure of the effect of increasing x_1 from -1 to $+1$. Runs 3 and 4 also have the same x_2 value so the difference $(22 - 16) = 6$ is another estimate of the effect of changing x_1. The estimate of the **main effect** of factor A, temperature, is the average of these two estimates.

$$\text{Main effect Factor A} = \frac{(\bar{y}_2 - \bar{y}_1) + (\bar{y}_4 - \bar{y}_3)}{2}$$
$$= \frac{(20 - 12) + (22 - 16)}{2} = 7$$

Similarly, the main effect of factor B is the average of two estimates

$$\text{Main effect Factor B} = \frac{(\bar{y}_3 - \bar{y}_1) + (\bar{y}_4 - \bar{y}_2)}{2}$$
$$= \frac{(16 - 12) + (22 - 20)}{2} = 3$$

These estimates also have the interpretation as the average response on the face of the square with level $+1$ minus the average on the face of the square with level -1. Both interpretations for the estimate of the main effect of Factor A are illustrated in Figure 13.9 (a) and (b).

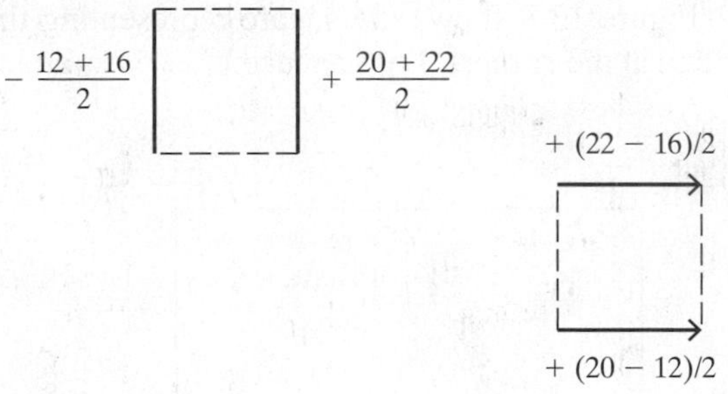

Figure 13.9
Two interpretations of estimate of main effect of Factor A from a 2^2 factorial experiment
(a) Difference of face means
(b) Mean of differences top and bottom

Next, consider the increase in yield $(22 - 16) = 6$ at the high level of Factor B and the increase $(20 - 12) = 8$ at the low level. If the two factors, temperature and pH, do not interact with each other, these two increases should be approximately equal. To estimate the AB **interaction effect**, we divide their difference by 2.

$$\text{AB interaction} = \frac{(\bar{y}_4 - \bar{y}_3) - (\bar{y}_2 - \bar{y}_1)}{2}$$
$$= \frac{(22 - 16) - (20 - 12)}{2} = -1$$

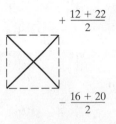

Figure 13.10
Geometric interpretation of estimate of AB interaction

Note the same answer can be obtained using the difference of increases due to second factor $[(22 - 20) - (16 - 12)]/2 = -1$. A second interpretation in terms of average response on the diagonals is illustrated in Figure 13.10.

An important connection between the plus and minus signs of the estimators becomes apparent by augmenting the design for x_1 and x_2 to include a column for the product term $x_1 x_2$. Notice that the signs in the x_1, x_2 and $x_1 x_2$ columns in the following table are the signs for the coefficients of the $\bar{y}_i$'s for estimating the corresponding effects.

Factor A x_1	Factor B x_2	Interaction $x_1 x_2$	Mean
-1	-1	1	$\bar{y}_1$
1	-1	-1	$\bar{y}_2$
-1	1	-1	$\bar{y}_3$
1	1	1	$\bar{y}_4$

We summarize the estimates of the effects.

Main effect Factor A: $\dfrac{(\bar{y}_2 - \bar{y}_1) + (\bar{y}_4 - \bar{y}_3)}{2}$

Main effect Factor B: $\dfrac{(\bar{y}_3 - \bar{y}_1) + (\bar{y}_4 - \bar{y}_2)}{2}$

AB interaction: $\dfrac{(\bar{y}_4 - \bar{y}_3) - (\bar{y}_2 - \bar{y}_1)}{2}$

Confidence intervals for the effects

To obtain confidence intervals, we introduce model assumptions that are extensions of the two independent samples case. When the number of replicates $r > 1$,

Model Assumptions

1. $Y_{i1}, Y_{i2}, \ldots, Y_{ir}$ are independent and distributed as $N(\mu_i, \sigma)$ for $i = 1, 2, 3, 4$.
2. The four random samples are independent.

The assumptions state that the four normal populations have a common variance.
    We develop an estimate of standard error of an estimator of an effect by first noting that all estimators are of the form

$$\frac{1}{2} \left[\pm \bar{Y}_1 \pm \bar{Y}_2 \pm \bar{Y}_3 \pm \bar{Y}_4 \right]$$

Each sample mean is the average of r independent observations so that $Var(\bar{Y}_i) = \sigma^2/r$. Then, because the means $\bar{Y}_1 \bar{Y}_2, \bar{Y}_3$ and $\bar{Y}_4$ are independent, the variances add.

$$Var(\text{ estimator effect }) = \frac{1}{4} \left[\frac{\sigma^2}{r} + \frac{\sigma^2}{r} + \frac{\sigma^2}{r} + \frac{\sigma^2}{r} \right] = \frac{\sigma^2}{r}$$

To estimate σ^2, we extend the concept of pooling connected with the two sample t statistic. Expressing the observations as random variables, the treatment with $x_1 = -1$ and $x_2 = -1$ contributes $(r-1)S_1^2 = \sum_{j=1}^{r} (Y_{1j} - \bar{Y}_1)^2$ to the pooled estimate of variance. Similarly, the other three treatments contribute $(r-1)S_2^2$, $(r-1)S_3^2$, and $(r-1)S_4^2$. The pooled estimate of σ^2 is the sum of these four contributions divided by the number of degrees of freedom $(r - 1) + (r - 1) + (r - 1) + (r - 1) = 4(r - 1)$.

$$s^2 = \frac{\sum_{i=1}^{4} \sum_{j=1}^{r} (Y_{ij} - \overline{Y}_i)^2}{4(r - 1)} = \frac{1}{4}\left(s_1^2 + s_2^2 + s_3^2 + s_4^2\right)$$

(See Theorem 6.5 and Example 14, Chapter 6, that relate the numerator to a χ^2 distribution and that leads to the t distribution for making inferences about effect.)

In our example of growing crystals, the observed variance s is

$$s^2 = \frac{1}{4}(4 + 4 + 1 + 1 + 1 + 1 + 4 + 4) = 5$$

A 95% confidence interval for an effect based on r replications of a 2^2 design.

$$\text{Estimated effect} \pm \sqrt{\frac{s^2}{r}}\ t_{0.025}$$

where $s^2 = \frac{1}{4}(s_1^2 + s_2^2 + s_3^2 + s_4^2)$ and $t_{0.025}$ is based on $4(r - 1)$ degrees of freedom.

We can now summarize the crystal growing experiment where $r = 2$ so the degrees of freedom $= 4$ and $t_{0.025} = 2.776$. Also, $s^2 = 5$ and $t_{0.025}\sqrt{s^2/2} = 2.776\sqrt{5/2} = 4.39$.

temperature effect:	7 ± 4.39	or	$(2.61, 11.39)$
pH effect:	3 ± 4.39	or	$(-1.39, 7.39)$
temperature $\times$ pH interaction:	-1 ± 4.39	or	$(-5.39, 3.39)$

The confidence intervals have not only revealed that only the main effect of temperature is non-zero but have indicated the size of the effect when temperature is increased from $300°$ to $350°$.

Whenever the interaction effect is judged to be significant, we *cannot* conclude that either main effect is not significant. In the presence of interaction, the two factors must be considered jointly. The proper summary is the two-way table of means

	Factor B	
Factor A	Low	High
Low	$\overline{y}_1$	$\overline{y}_3$
High	$\overline{y}_2$	$\overline{y}_4$

or, equivalently, the square with the average responses in Figure 13.8.

2^3 Design

We now turn to graphic displays for the 2^3 factorial design. First, we record the eight possible experimental conditions in the **standard order** for the 2^3 factorial design using the coded values -1 and $+1$.

	Design	
x_1	x_2	x_3
-1	-1	-1
1	-1	-1
-1	1	-1
1	1	-1
-1	-1	1
1	-1	1
-1	1	1
1	1	1

The first column alternates signs between -1 and $+1$ starting with -1, while the second column alternates doubles starting with two -1's. The third column alternates four rows at a time starting with four -1's.

In order to help expand the company's market for a plastic wrapping material, engineers were asked to improve the opacity. They felt that three factors—rate of extrusion, amount of an additive, and nozzle setting—might have an effect. Two levels were selected for each factor, and a 2^3 factorial experiment produced the observations:

Factor A Rate	Factor B Amount Additive	Factor C Nozzle Setting	Rep. 1	Rep. 2
-1	-1	-1	4.5	4.1
1	-1	-1	3.8	3.4
-1	1	-1	3.1	4.3
1	1	-1	7.2	6.8
-1	-1	1	5.4	5.0
1	-1	1	4.5	4.9
-1	1	1	4.2	5.4
1	1	1	7.3	6.9

A cube representing the factors is shown in Figure 13.11. At the corners of the cube, we have attached the mean response for that set of experimental conditions. For instance, $(7.3 + 6.9)/2 = 7.1$ appears at the upper right corner of the front face. It is clear from Figure 13.11 that the response increases as both factor A and factor B are simultaneously changed from their low to high levels. This is an interaction effect since just changing one of the factors does not always increase the response. Factor C, nozzle setting, also seems to have an effect.

Estimates of effects

The estimate of the main effect of A is the average of the four $\bar{y}_i$'s on the front face minus the average of the four on the back face.

$$\begin{aligned}
\text{main effect Factor } A &= \frac{1}{4}\left(\bar{y}_2 + \bar{y}_4 + \bar{y}_6 + \bar{y}_8\right) - \frac{1}{4}\left(\bar{y}_1 + \bar{y}_3 + \bar{y}_5 + \bar{y}_7\right) \\
&= \frac{-\bar{y}_1 + \bar{y}_2 - \bar{y}_3 + \bar{y}_4 - \bar{y}_5 + \bar{y}_6 - \bar{y}_7 + \bar{y}_8}{4} \\
&= \frac{1}{4}\left(-4.3 + 3.6 - 3.7 + 7.0 - 5.2 + 4.7 - 4.8 + 7.1\right) \\
&= 1.1
\end{aligned}$$

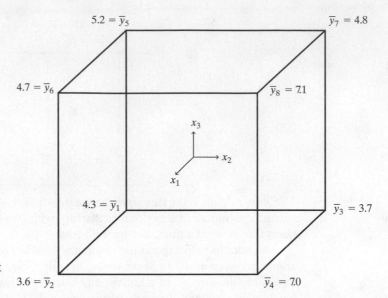

Figure 13.11

Visual display of 2^3 experiment with replication

Alternatively, this estimate is the average of the four increases along the edges where a factor is changed from its low to high level. (See Figure 13.12(a)). The estimates of the other two main effects have similar interpretations.

$$\text{main effect Factor } B = \frac{-\bar{y}_1 - \bar{y}_2 + \bar{y}_3 + \bar{y}_4 - \bar{y}_5 - \bar{y}_6 + \bar{y}_7 + \bar{y}_8}{4}$$

$$= 1.2$$

$$\text{main effect Factor } C = \frac{-\bar{y}_1 - \bar{y}_2 - \bar{y}_3 - \bar{y}_4 + \bar{y}_5 + \bar{y}_6 + \bar{y}_7 + \bar{y}_8}{4}$$

$$= 0.8$$

The AB interaction is estimated as the average of the interaction on the top face and the interaction on the bottom face. Figure 13.12 gives the signs for the $\bar{y}_i$. We estimate

$$AB \text{ interaction} = \frac{(\bar{y}_1 - \bar{y}_2 - \bar{y}_3 + \bar{y}_4)/2 + (\bar{y}_5 - \bar{y}_6 - \bar{y}_7 + \bar{y}_8)/2}{2}$$

$$= \frac{1}{4}\left[\bar{y}_1 - \bar{y}_2 - \bar{y}_3 + \bar{y}_4 + \bar{y}_5 - \bar{y}_6 - \bar{y}_7 + \bar{y}_8\right]$$

$$= \frac{1}{4}[4.3 - 3.6 - 3.7 + 7.0 + 5.2 - 4.7 - 4.8 + 7.1]$$

$$= 1.7$$

The combination of signs for the 8 means are obtained in a similar manner for the other 2 factor interactions.

$$AC \text{ interaction} = \frac{1}{4}\left[\bar{y}_1 - \bar{y}_2 + \bar{y}_3 - \bar{y}_4 - \bar{y}_5 + \bar{y}_6 - \bar{y}_7 + \bar{y}_8\right]$$

$$= \frac{1}{4}[4.3 - 3.6 + 3.7 - 7.0 - 5.2 + 4.7 - 4.8 + 7.1]$$

$$= -0.2$$

$$BC \text{ interaction} = \frac{1}{4}\left[\bar{y}_1 + \bar{y}_2 - \bar{y}_3 - \bar{y}_4 - \bar{y}_5 - \bar{y}_6 + \bar{y}_7 + \bar{y}_8\right]$$

$$= \frac{1}{4}[4.3 + 3.6 - 3.7 - 7.0 - 5.2 - 4.7 + 4.8 + 7.1]$$

$$= -0.2$$

The three-factor interaction ABC is a measure of the difference of the AB interaction on the top faces and the interaction on the bottom face. (See Figure 13.12.) That is, it quantifies the influence of C on the AB interaction. (There is a symmetry here, and, up to a minus sign, the same result is obtained starting with any two-factor interaction.)

$$ABC \text{ interaction} = \frac{(\bar{y}_5 - \bar{y}_6 - \bar{y}_7 + \bar{y}_8)/2 - (\bar{y}_1 - \bar{y}_2 - \bar{y}_3 + \bar{y}_4)/2}{2}$$

$$= \frac{1}{4}\left[-\bar{y}_1 + \bar{y}_2 + \bar{y}_3 - \bar{y}_4 + \bar{y}_5 - \bar{y}_6 - \bar{y}_7 + \bar{y}_8\right]$$

$$= \frac{1}{4}[-4.3 + 3.6 + 3.7 - 7.0 + 5.2 - 4.7 - 4.8 + 7.1]$$

$$= -0.3$$

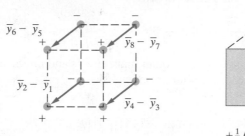
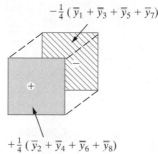

(a) Main effect of A

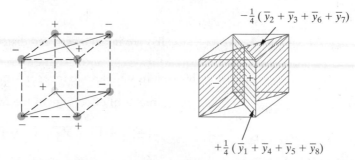

(b) AB interaction

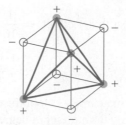

(c) ABC interaction

Figure 13.12
The signs for estimating effects in a 2^3 design

The geometrical interpretation of the estimates is given in Figure 13.13.

An alternative method of determining the sign of any $\bar{y}_i$ arises when we augment the design by including additional columns for products of the original columns.

Rate x_1	Design Additive x_2	Nozzle x_3	$x_1 x_2$	$x_1 x_3$	$x_2 x_3$	$x_1 x_2 x_3$	Rep. 1	Rep. 2	Mean
−1	−1	−1	1	1	1	−1	4.5	4.1	4.3
1	−1	−1	−1	−1	1	1	3.8	3.4	3.6
−1	1	−1	−1	1	−1	1	3.1	4.3	3.7
1	1	−1	1	−1	−1	−1	7.2	6.8	7.0
−1	−1	1	1	−1	−1	1	5.4	5.0	5.2
1	−1	1	−1	1	−1	−1	4.5	4.9	4.7
−1	1	1	−1	−1	1	−1	4.2	5.4	4.8
1	1	1	1	1	1	1	7.3	6.9	7.1

Notice that again the signs of the columns are the coefficients of the linear combination of means. The value of each linear combination, when divided by 2^{3-1}, is the estimated effect when changing from the low to high values of the input variables. For the main effect of rate,

$$\frac{1}{2^{3-1}} \left(-4.3 + 3.6 - 3.7 + 7.0 - 5.2 + 4.7 - 4.8 + 7.1 \right) = 1.1$$

The estimates of the other main effects and interactions are similarly obtained using the appropriate columns of signs.

Confidence intervals for the effects

To obtain confidence intervals, we must specify the model assumptions. When the number of replicates $r > 1$,

Model Assumptions

1. $Y_{i1}, Y_{i2}, \ldots, Y_{ir}$ are independent and distributed as $N(\mu_i, \sigma)$ for $i = 1, 2, \ldots, 8$.

2. The eight random samples are independent.

All seven estimates of effects are of the form

$$\frac{1}{4}[\pm \bar{Y}_1 \pm \bar{Y}_2 \pm \bar{Y}_3 \pm \bar{Y}_4 \pm \bar{Y}_5 \pm \bar{Y}_6 \pm \bar{Y}_7 \pm \bar{Y}_8]$$

and $Var(\bar{Y}_i) = \sigma^2/r$ for $i = 2, \ldots, 8$. Then, because the means are independent,

$$\text{Var (estimator effect)} = \frac{1}{16} \left(\frac{\sigma^2}{r} + \frac{\sigma^2}{r} + \frac{\sigma^2}{r} + \frac{\sigma^2}{r} + \frac{\sigma^2}{r} + \frac{\sigma^2}{r} + \frac{\sigma^2}{r} + \frac{\sigma^2}{r} \right)$$

$$= \frac{1}{2} \frac{\sigma^2}{r}$$

To estimate σ^2, we pool the eight contributions $(r-1)S_i^2 = \sum_{j=1}^{r}(Y_{ij} - \bar{Y}_i)^2$ and divide by the number of degrees of freedom $8(r-1)$. (See Theorem 6.5 and Example 14, Chapter 6)

> **A 95% confidence interval for an effect in a 2^3 design,**
> Based on r replicates,
>
> $$\text{Estimated effect} \pm \sqrt{\frac{s^2}{2r}}\ t_{0.025}$$
>
> where $s^2 = \frac{1}{8}(s_1^2 + s_2^2 + s_3^2 + s_4^2 + s_5^2 + s_6^2 + s_7^2 + s_8^2)$ and $t_{0.025}$ is based on $8(r-1)$ degrees of freedom.

In our example

$$s^2 = \frac{1}{8}(0.08 + 0.08 + 0.72 + 0.08 + 0.08 + 0.08 + 0.72 + 0.08)$$

$$= \frac{1}{8}(1.92) = 0.24$$

and $t_{0.025} = 2.306$ for $(r-1)2^3 = 8$ degrees of freedom. The half length of the confidence interval is $t_{0.025}\sqrt{s^2/(2r)} = 2.306\sqrt{0.24/4} = 0.56$.

Naming the factors according to our application, the resulting 95% individual confidence intervals are

rate effect:	$1.1 \pm 0.56,$	or	0.54	to	1.66
additive effect:	$1.2 \pm 0.56,$	or	0.64	to	1.76
nozzle effect:	$0.8 \pm 0.56,$	or	0.24	to	1.36
rate × additive interaction:	$1.7 \pm 0.56,$	or	1.14	to	2.26
rate × nozzle interaction:	$-0.2 \pm 0.56,$	or	-0.76	to	0.36
additive × nozzle interaction:	$-0.2 \pm 0.56,$	or	-0.76	to	0.36
rate × additive × nozzle interaction:	-0.3 ± 0.56	or	-0.86	to	0.26

Here the confidence interval for the three-factor interaction covers 0, so we neglect this interaction. However, we are 95% confident that the interval from 1.14 to 2.26 contains the rate × additive interaction, so we cannot interpret the rate and additive factors individually. Factor C, nozzle setting, is not involved in any significant interaction. With 95% confidence, we conclude that using the high nozzle setting will increase the mean opacity of the plastic wrap by 0.24 to 1.36 units.

Blocking in a 2^3 design

It is always desirable to conduct the 8 runs of a 2^3 factorial design under conditions that are as homogeneous as possible except for the settings of the 3 factors. However, limitations of time, space, equipment or available people may make it impossible to perform all 8 runs under homogeneous conditions. For instance, if large ceramic parts need to be baked, the oven may only hold 4 at one time. In our example of a

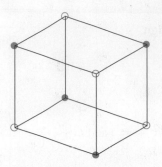

Figure 13.13
A 2^3 experiment arranged in 2 blocks of 4 runs; the solid circles have $x_1x_2x_3 = -1$

2^3 design for investigating factors that influence opacity, the time to manufacture the large rolls of plastic could be a limiting condition. Suppose a maximum of 4 runs can be conducted during one day. Consequently, the experiment needs to be divided up so 4 runs can be conducted one day and the other 4 the next day. How should the 4 runs for the first day be selected from among the total of 8 runs?

The answer lies in the fact that the three-factor interaction can often be neglected. Figure 13.13 shows how the 2^3 design can be divided into two groups of runs or **blocks**, each consisting of four runs. The 4 runs in each block all have the same sign for the three-factor interaction term $x_1x_2x_3$.

The 4 runs in the first block, with $x_1 \, x_2 \, x_3 = -1$, are

Block I						
x_1	x_2	x_3	x_1x_2	x_1x_3	x_2x_3	$x_1x_2x_3$
-1	-1	-1	1	1	1	-1
1	1	-1	1	-1	-1	-1
1	-1	1	-1	1	-1	-1
-1	1	1	-1	-1	1	-1

Those in the second block are

Block II						
x_1	x_2	x_3	x_1x_2	x_1x_3	x_2x_3	$x_1x_2x_3$
1	-1	-1	-1	-1	1	1
-1	1	-1	-1	1	-1	1
-1	-1	1	1	-1	-1	1
1	1	1	1	1	1	1

This division of the 8 runs into 2 blocks tends to neutralize the effect of differences in uncontrolled variables between the 2 days on which the experiment is conducted. Suppose that, due to some uncontrollable variable, all the runs on the second day were d units higher than if they were performed on the first day. Whatever the value of d, it will cancel out in the estimation of the 3 main effects and the two-factor interactions. To see this, we inspect the appropriate column in the tables of signs for each of the 2 blocks. The column corresponding to each of these effects, in each block, has 2 plus signs and 2 minus signs, so the d term cancels. Alternatively, from the geometric representation, each face of the cube has 2 solid dots and 2 white dots. Consequently, the average over any face is unaffected by the additive effect d and hence the estimates of the main effects are unaffected. The diagonal contrasts for a two-factor interaction have the same property on each face.

To be able to split the experiment into 2 blocks, we did have to give up information about the three-factor interaction. We deliberately confused or **confounded** the three-factor interaction with the day-to-day differences. Where the three-factor interaction is unimportant, this choice ensures that the main effects and the two-factor interactions can be estimated more precisely than would be the case if the 8 runs had to be conducted over a two-day period under less homogeneous conditions.

Exercises

13.10 As a preliminary step in optimizing the coating process of iron oxide nanoparticles engineers explored the effects, of two factors each having two levels on the response $y = $ increase in particle size (%).

Factor	Low level	High Level
Factor A molecular weight of chitosan	low	high
Factor B concentration ratio of chitosan-TPP	2:1	6:1

Given the following observations (Source: Replicate 1 extracted from the design in S. Honary et al. (2013), *International Nano Letters*, **3**(48). doi:10.1186/2228-5326-3-48):

x_1	x_2	Rep. 1	Rep. 2
−1	−1	96.0	90.8
1	−1	91.0	93.6
−1	1	76.0	88.4
1	1	223.8	214.2

(a) Attach the sample means at the corners of a square. Comment on any obvious pattern.

(b) Obtain the point estimates of the effects and 95% confidence intervals.

13.11 Shape memory alloys can undergo a reversible phase transformation. These materials display dramatic shape memory temperature-induced deformations that are recoverable. Investigators want to evaluate the influence of two factors

Factor A: Temperature at levels 350 °C and 450°C

Factor B: Time at levels 1 h and 5 h.

on the phase transformation temperature.

Given the following observations (Source: Replicate 1 extracted from the design in W. de Castro and G. Anselmo (2013), A Factorial design study of Ageing heat treatment influence on phase transformation of Ni-44.8wt%Ti Alloy, *22nd International Congress of Mechanical Engineering* (COBEM 2013), 907–911):

Factor A	Factor B	Rep. 1	Rep. 2
−1	−1	31.7	31.3
1	−1	33.5	33.1
−1	1	33.9	34.5
1	1	38.8	39.6

(a) Attach the sample means at the corners of a square. Comment on any obvious pattern.

(b) Obtain the point estimates of the effects and 95% confidence intervals.

13.12 Given the following observations

Factor A	Factor B	Rep. 1	Rep. 2
−1	−1	14	6
1	−1	15	21
−1	1	11	17
1	1	24	14

(a) Attach the sample means at the corners of a square. Comment on any obvious pattern.

(b) Obtain the point estimates of the effects and 95% confidence intervals.

13.13 Two factors are thought to influence the deposition rate (seconds) for a pulse laser to deposit one monolayer of material. Initially, a 2^2 design was run with two factors: spot size 50 mm or 60 mm and laser energy 1.5 J/cm^2 or 2.0 J/cm^2.

Spot Size	Laser Energy	Rep. 1	Rep. 2
−1	−1	8.34	7.44
1	−1	5.20	4.96
−1	1	7.01	7.09
1	1	4.45	4.73

Summarize the experiment according to the visual procedure. Interpret the effects based on confidence intervals.

13.14 Tomatoes have one of highest production volumes in the world and drying is one major process for preservation. Color is an important quality index for consumers. Three factors of storage, each at two levels, are considered in a replicated 2^3 design.

Factor	Low level	High level
A: Storage temperature (° C)	4	20
B: Packaging	Vacuum (≤ 40 mbar)	Normal (1023 mbar)
C: Storage time(month)	0	9

The response is color as measured by a redness index. (Source: B. Akdeniz, et al. (2012), Use of factorial experimental design for analyzing the effect of storage conditions on quality of sum-dried tomatoes, *Scientific Research and Essays*, **7**(4), 477–489.)

Factor A	Factor C	Factor C	Rep. 1	Rep. 2
−1	−1	−1	2.38	2.40
1	−1	−1	2.40	2.38
−1	1	−1	2.38	2.40
1	1	−1	2.38	2.40
−1	−1	1	2.40	2.42
1	−1	1	2.29	2.31
−1	1	1	1.94	1.94
1	1	1	1.92	1.93

(a) Attach the sample means at the corners of a cube. Comment on any obvious pattern.

(b) Obtain the point estimates of the effects and 95 % confidence intervals.

13.15 An engineering student wanted to know which factors influence the time (in seconds) for his car to go from 0 to 30 to 0 miles per hour. Factor A was the launch, which was either no wheel spin or dropping the clutch at 2,500 rpm. Factor B is either stopping with transmission in neutral or in second gear. Factor C is the air conditioning off or on.

Factor Launch	Factor B Transmission	Factor C A/C	Rep. 1	Rep. 2
−1	−1	−1	9.43	9.34
1	−1	−1	8.80	8.52
−1	1	−1	9.17	9.15
1	1	−1	8.36	8.43
−1	−1	1	9.87	9.66
1	−1	1	8.81	8.92
−1	1	1	9.56	8.94
1	1	1	8.40	8.46

Summarize the experiment according to the visual procedure. Interpret the effects based on confidence intervals.

13.16 The effect on engine wear of oil viscosity, temperature, and a special additive was tested using a 2^3 factorial design. Given the following results from the experiment,

Factor A Viscosity	Factor B Temperature	Factor C Additive	Rep. 1	Rep. 2
−1	−1	−1	3.7	4.1
1	−1	−1	4.6	5.0
−1	1	−1	3.1	2.7
1	1	−1	3.4	3.8
−1	−1	1	3.4	3.6
1	−1	1	5.3	4.9
−1	1	1	2.4	3.2
1	1	1	4.7	4.1

summarize the experiment according to the visual procedure given in Section 13.3. Interpret the effects based on the confidence intervals.

13.17 Two analysts A and B grew bacterial cultures using samples of sewage effluent and of clean stream water. The bacterial cultures were grown on two media C and CP. The design was repeated 3 times. The y values given below are the logarithms of the measured bacterial populations.

Source	Analyst	Medium	Rep. 1	Rep. 2	Rep. 3
Effluent	A	C	3.54	3.79	3.40
Stream	A	C	1.85	1.76	1.72
Effluent	B	C	3.81	3.82	3.79
Stream	B	C	1.72	1.75	1.55
Effluent	A	CP	3.63	3.67	3.71
Stream	A	CP	1.60	1.74	1.72
Effluent	B	CP	3.86	3.86	4.08
Stream	B	CP	2.05	1.51	1.70

Analyze this experiment using the visual procedure and determine if there is a difference between the two analysts.

13.18 The response variable Y_{ij} in a 2^2 design can also be expressed as a regression model

$$Y_{ij} = \mu + \beta_1 x_1 + \beta_2 x_2 + \beta_{12} x_1 x_2 + \varepsilon_{ij}$$

where the ε_{ij} are independent normal random variables and each has mean 0 and variance σ^2.

Because β_1 is a regression coefficient, it quantifies the change in the expected response when x_1 is changed by one unit. The effects are calculated on a change from -1 to 1 or 2 units.

(a) Obtain the expected values of $\overline{Y}_1$, $\overline{Y}_2$, $\overline{Y}_3$, and $\overline{Y}_4$.

(b) Show that the expected value of the main effect of Factor A is $2\beta_1$.

(c) Show that the expected value of the AB interaction effect is $2\beta_{12}$.

13.4   Response Surface Analysis

The aim of a response surface analysis is to use designed experiments to obtain an optimal response. The two-level factorial designs help experimenters locate regions where the response is a maximum, or, if desired, a minimum. Once the general region is located, another experiment should be conducted that allows for the estimation of a quadratic surface. Then it is usually possible to more accurately determine the best setting of the factors. We will illustrate the general idea with an example involving two factors. The **response surface** is then the surface traced out by the expected value of the response as the values of the two variables are changed.

We choose an experimental design which is composed of (i) a square and (ii) star plus center point. These two components are illustrated in Figure 13.14.

Figure 13.14
Design for fitting a quadratic surface in two variables

Notice that, with only two variables, the design is the same as a 3×3 factorial design. But with three factors, the cube has 8 points and there are 6 star points plus 1 center point, making a total of 15, not $27 = 3 \times 3 \times 3$ runs.

With only two factors, we fit the quadratic model in two variables x_1 and x_2:

$$E(Y) = \beta_0 + \beta_1 x_1 + \beta_2 x_2 + \beta_{11} x_1^2 + \beta_{22} x_2^2 + \beta_{12} x_1 x_2$$

which provides an approximation to the response surface.

EXAMPLE 4   **A response surface design to maximize yield**

A compound is produced for a coating process. It is added to an otherwise fixed recipe and the coating process is completed. Yield is the response variable. Two factors x_1 and x_2 have been identified as "vital," and they are

Factor (x_i)	Range
Additive amount	0–70 gm/kg
Reaction temperature	100–180°C

The goal of the experiment is to improve the yield. The business requires that yield be greater than 95%.

An experimental design, square plus star and center points, was implemented and the response yield measured.

Run	Additive	Temperature	Yield
1	0	100	81
2	70	100	65
3	35	140	92
4	0	180	50
5	70	180	75
6	70	140	75
7	0	140	68
8	35	100	90
9	35	180	77

(*Courtesy of Asit Banerjee*)

Estimate the response surface and suggest a good region in which to operate.

Solution   *MINTAB* software (see Exercise 13.20) provides the regression analysis

```
Estimated Regression Coefficients for Yield

Term                        Coef   SE Coef          T        P
Constant                 60.2639   18.5242      3.253    0.047
Additive                  0.0417    0.1421      0.293    0.788
Temperature               0.5354    0.2714      1.972    0.143
Additive*Additive        -0.0141    0.0013    -11.269    0.001
Temperature*Temperature  -0.0033    0.0010     -3.467    0.040
Additive*Temperature      0.0073    0.0008      9.424    0.003

S = 2.175    R-Sq = 98.9%
```

The estimated response surface is

$$\hat{y} = 60.2639 + 0.0417\,x_1 + 0.5354\,x_2 - 0.0141\,x_1^2 - 0.0033\,x_2^2 + 0.0073\,x_1\,x_2$$

It is usually not reasonable to drop a linear term when the associated square term is in the response surface model. Figure 13.15 presents a contour and 3D surface plot. There is a small region, near $x_1 = 33$ and $x_2 = 117$, where the estimated yield is nearly maximum.   ∎

The book by Box and Draper, listed in the bibliography, provides a comprehensive introduction to response surface methodology.

Exercises

13.19   Refer to the Example 4. Use calculus to obtain the location of the estimated maximum yield when all terms are included in the model.

13.20   *MINITAB* **response surface analysis**
We illustrate the commands for the coating data in Example 4 where yield is the response. Start with the Run, Additive, Temperature, Yield in C1–C4.

Surface plot of yield vs. temperature, additive

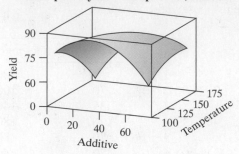

Contour plot of yield vs. temperature, additive

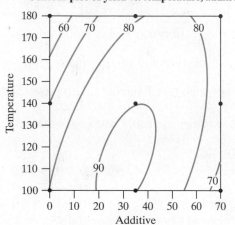

Figure 13.15
Estimated response surface for yield

Dialog box:

Stat> DOE > Response Surface. Click **Define Custom Response . . .**
Type *Additive* and *Temperature* in **Factors**. Click **High/Low** and
Type 0 and 70 for Additive and 100 and 180 for Temperature.
Click **OK**. Click **OK**.

Stat> DOE > Response Surface. Click **Analyze Response Surface Design**
Type *Yield* in **Response**. Click **Uncoded units**. Choose **Terms**
and then **full Quadratic**. Click **OK**. Click **OK**.

The graph of the estimated response surface is produced by the commands

Stat> DOE > Response Surface > Contour/Surface Plots

Click **Surface plot** and then fill in choices similar to the analyze step shown above.
    Repeat the analysis in the example after adding one more center point where the
measured yield is 93%.

13.21 Refer to Exercise 13.20. In Example 4, the experimenters also obtained the nine
responses for adhesion. The business wants adhesion greater than 45 grams.

Adhesion   10   48   41   40   39   44   24   31   44

Repeat the analysis in Example 4 but change the response to adhesion.

13.22 Is there a region within the experimental region where estimated adhesion is greater than
45 grams? Construct a contour plot to show this region. Note that *MINITAB* does keep
nonsignificant terms. Refer to Exercise 13.21.

Do's and Don'ts

Do's

1. Whenever possible, randomize the assignment of treatments in a factorial design.
2. Always create a graphic presentation of the data from 2^2 and 2^3 designs, along with confidence intervals for the main effects and interactions.

Don'ts

1. Don't routinely accept the analysis of the factorial design presented in a computer output. Instead, inspect the residuals for outliers and moderate to severe lack of normality. A normal-scores plot is useful if there are more than 20 or so residuals. It may suggest a transformation.

Review Exercises

13.23 A rocket-launcher experiment is poorly designed if not replicated, since it would be impossible to test whether there is an interaction between launchers and rocket fuels. Suppose that two replicates are performed, with the following results obtained for the range (nautical miles).

		Fuel I	Fuel II	Fuel III	Fuel IV
Rep. 1	Launcher X	62.5	49.3	33.8	43.6
	Launcher Y	40.4	39.7	47.4	59.8

		Fuel I	Fuel II	Fuel III	Fuel IV
Rep. 2	Launcher X	66.8	51.1	40.1	49.2
	Launcher Y	47.7	38.2	50.6	64.3

Perform an appropriate analysis of variance, and test for the presence of an interaction.

13.24 A study was conducted to measure the effect of 3 different meat tenderizers on the weight loss of steaks having the same initial (precooked) weights. The effects of cooking temperatures and cooking times also were measured by performing a $3 \times 2 \times 2$ factorial experiment in 3 replicates. The results are as follows:

Tenderizer	Cooking Time (minutes)	Cooking Temperature (°F)	Weight Loss (ounces)		
			Rep. 1	Rep. 2	Rep. 3
A	20	350	1.5	1.3	1.4
A	20	400	1.6	1.4	1.5
A	30	350	1.7	1.8	1.7
A	30	400	1.8	1.9	2.0
B	20	350	1.9	2.1	2.0
B	20	400	2.2	2.4	2.5
B	30	350	2.6	2.3	2.4
B	30	400	2.6	2.7	2.5
C	20	350	0.9	0.8	0.8
C	20	400	1.1	1.0	0.9
C	30	350	0.8	0.9	1.0
C	30	400	1.2	1.0	1.1

Present the results in an analysis of variance table and interpret the experiment.

13.25 An experiment was conducted to determine the effects of certain alloying elements on the ductility of a metal, and the following results were obtained:

			Breaking Strength (ft-lb)		
Nickel	Carbon	Manganese	Rep. 1	Rep. 2	Rep. 3
0.0%	0.3%	0.5%	36.7	39.6	38.2
0.0	0.3	1.0	47.5	43.5	45.9
0.0	0.6	0.5	40.6	36.8	36.0
0.0	0.6	1.0	41.1	45.8	46.4
4.0	0.3	0.5	37.8	32.7	31.6
4.0	0.3	1.0	34.2	37.2	36.5
4.0	0.6	0.5	39.5	41.7	39.1
4.0	0.6	1.0	46.4	43.7	49.4

Perform an appropriate analysis of variance and interpret the results.

13.26 Given the two replicates of a 2×3 factorial experiment, calculate the analysis of variance tables using the formulas on page 418.

Factor A	Factor B	Rep. 1	Rep. 2
1	1	29	35
1	2	15	17
1	3	14	22
2	1	15	13
2	2	27	25
2	3	16	24

13.27 Given the following observations,

Factor A	Factor B	Rep. 1	Rep. 2
−1	−1	11	7
1	−1	16	14
−1	1	10	12
1	1	17	13

summarize the experiment according to the visual procedure given in Section 13.3. Interpret the effects based on the confidence intervals.

13.28 With reference to the example on page 433, suppose a third replicate

Temperature	pH	Rep. 3
300	2	9
350	2	23
300	3	13
350	3	25

is run. Analyze the experiment, using all 3 replicates, according to the visual procedure given in Section 13.3. Interpret the effects based on the confidence intervals.

13.29 Trouble was being experienced by a new high-tech machine for joining two pieces of sheet metal. The two factors considered first are the pressure (low/high) and temperature of the pump low/high. The response is the diameter (mm) of a button-shaped joint which is an indirect measure of strength.

Pressure	Temperature	Rep. 1	Rep. 2
−1	−1	8	12
1	−1	16	22
−1	1	5	11
1	1	16	21

Summarize the experiment according to the visual procedure. Interpret the effects based on confidence intervals.

13.30 A mechanical engineer studied the mechanical properties of a resin under different running conditions for injection-molding equipment. The response is tensile strength (coded units). The factors are supercritical fluid levels (wt%), shot size (mm), and injection speed (%).

Factor A Content (wt%)	Factor B Size (mm)	Factor C Speed (%)	Rep. 1	Rep. 2
−1	−1	−1	41.8	42.2
1	−1	−1	44.5	43.9
−1	1	−1	56.5	56.3
1	1	−1	57.3	56.5
−1	−1	1	43.4	42.7
1	−1	1	42.5	43.1
−1	1	1	56.5	55.3
1	1	1	56.5	55.6

Summarize the experiment according to the visual procedure. Interpret the effects based on confidence intervals.

13.31 Given the following results from a 2^3 factorial experiment,

Factor A	Factor B	Factor C	Rep. 1	Rep. 2
−1	−1	−1	13.8	14.6
1	−1	−1	10.8	8.4
−1	1	−1	9.0	9.8
1	1	−1	10.1	10.9
−1	−1	1	14.4	13.6
1	−1	1	6.2	8.6
−1	1	1	7.7	7.9
1	1	1	9.0	8.2

summarize the experiment according to the visual procedure given in Section 13.3. Interpret the effects based on the confidence intervals.

13.32 The total sum of squares is given by $\sum_{i=1}^{k} \sum_{j=1}^{r} (y_{ij} - \bar{y})^2$ where the overall mean $\bar{y} = \sum_{i=1}^{k} \sum_{j=1}^{r} (y_{ij} / n)$. With reference to Exercise 13.27, show that the total sum of squares can be expressed as the sum of squares due to each of the treatments SSA, SSB, and $SSAB$ plus the error of sum of squares.

This decomposition is the basis for the analysis variance and it is summarized in the first column of the ANOVA table.

[*Hint:* Sum of squares for any treatment in a 2^2 design is (estimated effect)^2 × r The error sum of squares is $\sum_{i=1}^{k} \sum_{j=1}^{r} (y_{ij} - \bar{y}_i)^2$]

13.33 With reference to the example of the 2^3 design on page ***, express the total sum of squares as the sum of the contributions from each of the seven treatments plus the error sum of squares.

This decomposition is the basis for the analysis variance and it is summarized in the first column of the ANOVA table.

[*Hint:* Sum of squares for any treatment in a 2^3 design is (estimated effect $)^2$ $2 \times r$. Refer to Exercise 13.32 for the total and error sum of squares.]

13.34 With reference to the Example 3 concerning improvements in the safety of an ignitor, the time to reach maximum pressure was also recorded. Two replicates were run of the factorial design and the times to reach maximum pressure recorded. Analyze the results of this experiment.

A	B	C	Time (milliseconds) Rep. 1	Rep. 2
Initiator 1	Powder	Mc 1	54.02	49.64
Initiator 1	Pellet	Mc 1	60.74	70.66
Initiator 1	Powder	Mc 2	37.56	43.72
Initiator 1	Pellet	Mc 2	46.40	42.04
Initiator 1	Powder	Mc 3	40.54	41.60
Initiator 1	Pellet	Mc 3	50.56	45.44
Initiator 1	Powder	Mc 4	47.08	44.28
Initiator 1	Pellet	Mc 4	47.68	60.22
Initiator 2	Powder	Mc 1	47.00	58.74
Initiator 2	Pellet	Mc 1	51.30	53.96
Initiator 2	Powder	Mc 2	59.82	45.66
Initiator 2	Pellet	Mc 2	52.20	57.82
Initiator 2	Powder	Mc 3	68.46	59.86
Initiator 2	Pellet	Mc 3	82.78	57.16
Initiator 2	Powder	Mc 4	75.14	90.82
Initiator 2	Pellet	Mc 4	82.38	75.38
Initiator 3	Powder	Mc 1	35.78	30.34
Initiator 3	Pellet	Mc 1	56.06	38.30
Initiator 3	Powder	Mc 2	20.60	22.18
Initiator 3	Pellet	Mc 2	21.38	21.50
Initiator 3	Powder	Mc 3	45.06	43.18
Initiator 3	Pellet	Mc 3	50.24	56.74
Initiator 3	Powder	Mc 4	44.54	69.96
Initiator 3	Pellet	Mc 4	63.42	49.66

13.35 Refer to Exercise 13.10 where the response is $y =$ increase in particle size. Besides the first replicate, the investigators also performed the experiments that form the star part of the design.

x_1	x_2	y
-1	0	116.0
$+1$	0	138.0
0	-1	78.0
0	$+1$	118.0
0	0	97.0

Using all 9 measurements, fit a response surface as in Example 4.

Key Terms

CHAPTER

14

NONPARAMETRIC TESTS

Most of the methods of inference that we have studied are based on the assumption that the observations come from normal populations. If this is the case, these methods extract all the information available in a sample, and they usually attain the best possible precision. However, since there are many situations where it is doubtful whether the assumption of normality can be met, statisticians have developed alternative techniques based on less stringent assumptions, which have become known as nonparametric tests.

Here, we present the sign test in Section 14.2; tests based on rank sums in Section 14.3; a rank-based correlation coefficient in Section 14.4; a test of randomness in Section 14.5; and goodness-of-fit tests in Section 14.6.

14.1 Introduction

In this chapter, we expand the choice of statistical methods available for inferences concerning one or more populations. The assumption of a normal population underlies most of the "standard methods" discussed in the previous chapters. Understandably, it is often difficult to verify this tentative assumption, especially when sample sizes are small. Here we introduce tests that depend only on order relationships among the observations. Consequently, much less has to be assumed about the form of the underlying populations. The main advantage of these **nonparametric tests** is that exact inferences can be made when the assumptions underlying the so-called standard methods cannot be met. When the normal assumption is met, the standard tests will have more power. However, asymmetry or other departures from normality will have no effect on the sampling distribution of any nonparametric statistic when the null hypothesis prevails. Moreover, their power is usually satisfactory even when the populations deviate from normality.

Also, nonparametric tests apply even when the choice of a particular numerical scale of measurement is arbitrary. Still, their strongest value is the fact that the level of significance is exact even when the populations are quite nonnormal.

14.2 The Sign Test

We now describe a simple nonparametric alternative to the one-sample t test, the paired-sample t test, and corresponding large-sample tests. The **sign test** applies when we sample a continuous symmetrical population, so that the probability of getting a sample value less than the mean and the probability of getting a sample value greater than the mean are both $\frac{1}{2}$. More generally, because symmetry is often difficult to verify with small or moderate sample sizes, we can formulate the hypotheses in terms of the population median $\widetilde{\mu}$. To test the null hypothesis $\widetilde{\mu} = \widetilde{\mu}_0$ against an appropriate alternative on the basis of a random sample of size n, we replace each sample value greater than $\widetilde{\mu}_0$ with a plus sign and each sample value less than $\widetilde{\mu}_0$ with a minus sign. We then test the null hypothesis that these plus and minus signs are the outcomes of binomial trials with $p = \frac{1}{2}$. If a sample value equals $\widetilde{\mu}_0$, which may well happen since the values of continuous random variables are virtually always rounded, it is discarded.

To perform this kind of test when the sample is small, we refer directly to a table of binomial probabilities such as Table 1 at the end of the book; when the sample is large we use the test described in Section 10.2.

EXAMPLE 1   **Conducting a sign test**

The following data constitute a random sample of 15 measurements of the octane rating of a certain kind of gasoline:

99.0	102.3	99.8	100.5	99.7	96.2	99.1	102.5
103.3	97.4	100.4	98.9	98.3	98.0	101.6	

Test the null hypothesis $\widetilde{\mu} = 98.0$ against the alternative hypothesis $\widetilde{\mu} > 98.0$ at the 0.01 level of significance.

Solution   Since one of the sample values equals 98.0 and must be discarded, the sample size for the sign test is only $n = 14$.

1. *Null hypothesis*: $\widetilde{\mu} = 98.0$ ($p = \frac{1}{2}$)
   Alternative hypothesis: $\widetilde{\mu} > 98.0$ ($p > \frac{1}{2}$)
2. *Level of significance*: $\alpha = 0.01$
3. *Criterion*: The criterion may be based on the number of plus signs or the number of minus signs. Using the number of plus signs, denoted by x, reject the null hypothesis if the probability of getting x or more plus signs is less than or equal to 0.01.
4. *Calculations*: Replacing each value greater than 98.0 with a plus sign and each value less than 98.0 with a minus sign, the 14 sample values yield

$$+ \quad + \quad + \quad + \quad + \quad - \quad + \quad + \quad + \quad - \quad + \quad + \quad + \quad +$$

Thus $x = 12$ and Table 1 shows that for $n = 14$ and $p = 0.50$ the probability of $X \geq 12$ is $1 - 0.9935 = 0.0065$.
5. *Decision*: Since 0.0065 is less than 0.01, the null hypothesis must be rejected; we conclude that the median octane rating of the given kind of gasoline exceeds 98.0. ∎

The sign test can also be used as a nonparametric alternative to the paired-t test or the corresponding large-sample test. In such problems, each pair of sample values is replaced with a plus sign if the first is greater than the second, with a minus sign if the first value is smaller than the second, or it is discarded if the two values are equal. The procedure is the same as before. Let $\widetilde{\mu}_D$ denote the median of the differences.

EXAMPLE 2   **A sign test of the effectiveness of a safety program**

With reference to Example 12, Chapter 8, which deals with the effectiveness of an industrial safety program, use the sign test at the 0.051 level of significance to test whether the safety program is effective.

Solution   1. *Null hypothesis*: $\widetilde{\mu}_D = 0 \quad \left(p = \dfrac{1}{2} \right)$

   Alternative hypothesis: $\widetilde{\mu}_D > 0 \quad \left(p > \dfrac{1}{2} \right)$

2. *Level of significance*: $\alpha = 0.05$
3. *Criterion*: If x is the number of plus signs, reject the null hypothesis if the probability of getting x or more plus signs is less than or equal to 0.05.
4. *Calculations*: Replacing each pair of values with a plus sign if the first value is greater than the second or with a minus sign if the first value is smaller than the second, the 10 sample pairs yield

$$+ \quad + \quad + \quad + \quad - \quad + \quad + \quad + \quad + \quad +$$

   Thus $x = 9$ and Table 1 shows that for $n = 10$ and $p = 0.50$ the probability of $X \geq 9$ is $1 - 0.9893 = 0.0107$.
5. *Decision*: Since 0.0107 is less than 0.05, the null hypothesis must be rejected; we conclude that the safety program is effective.   ∎

14.3   Rank-Sum Tests

In this section we shall introduce two tests based on **rank sums**—the **U test** will be presented as a nonparametric alternative to the two-sample t test, and the **H test** will be presented as a nonparametric alternative to the one-way analysis of variance, which we studied in Chapter 12. In other words, the H test serves to test the null hypothesis that k samples come from identical populations against the alternative that the populations are not identical.

To illustrate how the U test (also called the **Wilcoxon test** or the **Mann-Whitney test**, named after the statisticians who contributed to its development) is performed, suppose that in a study of sedimentary rocks, the following diameters (in millimeters) were obtained for two kinds of sand:

Sand I:	0.63	0.17	0.35	0.49	0.18	0.43	0.12	0.20
	0.47	1.36	0.51	0.45	0.84	0.32	0.40	
Sand II:	1.13	0.54	0.96	0.26	0.39	0.88	0.92	0.53
	1.01	0.48	0.89	1.07	1.11	0.58		

The problem is to decide whether the two populations are the same or if one is more likely to produce larger observations than the other. Let X_1 be a random variable having the first distribution and X_2 a random variable having the second distribution. If $P(a < X_1) \leq P(a < X_2)$ for all a, with strict inequality for some a, we say that the second population (distribution) is **stochastically larger** than the first population (distribution). We formulate one-sided hypotheses in terms of this stochastic order relation.

We begin the U test by ranking the data jointly, as if they comprise one sample, in an increasing order of magnitude, and for our data we get

0.12	0.17	0.18	0.20	0.26	0.32	0.35	0.39	0.40	0.43
I	I	I	I	II	I	I	II	I	I
0.45	0.47	0.48	0.49	0.51	0.53	0.54	0.58	0.63	0.84
I	I	II	I	I	II	II	II	I	I
0.88	0.89	0.92	0.96	1.01	1.07	1.11	1.13	1.36	
II	II	II	II	II	II	II	II	I	

Note that, for each value, we indicate whether it is a measurement of Sand I or Sand II. Assigning the data in this order the ranks $1, 2, 3, \ldots$, and 29, we find that the values of the first sample (Sand I) occupy ranks 1, 2, 3, 4, 6, 7, 9, 10, 11, 12, 14, 15, 19, 20, and 29, while those of the second sample (Sand II) occupy ranks 5, 8, 13, 16, 17, 18, 21, 22, 23, 24, 25, 26, 27, and 28. There are no ties here among values belonging to different samples, but if there were, we would assign to each of the tied observations the mean of the ranks which they jointly occupy. For instance, if the third and fourth values are identical we would assign each the rank

$$\frac{3+4}{2} = 3.5$$

and if the ninth, tenth, and eleventh values are identical we would assign each the rank

$$\frac{9 + 10 + 11}{3} = 10$$

The null hypothesis we want to test is that the two samples come from identical populations, and it stands to reason that, in that case, the means of the ranks assigned to the values of the two samples should be more or less the same. Instead of the means, we can also compare the sums of the ranks assigned to the values of the two samples, suitably accounting for a possible difference in their size. For our two samples, the sums of the ranks are $W_1 = 162$ and $W_2 = 273$, and it remains to be seen whether their difference is large enough to reject the null hypothesis.

When the use of rank sums was first proposed as a nonparametric alternative to the two sample t test, the decision was based on W_1 or W_2, but now the decision is based on either of the related statistics

U_1 and U_2 statistics or

$$U_1 = W_1 - \frac{n_1 (n_1 + 1)}{2}$$

$$U_2 = W_2 - \frac{n_2 (n_2 + 1)}{2}$$

or on the statistic U which equals the smaller of the two. The sizes of the two samples are n_1 and n_2, and as it does not matter how we number the samples, we shall use here the statistic U_1.[1]

Under the null hypothesis that the two samples come from identical populations, it can be shown that the mean and the variance of the sampling distribution of U_1 are

Mean and variance of U_1 statistic and

$$\mu_{U_1} = \frac{n_1 n_2}{2}$$

$$\sigma_{U_1}^2 = \frac{n_1 n_2 (n_1 + n_2 + 1)}{12}$$

If there are ties in rank, these formulas provide only approximations, but if the number of ties is small, these approximations will generally be good.

[1] The tests based on U_1 and U_2 are equivalent to those based on W_1 or W_2, but they have the advantage that they lend themselves more readily to the construction of tables of critical values. Not only do U_1 and U_2 take on values on the interval from 0 to $n_1 n_2$—indeed, their sum is always equal to $n_1 n_2$—but their sampling distributions are symmetrical about $\frac{n_1 n_2}{2}$.

Since numerical studies have shown that the sampling distribution of U_1 can be approximated closely by a normal distribution when n_1 and n_2 are both greater than 8, the test of the null hypothesis that the two samples come from identical populations can be based on

Statistic for large sample U test

$$Z = \frac{U_1 - \mu_{U_1}}{\sigma_{U_1}}$$

which is a random variable having approximately the standard normal distribution. For small samples, we can base the test on special tables; for instance, on those in the book by Johnson and Bhattacharyya, listed in the bibliography.

Note that when we test the null hypothesis—the two samples come from identical populations—against the alternative hypothesis

population 2 is stochastically larger than population 1

we reject the null hypothesis, if $Z < -z_\alpha$, since small values of U_1 correspond to small values of W_1; correspondingly, if the alternative hypothesis is

population 1 is stochastically larger than population 2

we reject the null hypothesis, if $Z > z_\alpha$, since large values of U_1 correspond to large values of W_1.

EXAMPLE 3   Conducting the Wilcoxon test with large samples

With reference to the grain-size data on page 456, use the U test at the 0.01 level of significance to test the null hypothesis that the two samples come from identical populations against the alternative hypothesis that the populations are not identical.

Solution

1. *Null hypothesis*: Populations are identical.
   Alternative hypothesis: The populations are not identical.

2. *Level of significance*: $\alpha = 0.01$

3. *Criterion*: Reject the null hypothesis if $Z < -2.575$ or $Z > 2.575$, where Z is given by the formula above.

4. *Calculations*: Since $n_1 = 15$, $n_2 = 14$, and we have already shown that $W_1 = 162$, we find that

$$U_1 = 162 - \frac{15 \cdot 16}{2} = 42$$

$$\mu_{U_1} = \frac{15 \cdot 14}{2} = 105$$

and

$$\sigma_{U_1}^2 = \frac{15 \cdot 14 \cdot 30}{12} = 525$$

and it follows that

$$z = \frac{42 - 105}{\sqrt{525}} = -2.75$$

5. *Decision*: Since $z = -2.75$ is less than -2.575, the null hypothesis must be rejected at $\alpha = 0.01$. The P-value $= 0.0060$ and we conclude that there is a strong evidence of difference in the populations of grain size. ∎

The *H* **test**, or **Kruskal-Wallis test**, is a generalization of the *U* test in that it enables us to test the null hypothesis that *k* independent random samples come from identical populations. As in the *U* test, all the observations are ranked jointly, and if R_i is the sum of the ranks occupied by the n_i observations of the *i*th sample and $n_1 + n_2 + \cdots + n_k = n$, the test is based on the statistic

Statistic for *H* test

$$H = \frac{12}{n(n+1)} \sum_{i=1}^{k} \frac{R_i^2}{n_i} - 3(n+1)$$

When $n_i > 5$ for all *i* and the null hypothesis is true, the sampling distribution of the *H* statistic is well approximated by the chi square distribution with $k - 1$ degrees of freedom. There exist special tables of critical values for the *H* test for selected small values of the n_i and *k*.

EXAMPLE 4 **Conducting an *H* test to compare three methods**

An experiment, designed to compare three methods for preventing corrosion, yielded the following maximum depths of pits (in thousandths of an inch) in pieces of wire subjected to the respective treatments:

Method A: 77 54 67 74 71 66
Method B: 60 41 59 65 62 64 52
Method C: 49 52 69 47 56

Use the 0.05 level of significance to test the null hypothesis that the three samples come from identical populations.

Solution 1. *Null hypothesis*: Populations are identical.
   Alternative hypothesis: The populations are not all equal.

2. *Level of significance*: $\alpha = 0.05$

3. *Criterion*: Reject the null hypothesis if $H > 5.991$, the value of $\chi^2_{0.05}$ for 2 degrees of freedom, where *H* is given by the formula above.

4. *Calculations*: Ranking these measurements jointly from smallest to largest, we find that those of the first sample occupy ranks 6, 13, 14, 16, 17, and 18; those of the second sample occupy ranks 1, 4.5, 8, 9, 10, 11, and 12; and those of the third sample occupy ranks 2, 3, 4.5, 7, and 15. Thus, $R_1 = 84$, $R_2 = 55.5$, $R_3 = 31.5$, and substitution into the formula for *H* yields

$$H = \frac{12}{18 \cdot 19} \left(\frac{84^2}{6} + \frac{55.5^2}{7} + \frac{31.5^2}{5} \right) - 3 \cdot 19$$

$$= 6.7$$

5. *Decision*: Since $H = 6.7$ exceeds 5.991, the null hypothesis must be rejected at $\alpha = 0.05$. The *P*-value $= P(\chi^2_2 > 6.7) = 0.035$ and we conclude that the three preventive methods against corrosion are not equally effective. ∎

14.4 Correlation Based on Ranks

In Chapter 11, we introduced Pearson's product moment correlation coefficient as a measure of association. An alternative measure, called **Spearman's rank-correlation** or the **rank-correlation coefficient**, is analogous to Pearson's correlation, r, except that Spearman replaces the observations with their ranks:

$$r_S = \frac{\sum_{i=1}^{n}(R_i - \overline{R})(S_i - \overline{S})}{\sqrt{\sum_{i=1}^{n}(R_i - \overline{R})^2}\sqrt{\sum_{i=1}^{n}(S_i - \overline{S})^2}}$$

where R_i is the rank of x_i among the x's and S_i is the rank of y_i among the y's. Because each of the ranks, 1, 2, ..., n, must occur exactly once in the set $R_1, R_2, \ldots, R_n$, it can be shown that $\overline{R} = \overline{S} = (n+1)/2$ and $\sum_{i=1}^{n}(R_i - \overline{R})^2 = n(n^2 - 1)/12 = \sum_{i=1}^{n}(S_i - \overline{S})^2$.

Spearman's rank correlation coefficient

> The rank-correlation coefficient
>
> $$r_S = \frac{\sum_{i=1}^{n}\left(R_i - \frac{n+1}{2}\right)\left(S_i - \frac{n+1}{2}\right)}{n(n^2 - 1)/12} = \frac{\sum_{i=1}^{n}R_iS_i - n(n+1)^2/4}{n(n^2 - 1)/12}$$
>
> 1. $-1 \le r_S \le 1$
> 2. Values of r_S near 1 indicate a tendency of large values for X and Y to be paired together. An r_S near -1 indicates the opposite relationship.
> 3. r_S is a measure of a monotone increasing/decreasing relationship that is not necessarily linear.

When ties are present, assign the average of the corresponding ranks to each tied observation. It can be shown that

Large sample statistic for testing independence

> if X and Y are independent, then
>
> $$Z = \sqrt{n}\, r_S \qquad \text{is approximately distributed as standard normal}$$
>
> provided the sample size is large.

EXAMPLE 5   **Rank correlation of before and after plant safety**

Refer to Example 12 of Chapter 8 concerning losses of worker-hours before and after safety programs in 10 industrial plants. Calculate r_S.

Solution   The two sets of ranks are

$$R_i = \text{rank}(x_i): 5 \quad 8 \quad 6 \quad 10 \quad 3 \quad 7 \quad 9 \quad 4 \quad 2 \quad 1$$
$$S_i = \text{rank}(y_i): 5 \quad 8 \quad 6 \quad 10 \quad 4 \quad 7 \quad 9 \quad 3 \quad 2 \quad 1$$

We calculate $\sum_{i=1}^{10} R_iS_i = 384$, $(n+1)/2 = 5.5$, and $n(n^2 - 1) = 990$ so

$$r_S = \frac{\sum_{i=1}^{n}R_iS_i - n(n+1)^2/4}{n(n^2 - 1)/12} = \frac{384 - 10(5.5)^2}{990/12} = 0.988$$

This large positive value indicates strong association along an increasing curve. Since $\sqrt{n}\, r_S = 3.12$, there is strong evidence against independence.   ■

Exercises

14.1 In a laboratory experiment, 18 determinations of the coefficient of friction between leather and metal yielded the following results: 0.59, 0.56, 0.49, 0.55, 0.65, 0.55, 0.51, 0.60, 0.56, 0.47, 0.58, 0.61, 0.54, 0.68, 0.56, 0.50, 0.57, and 0.53. Use the sign test at the 0.05 level of significance to test the null hypothesis $\tilde{\mu} = 0.55$ against the alternative hypothesis $\tilde{\mu} \neq 0.55$.

14.2 The quality-control department of a large manufacturer obtained the following sample data (in pounds) on the breaking strength of a certain kind of 2-inch cotton ribbon: 153, 159, 144, 160, 158, 153, 171, 162, 159, 137, 159, 159, 148, 162, 154, 159, 160, 157, 140, 168, 163, 148, 151, 153, 157, 155, 148, 168, 152, and 149. Use the sign test at the 0.01 level of significance to test the null hypothesis $\tilde{\mu} = 150$ against the alternative hypothesis $\tilde{\mu} > 150$.

14.3 With reference to Exercise 2.12, which pertained to the ignition times of certain upholstery materials, use the sign test at the 0.01 level of significance to test the null hypothesis $\tilde{\mu} = 6.50$ seconds against the alternative hypothesis $\tilde{\mu} < 6.50$ seconds.

14.4 The following are the number of speeding tickets issued by 2 police officers on 17 days: 7 and 10, 11 and 13, 14 and 14, 11 and 15, 12 and 9, 6 and 10, 9 and 13, 8 and 11, 10 and 11, 11 and 15, 13 and 11, 7 and 10, 8 and 8, 11 and 12, 9 and 14, 10 and 9, 13 and 16. Use the sign test at the 0.05 level of significance to test the null hypothesis that on the average the 2 police officers issue equally many speeding tickets per day against the alternative hypothesis that the second police officer tends to issue more than the first.

14.5 Comparing 2 kinds of emergency flares, a consumer testing service obtained the following burning times (rounded to the nearest tenth of a minute):

Brand C:	19.4	21.5	15.3	17.4	16.8	16.6	20.3	22.5	21.3
	23.4	19.7	21.0						

Brand D:	16.5	15.8	24.7	10.2	13.5	15.9	15.7	14.0	12.1
	17.4	15.6	15.8						

Use the U test at the 0.01 level of significance to check whether it is reasonable to say that the populations of burning times of the two kinds of flares are identical.

14.6 The following are the self-reported times (hours for month), spent on homework, by random samples of juniors in two different majors.

Major 1:	63	72	29	58	81	65	79	57	40	76	47	55	60
Major 2:	41	32	26	43	78	49	39	56	15	54	8	66	64

Use the U test at the 0.05 level of significance to test whether or not students from the 2 groups devote the same amounts of time to homework.

14.7 The following are data on the breaking strength (in pounds) of 2 kinds of material:

Material 1:	144	181	200	187	169	171	186	194	176	182
	133	183	197	165	180	198				

Material 2:	175	164	172	194	176	198	154	134	169	164
	185	159	161	189	170	164				

Use the U test at the 0.05 level of significance to test the claim that the strength of Material 1 is stochastically larger than the strength of Material 2.

14.8 A company that processes health claims maintains three centers. Software was installed so they could monitor non-business internet usage by their employees. Initially, six employees were randomly selected from each of three service centers and the number of hours of non-business internet usage recorded.

Service Center 1	4.1	10.4	2.2	5.7	3.8	12.3
Service Center 2	7.9	5.4	13.1	7.7	8.3	9.8
Service Center 2	6.9	9.3	11.2	1.9	13.8	7.3

Use the H test at the 0.05 level of significance to test the null hypothesis that the 3 samples come from identical populations.

14.9 So-called Franklin tests were performed to determine the insulation properties of grain-oriented silicon steel specimens that were annealed in five different atmospheres with the following results:

Atmosphere	Test Results (amperes)						
1	0.58	0.61	0.69	0.79	0.61	0.59	
2	0.37	0.37	0.58	0.40	0.28	0.44	0.35
3	0.29	0.19	0.34	0.17	0.29	0.16	
4	0.81	0.69	0.75	0.72	0.68	0.85	0.57 0.77
5	0.26	0.34	0.29	0.47	0.30	0.42	

Use the H test at the 0.05 level of significance to decide whether or not these 5 samples can be assumed to come from identical populations.

14.10 A panel of 7 experts was asked to rate each of 3 industries on the likelihood that technological changes would produce improvement in environmental pollution over the next 10 years. Their ratings (in the form of judgmental probabilities) are as follows:

	Industry		
Expert	A	B	C
1	0.15	0.75	0.10
2	0.30	0.60	0.20
3	0.20	0.80	0.30
4	0.00	0.50	0.25
5	0.10	0.55	0.15
6	0.25	0.70	0.35
7	0.40	0.95	0.45

Calculate the rank correlation coefficient, r_S.

(a) Using industries A and B.

(b) Using industries A and C.

14.5 Tests of Randomness

When we discussed random sampling in Chapter 6, we gave several methods which provide some assurance in advance that the selected sample will be random. In some situations however, we have no control over the way in which the data are selected. Then, it is useful to have a technique for testing whether the selected sample may be looked upon as random. One such technique is based on the order in which the sample values were obtained. More specifically, it is based on the number of **runs** exhibited in the sample results.

Given a sequence of two symbols, such as H and T (which might represent the occurrence of heads and tails in repeated tosses of a coin), a run is defined as a succession of identical symbols contained between different symbols or none at all. For example, the sequence

$$\underline{T\,T}\ \ \underline{H\,H}\ \ \underline{T\,T}\ \ \underline{H\,H\,H}\ \ \underline{T}\ \ \underline{H\,H\,H}\ \ \underline{T\,T\,T\,T}\ \ \underline{H\,H\,H}$$

contains 8 runs, as indicated by the underlines. The total number of runs in a sequence of n trials often serves as an indication that the arrangement is not random. For instance, if there had been only two runs consisting of 10 heads followed by 10 tails, we should suspect that the probability of a success did not remain constant from trial to trial. On the other hand, had the sequence of 20 tosses consisted of alternating heads and tails, we might have suspected that the trials were not independent. In either case, there are grounds to suspect a lack of randomness. Note that our suspicion is not aroused by the numbers of H's and T's, but by the order in which they appeared.

If a sequence contains n_1 symbols of one kind and n_2 of another kind (and neither n_1 nor n_2 is less than 10), the sampling distribution of the **total number of runs**, u, can be approximated closely by a normal distribution with

Mean and standard deviation of u

$$\mu_u = \frac{2n_1 n_2}{n_1 + n_2} + 1 \quad \text{and} \quad \sigma_u = \sqrt{\frac{2n_1 n_2 (2n_1 n_2 - n_1 - n_2)}{(n_1 + n_2)^2 (n_1 + n_2 - 1)}}$$

Thus, the test of the null hypothesis that the arrangement of the symbols (and, hence, the sample) is random can be based on the statistic

Statistic for tests of randomness

$$Z = \frac{u - \mu_u}{\sigma_u}$$

which has approximately the standard normal distribution. Special tables are available for performing the test when n_1, n_2, or both are small.

EXAMPLE 6 **Conducting a test for randomness**

The following is the arrangement of defective, d, and nondefective, n, pieces produced in the given order by a certain machine:

$$n\,n\,n\,n\,n \;\; d\,d\,d\,d \;\; n\,n\,n\,n\,n\,n\,n\,n\,n \;\; d\,d \;\; n\,n \;\; d\,d\,d\,d$$

Test for randomness at the 0.01 level of significance.

Solution 1. *Null hypothesis*: Arrangement is random.
   Alternative hypothesis: Arrangement is not random.

2. *Level of significance*: $\alpha = 0.01$

3. *Criterion*: Reject the null hypothesis if $Z < -2.575$ or $Z > 2.575$, where Z is given by the above formula.

4. *Calculations*: Since $n_1 = 10$, $n_2 = 17$, and $u = 6$, we get

$$\mu_u = \frac{2 \cdot 10 \cdot 17}{10 + 17} + 1 = 13.59$$

$$\sigma_u = \sqrt{\frac{2 \cdot 10 \cdot 17 (2 \cdot 10 \cdot 17 - 10 - 17)}{(10 + 17)^2 (10 + 17 - 1)}} = 2.37$$

and

$$z = \frac{6 - 13.59}{2.37} = -3.20$$

5. *Decision*: Since $z = -3.20$ is less than -2.575, the null hypothesis must be rejected. We conclude that the arrangement is not random. The small P-value $= .0014 = P(Z < -3.20) + P(Z > 3.20)$ strengthens the conclusion. Indeed, the total number of runs is much smaller than expected and there is a strong indication that the defective pieces appear in clusters or groups. The reason for this will have to be uncovered by an engineer who is familiar with the process. ∎

The run test can be used also to test the randomness of samples consisting of numerical data by counting **runs above and below the median**. Denoting an observation exceeding the median of the sample by the letter a and an observation less than the median by the letter b, we can use the resulting sequence of a's and b's to test for randomness by the method just indicated. A frequent application of this test is in quality control, where the means of successive small samples are exhibited on a graph in chronological order. The run test can then be used to check whether there might be a trend in the data. If so, it may be possible to adjust a machine setting or some other process variable before any serious damage occurs.

EXAMPLE 7   **Testing for too many changes**

An engineer is concerned about the possibility that too many changes are being made in the settings of an automatic lathe. Given the following mean diameters (in inches) of 40 successive shafts turned on the lathe

0.261	0.258	0.249	0.251	0.247	0.256	0.250	0.247	0.255	0.243
0.252	0.250	0.253	0.247	0.251	0.243	0.258	0.251	0.245	0.250
0.248	0.252	0.254	0.250	0.247	0.253	0.251	0.246	0.249	0.252
0.247	0.250	0.253	0.247	0.249	0.253	0.246	0.251	0.249	0.253

use the 0.01 level of significance to test the null hypotheses of randomness against the alternative that there is a frequently alternating pattern.

Solution   1. *Null hypothesis*: Arrangement of sample values is random.
Alternative hypothesis: There is a frequently alternating pattern.

2. *Level of significance*: $\alpha = 0.01$

3. *Criterion*: Reject the null hypothesis if $Z > 2.33$, where Z is given by the formula on page 463 for the total number of runs above and below the median.

4. *Calculations*: The median of the 40 measurements is 0.250, so that we get the following arrangement of values above and below 0.250:

$a\ a\ b\ a\ b\ a\ b\ a\ b\ a\ b\ a\ a\ b\ a\ b\ a\ b\ a\ a\ b\ b\ a\ a\ b\ a\ a\ b\ b\ a\ b\ a\ b\ b\ a\ b\ a\ b\ a$

Thus, $n_1 = 19$, $n_2 = 16$, and $u = 27$, so that

$$\mu_u = \frac{2 \cdot 19 \cdot 16}{35} + 1 = 18.37$$

$$\sigma_u = \sqrt{\frac{2 \cdot 19 \cdot 16(2 \cdot 19 \cdot 16 - 19 - 16)}{(19 + 16)^2(19 + 16 - 1)}} = 2.89$$

and

$$z = \frac{27 - 18.37}{2.89} = 2.98$$

5. *Decision*: Since $z = 2.98$ exceeds 2.33, the null hypothesis of randomness must be rejected. The number of runs is much larger than one might expect due to chance, so it is reasonable to conclude that the lathe is being adjusted too often. The P-value $= .0014 = P(Z > 2.98)$ strengthens this conclusion.   ∎

14.6   The Kolmogorov-Smirnov and Anderson-Darling Tests

The **Kolmogorov-Smirnov tests** are nonparametric tests for differences between cumulative distributions. The one sample test concerns the agreement between an observed, or empirical, cumulative distribution of sample values and a specified continuous distribution function; thus, it is a test of goodness of fit. The two sample test concerns the agreement between two observed cumulative distributions; it tests the hypothesis whether two independent samples come from identical continuous distributions, and it is sensitive to population differences with respect to location, dispersion, or skewness.

The Kolmogorov-Smirnov one sample test is generally more efficient than the chi square test for goodness of fit for small samples, and it can be used for very small samples where the chi square test does not apply. It must be remembered, however, that the chi square test of Section 10.5 can be used in connection with discrete distributions, whereas the Kolmogorov-Smirnov test cannot.

The one sample test is based on the maximum absolute difference D between the values of the empirical cumulative distribution of a random sample of size n and a specified theoretical cumulative distribution. To determine whether this difference is larger than can reasonably be expected for a given level of significance, we obtain a P-value (see Exercise 14.15).

EXAMPLE 8   **Using the Kolmogorov-Smirnov test for uniformity**

It is desired to check whether pinholes in electrolytic tin plate are uniformly distributed across a plated coil on the basis of the following distances in inches of 10 pinholes from one edge of a long strip of tin plate 30 inches wide:

$$4.8 \quad 14.8 \quad 28.2 \quad 23.1 \quad 4.4 \quad 28.7 \quad 19.5 \quad 2.4 \quad 25.0 \quad 6.2$$

Test the null hypothesis at the 0.05 level of significance.

Solution   1. *Null hypothesis*:

$$F(x) = \begin{cases} 0 & \text{for } x \leq 0 \\ \dfrac{x}{30} & \text{for } 0 < x < 30 \\ 1 & \text{for } x \geq 30 \end{cases}$$

where x is the distance of a pinhole from the edge.
Alternative hypothesis: The pinholes are not uniformly distributed across the tin plate.

2. *Level of significance*: $\alpha = 0.05$

3. *Criterion*: Reject the null hypothesis if D is large, where D is the maximum difference between the empirical cumulative distribution and the cumulative distribution assumed under the null hypothesis.

4. *Calculations*: Plotting the 2 cumulative distributions as in Figure 14.1, we find that the difference is greatest at $x = 6.2$, and that its value is

$$D = 0.40 - \frac{6.2}{30} = 0.193$$

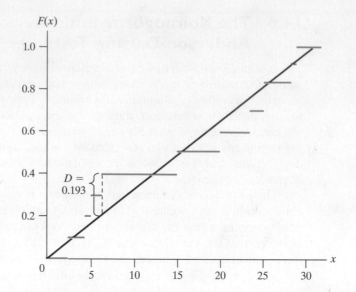

Figure 14.1
Diagram for Kolmogorov-
Smirnov test

5. *Decision*: We use the software R to perform the test and calculate the P-value (see Exercise 14.14). The resulting P-value

$$P(\,|\,D\,| > 0.193\,) = 0.783$$

is so large that the null hypothesis (that the pinholes are uniformly distributed across the tin plate) cannot be rejected. ∎

Despite their intuitive appeal, the Kolmogorov-Smirnov tests do not have good power. Differences in the tails can be easier to detect if the difference between the empirical cumulative distribution, F_n, and F is divided by $\sqrt{F(x)(1 - F(x))}$. In particular, the **Anderson-Darling test** is based on the large values of the statistic

$$A^2 = n \int_{-\infty}^{\infty} [\,F_n(x) - F(x)\,]^2 \, \frac{1}{F(x)(1 - F(x))} \, f(x)\,dx$$

At first sight the numerical calculation of this statistic looks difficult. But, it can be shown that, for continuous distributions

$$A^2 = - \left[\sum_{i=1}^{n} (2i - 1)(\ln(u_i) + \ln(1 - u_{n+1-i})) \right] \Big/ n - n$$

where $u_i = F(x_{(i)})$ is the value of the theoretical cumulative distribution at the ith largest observation $x_{(i)}$.

The null hypothesis is rejected for large values of the statistic A^2. As a guideline, the large sample 10%, 5%, and 1% points are 1.933, 2.492, and 3.878. It has been suggested that these critical values are quite accurate even for samples as small as 10.

EXAMPLE 9   **Evaluating the Anderson-Darling statistic**

With reference to the preceding example, evaluate the Anderson-Darling statistic A^2.

Solution   The smallest observation is 2.4 so $u_1 = F(2.4) = 2.4/30 = 0.08000$. Continuing, the ordered values of the observations and the u_i are

2.4	4.4	4.8	6.2	14.8	19.5	23.1	25.0	28.2	28.7
0.08000	0.14667	0.16000	0.20667	0.49333	0.65000	0.77000	0.83333	0.94000	0.95667

Therefore,

$$A^2 = -[(2 \times 1 - 1)(\ln(0.080000) + \ln(1 - 0.95667)) + \cdots$$
$$+ (2 \times 10 - 1)(\ln(0.95667) + \ln(1 - 0.08000))]/10 - 10$$
$$= 0.5267$$

According to the large sample critical value, we fail to reject the null hypothesis that the distribution of pin holes is uniform, with $\alpha = 0.05$. ∎

Exercises

14.11 The following arrangement indicates whether 60 consecutive cars which went by the toll booth of a bridge had local plates, L, or out-of-state plates, O:

$$L L O L L L L O O L L L L O L O O L L L L O L O O L L L L L$$
$$O L L L O L O L L L L O O L O O O O L L L L O L O O L L L O$$

Test at the 0.05 level of significance whether this arrangement of L's and O's may be regarded as random.

14.12 The following are the number of defective pieces turned out by a machine during 24 consecutive shifts: 15, 11, 17, 14, 16, 12, 19, 17, 21, 15, 17, 19, 21, 14, 22, 16, 19, 12, 16, 14, 18, 17, 24, and 13. Test for randomness at the 0.01 level of significance.

14.13 The following are 50 consecutive downtimes of a machine (in minutes) which were observed during a certain period of time: 22, 29, 32, 25, 33, 34, 38, 34, 29, 25, 27, 33, 34, 28, 39, 41, 24, 31, 34, 29, 34, 25, 30, 37, 40, 39, 35, 24, 32, 43, 44, 34, 40, 38, 39, 43, 46, 34, 39, 45, 42, 39, 54, 50, 38, 41, 43, 46, 52, and 55. Use the method of runs above and below the median and the 0.05 level of significance to test the null hypothesis of randomness against the alternative that there is a trend.

14.14 The P-value on page 466 was calculated using the R software command

```
ks.test(x, "punif", 0,30, alternative = "t")
```

The following are 15 measurements of the boiling point of a silicon compound (in degrees Celsius):

  166 141 136 154 170 162 155 146 183 157 148 132 160 175   150

(a) Use the Kolmogorov-Smirnov test at the 0.01 level of significance to test the null hypothesis that the boiling points come from a normal population with $\mu = 160$ degrees Celsius and $\sigma = 10$ degrees Celsius. Use the R software commands

```
y = c (166,141,136,154,170,162,155,146,183,157,148,132,160,175,150)
ks.test(y, "pnorm", m = 160, sd = 10)
```

(b) Calculate the Anderson-Darling statistic.

14.15 In a vibration study, certain airplane components were subjected to severe vibrations until they showed structural failures. Given the following failure times (in minutes), test whether they can be looked upon as a sample from an exponential population with the mean $\mu = 10$:

1.5	10.3	3.6	13.4	18.4	7.7	24.3	10.7	8.4
15.4	4.9	2.8	7.9	11.9	12.0	16.2	6.8	14.7

Use the Kolmogorov-Smirnov test with a 0.05 level of significance. Refer to Exercise 14.14, but use the R command

```
ks.test(x, "pweibull", shape = 1, scale = 10)
```

Do's and Don'ts

Do's

1. When comparing the locations of two samples and the sample sizes are small, consider applying the Wilcoxon test. This test does not require the population to be normal, an assumption difficult to check with small sample sizes.

Don'ts

1. Don't apply the Wilcoxon test if the dot diagrams of the two samples suggest very different amounts of variation as well a different locations. It will not tell the whole story about differences between the two populations.

2. Don't routinely apply nonparametric rank tests without confirming that the observations are independent. Even moderate time dependence in the observations can seriously affect the level of significance of a rank test. Rank tests are not distribution-free in the presence of dependence. Also, when sampling without replacement from a finite population of size N, you need to account for dependence when the sample size becomes about as large as 5% to 10% of the population.

Review Exercises

14.16 According to Einstein's theory of relativity, light should bend when it passes through a gravitational field. This was first tested experimentally in 1919 when photographs were taken of stars near the sun during a total eclipse and again when the sun had moved to another part of the sky. These eclipse pictures should show the stars displaced outward from the position of the sun. The direction, in the first of 2 coordinate axes, predicted by the theory was matched by the observed direction for 6 out of 7 stars. Record + for a match and − for a mismatch. Guessing would give probability $\frac{1}{2}$ of a match. Use the sign test with level 0.063 to support the claim that the theory holds with respect to matching the direction of displacement.

14.17 Referring to Exercise 12.6, use the U statistic at the 0.05 level of significance to test whether weight loss using lubricant A tends to be less than the loss using lubricant B.

14.18 To find the best arrangement of instruments on a control panel of an airplane, 2 different arrangements were compared by simulating an emergency condition and measuring the reaction time required to correct the condition. The reaction times (in tenths of a second) of 20 pilots (randomly assigned to the 2 different arrangements) were as follows:

Arrangement 1:   8   15   10   13   17   10   9   11   12   15
Arrangement 2:   12   7   13   8   14   6   16   7   10   9

Use the U test at the 0.05 level of significance to check the claim that the second arrangement is better than the first.

14.19 The following are the miles per gallon which a test driver got for 10 tankfuls each of 3 brands of gasoline:

Brand 1:   22   25   32   18   23   15   30   27   19   23
Brand 2:   19   22   18   29   28   32   17   33   28   20
Brand 3:   30   29   25   24   15   27   30   27   18   32

Use the H test at the 0.05 level of significance to test whether there is a difference in the performance of the three brands of gasoline.

14.20 To test whether radio signals from deep space contain a message, an interval of time could be subdivided into a number of very short intervals and it could then be determined whether the signal strength exceeded a certain level (background noise) in each short interval. Suppose that the following is part of such a record, where H denotes a high signal strength and L denotes that the signal strength does not exceed a given noise level.

$$L\,L\,H\,L\,H\,L\,H\,L\,H\,H\,H\,L\,H\,H\,H$$
$$L\,H\,H\,H\,L\,H\,L\,H\,L\,H\,L\,L\,L\,L$$

Test this sequence for randomness (using the 0.05 level of significance) and ascertain whether it is reasonable to assume that the signal contains a message.

14.21 The total number of retail stores opening for business and also quitting business within the calendar years 1983–2015 in a large city were

108 103 109 107 125 142 147 122 116 153 144
162 143 126 145 129 134 137 143 150 148 152
125 106 112 139 132 122 138 148 155 146 158

Making use of the fact that the median is 138, test at the 0.05 level of significance whether there is a significant trend.

14.22 When two populations have the same probability density function, each outcome of n_1 ranks for the first sample, out of the possible values $1, 2, \ldots, n_1 + n_2$, is equally likely.

(a) Write out all of the possible outcomes when $n_1 = 3 = n_2$.

(b) Evaluate U_1 at each of the outcomes and construct its probability distribution.

14.23 With reference to Example 2, Chapter 2, use the U statistic to test the null hypothesis of equality versus the alternative that the distribution of copper content from the first heat is stochastically larger than the distribution for the second heat. Following the approach in Exercise 14.22, it can be shown that the exact distribution gives $P(U_i \geq 19) = 0.033$. Use this as the level of significance.

14.24 The difference between the observed flux and the theoretical value was observed at 20 points within a reactor. The values were

2 −2 −4 −6 −3 −6 3 −5 2 6
8 5 3 9 7 3 2 −1 −3 −1

Use a sign test at the 0.036 level to test the null hypothesis $\tilde{\mu} = 0$ versus the alternative hypothesis $\tilde{\mu} \neq 0$.

14.25 With reference to Exercise 14.24, test for randomness with level 0.05.

14.26 Survival times (days) of fuel rods in a nuclear reactor are as follows:

16 11 24 18 31 15 12 21

Test at the 0.01 level of significance whether these data are consistent with the assumption of a log-normal distribution of survival times. Use the Kolmogorov-Smirnov test and see Exercise 14.14.

Key Terms

Anderson-Darling test 466	Rank-correlation coefficient 459	Stochastically larger 456
H test 459	Rank sums 456	Total number of runs 463
Kolmogorov-Smirnov test 465	Run 462	U test 456
Kruskal-Wallis test 459	Runs above and below the median 464	Wilcoxon test 456
Mann-Whitney test 456	Sign test 454	
Nonparametric tests 454	Spearman's rank correlation 459	

THE STATISTICAL CONTENT OF QUALITY-IMPROVEMENT PROGRAMS

Although there is a tendency to think of monitoring quality as a recent development, there is nothing new about the basic idea of making a quality product characterized by a high degree of uniformity. For centuries skilled artisans have striven to make products distinctive through superior quality, and once a standard of quality was achieved, to eliminate insofar as possible all variability between products that were nominally alike.

What is new in **quality improvement** is the idea that a product is never good enough and should be continually improved. This concept, honed to a fine edge in Japan, created a crisis in the international marketplace for firms that did not follow suit. In quality-improvement programs, the emphasis is on employing designed experiments to improve the product in the design, production, and assembly stages rather than in futile attempts to inspect quality into a product after it is produced. We introduce these ideas in Sections 15.1 through 15.3.

Three special techniques of (statistical) **quality assurance** are also treated in this chapter—quality control is discussed in Sections 15.4 through 15.6, and the establishment of tolerance limits in Section 15.7. Note that the word *quality*, when used technically as in this discussion, refers to some measurable or countable property of a product, the breaking strength of a nano-circuit board, the number of imperfections in a piece of cloth, the potency of a drug, and so forth.

15.1   Quality-Improvement Programs

What is a quality-improvement program? To answer this question, we present a scenario of what happens when action is taken to improve quality. In the context of a machine tooling operation, we first plot the fraction of defective pieces per day [see Figure 15.1(a)] for each day over a 5-week period. This plot reveals stable variation about a value of nearly 15% defective pieces. That is, the process is predictable. We can estimate the mean by the average over days and we can also estimate the amount of variation (see Section 15.4). The fact that the process is stable does not make it good! It is turning out too many defective pieces.

Once it is realized that the process needs improvement, action can be taken. Data collected on several possible sources of variation are displayed in the Pareto diagram (see Section 2.1) in Figure 15.2. The cumulative percentages are given by the broken-line curve and are read from the right-hand scale.

Based on these data, it was decided to give the operators more training on the use of the machine. The record for the next 5 weeks of the daily fraction defectives, after the training, is plotted in Figure 15.1(b). The new process also appears stable but this time about a lower mean. Should we be satisfied? No. The central precept

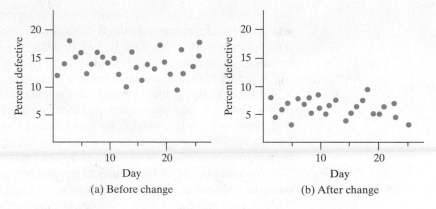

Figure 15.1

Fraction defective per day

(a) Before change

(b) After change

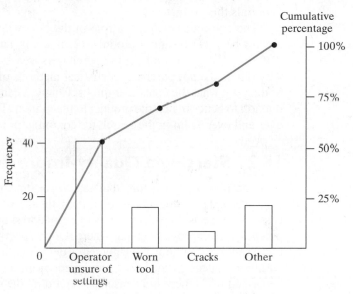

Figure 15.2

Pareto diagram of defects

of quality improvement calls for the process of improvement to be continual. Maybe the gains will be smaller at each progressive stage, but efforts must be continued to reduce the amount of variation and the proportion of defectives. Further substantial improvements will come only by taking action on the system. However, since the process is stable, the effects of change can be observed. Engineers can make innovations to improve the process. The two-level factorial designs discussed in the Chapter 13 are particularly relevant.

There is some folklore that high quality and low production costs are incompatible. But time and again it is the cost of reworking bad products that is a major component of production cost. It really is low quality that results in high costs. Besides the high costs of reworking pieces to make them usable, there are also high costs associated with lost customers who were sold inferior products.

The transformation to quality production in Japan, starting in 1950, created a new economic age leading to a crisis in the 1980s for American businesses. Briefly, Japanese merchandise of the time was known to be shoddy. Several highly placed engineers in Japan studied the literature on quality control produced at Bell Laboratories by Walter Shewhart and others. W. Edwards Deming (1900–1993) was brought in by the Japanese as a foreign expert. Unlike in America, where the applications of statistical methods to problems of quality fell far below their wartime successes because managers did not fully appreciate them, in Japan the top managers came with their engineers to learn about the techniques. What followed was company- and industrywide commitments to improve quality through education, which included statistical methods. This transformation has taken many Japanese companies to world market leadership. In order to compete in the international marketplace, other companies and countries have also had to stress quality improvement.

A theory of management for product or service improvement was pioneered by W. Edwards Deming. It contains concise statements of the elements of the transformation that must take place. Deming summed up his ideas on the transformation of American industry in 14 points for management (see book by Deming, listed in the bibliography). They apply not only to manufacturing but also service industries, and pertain to organizations of all sizes, large and small.

The main thrust of the statistical approach to quality improvement is that, in order to improve quality, it is better to work upstream on the processes. That is, build quality into the product by concentrating on the equipment, components, and materials that go into making it.

The consumer also has a role in the new way of quality improvement. It has always been (1) design a product, (2) make it, and (3) market it. Now, there is a new fourth step, (4) find out the purchasers' reactions to the product. Also find out why others did not purchase. Statistical methods of sampling will provide a way of finding out what the consumer thinks. Changes can then be made in design and production to better match the product to the market. These four steps must be repeated over and over again in the search for continual product improvement.

15.2 Starting a Quality-Improvement Program

It is the prevailing wisdom that top managers must be involved in any quality-improvement program. Once committed, they must take action and select initial processes to serve as flagship projects. It is good to start with processes that have a large potential for improvement and where the prospect for large financial gains is greatest. Even though that is management's decision, the most successful programs start with committees formed with employees from all levels. More enthusiasm can be generated when there is a consensus regarding the selection of the process. A modified Delphi technique can help groups reach unanimity. Each person writes down his or her top three choices. With 3 points for first place, 2 for second, and 1 for third, the totals for each candidate process are tabulated for all to see. Perhaps after some discussion, each person votes again and the process continues until a unanimous choice is reached.

Suppose the process selected concerns piston rings. The first step is to collect data. We will talk to those who run the process about causes and types of defects, but to start we want fresh data on all of the defectives that occur over a period of two weeks.

Defect	Number
Height	30
Diameter	14
Cracks	4
Scratches	2
Other	5

This information is presented as a Pareto diagram in Figure 15.3. We see that 30 out of 55 defective rings have incorrect heights.

To proceed, we gather the engineers, supervisors, and operators who make the rings for a brainstorming session. They construct a list of the possible causes for the variation in height. These may be graphically displayed in a

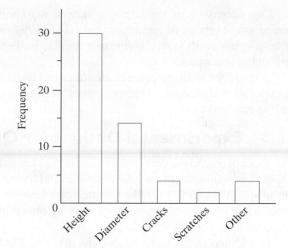

Figure 15.3
Pareto diagram for piston rings

cause-and-effect diagram. It arranges causes, and causes of causes, as shown in Figure 15.4. The cause-and-effect diagram, which resembles the skeleton of a fish, starts with a central horizontal line for a major problem such as incorrect height. Major factors that affect height are listed on diagonal lines attached to the central horizontal line. Factors that affect the major factors, such as cooling time affects tempering, are labeled on horizontal lines connected to the diagonal lines. To proceed further, action must be taken on the system.

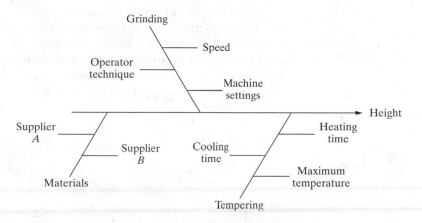

Figure 15.4
Cause-and-effect diagram for piston rings

A two-level factorial design is run with the suppliers as Factor *A* and two speeds for the grinder as Factor *B*. The response, the number of defectives out of 200 rings made at each condition, is recorded in Figure 15.5. The results confirm one of Deming's points: Work with a single supplier.

If the process is stable with the material from supplier 1, then it is time to make another Pareto diagram and continue the cycle of improvement.

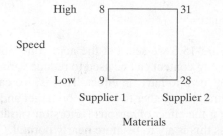

Figure 15.5
Number of defective rings from a 2^2 design

One outgrowth of the Japanese way of working together has been the forma-
tion of small groups of employees called *quality circles*. These groups, consisting
of employees at all levels, meet on a regular basis to discuss ways of continually
improving processes.

With all the workers given some statistical training and engineers some training
in experimental design, all the processes within the company can receive attention
and be improved.

15.3 Experimental Designs for Quality

The modern emphasis, developed in Japan, has been to build quality into the product
rather than waiting until the end of the line and trying to inspect bad quality out. The
job of quality becomes a full-time job for everyone in the company, working as a
team. They must learn about the process by observing and conducting statistically
designed experiments.

In addition to the factorial designs discussed in Chapter 13, the Japanese, and
Professor Genichi Taguchi in particular, have introduced good engineering ideas
to produce new design procedures. Two of his major contributions involve using
designed experiments to

**(a) select one input variable to minimize variation while another input
variable holds the response on target;**

**(b) create products that are not sensitive to variations in their components or
environmental conditions.**

To illustrate the minimization of variation concept, suppose a 2^3 factorial design is
run to study the effects of initial concentration of Acid A, rag content, and digester
time on the tear strength of writing paper. Rather than summarize the experiment in
terms of the means $\bar{y}_i$ at each experimental condition, the statistician G. E. P. Box
suggests a chart of the individual values. This graph portrays both the level with
respect to the indicated target and the amount of variation.

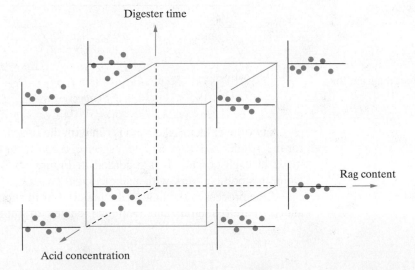

Figure 15.6

The summary of a design to
study the effect on both mean
response and variation

From Figure 15.6 we see that the acid concentration can influence the mean
level, whereas rag content can be used to reduce variation. The third factor, digester
time, does not seem to have an effect. That is, we can manipulate rag content and
initial concentration together to both be on target and to reduce variation.

In studying the effect of various factors on variance, it is usually better to consider $\ln(s^2)$, as this is apt to be more nearly normal.

EXAMPLE 1   **Estimating the factor effects on variance**

Investigators replicated a 2^3 experiment three times to study the effects of $x_1 =$ type of solvent, $x_2 =$ time in oven, and $x_3 =$ temperature of oven on the tensile strength of test specimens of synthetic fiber to be used in carpeting. The main effects and interactions would be estimated as described in Section 13.3. Here, for each run, we have computed the variance of the 3 responses and placed them in the second to last column of the table. The values of their natural logarithms are given in the last column. Determine if any of the factors have an influence on the variance.

	Design			
Solvent	Time	Temp.		
x_1	x_2	x_3	s_i^2	$\ln(s_i^2)$
-1	-1	-1	2,048	7.6246
1	-1	-1	2,813	7.9420
-1	1	-1	800	6.6846
1	1	-1	1,352	7.2093
-1	-1	1	2,113	7.6559
1	-1	1	1,568	7.3576
-1	1	1	882	6.7822
1	1	1	1,013	6.9207

Solution   Multiplying the column of ln(variance) by the appropriate column and dividing by 4, we obtain the estimated effects on ln(variance);

$$solvent: \frac{-7.6246 + 7.9420 - 6.6846 + \cdots + 6.9207}{4} = 0.171$$

$$time: -0.746 \quad temp: -0.186 \quad solvent \times time: 0.161$$
$$solvent \times temp: -0.250 \quad time \times temp: -0.090$$
$$solvent \times time \times temp: 0.057$$

Because there is no replication of values for s_i^2, we cannot compute an F statistic. Instead we create a **half-normal plot**. Plotting the absolute values of the estimated effects in Figure 15.7, we see that the second factor, B (time in oven), seems important. It is far above the straight-line pattern formed by the small estimated effects. If the oven time is kept at its higher level, the variance will be reduced.

Figure 15.7
A half-normal plot of the
| estimated effect | for $\ln(s_i^2)$

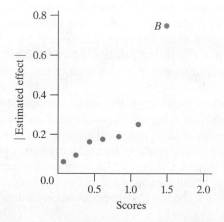

We caution that 7 may be too few effects in order to read patterns from the half-normal plot, so this tentative conclusion needs to be verified by further experimentation. ∎

To illustrate the idea of making products that are insensitive to variation, suppose that the output voltage of a circuit is related to the value of a resistor, as in Figure 15.8. It is possible to exploit the *nonlinear* relationship to obtain a more stable output. Even if the resistors have a 10% tolerance (the nominal 200-ohm resistor would vary between 180 and 220, whereas the 400-ohm resistor would vary between 360 and 440), it would be better to use the higher value resistor because the variation in the response, output voltage, is much smaller. Then, as in Figure 15.6, we could seek a device that would bring the output voltage into the desired range.

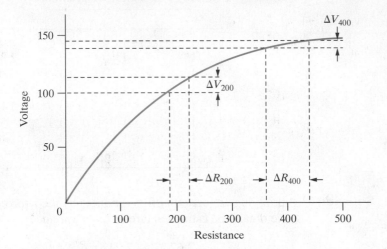

Figure 15.8
A nonlinear relation between output voltage and a resistance can be exploited to reduce variation

G. Taguchi, Chowdhury and Y. Wu, in the reference listed in the bibliography, give an example of making a product insensitive to environmental conditions. The Ina Tile company had experienced large amounts of scrap because of temperature variations within their kiln. Rather than immediately buy a new, expensive kiln, they experimented with new recipes for making the tiles. After running a fractional factorial design with 7 factors, it was found that only lime content was important. As suggested by the Pareto rule, which postulates a vital few and many trivial elements, only 1 out of 7 variables had an effect. A low-cost solution was found and the problem surmounted by increasing the lime content.

15.4 Quality Control

It may surprise some persons to learn that two apparently identical parts made under carefully controlled conditions, from the same batch of raw material, and only seconds apart by the same machine, can nevertheless be different in many respects. Indeed, any manufacturing process, however good, is characterized by a certain amount of variability, which is of a random nature and which cannot be completely eliminated.

Usually manufacturing processes go through several stages of development before actual production begins. Assessments must be made to determine whether the process can produce units that meet engineering specifications. If a characteristic is nearly normally distributed, its natural variation is within plus or minus 3 standard deviations of its mean. A typical baseline assessment is to determine if this interval of length 6σ is within the specification limits. Process capability can be quantified on this basis. Let LSL be the lower specification limit and USL the upper specification limit for the process. Then process potential can be assessed from a **process capability index**, which is estimated by

$$\widehat{C}_p = \frac{USL - LSL}{6\,s}$$

where s is the standard deviation obtained by measuring a sample of units.

Because we must deal with the estimated capability index, $\widehat{C}_p$, practitioners suggest that a value of at least 1.33 is required before an ongoing process is deemed capable.

When the process mean is not centered between the specification limits, the closest specification may be most important. An alternative process capability index having estimated value

$$\widehat{C}_{pk} = \frac{\min(\bar{x} - LSL,\ \ USL - \bar{x})}{3\,s}$$

takes this distance into account.

EXAMPLE 2   **Calculating the process capability index**

The specification limits on a valve diameter (mm) are $LSL = 10.98$ and $USL = 11.01$. Measurements on 80 valves gave $\bar{x} = 10.991$ and $s = 0.0035$. Estimate the process capability indices C_p and C_{pk}.

Solution

$$\widehat{C}_p = \frac{USL - LSL}{6\,s} = \frac{11.01 - 10.98}{6\,(0.0035)} = 1.43$$

so the process would be judged to be capable.

$$\widehat{C}_{pk} = \frac{\min(\bar{x} - LSL,\ \ USL - \bar{x})}{3\,s}$$

$$= \frac{\min(10.991 - 10.98,\ \ 11.01 - 10.991)}{3\,(0.0035)} = 1.048$$

This second index is substantially smaller than the first because the mean is off-center.   ∎

According to the ideas of quality improvement, getting the six-sigma interval within specifications is just a first step. Further improvements can lead to tighter specifications and the production of better units. However, before any assessment of capability can be made, the process must be made stable or in control.

When the variability present in a production process is confined to chance variation, the process is said to be in a state of **statistical control**. Such a state is usually attained by finding and eliminating trouble of the sort causing another kind of variation, called **assignable variation**, which may be due to poorly trained operators, poor-quality raw materials, faulty machine settings, worn parts, and the like. Since manufacturing processes are rarely free from trouble of this kind, it is important to have some systematic method of detecting serious deviations from a state of statistical control when, or if possible *before*, they actually occur. It is to this end that **control charts** are principally used.

In what follows, we shall differentiate between **control charts for measurements** and **control charts for attributes**, depending on whether the observations with which we are concerned are measurements or count data (say, the numbers of defectives in a sample of a given size). In either case, a control chart consists of a **central line** (see Figure 15.9) corresponding to the *average* quality at which the

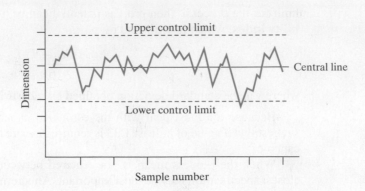

Figure 15.9
Control chart

process is to perform, and lines corresponding to the **upper and lower control limits**. These limits are chosen so that values falling between them can be attributed to chance, while values falling beyond them are interpreted as indicating a lack of control. By plotting the results obtained from samples taken periodically at frequent intervals, it is possible to check by means of such a chart whether the process is under control, or whether trouble of the sort indicated above has entered the process. When a sample point falls beyond the control limits, one looks for trouble, but even if the point falls between the control limits, a trend of some other systematic pattern may serve notice that action should be taken to avoid serious trouble.

The ability to read control charts and to determine from them just what corrective action should be taken is a matter of experience and highly developed judgment.

A quality-control engineer must not only understand the statistical foundation of the subject but must also be thoroughly acquainted with the processes themselves. The engineering and managerial aspects of quality control (and quality assurance in general), which nowadays includes incoming raw materials, outgoing products, and in-process control, constitute an extensive subject in themselves. In the following sections we present only the statistical aspects of the subject.

15.5   Control Charts for Measurements

When dealing with measurements, it is customary to exercise control over the average quality of a process as well as its variability. The first goal is accomplished by plotting the means of periodic samples on a **control chart for means**, called an **X-bar** or $\bar{x}$ **chart**. Variability is controlled by plotting the sample ranges or standard deviations, respectively, on an **R chart**, or a **σ chart**, depending on which statistic is used to estimate the population standard deviation.

If the process mean and standard deviation, μ and σ, are known, and it is reasonable to treat the measurements as samples from a normal population, we can assert with probability $1 - \alpha$ that the mean of a random sample of size n will fall between

$$\mu - z_{\alpha/2}\,\frac{\sigma}{\sqrt{n}} \quad \text{and} \quad \mu + z_{\alpha/2}\,\frac{\sigma}{\sqrt{n}}$$

These two limits on $\bar{x}$ provide upper and lower control limits, and, under the given assumptions, they enable the quality-control engineer to determine whether or not to make an adjustment in the process.

In actual practice, μ and σ are usually unknown and it is necessary to estimate their values from a large sample (or samples) taken while the process is in control. For this reason and because there may be no assurance that the measurements can be treated as samples from a normal population, the $(1 - \alpha)100\%$ confidence level associated with the control limits is only approximate, and such probability limits are seldom used in practice. Instead, it is common industrial practice to use **three-sigma limits** obtained by substituting 3 for $z_{\alpha/2}$. With three-sigma limits one usually is highly confident that the process will not be declared out of control when, in fact, it is actually in control.

If there exists a long history of a process in good control, μ and σ can be estimated from past data practically without error. Thus, the central line of an $\bar{x}$ chart is given by μ, and the upper and lower three-sigma control limits are given by $\mu \pm A\sigma$, where $A = 3/\sqrt{n}$ and n is the size of each sample.[1] For convenience, values of A for $n = 2, 3, \ldots,$ and 15 are given in Table 8W. The use of a constant sample size n simplifies the maintenance and interpretation of an $\bar{x}$ chart, but as the reader will observe in Exercise 15.4, this restriction is not absolutely necessary.

In the more common case where the population parameters are *unknown*, it is necessary to estimate these parameters on the basis of preliminary samples. For this purpose, it is usually desirable to obtain the results of 20 or 25 consecutive samples taken when the process is in control. If k samples are used, each of size n, we shall denote the mean of the ith sample by $\bar{x}_i$, and the grand mean of the k sample means by $\bar{\bar{x}}$, that is,

Grand mean of sample means

$$\bar{\bar{x}} = \frac{1}{k} \sum_{i=1}^{k} \bar{x}_i$$

The process variability σ can be estimated either from the standard deviations or the ranges of the k samples. Since the sample size commonly used in connection with control charts for measurements is small, there is usually very little loss of efficiency in estimating σ from the sample ranges. (For an example where sample standard deviations are used in this connection, see Exercise 15.5.) Denoting the range of the ith sample by R_i, we shall thus make use of the statistic

Mean sample range

$$\bar{R} = \frac{1}{k} \sum_{i=1}^{k} R_i$$

Since $\bar{\bar{x}}$ provides an unbiased estimate of the population mean μ, the central line for the $\bar{x}$ chart is given by $\bar{\bar{x}}$. The statistic $\bar{R}$ does not provide an unbiased estimate of σ, but multiplying $\bar{R}$ by the constant A_2, we obtain an unbiased estimate of $3\sigma/\sqrt{n}$. The constant multiplier A_2, tabulated in Table 8W for various values of n, depends on the assumption that the measurements constitute a sample from a normal population. Thus, the central line and the upper and lower three-sigma control limits, UCL and LCL, for an $\bar{x}$ chart (with μ and σ estimated from past data) are given by

Control-chart values for an $\bar{x}$ chart

$$
\begin{aligned}
central\ line &= \bar{\bar{x}} \\
UCL &= \bar{\bar{x}} + A_2 \bar{R} \\
LCL &= \bar{\bar{x}} - A_2 \bar{R}
\end{aligned}
$$

---

[1]Throughout this chapter we depart somewhat from the customary quality-control notation in order to be consistent with the more widely accepted statistical notation used elsewhere in this book. (For instance, in quality control it is customary to denote the sample mean and standard deviation by $\bar{x}$ and σ and the corresponding population parameters by $\bar{x}'$ and σ'.)

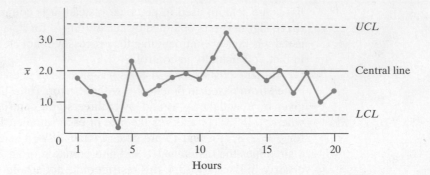

Figure 15.10
$\bar{x}$ chart

An example of this kind of control chart for the mean is shown in Figure 15.10.

In controlling a process, it may not be enough to monitor the population mean. Although an increase in process variability may become apparent from increased fluctuations of the $\bar{x}$'s, a more sensitive test of shifts in process variability is provided by a separate control chart, an R chart based on the sample ranges or a σ chart based on the sample standard deviations. An example of the latter may be found in Exercise 15.5.

The central line and control limits of an R chart are based on the distribution of the range of samples of size n from a normal population. As we observed on page 281, the mean and the standard deviation of this sampling distribution are given by $d_2\sigma$ and $d_3\sigma$, respectively, when σ is known. Thus, three-sigma control limits for the range are given by $d_2\sigma \pm 3d_3\sigma$, and the complete set of control-chart values for an R chart (with σ known) is given by

Control-chart values for an R chart (σ known)

$$\begin{aligned} central\ line &= d_2\sigma \\ UCL &= D_2\sigma \\ LCL &= D_1\sigma \end{aligned}$$

Here $D_1 = d_2 - 3d_3$ and $D_2 = d_2 + 3d_3$, and values of these constants can be found in Table 8W for various values of n.

If σ is unknown, it is estimated from past data as previously described, and the control-chart values for R chart (with σ unknown) are as follows:

Control-chart values for an R chart (σ unknown)

$$\begin{aligned} central\ line &= \overline{R} \\ UCL &= D_4\overline{R} \\ LCL &= D_3\overline{R} \end{aligned}$$

Here $D_3 = D_1/d_2$ and $D_4 = D_2/d_2$, and values of these constants can also be found in Table 8W for various values of n.

To illustrate the construction of an $\bar{x}$ chart and an R chart, suppose a manufacturer of a certain bearing knows from a preliminary record of 20 hourly samples of size 4 that, for the diameters of these bearings, $\bar{\bar{x}} = 0.9752$ and $\overline{R} = 0.0002$. Coding her data by means of the expression

$$\frac{x - 0.9750}{0.0001}$$

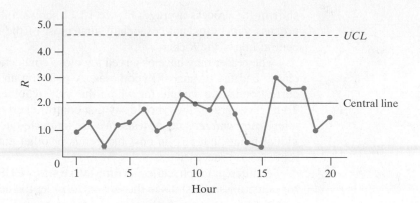

Figure 15.11
R chart

that is, expressing each measurement as a deviation from 0.9750 in 0.0001 inch, she obtains

$\bar{x}$ Chart (coded)		R Chart (coded)	
central line	$\bar{\bar{x}} = 2.0$	central line	$\bar{R} = 2.0$
UCL	$\bar{\bar{x}} + A_2\bar{R} = 3.5$	UCL	$D_4\bar{R} = 4.6$
LCL	$\bar{\bar{x}} - A_2\bar{R} = 0.5$	LCL	$D_3\bar{R} = 0.0$

The values of $A_2 = 0.729$, $D_3 = 0$, and $D_4 = 2.282$ for samples of size 4 were obtained from Table 8W. Graphically, these control charts are shown in Figures 15.10 and 15.11, where we have also indicated the results subsequently obtained in the following 20 samples:

Hour	Coded Sample Values				$\bar{x}$	R
1	1.7	2.2	1.9	1.2	1.75	1.0
2	0.8	1.5	2.1	0.9	1.32	1.3
3	1.0	1.4	1.0	1.3	1.18	0.4
4	0.4	−0.6	0.7	0.2	0.18	1.3
5	1.4	2.3	2.8	2.7	2.30	1.4
6	1.8	2.0	1.1	0.1	1.25	1.9
7	1.6	1.0	1.5	2.0	1.52	1.0
8	2.5	1.6	1.8	1.2	1.78	1.3
9	2.9	2.0	0.5	2.2	1.90	2.4
10	1.1	1.1	3.1	1.6	1.72	2.0
11	1.7	3.6	2.5	1.8	2.40	1.9
12	4.6	2.8	3.5	1.9	3.20	2.7
13	2.6	2.8	3.2	1.5	2.52	1.7
14	2.3	2.1	2.1	1.7	2.05	0.6
15	1.9	1.6	1.8	1.4	1.68	0.5
16	1.3	2.0	3.9	0.8	2.00	3.1
17	2.8	1.5	0.6	0.2	1.28	2.6
18	1.7	3.6	0.9	1.5	1.92	2.7
19	1.6	0.6	1.0	0.8	1.00	1.0
20	1.7	1.0	0.5	2.2	1.35	1.7

Inspection of Figure 15.10 shows that only one of the points falls outside of the control limits, but it also shows that there may nevertheless have been a downward shift in the process average. Figure 15.11 shows a definite downward shift in the process variability; note especially that most of the sample ranges fall below the central line of the R chart.

The reader may have observed the close connection between the use of control charts and the testing of hypotheses. A point on an $\bar{x}$ chart that is out of control corresponds to a sample for which the null hypotheses that $\mu = \mu_0$ is rejected. To be more precise, we should say that control-chart techniques provide sequential, *temporally ordered* sets of tests. We are interested not only in the position of individual points, but also in possible trends or other patterns exhibited by the points representing successive samples.

A different graph, called a **cumulative sum (CUSUM) chart**, is more effective for detecting small shifts in the mean. Consider the deviations *observation − target value*. In the context of our example, the target value should be 2.00. To construct the CUSUM chart, plot the **CUSUM statistic** versus time order

$$S_1 = (1.75 - 2.00) = -0.25 \text{ versus } 1$$
$$S_2 = (1.75 - 2.00) + (1.32 - 2.00) = -0.93 \text{ versus } 2$$
$$S_3 = (-0.25) + (-0.68) + (1.18 - 2.00) = -1.75 \text{ versus } 3$$
$$\vdots$$

In Figure 15.12 we see evidence of an initial constant downward trend, masked somewhat by random variation. This behavior indicates that the level of the process is a fixed amount below the target value of 2.00. At observation 11 there is a distinct shift up in level, as if a change has been made, but after a few hours a shift to a level below 2.0 appears to occur.

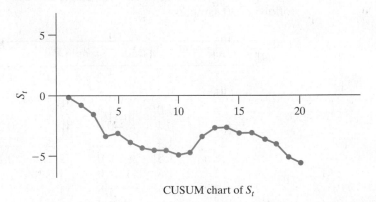

Figure 15.12

A CUSUM plot

CUSUM chart of S_t

R. Crosier proposed a two-sided CUSUM scheme, which first updates the previous CUSUM by a new observation. Depending on the updated value of this sum, the new value of the CUSUM is either set equal to zero or the CUSUM is shrunk toward zero. This modification reduces the chance of raising a false alarm.

In particular, **Crosier's** two-sided **CUSUM** starts with $S_0 = 0$. At each step, the tentative absolute value of the sum $C_n = |S_{n-1} + (X_n - a)|$ is first calculated. Then the next value of the statistic S_n is defined as

$$S_n = \begin{cases} 0, & \text{if } C_n \leq ks \\ (S_{n-1} + X_n - a)(1 - ks/C_n), & \text{otherwise} \end{cases}$$

If μ_0 denotes the in-control mean and μ_1 a value that should be detected quickly, the constant k can be set to one-half of the specified mean-shift (expressed in standard deviations).

$$k = \frac{1}{2} \frac{(\mu_1 - \mu_0)}{\sigma}$$

Crosier's scheme signals that the mean has shifted when

$$S_n \geq hs \text{ (increase)} \quad \text{or} \quad S_n \leq -hs \text{ (decrease)}$$

where h is a specified constant. See R. Johnson and R. Li, listed in the bibliography, for examples and more details on CUSUM statistics.

15.6   Control Charts for Attributes

Although more complete information can usually be gained from measurements made on a finished product, it is often quicker and cheaper to check the product against specifications on an "attribute" or "go, no-go" basis. For example, in checking the diameter and eccentricity of a ball bearing it is far simpler to determine whether it will pass through circular holes cut in a template than to make several measurements of the diameter with a micrometer. In this section we discuss two fundamental kinds of control charts used in connection with attribute sampling, the **fraction-defective chart**, also called a **p chart**, and the **number-of-defects chart**, also called a **c chart**. To clarify the distinction between "number of defective" and "number of defects," note that a unit tested can have several **defects**, whereas on the other hand, it is either defective or it is not. In many applications a unit is referred to as **defective** if it has at least one defect.

Control limits for a fraction-defective chart are based on the sampling theory for proportions introduced in Section 10.1 and on the normal curve approximation to the binomial distribution. Thus, if a standard is given—that is, if the fraction defective should take on some preassigned value p—the central line is p and three-sigma control limits for the fraction defective in random samples of size n are given by

$$p \pm 3\sqrt{\frac{p(1-p)}{n}}$$

If no standard is given, which is more frequently the case in actual practice, p will have to be estimated from past data. If k samples are available, d_i is the number of defectives in the ith sample, and n_i is the number of observations in the ith sample, it is customary to estimate p as the proportion of defectives in the combined sample, namely, as

Proportion of defectives in combined sample

$$\overline{p} = \frac{d_1 + d_2 + \cdots + d_k}{n_1 + n_2 + \cdots + n_k}$$

Control-chart values for a fraction-defective chart

$$central\ line = \overline{p}$$

$$UCL = \overline{p} + 3\sqrt{\frac{\overline{p}(1-\overline{p})}{n}}$$

$$LCL = \overline{p} - 3\sqrt{\frac{\overline{p}(1-\overline{p})}{n}}$$

Note that if p is small, as is often the case in practice, substitution in the formula for the lower control limit might yield a negative number. When this occurs, it is customary to regard the lower control limit as if it were zero and, in effect, to use only the upper control limit. Another complication that can arise if p is small is that the binomial distribution may not be adequately approximated by the normal distribution. Generally speaking, the use of the above control limits for p charts is unrealistic whenever n and p are such that the underlying binomial (or hypergeometric) distribution cannot be approximated by a normal curve (see page 138). In such cases it is best to use an upper control limit obtained directly from a table of binomial probabilities, or, perhaps use the Poisson approximation to the binomial distribution.

As an illustration of a p chart, suppose that it is desired to control the output of a certain integrated circuit production line to maintain a yield of 60 percent, that is, a proportion defective of 40 percent. To this end, daily samples of 100 units are checked to electrical specifications, with the following results:

Date	Number of Defectives	Date	Number of Defectives	Date	Number of Defectives
3-12	24	3-26	44	4-09	23
3-13	38	3-27	52	4-10	31
3-16	62	3-30	45	4-13	26
3-17	34	3-31	30	4-14	32
3-18	26	4-01	34	4-15	35
3-19	36	4-02	33	4-16	15
3-20	38	4-03	22	4-17	24
3-23	52	4-06	34	4-20	38
3-24	33	4-07	43	4-21	21
3-25	44	4-08	28	4-22	16

Since the standard is given as $p = 0.40$, the control-chart values are

$$central\ line = 0.40$$

$$UCL = 0.40 + 3\sqrt{\frac{(0.40)(0.60)}{100}} = 0.55$$

$$LCL = 0.40 - 3\sqrt{\frac{(0.40)(0.60)}{100}} = 0.25$$

The corresponding control chart with points for the 30 sample fractions defective is shown in Figure 15.13, and it exhibits some interesting characteristics. Note that there is only 1 point out of control on the high side, but there are 7 points out of control on the low side. Most of these 7 low points occurred after April 1, and there appears to be a general downward trend. In fact, there is an unbroken run of 11 points below the central line after April 7. It would appear from this chart that the yield is not yet stabilized and that the process is potentially capable of maintaining a yield well above the nominal 60 percent value.

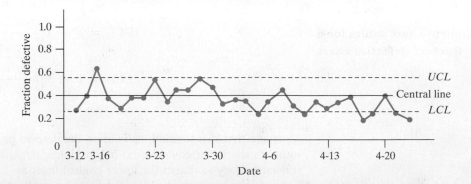

Figure 15.13
p chart

Equivalent to the p chart for the fraction defective is the control chart for the number of defectives. Instead of plotting the fraction defective in a sample of size n, one plots the number of defectives. The control-chart values for this kind of chart are obtained by multiplying the above values for the central line and the control-limits by n. Thus, if p is estimated by $\overline{p}$, the control-chart values for a **number-of-defectives chart** are as follows:

Control-chart values for a number-of-defectives chart

$$central\ line = n\overline{p}$$
$$UCL = n\overline{p} + 3\sqrt{n\overline{p}(1 - \overline{p})}$$
$$LCL = n\overline{p} - 3\sqrt{n\overline{p}(1 - \overline{p})}$$

There are situations where it is necessary to control the number of defects in a unit of product, rather than the fraction defective or the number of defectives. For example, in the production of carpeting, it is important to control the number of defects per hundred yards; in the production of newsprint one may wish to control the number of defects per roll. These situations are similar to the one described in Section 4.7, which led to the Poisson distribution. Thus, if c is the number of defects per manufactured unit, c is taken to be a value of a random variable having the Poisson distribution.

It follows that the center line for a number-of-defects chart is the parameter λ of the corresponding Poisson distribution, and the three-sigma control limits can be based on the fact that the standard deviation of this distribution is $\sqrt{\lambda}$. If λ is unknown, that is, if no standard is given, its value is usually estimated from at least 20 values of c observed from past data. If k is the number of units of product available for estimating λ, and if c_i is the number of defects in the ith unit, then λ is estimated by

Mean number of defects

$$\overline{c} = \frac{1}{k} \sum_{i=1}^{k} c_i$$

and the control-chart values for the c **chart**, or **number-of-defects chart**, are

Control-chart values for a number-of-defects chart

$$central\ line = \overline{c}$$
$$UCL = \overline{c} + 3\sqrt{\overline{c}}$$
$$LCL = \overline{c} - 3\sqrt{\overline{c}}$$

To illustrate this kind of control chart, suppose that it is known from past experience that on the average an aircraft assembly made by a certain company has $\overline{c} = 4$ missing rivets. The corresponding control chart for the number of missing rivets is shown in Figure 15.14, on which we have also plotted the results of inspections that revealed 4, 6, 5, 1, 2, 3, 5, 7, 1, 2, 2, 4, 6, 5, 3, 2, 4, 1, 8, 4, 5, 6, 3, 4, and 2 missing rivets in 25 assemblies.

For any $\overline{x}$ chart, p chart, or c chart, one point outside of the control limits is enough to suggest that a special cause has influenced the process. It is now common

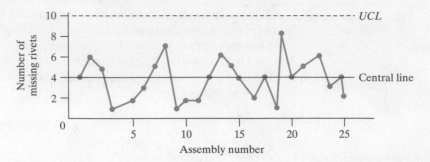

Figure 15.14
c chart

practice to increase the sensitivity of the chart to detect special causes by adding other tests for special causes. Two of these extra tests merit special mention. In addition to one point outside of the control limits, a chart will signal a special cause if either

---

Additional tests for special causes

1. There are 9 points in a row on the same side of the central line.
2. There are 6 points in a row, all increasing or all decreasing.

As mentioned above, there is a run of 11 points below the central line in Figure 15.13.

Exercises

15.1 A plastics manufacturer extrudes blanks for use in the manufacture of eyeglass temples. Specifications require that the thickness of these blanks have $\mu = 0.150$ inch and $\sigma = 0.002$ inch.

(a) Use the specifications to calculate a central line and three-sigma control limits for an $\bar{x}$ chart with $n = 5$.

(b) Use the specifications to calculate a central line and three-sigma control limits for an R chart with $n = 5$.

(c) Plot the following means and ranges, obtained in 20 successive random samples of size 5, on charts based on the control-chart constants obtained in parts (a) and (b), and discuss the process.

Sample	$\bar{x}$	R	Sample	$\bar{x}$	R
1	0.152	0.004	11	0.149	0.003
2	0.147	0.006	12	0.153	0.004
3	0.153	0.004	13	0.150	0.005
4	0.153	0.002	14	0.152	0.001
5	0.151	0.003	15	0.149	0.003
6	0.148	0.002	16	0.146	0.002
7	0.149	0.006	17	0.154	0.004
8	0.144	0.001	18	0.152	0.005
9	0.149	0.003	19	0.151	0.002
10	0.152	0.005	20	0.149	0.004

15.2 Calculate $\bar{\bar{x}}$ and $\bar{R}$ of the data of part (c) of Exercise 15.1, and use these values to construct the central lines and three-sigma control limits for new $\bar{x}$ and R charts to be used in the control of the thickness of the extruded plastic blanks.

15.3 The following data give the means and ranges of 25 samples, each consisting of 4 compression test results on steel forgings, in thousands of pounds per square inch:

Sample	1	2	3	4	5	6	7	8
$\bar{x}$	45.4	48.1	46.2	45.7	41.9	49.4	52.6	54.5
R	2.7	3.1	5.0	1.6	2.2	5.7	6.5	3.6

Sample	9	10	11	12	13	14	15	16
$\bar{x}$	45.1	47.6	42.8	41.4	43.7	49.2	51.1	42.8
R	2.5	1.0	3.9	5.6	2.7	3.1	1.5	2.2

Sample	17	18	19	20	21	22	23	24	25
$\bar{x}$	51.1	52.4	47.9	48.6	53.3	49.7	48.2	51.6	52.3
R	1.4	4.3	2.2	2.7	3.0	1.1	2.1	1.6	2.4

(a) Use these data to find the central line and control limits for an $\bar{x}$ chart.

(b) Use these data to find the central line and control limits for an R chart.

(c) Plot the given data on $\bar{x}$ and R charts based on the control-chart constants computed in parts (a) and (b), and interpret the results.

(d) Using runs above and below the central line (similar to runs above and below the median discussed on page 464), test at a level of significance of 0.05 whether there is a trend in the $\bar{x}$ values.

(e) Would it be reasonable to use the control limits found in this exercise in connection with subsequent compression test measurements from the same process? Why or why not?

15.4 Reverse-current readings (in nanoamperes) are made at the location of a transistor on an integrated circuit. A sample of size 10 is taken every half hour. Since some of the units may prove to be "shorts" or "opens," it is not always possible to obtain 10 readings. The following table shows the number of readings made at the end of each half-hour interval during an 8-hour shift, and the mean reverse currents obtained:

Sample	1	2	3	4	5	6	7	8
n	10	6	9	8	8	10	7	9
$\bar{x}$	12.5	11.1	10.2	11.6	21.9	12.3	9.7	15.6

Sample	9	10	11	12	13	14	15	16
n	7	8	10	9	7	8	9	10
$\bar{x}$	16.7	9.8	11.6	17.2	10.1	9.5	13.1	14.2

(a) Find the central line for an $\bar{x}$ chart by taking the weighted mean of the 16 $\bar{x}$'s, weighting each value with the size of the corresponding sample.

(b) Construct a table showing the central line in part (a) and three-sigma control limits corresponding to $n = 6, 7, 8, 9$, and 10. Use $R = 4.0$, a value based on prior data.

(c) Plot the data on a control chart like the one in Figure 15.15 and interpret the results.

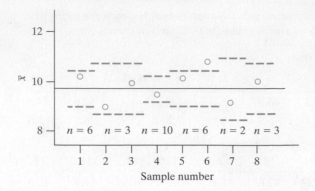

Figure 15.15
Exercise 15.4

15.5 If the sample standard deviations instead of the sample ranges are used to estimate σ, the control limits for the resulting $\bar{x}$ chart are given by $\bar{\bar{x}} \pm A_1 \bar{s}$, where $\bar{s}$ is the mean of the sample standard deviations obtained from given data, and A_1 can be found in Table 8W. Note that in connection with problems of quality control the sample standard deviation is defined using the divisor n instead of $n-1$. The corresponding R chart is replaced by a σ chart, having the central line $c_2 \bar{s}$ and the lower and upper control limits $B_3 \bar{s}$ and $B_4 \bar{s}$, where c_2, B_3, and B_4 can be obtained from Table 8W.

(a) Construct an $\bar{x}$ chart and a σ chart for 20 samples of size 3 which had $\bar{x}$ equal to 21.2, 19.4, 20.4, 20.4, 20.4, 19.0, 20.3, 21.1, 21.6, 22.1, 24.4, 23.9, 24.9, 24.1, 21.8, 19.5, 20.3, 22.5, 23.4, 23.3, and s is equal to 2.0, 0.8, 1.1, 0.9, 1.0, 0.3, 1.3, 2.0, 0.8, 1.0, 1.5, 1.0, 1.5, 0.8, 1.3, 2.9, 4.3, 1.2, 0.3, 3.1.

(b) Would it be reasonable to use these control limits for subsequent data? Why?

15.6 In order to establish control charts for a boring process, 30 samples of five measurements of the inside diameters are taken, and the results are $\bar{\bar{x}} = 1.317$ inches and $\bar{s} = 0.002$ inch. Using the method of Exercise 15.5, construct an $\bar{x}$ chart for $n = 5$ and on it plot the following means obtained in 25 successive samples: 1.328, 1.330, 1.321, 1.325, 1.332, 1.340, 1.327, 1.321, 1.324, 1.325, 1.329, 1.326, 1.330, 1.324, 1.328, 1.322, 1.326, 1.327, 1.329, 1.325, 1.324, 1.329, 1.330, 1.321, and 1.329. Discuss the results.

15.7 Suppose that with the samples of Exercise 15.6, it is desired to establish control also over the variability of the process. Using the method of Exercise 15.5 and the values of $\bar{\bar{x}}$ and $\bar{s}$ given in Exercise 15.6, calculate the central line and control limits for a σ chart with $n = 5$.

15.8 Thirty-five successive samples of 100 castings each, taken from a production line, contained, respectively, 3, 3, 5, 3, 5, 0, 3, 2, 3, 5, 6, 5, 9, 1, 2, 4, 5, 2, 0, 10, 3, 6, 3, 2, 5, 6, 3, 3, 2, 5, 1, 0, 7, 4, and 3 defectives. If the fraction defective is to be maintained at 0.02, construct a p chart for these data and state whether or not this standard is being met.

15.9 The data of Exercise 15.8 may be looked upon as evidence that the standard of 2% defectives is being exceeded.

(a) Use the data of Exercise 15.8 to construct new control limits for the fraction defective.

(b) Using the control limits found in part (a), continue the control of the process by plotting the following data on the number of defectives obtained in 20 subsequent samples of size $n = 100$: 2, 4, 2, 4, 7, 5, 3, 2, 2, 3, 5, 6, 4, 5, 8, 0, 5, 5, 4, and 2.

15.10 The specifications for a certain mass-produced valve prescribe a testing procedure according to which each valve can be classified as satisfactory or unsatisfactory (defective). Past experience has shown that the process can perform so that $\bar{p} = 0.03$. Construct a three-sigma control chart for the number of defectives obtained in samples of size 100, and on it plot the following numbers of defectives obtained in such samples randomly selected from 30 successive half-days of production: 3, 4, 2, 1, 5, 2, 1, 2, 3, 1, 3, 2, 2, 2, 1, 1, 2, 0, 4, 3, 1, 0, 2, 4, 0, 1, 5, 7, 3, and 2.

15.11 The standard for a process producing tin plate in a continuous strip is 5 defects in the form of pinholes or visual blemishes per 100 feet. Based on the following set of 25 observations, giving the number of defects per 100 feet, can it be concluded that the process is in control to this standard?

Inspection number	1	2	3	4	5	6	7	8	9	10	11	12
Number of defects	3	2	2	4	4	4	6	4	1	7	5	5

Inspection number	13	14	15	16	17	18	19	20	21	22	23	24	25
Number of defects	4	6	6	9	5	2	6	5	11	6	6	8	2

15.12 A process for the manufacturer of 4-by-8-foot woodgrained panels has performed in the past with an average of 2.7 imperfections per 100 panels. Construct a chart to be used in the inspection of the panels and discuss the control if 25 successive 100-panel lots contained, respectively, 4, 1, 0, 3, 5, 3, 5, 4, 1, 4, 0, 1, 4, 2, 3, 7, 4, 2, 1, 3, 0, 2, 6, 1, and 3 imperfections.

15.7   Tolerance Limits

Any process that is stable has natural limits of its own. A random sample of any quality characteristic can then lead to **tolerance limits** that locate a reasonably high proportion of the values of the quality characteristic being produced.

Suppose, long experience with a product strongly suggests that a certain dimension is normally distributed with the mean μ and the standard deviation σ. Then, it is easy to construct limits between which we can expect to find any given proportion P of the population. For $P = 0.90$, we have the tolerance limits $\mu \pm 1.645\,\sigma$, and for $P = 0.95$ we have $\mu \pm 1.96\,\sigma$, as can easily be verified from Table 3.

In most practical situations the true values of μ and σ are not known, and tolerance limits must be based on the mean $\overline{X}$ and the standard deviation S of a random sample. Whereas $\mu \pm 1.96\,\sigma$ are limits including 95% of a normal population, the same cannot be said for the limits $\overline{X} \pm 1.96\,S$. These limits are random variables and they may or may not include a given proportion of the population. Nevertheless, it is possible to determine a constant K so that *one can assert with $(1 - \alpha)100\%$ confidence that the proportion of the population contained between $\overline{x} - Ks$ and $\overline{x} + Ks$ is at least P*.

Data: A random sample for a normal population X_1, X_2, ..., X_n

Given n, a specified confidence level $(1 - \alpha)100\%$, and population proportion P, let K be determined from Table 9W(a), Appendix B.

Tolerance limits: $\overline{x} \pm K s$

With $(1 - \alpha)\%$ confidence, the interval contains at least proportion P of the population.

Table 9W(a), Appendix B, gives the value of K for $P = 0.90, 0.95$, and 0.99, with 95% or 99% levels of confidence, and selected values of n from 2 to 1,000.

EXAMPLE 3   A tolerance interval for the free length of springs

A manufacturer produces compression springs in very large lots. It is helpful to find an interval that locates a large majority of the free lengths of these springs. A sample of size $n = 100$ yields $\overline{x} = 1.507$ and $s = 0.004$ inch.

With 99% confidence determine an interval that contains a minimum proportion $P = 0.95$ of all springs that will be produced. Assume that the distribution of free length is normal.

Solution   From Appendix B, Table 9W(a), with $n = 100$, $1 - \alpha = 0.99$, and $P = 0.95$, we find $K = 2.335$. The resulting tolerance interval is

$$\overline{x} \pm K s = 1.507 \pm 2.335(0.004) \qquad \text{or} \qquad (1.497, 1.517)$$

We assert, with 99% confidence, that at least 95 % of springs have free lengths from 1.497 to 1.517 inches.

Note that, in problems like this, the lower tolerance limit is rounded *down* and the upper tolerance limit is rounded *up*.   ∎

To avoid confusion, let us also point out that there is an essential difference between confidence limits and tolerance limits. Whereas confidence limits are used to estimate a parameter of a population, tolerance limits are used to indicate between what limits one can find a certain proportion of a population. This distinction is emphasized by the fact that when n becomes large the length of a confidence interval approaches zero, while the tolerance limits will approach the corresponding values for the population. Thus, for large n, K approaches 1.96 in the columns for $P = 0.95$ in Table 9W(a).

The situation for one-sided tolerance bounds is different. In the context of strength of materials, it is the weaker specimens that break. Consequently, it is important for engineers to have an accurate estimate of the lower tail of the population of strengths. Recently engineers have realized that it is wiser to set specifications for strength in terms of a lower percentile η_β rather than the mean μ. It is the weaker specimens, not those of average strength, which break. The lumber industry and many space-age materials groups specify that a 95% one-sided confidence bound be calculated for the fifth percentile $\eta_{0.05}$. That is, a lower bound $L(x_1, x_2, \ldots, x_n)$ is calculated from the observations and, prior to taking the observations,

$$P[L(X_1, X_2, \ldots, X_n) < \eta_{0.05}] = 0.95$$

But this one-sided confidence bound is just a **one-sided tolerance bound**, since the event *the bound $L(x_1, x_2, \ldots, x_n)$ is less than the population 0.05 point $\eta_{0.05}$* is the same as the event *at least 95% of the population is above $L(x_1, x_2, \ldots, x_n)$*.

For normal populations

$$L(x_1, x_2, \ldots, x_n) = \bar{x} - Ks$$

where K can be obtained from Table 9W(b).

EXAMPLE 4   **Calculating a lower tolerance bound for the strength of cardboard**

The cardboard industry is considering new standards for the cardboard used in boxes. One test involves placing weight on the box until it bursts. The burst strengths, in pounds per square inch, for 40 boxes are

210	234	216	232	262	183	227	197
248	218	256	218	244	259	263	185
218	196	235	223	212	237	275	240
217	263	240	247	253	269	231	254
248	261	268	262	247	292	238	215

Obtain a 95% tolerance bound that will be less than proportion 0.95 of the population of burst strengths.

Solution   By computer, we determine that $\bar{x} = 237.32$ and $s = 25.10$. From Table 9W(b), $K = 2.125$, so

$$L = \bar{x} - Ks = 237.32 - 2.125(25.10) = 183.93$$

which is rounded down to 183.

We are 95% confident that at least a proportion 0.95 of the population of burst strengths, for cardboard boxes, is above 183 psi.

In Exercise 15.25 you are asked to verify that the strength measurements fail to exhibit departures from normality.   ∎

Exercises

15.13 To check the strength of carbon steel for use in chain links, the yield stress of a random sample of 25 pieces was measured, yielding a mean and a standard deviation of 52,800 psi and 4,600 psi, respectively. Establish tolerance limits with $\alpha = 0.05$ and $P = 0.99$, and express *in words* what these tolerance limits mean.

15.14 In a study designed to determine the number of turns required for an artillery-shell fuse to arm, 75 fuses, rotated on a turntable, averaged 38.7 turns with a standard deviation of 4.3 turns. Establish tolerance limits for which one can assert with 99% confidence that *at least 95%* of the fuses will arm within these limits.

15.15 In a random sample of 40 piston rings chosen from a production line, the mean edge width was 0.1063 inch, and the standard deviation was 0.0004 inch.

(a) Between what limits can it be said with 95% confidence that at least 90% of the edge widths of piston rings produced by this production line will lie?

(b) Find 95% confidence limits for the true mean edge width, and explain the difference between these limits and the tolerance limits found in part (a).

15.16 *Nonparametric tolerance limits* can be based on the extreme values in a random sample of size n from any continuous population. The following equation relates the quantities n, P, and α, where P is the minimum proportion of the population contained between the smallest and the largest observations with $(1 - \alpha)100\%$ confidence:

$$n\,P^{n-1} - (n-1)\,P^n = \alpha$$

An approximate solution for n is given by

$$n = \frac{1}{2} + \frac{1+P}{1-P} \cdot \frac{\chi_\alpha^2}{4}$$

where χ_α^2 is the value of chi square for 4 degrees of freedom that corresponds to a right-hand tail area α.

(a) How large a sample is required to be 95% certain that at least 90% of the population will be included between the extreme values of a sample?

(b) With 95% confidence, at least what proportion of the population can be expected to be included between the extreme values of a sample of size 100?

Do's and Don'ts
Do's
1. Make sure the process is operating in a stable manner before calculating a central line and limits for a control chart. There should be no trends over time, either in location or amount of variation.
2. Continue to improve any product or service by finding ways to reduce variation.
Don'ts
1. Don't forget to check for dependence between adjacent values that are plotted on a control chart. When means are plotted, you might graph the adjacent pairs $(\bar{x}_i, \bar{x}_{i-1})$. Even moderate correlation between adjacent points can greatly deteriorate the performance of control charts.

Review Exercises

15.17 The specifications require that the weight of castings have $\mu = 4.1$ ounces and $\sigma = 0.05$ ounce.

(a) Use the specifications to calculate a central line and three-sigma control limits for an $\bar{x}$ chart with $n = 5$.

(b) Use the specifications to calculate a central line and three-sigma control limits for an R chart with $n = 5$.

(c) Plot the following means and ranges, obtained in 20 successive random samples of size 5, on charts based on the control-chart constants obtained in part (a) and (b), and discuss the process.

Sample	$\bar{x}$	R	Sample	$\bar{x}$	R
1	4.24	0.09	11	4.20	0.21
2	4.18	0.12	12	4.25	0.20
3	4.26	0.14	13	4.25	0.17
4	4.21	0.24	14	4.21	0.07
5	4.22	0.15	15	4.19	0.16
6	4.18	0.28	16	4.23	0.16
7	4.23	0.06	17	4.27	0.19
8	4.19	0.15	18	4.22	0.20
9	4.21	0.09	19	4.20	0.12
10	4.18	0.15	20	4.19	0.16

15.18 Calculate $\bar{\bar{x}}$ and $\bar{R}$ for the data of part (c) of Exercise 15.17, and use these values to construct the central lines and three-sigma control limits for new $\bar{x}$ and R charts to be used in the control of the weight of the castings.

15.19 Twenty-five successive samples of 200 switches, each taken from a production line, contained, respectively, 6, 7, 13, 7, 0, 9, 4, 6, 0, 4, 5, 11, 6, 18, 1, 4, 9, 8, 2, 17, 9, 12, 10, 5, and 4 defectives. If the fraction of defectives is to be maintained at 0.02, construct a p chart for these data and state whether or not this standard is being met.

15.20 The data of Exercise 15.19 may be looked upon as evidence that the standard of 2% defectives is being exceeded.

(a) Use the data of Exercise 15.19 to construct new control limits for the fraction defective.

(b) Using the limits found in part (a), continue the control of the process by plotting the following data on the next ten samples of size $n = 200$: 4, 7, 5, 3, 8, 3, 1, 4, 3, 9.

15.21 A process for the manufacture of film has performed in the past with an average of 0.8 imperfections per 10 linear feet.

(a) Construct a chart to be used in the inspection of 10-foot sections.

(b) Discuss the control if 20 successive 10-foot sections contained, respectively, 1, 0, 0, 1, 3, 1, 2, 1, 0, 2, 1, 3, 0, 0, 1, 1, 2, 0, 4 and 1 imperfections.

15.22 With reference to the aluminum alloy strength data on page 19, obtain two-sided 95% tolerance limits on the proportion $P = 0.90$ of the population of strengths.

15.23 With reference to the interrequest time data on page 19, obtain 95% tolerance limits on the proportion $P = 0.90$ of the population of interrequest times. Take logs, use the normal theory approach, and then transform back to the original scale.

15.24 With reference to the discussion on page 482, calculate the CUSUM using 2.25 in place of 2.00 as the centering value. Also make the CUSUM chart.

15.25 With reference to Example 4.

(a) verify the calculation of the tolerance bound L;

(b) if the confidence is decreased to 90%, calculate the new tolerance bound (use $K = 2.010$);

(c) check the cardboard strength data for departures from normality using a normal-score plot.

15.26 Explain, from the perspective of quality improvement programs, why the $\bar{x}$, R, and fraction defective charts should be used to listen to the process and observe its natural variability, at any stage, rather than for the long-run control of the process.

15.27 A critical width dimension on an integrated circuit board was measured on 100 boards. The ordered measurements are:

2.500	2.502	2.502	2.502	2.503	2.503	2.503	2.504	2.504
2.504	2.504	2.504	2.504	2.504	2.504	2.504	2.504	2.504
2.504	2.504	2.504	2.504	2.504	2.505	2.505	2.505	2.505
2.505	2.505	2.505	2.505	2.505	2.505	2.505	2.505	2.505
2.505	2.506	2.506	2.506	2.506	2.506	2.506	2.506	2.506
2.506	2.506	2.506	2.506	2.506	2.506	2.506	2.506	2.507
2.507	2.507	2.507	2.507	2.507	2.507	2.507	2.507	2.507
2.507	2.508	2.508	2.508	2.508	2.508	2.508	2.508	2.508
2.508	2.508	2.508	2.508	2.508	2.508	2.509	2.509	2.509
2.509	2.509	2.509	2.509	2.509	2.509	2.509	2.509	2.509
2.509	2.509	2.509	2.509	2.509	2.509	2.509	2.510	2.511
2.511								

Given the specification limits $LSL = 2.496$ and $USL = 2.516$, evaluate the process capability by determining the estimates (a) $\widehat{C}_p$ and (b) $\widehat{C}_{pk}$.

15.28 The following are the number of pounds per day shipped by a trucking company.

222,415	140,670	396,868	240,678	101,786	166,217	177,900
349,900	131,100	465,800	417,700	305,600	264,500	224,400
360,400	211,600	378,200	285,400	166,100	230,900	593,300
214,200	147,800	119,510	159,200	353,200	408,300	275,100
254,100	423,500	324,800	304,500	298,600	202,200	

It is suggested that the shipments be treated as a process. Because these data are not symmetrically distributed, you could try a transformation. For the choice of the fourth root of weight, set the specification limits LSL and USL symmetrically about 22.6 $(pounds)^{1/4}$, so that the estimated capability index $\widehat{C}_{pk}$ is 1.5.

Key Terms

16

APPLICATION TO RELIABILITY AND LIFE TESTING

The task of designing and supervising the manufacture of a product has been made increasingly difficult by rapid strides in the sophistication of modern products and the severity of the environmental conditions under which they must perform. No longer can an engineer be satisfied if the operation of a product is technically feasible, or if it can be made to work under optimum conditions. In addition to such considerations as cost and ease of manufacture, increasing attention must now be paid to size and weight, ease of maintenance, and reliability. The magnitude of the problem of maintainability and reliability is illustrated by surveys which have uncovered the fact that a high percentage of space-age electronic equipment is inoperative. Military surveys have further shown that maintenance and repair expenses for electronic equipment often exceed the original cost of procurement, even during the first year of operation.

In Section 16.1, we define the concept of reliability. In Section 16.2, we discuss and apply special probability distributions to the calculation of reliabilities. In Sections 16.3 and 16.4, some theory and applications relating to testing products for useful lifetime are introduced.

16.1 Reliability

The problem of assuring and maintaining reliability has many facets, including original-equipment design, control of quality during production, acceptance inspection, field trials, life testing, and design modifications. To complicate matters further, reliability competes directly or indirectly with a host of other engineering considerations, chiefly cost, complexity, size and weight, and maintainability. In spite of its complicated engineering aspects, it is possible to give a relatively simple mathematical definition for reliability. To motivate this definition, we can call the reader's attention to the fact that a product may function satisfactorily under one set of conditions but not under other conditions. Also satisfactory performance for one purpose does not assure adequate performance for another purpose. For example, a microchip perfectly satisfactory for use in a home audio system may be entirely unsatisfactory for use in the airborne guidance system of a missile. Accordingly, we shall define **reliability** of any unit in terms of the probability it will operate successfully under specified environmental conditions.

Reliability of a unit

The **reliability** of a unit is probability that it will function within specified limits for at least a specified period of time under specified environmental conditions.

Thus, the reliability of a standard-equipment automobile tire is close to unity for 10,000 miles of normal operation on a passenger car, but it is virtually zero for use at the Indianapolis 500.

Since reliability is defined as a probability, the theoretical treatment of this subject is based essentially on the material introduced in the early chapters of this book. Thus, the rules of probability introduced in Chapter 3 can be applied directly to the calculation of the reliability of a complex system, if the reliabilities of the individual components are known. (Estimates of the reliabilities of the individual components are usually obtained from statistical life tests, such as those discussed in Sections 16.3 and 16.4.)

Many systems can be considered to be series or parallel systems, or a combination of both. A **series system** is one in which all components are so interrelated that the entire system will fail if any one of its components fails; a **parallel system** is one that will fail only if all of its components fail.

Let us first discuss a system of n components connected in series, and let us suppose that the components are independent, namely, that the performance of any one part does not affect the reliability of the others. Under these conditions, the probability that the system will function is given by the special rule of multiplication for probabilities, and we have

Product law of reliabilities

$$R_S = \prod_{i=1}^{n} R_i$$

where R_i is the reliability of the ith component and R_S is the reliability of the series system. This simple **product law of reliabilities**, applicable to series systems of independent components, vividly demonstrates the effect of increased complexity on reliability.

EXAMPLE 1 **Calculating the reliability of a series system**

A system consists of 5 independent components in series, each having a reliability of 0.970. What is the reliability of the system? What happens to the system reliability if its complexity is increased so that it contains 10 similar components?

Solution The reliability of the 5-component system is

$$(0.970)^5 = 0.859$$

Increasing system complexity to 10 components will decrease the system reliability to

$$(0.970)^{10} = 0.737$$

Looking at the effect of increasing complexity in another way, we find that each of the components in the 10-component system would require a reliability of 0.985, instead of 0.970, for the 10-component system to have a reliability equal to that of the original 5-component system. ∎

One way to increase the reliability of a system is to replace certain components by several similar components connected in parallel. If a system consists of n independent components connected in parallel, it will fail to function only if all n components fail. Thus, if $F_i = 1 - R_i$ is the "unreliability" of the ith component, we can again apply the special rule of multiplication for probabilities to obtain

$$F_P = \prod_{i=1}^{n} F_i$$

where F_P is the unreliability of the parallel system and $R_P = 1 - F_P$ is the reliability of the parallel system. Thus, for parallel systems, we have a **product law of unreliabilities** analogous to the product law of reliabilities for series systems. Writing this law in another way, we get

Product law of unreliabilities

$$R_P = 1 - \prod_{i=1}^{n} (1 - R_i)$$

for the reliability of a parallel system.

EXAMPLE 2   **Calculating reliability for a complex system**

The two basic formulas for the reliability of series and parallel systems can be used in combination to calculate the reliability of a system having both series and parallel parts. To illustrate such a calculation, consider the system diagramed in Figure 16.1, which consists of eight components having the reliabilities shown in that figure. Find the reliability of this system.

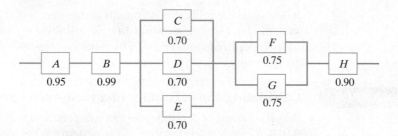

Figure 16.1
System reliability

Solution   The parallel assembly C, D, E can be replaced by an equivalent component C' having the reliability $1 - (1 - 0.70)^3 = 0.973$, without affecting the overall reliability of the system. Similarly, the parallel assembly F, G can be replaced by a single component F' having the reliability $1 - (1 - 0.75)^2 = 0.9375$. The resulting series system A, B, C', F', H, equivalent to the original system, has the reliability

$$(0.95)(0.99)(0.973)(0.9375)(0.90) = 0.772 \qquad \blacksquare$$

16.2 Failure-Time Distribution

According to the definition of reliability given in the preceding section, the reliability of a system or a component will often depend on the length of time it has been in service. Thus, of fundamental importance in reliability studies is the **failure-time distribution**. Specifically, this is, the distribution of the time to failure of a component under given environmental conditions. A useful way to characterize this distribution is by means of its associated **instantaneous failure rate**. To develop this concept, first let $f(t)$ be the probability density of the time to failure of a given component. The probability that the component will fail between times t and $t + \Delta t$ is approximately $f(t) \cdot \Delta t$. Then, the probability that the component will fail on the interval from 0 to t is given by

$$F(t) = \int_0^t f(x)\,dx$$

and the **reliability function**, expressing the probability that it survives to time t, is given by

$$R(t) = 1 - F(t)$$

We can then express the probability that the component will fail in the interval from t to $t + \Delta t$ as $F(t + \Delta t) - F(t)$, and the conditional probability of failure in this interval, *given that the component survived to time t*, is expressed by

$$\frac{F(t + \Delta t) - F(t)}{R(t)}$$

Dividing by Δt, we find that the average rate of failure in the interval from t to $t + \Delta t$, given that the component survived to time t, is

$$\frac{F(t + \Delta t) - F(t)}{\Delta t} \cdot \frac{1}{R(t)}$$

Taking the limit as $\Delta t \to 0$, we then get the instantaneous failure rate, or simply the **failure rate or hazard rate**

$$Z(t) = \frac{F'(t)}{R(t)}$$

where $F'(t)$ is the derivative of $F(t)$ with respect to t. Finally, observing that $f(t) = F'(t)$ (see page 127), we get the relation

General equation for failure-rate function

$$Z(t) = \frac{f(t)}{R(t)} = \frac{f(t)}{1 - F(t)}$$

The **failure-rate function** expresses the failure rate in terms of the failure-time distribution.

A failure-rate curve that is typical of many manufactured items is shown in Figure 16.2. The curve is conveniently divided into three parts. The first part is characterized by a decreasing failure rate and it represents the period during which poorly manufactured items are weeded out. (It is common in the electronics industry to burn in components prior to actual use in order to eliminate any early failures.) The second part, which is often characterized by a constant failure rate, is normally regarded as the period of useful life during which only chance failures occur. The

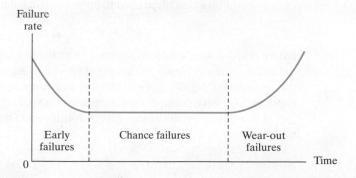

Figure 16.2
Typical failure-rate curve

third part is characterized by an increasing failure rate, and it is the period during which components fail primarily because they are worn out. Note that the same general failure-rate curve is typical of human mortality, where the first part represents infant mortality, and the third part corresponds to old-age mortality.

Let us now derive an important relationship expressing the failure-time density in terms of the failure-rate function. Making use of the fact that $R(t) = 1 - F(t)$ and, hence, that $F'(t) = -R'(t)$, we can write

$$Z(t) = -\frac{R'(t)}{R(t)} = -\frac{d\,[\,\ln R(t)\,]}{dt}$$

Solving this differential equation for $R(t)$, we obtain

$$R(t) = e^{\displaystyle -\int_0^t Z(x)\,dx}$$

and, making use of the relation $f(t) = Z(t) \cdot R(t)$, we finally get

General equation for failure-time distribution

$$f(t) = Z(t) \cdot e^{\displaystyle -\int_0^t Z(x)\,dx}$$

As illustrated in Figure 16.2, it is often assumed that the failure rate is constant during the period of useful life of a component. Denoting this constant failure rate by α, where $\alpha > 0$, and substituting α for $Z(t)$ in the formula for $f(t)$, we obtain

$$f(t) = \alpha \cdot e^{-\alpha t}, \quad t > 0$$

Thus, we have an **exponential failure-time distribution** when it can be assumed that the failure rate is constant. For this reason, the assumption of constant failure rates is sometimes also called the *exponential assumption*. The time to failure also has an interpretation as a waiting time. If a component which fails is immediately replaced with a new one having the same constant failure rate α and the occurrence of failures follows a Poisson process, then the waiting times have this exponential distribution according to results in Section 5.7. As we observed on page 147, the mean waiting time between successive failures is $1/\alpha$, or the reciprocal of the failure rate. Thus, the constant $1/\alpha$ is often referred to as the **mean time between failures (MTBF)**.

There are situations in which the assumption of a constant failure rate is not realistic, and in many of these situations one assumes instead that the failure-rate function increases or decreases smoothly with time. In other words, it is assumed that there are no discontinuities or turning points. This assumption would be consistent with either the initial or the last stage of the failure-rate curve shown in Figure 16.2.

A useful function often used to approximate such failure-rate curves is given by

$$Z(t) = \alpha \beta t^{\beta - 1}, \quad t > 0$$

where α and β are positive constants. Note the generality of this function: If $\beta < 1$, the failure rate *decreases* with time; if $\beta > 1$, it *increases* with time; and if $\beta = 1$, the failure rate equals α. Note that the assumption of a constant failure rate, the exponential assumption, is thus included as a special case.

If we substitute the above expression for $Z(t)$ into the formula for $f(t)$ above, we obtain

$$f(t) = \alpha \beta t^{\beta - 1} e^{-\alpha t^{\beta}}, \quad t > 0$$

where α and β are positive constants. This density, or distribution, is the Weibull distribution, introduced in Section 5.9, and we discuss its application to problems of life testing in Section 16.4.

Exercises

16.1 A holiday wreath has 8 light bulbs connected in series. Determine the reliability of each bulb if there were a 95% chance of the string's lighting after a year's storage?

16.2 A system consists of 5 identical components connected in parallel. Determine the reliability of each component if the overall reliability of the system is to be 0.96?

16.3 A system consists of 6 components connected as in Figure 16.3. Find the overall reliability of the system, given that the reliabilities of A, B, C, D, E, and F are, respectively, 0.95, 0.80, 0.90, 0.99, 0.90, and 0.85.

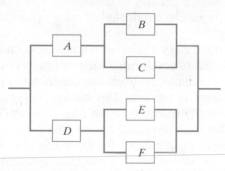

Figure 16.3   System for Exercise 16.3

16.4 Suppose that the flight of an aircraft is regarded as a system having the three main components A (aircraft), B (pilot), and C (airport). Suppose, furthermore, that component B can be regarded as a parallel subsystem consisting of B_1 (captain), B_2 (first officer), and B_3 (flight engineer); and C is a parallel subsystem consisting of C_1 (scheduled airport) and C_2 (alternate airport). Under given flight conditions, the reliabilities of components A, B_1, B_2, B_3, C_1, and C_2 (defined as the probabilities that they can contribute to the successful completion of the scheduled flight) are, respectively, 0.9999, 0.9995, 0.999, 0.20, 0.95, and 0.85.

(a) What is the reliability of the system?

(b) What is the effect on system reliability of having a flight engineer who is also a trained pilot, so that the reliability of B_3 is increased from 0.20 to 0.99?

(c) If the flight crew did not have a first officer, what then would be the effect of increasing the reliability of B_3 from 0.20 to 0.99?

(d) What is the effect of adding a second alternate landing point, C_3, with reliability 0.80?

16.5 In some reliability problems we are concerned only with initial failures, treating a component as if (for all practical purposes) it never fails, once it has survived past a certain time $t = \alpha$. In a problem like this, it may be reasonable to use the failure rate

$$Z(t) = \begin{cases} \beta \left(1 - \dfrac{t}{\alpha} \right) & \text{for } 0 < t < \alpha \\ 0 & \text{elsewhere} \end{cases}$$

(a) Find expressions for $f(t)$ and $F(t)$.

(b) Show that the probability of an initial failure is given by

$$1 - e^{-\alpha\beta/2}$$

16.6 As indicated in the text, one often distinguishes between initial failures, random failures during the useful life of the product, and wear-out failures. For a given product, suppose the probability of an initial failure (a failure prior to time $t = \alpha$) is θ_1, the probability of a wear-out failure (a failure beyond time $t = \beta$) is θ_2, and that for the interval $\alpha \le t \le \beta$ the failure-time density is given by

$$f(t) = \frac{1 - \theta_1 - \theta_2}{\beta - \alpha}$$

(a) Find an expression for $F(t)$ for the interval $\alpha \le t \le \beta$.

(b) Show that for the interval $\alpha \le t \le \beta$, the failure rate is given by

$$Z(t) = \frac{1 - \theta_1 - \theta_2}{(\beta - \alpha)(1 - \theta_1) - (1 - \theta_1 - \theta_2)(t - \alpha)}$$

(c) Suppose that the failure of a digital television set is considered to be an initial failure if it occurs during the first 100 hours of usage and a wear-out failure if it occurs after 15,000 hours. Assuming that the model given in this exercise holds and that θ_1 and θ_2 equal 0.05 and 0.75, respectively, sketch the graph of the failure-rate function from $t = 100$ to $t = 15,000$ hours.

16.7 An integrated-circuit chip has a constant failure rate of 0.02 per thousand hours.

(a) What is the probability that it will operate satisfactorily for at least 20,000 hours?

(b) What is the 5,000-hour reliability of a component consisting of 4 such chips connected in series?

16.8 After burn-in, the lifetime of a solar cell is modeled as an exponential distribution with failure rate $\alpha = 0.0005$ failures per day.

(a) What is the probability that the cell will fail within the first 365 days that it is in operation?

(b) What is the probability that two such cells, operating independently, will both survive the first 365 days they are in operation?

16.9 If a component has the Weibull failure-time distribution with the parameters $\alpha = 0.005$ per hour and $\beta = 0.80$, find the probability that it will operate successfully for at least 5,000 hours.

16.3   The Exponential Model in Life Testing

An effective and widely used method of handling problems of reliability is that of **life testing**. For the purpose of such tests, a random sample of n components is selected from a lot, put on test under specified environmental conditions, and the times to failure of the individual components are observed. If each component that fails is immediately replaced by a new one, the resulting life test is called a **replacement test**; otherwise, the life test is called a **nonreplacement test**. Whenever the mean lifetime of the components is so large that it is not practical, or economically feasible, to test each component to failure, the life test may be **truncated**, after a fixed period of time has elapsed. Alternatively, it may be terminated after the first r failures have occurred ($r \leq n$).

A special method often used when early results are required in connection with very high reliability components is that of **accelerated life testing**. In an accelerated life test the components are put on test under environmental conditions far more severe than those normally encountered in practice. This causes the components to fail more quickly, and it can drastically reduce both the time required for the test and the number of components that must be tested. Accelerated life testing can be used to compare two or more types of components for the purpose of obtaining a rapid assessment of which one is the most reliable. Sometimes, preliminary experimentation is carried out to determine the relationship between the proportion of failures that can be expected under nominal conditions and under various levels of accelerated environmental conditions. The methods of Sections 11.3 and 13.2 can be applied in this connection to determine "derating curves," relating the reliability of the component to the severity of the environmental conditions under which it is to operate.

In the remainder of this section we shall assume that the exponential model holds, namely, that the failure-time distribution of each component is given by

$$f(t) = \alpha \cdot e^{-\alpha t} \qquad t > 0, \quad \text{where } \alpha > 0$$

In what follows, we shall assume that n components are put on test, life testing is discontinued after a fixed number, r ($r \leq n$), of components have failed, and that the observed failure times are $t_1 \leq t_2 \leq \cdots \leq t_r$. We shall be concerned with estimating and testing hypotheses about the mean life of the component, namely, $\mu = 1/\alpha$.

It can be shown (see the reference to Lawless in the bibliography) that unbiased estimates of the mean life of the component are given by

Estimate of mean life

$$\widehat{\mu} = \frac{T_r}{r}$$

where T_r is the accumulated life of test until the rth failure occurs, and hence

Accumulated life to r failures (nonreplacement test)

$$T_r = \sum_{i=1}^{r} t_i + (n - r)t_r$$

for nonreplacement tests and

Accumulated life to r failures (replacement test)

$$T_r = n\,t_r$$

if the test is *with replacement*. Note that if the test is without replacement and $r = n$, $\hat{\mu}$ is simply the mean of the observed times to failure.

To make inferences concerning the mean life μ of the component, we use the fact that $2T_r/\mu$ is a value of a random variable having the chi square distribution with $2r$ degrees of freedom. With the appropriate expression substituted for T_r, this is true regardless of whether the test is conducted with or without replacement. Thus, in either case a two-sided $(1 - \alpha)100\%$ confidence interval for μ is given by

Confidence interval for mean life

$$\frac{2T_r}{\chi^2_{\alpha/2}} < \mu < \frac{2T_r}{\chi^2_{1-\alpha/2}}$$

where $\chi^2_{1-\alpha/2}$ and $\chi^2_{\alpha/2}$ cut off the left- and right-hand tails of area $\alpha/2$ under the chi square distribution with $2r$ degrees of freedom. (See Exercise 16.15.)

Tests of the null hypothesis that $\mu = \mu_0$ can also be based on the sampling distribution of $2T_r/\mu$, using the appropriate expression for T_r depending on whether the test is with or without replacement. Thus, if the alternative hypothesis is $\mu > \mu_0$, we reject the null hypothesis at the level of significance α when $2T_r/\mu_0$ exceeds χ^2_α, or

Critical region for testing H_0: $\mu = \mu_0$ against H_1: $\mu > \mu_0$

$$T_r > \frac{1}{2}\mu_0\,\chi^2_\alpha$$

where χ^2_α, to be determined for $2r$ degrees of freedom, is as defined on page 198. In Exercises 16.10 and 16.13 the reader is asked to construct and perform similar tests corresponding to the alternative hypotheses $\mu < \mu_0$ and $\mu \neq \mu_0$.

EXAMPLE 3   **Obtaining a confidence interval for mean life**

Suppose that 50 units are placed on life test (without replacement) and the test is to be truncated after $r = 10$ of them have failed. We shall suppose, furthermore, that the first 10 failure times are 65, 110, 380, 420, 505, 580, 650, 840, 910, and 950 hours. Estimate the mean life of the component, and its failure rate, and calculate a 90% confidence interval for μ.

Solution   Since $n = 50$, $r = 10$,

$$T_{10} = (65 + 100 + \cdots + 950) + (50 - 10)950$$
$$= 43,410 \text{ hours}$$

We estimate the mean life of the component as

$$\widehat{\mu} = \frac{43,410}{10} = 4,341 \text{ hours}$$

The failure rate α is estimated by $1/\widehat{\mu} = 0.00023$ failure per hour, or 0.23 failure per thousand hours. Using $\chi^2_{0.05} = 31.410$ and $\chi^2_{0.95} = 10.851$ for $2(10) = 20$ degrees of freedom, a 90% confidence interval for μ is given by

$$\frac{2(43,410)}{31.410} < \mu < \frac{2(43,410)}{10.851}$$

or

$$2,764 < \mu < 8,001 \qquad \blacksquare$$

EXAMPLE 4   **Testing hypotheses concerning mean life**

Using the data of the preceding example, test whether the failure rate is 0.40 failure per thousand hours against the alternative that the failure rate is less. Use the 0.05 level of significance.

Solution   1. *Null hypothesis*: $\mu = \dfrac{1,000}{0.40} = 2,500$ hours $= \mu_0$
   Alternative hypothesis: $\mu > 2,500$ hours

2. *Level of significance*: $\alpha = 0.05$

3. *Criterion*: Reject the null hypothesis if $T_r > \dfrac{1}{2} \mu_0 \chi^2_{0.05}$ where
   $\chi^2_{0.05} = 31.410$ is the chi square value for $2r = 20$ degrees of freedom.

4. *Calculations*: Substituting $\chi^2_{0.05} = 31.410$ and $\mu_0 = 2,500$, we find the critical value for this test to be

$$\frac{1}{2} \mu_0 \chi^2_{0.05} = \frac{1}{2}(2,500)(31.410) = 39,263$$

5. *Decision*: Since $T_{10} = 43,410$ exceeds the critical value, we must reject the null hypothesis, concluding that the mean lifetime exceeds 2,500 hours, or, equivalently, that the failure rate is less than 0.40 failure per thousand hours.   ■

Because of the simplicity of the statistical procedures, the exponential model is frequently considered. Before making inferences, it is imperative that this model be checked for adequacy. We recommend making a **total time on test plot**. Plot the total time on test until the ith failure, T_i, divided by the total time on test through the last (rth) observed failure, against i/r. If the population is exponential, we would expect to see a straight line along the 45-degree line. When this straight-line pattern occurs, we conclude that no violations of the exponential model are evident over the range of failure times. If the plot is a curve above the 45-degree line, the evidence favors an increasing hazard rate model.

We illustrate the total time on test plot using the data of Example 3. For $t_1 = 65$, we calculate the total time on test

$$T_1 = 65 + (50 - 1)65 = 3,250$$

Next, for $t_2 = 110$,

$$T_2 = 65 + 110 + (50 - 2)110 = 5,455$$

Continuing, we obtain all the values

3,250	5,455	18,415	20,295	24,205
27,580	30,660	38,830	41,770	43,410

so the total time on test until the last, $r = 10$, failure is $T_r = 43,410$. The first ratio $T_1/T_{10} = 3,250/43,410 = 0.0749$ is plotted against $1/10 = 0.10$. The ratios for all 10 failures are plotted in Figure 16.4. Over the range of failure times observed, the plot does not exhibit any marked departures from the assumed exponential model.

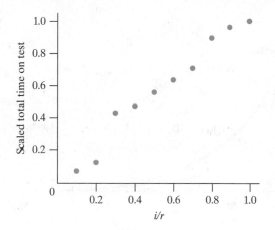

Figure 16.4
The total time on test plot for data in Example 3

16.4   The Weibull Model in Life Testing

Although life testing of components during the period of useful life is generally based on the exponential model, we have already pointed out that the failure rate of a component may not be constant throughout a period under investigation. In some instances the period of initial failure may be so long that the component's main use is during this period. However, the main purpose of most life testing is to determine the time to wear-out failure rather than chance failure of a critical component in a complex system. In such cases the exponential model generally does not apply, and it is necessary to consider a more general assumption for the failure rate.

As we observed earlier, the Weibull distribution may adequately describe the failure time of components when their failure rate either increases or decreases with time. It has the parameters α and β and its formula is given by

Weibull distribution

$$f(t) = \alpha \beta t^{\beta - 1} e^{-\alpha t^{\beta}} \qquad t > 0, \quad \text{where} \quad \alpha > 0, \quad \beta > 0$$

and it follows (see Exercise 16.20) that the reliability function associated with the **Weibull failure-time distribution** is given by

Weibull reliability function

$$R(t) = e^{-\alpha t^{\beta}}$$

We already showed on page 498 that the failure rate leading to the Weibull distribution is given by

**Weibull failure-rate
function**

$$Z(t) = \alpha \beta t^{\beta - 1}$$

The Weibull density can take a wide range of shapes depending primarily on the value of the parameter β. As illustrated in Figure 16.5, the Weibull curve is asymptotic to both axes and highly skewed to the right for values of β less than 1; it is identical to that of the exponential density for $\beta = 1$, and it is somewhat bell-shaped but skewed for values of β greater than 1.

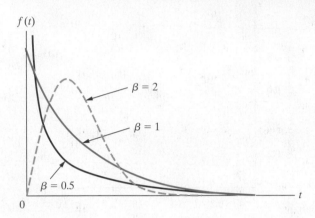

Figure 16.5
Weibull density functions
$(\alpha = 1)$

The mean of the Weibull distribution having the parameters α and β may be obtained by evaluating the integral

$$\mu = \int_0^\infty t \cdot \alpha \beta t^{\beta - 1} e^{-\alpha t^\beta} \, dt$$

Making the change of variable $u = \alpha t^\beta$, we get

$$\mu = \alpha^{-1/\beta} \int_0^\infty u^{1/\beta} e^{-u} \, du$$

Recognizing the integral as $\Gamma\left(1 + \dfrac{1}{\beta}\right)$, the gamma function evaluated at $1 + \beta^{-1}$, we find that the mean time to failure for the Weibull model is

**Mean time to failure
(Weibull model)**

$$\mu = \alpha^{-1/\beta} \, \Gamma\left(1 + \frac{1}{\beta}\right)$$

The reader will be asked to show in Exercise 16.21 that the variance of this distribution is given by

Variance of Weibull model

$$\sigma^2 = \alpha^{-2/\beta} \left\{ \Gamma\left(1 + \frac{2}{\beta}\right) - \left[\Gamma\left(1 + \frac{1}{\beta}\right)\right]^2 \right\}$$

Estimates of the parameters α and β of the Weibull distribution are somewhat difficult to obtain. The most widely accepted approach, the *maximum likelihood* method, maximizes the likelihood. Because the partial derivatives with respect to α and β must vanish at the maximum, the method selects the solution to these two equations as the estimates of α and β. If the lifetimes are **censored** at the rth failure (the test is terminated at the rth failure), or uncensored so $r = n$, the equations are

$$\frac{\sum_{i=1}^{r} t_i^{\beta} \ln t_i + (n - r) t_r^{\beta} \ln t_r}{\sum_{i=1}^{r} t_i^{\beta} + (n - r) t_r^{\beta}} - \frac{1}{\beta} - \frac{1}{r} \sum_{i=1}^{r} \ln t_i = 0$$

$$\alpha = \frac{1}{\frac{1}{r} \left[\sum_{i=1}^{r} t_i^{\beta} + (n - r) t_r^{\beta} \right]}$$

The first equation is solved for $\widehat{\beta}$ by numerical techniques. Then the second yields the estimate of $\widehat{\alpha}$. These are easy computer calculations.

If the lifetimes are time truncated at time T_0, the terms with a factor $n - r$ are modified by replacing each t_r by T_0.

A graphical method provides a check on the adequacy of the Weibull model. This method is based on the fact that the reliability function of the Weibull distribution can be transformed into a linear function of $\ln t$ by means of a double-logarithmic transformation. Taking the natural logarithm of $R(t)$, we obtain

$$\ln R(t) = -\alpha t^{\beta} \quad \text{or} \quad \ln \frac{1}{R(t)} = \alpha t^{\beta}$$

Again taking logarithms, we have

$$\ln \ln \frac{1}{R(t)} = \ln \alpha + \beta \cdot \ln t$$

and it can be seen that the right-hand side is linear in $\ln t$.

The usual experimental procedure is to place n units on life test and observe their failure times. If the ith failure occurs at time t_i, we estimate $F(t_i) = 1 - R(t_i)$ by the same method used for the normal-scores plot (see page 170), namely,

$$\widehat{F(t_i)} = \frac{i}{n + 1}$$

To construct a **Weibull plot**, we plot $\ln t_i$ versus

$$\ln \ln \frac{1}{1 - \widehat{F(t_1)}}$$

If the points do not fall reasonably close to a straight line, the assumption that the underlying failure-time distribution is of the Weibull type is contradicted.

A sample of 100 components is put on life test for 500 hours and the times to failure of the 12 components that failed during the test are as follows: 6, 21, 50, 84, 95, 130, 205, 260, 270, 370, 440, and 480 hours. Setting

$$x_i = \ln \ln \frac{1}{1 - \widehat{F(t_i)}}$$

$$y_i = \ln t_i$$

we obtain

$\widehat{F(t_i)}$	t_i	y_i	x_i
0.010	6	1.79	−4.61
0.020	21	3.04	−3.91
0.030	50	3.91	−3.50
0.040	84	4.43	−3.21
0.050	95	4.55	−2.98
0.059	130	4.87	−2.79
0.069	205	5.32	−2.63
0.079	260	5.56	−2.49
0.089	270	5.60	−2.37
0.099	370	5.91	−2.26
0.109	440	6.09	−2.16
0.119	480	6.17	−2.07

The points (x_i, y_i) are plotted in Figure 16.6, and it can be seen that they fall fairly close to a straight line. After checking the adequacy of the Weibull distribution, we obtain the maximum likelihood estimators defined on page 505. Our computer calculations yield $\widehat{\alpha} = 0.001505$ and $\widehat{\beta} = 0.7148$. It follows that the mean time to failure is estimated as

$$\widehat{\mu} = (0.001505)^{-1/0.7148}\, \Gamma \left(1 + \frac{1}{0.7148} \right)$$

which equals approximately 11,000 hours. Also, values of the failure-rate function may be obtained by substituting for t into

$$\widehat{Z(t)} = (0.001505)(0.7148)\, t^{0.7148-1} = 0.00108\, t^{-0.2852}$$

Since $\widehat{\beta} < 1$, the failure rate is decreasing with time. After 1 hour ($t = 1$), units are failing at the rate of 0.00108 unit per hour, and after 1,000 hours the failure rate has decreased to $0.00108\,(1000)^{-0.2852} = 0.00015$ unit per hour.

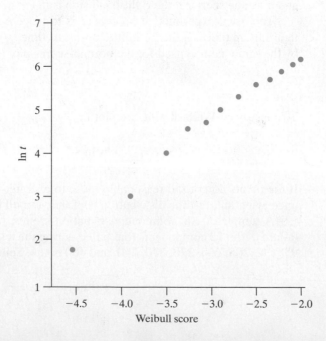

Figure 16.6

A Weibull plot of failure times

Exercises

16.10 Suppose that 50 units are put on life test, each unit that fails is immediately replaced, and the test is discontinued after 8 units have failed. If the eighth failure occurred at 760 hours, assuming an exponential model,

(a) construct a 95% confidence interval for the mean life of such units;

(b) test at the 0.05 level of significance whether or not the mean life is less than 10,000 hours.

16.11 In a nonreplacement life test, 35 space heaters were put into continuous operation, and the first 5 failures occurred after 250, 380, 610, 980, and 1,250 hours.

(a) Assuming the exponential model, construct a 99% confidence interval for the mean life of this kind of space heater.

(b) To check the manufacturer's claim that the mean life of these heaters is greater than 5,000 hours, test the null hypothesis $\mu = 5,000$ against an appropriate alternative, so that the burden of proof is put on the manufacturer. Use $\alpha = 0.05$.

16.12 With reference to the data in Exercise 16.11, make a total time on test plot.

16.13 To investigate the average time to failure of a certain weld subjected to continuous vibration, 7 welded pieces were subjected to specified frequencies and amplitudes of vibration and their times to failure were 211, 350, 384, 510, 539, 620, and 715 thousand cycles.

(a) Assuming the exponential model, construct a 95% confidence interval for the mean life (in thousands of cycles) of such a weld under the given vibration conditions.

(b) Assuming the exponential model, test the null hypothesis that the mean life of the weld under the given vibration conditions is 500,000 cycles against the two-sided alternative $\mu \neq 500,000$. Use the level of significance 0.10.

16.14 In life testing we are sometimes interested in establishing tolerance limits for the life of a component (see Section 15.7); in particular, we may be interested in a one-sided tolerance limit t^*, for which we can assert with a $(1-\alpha)100\%$ confidence that at least $100 \cdot P$ percent of the components have a life exceeding t^*. Using the exponential model, it can be shown that

$$ t^* = \frac{-2T_r(\ln P)}{\chi^2_\alpha} $$

where T_r is as defined on page 501 and the value of χ^2_α is to be obtained from Table 5W with $2r$ degrees of freedom.

(a) Using the data of Exercise 16.11, establish a lower tolerance limit for which one can assert with 95% confidence that it is exceeded by at least 80% of the lifetimes of the heaters.

(b) Using the data of Exercise 16.13, establish a lower tolerance limit for which one can assert with 99% confidence that it is exceeded by at least 90% of the lifetimes of the given welds.

16.15 Using the fact that $2T_r/\mu$ is a value of a random variable having the chi square distribution with $2r$ degrees of freedom, derive the confidence interval for μ given on page 501.

16.16 One hundred devices are put on life test and the times to failure (in hours) of the first 10 that fail are

7.0	14.1	18.9	31.6	52.8
80.0	164.5	355.4	451.0	795.1

Assuming a Weibull failure-time distribution, estimate the parameters α and β as well as the failure rate at 1,000 hours. How does this value of the failure rate compare with the value we would obtain if we assumed the exponential model?

16.17 A sample of 200 switches was placed on life test consisting of repeated on-off cycles. The test was terminated after the third failure. The first three failure times were 2,076, 3,667, and 9,102. Find a 95% *lower* confidence limit for the mean life, in number of cycles, of the switches. Use the exponential model.

16.18 A sample of 60 diaphragm valves, used in the control system of a chemical process, are placed on life test without replacement. The first 9 failures are observed after

3.6 6.9 9.5 15.7 27.3 41.2 81.7 178.3 227.1

hours. Using the Weibull model, estimate the mean life of this valve. How does this value compare with the mean life that would have been obtained under the exponential assumption?

16.19 Using the estimates of the parameters of the Weibull model obtained in Exercise 16.18, estimate the probability that this kind of diaphragm valve will perform satisfactorily for at least 150 hours.

16.20 Show that the reliability function associated with the Weibull failure-time distribution is given by

$$ R(t) = e^{-\alpha t^\beta} $$

16.21 Derive the formula for the variance of the Weibull distribution given on page 504.

Do's and Don'ts

Do's

1. Be aware that reliability analyses based on the exponential model can be quite misleading when components have an increasing failure rate.
2. When fitting an exponential model to data, it is good practice to also fit a Weibull distribution to see if its shape parameter is near one.

Don'ts

1. Don't routinely accept a reliability analysis of a system if the analysis is based on the independence of components. External shocks to the system can often cause multiple components to fail.
2. Don't routinely accept a reliability analysis of a system if the analysis is based on the assumption that the components have exponential distributions. It may be too optimistic since most real components eventually age and so eventually have increasing hazard rates.

Review Exercises

16.22 A system consists of 7 identical components connected in parallel. Determine the reliability of each component if the overall reliability of the system is to be 0.90?

16.23 A certain component has an exponential life distribution with a failure rate of $\alpha = 0.0045$ failures per hour.

(a) What is the probability that the component will fail during the first 250 hours it is in operation?

(b) What is the probability that two such components will both survive the first 100 hours of operation?

16.24 Fifteen assemblies are put on accelerated life test without replacement, and the test is truncated after 4 failures. If the first 4 failures occurred at 16.5, 19.2, 20.8, and 37.3 hours, assuming an exponential model,

(a) determine a 90% confidence interval for the failure rate of such assemblies under these accelerated conditions;

(b) test the null hypothesis that the failure rate is 0.004 failure per hour against the alternative that it is less than 0.004, using the 0.01 level of significance.

16.25 A sample of 300 high-reliability capacitors was placed on life test until the first four failures occurred and the test was then terminated. The first 4 failures were 3,582, 8,482, 8,921, and 16,303 hours. Find a 95% *lower* confidence limit for the mean life of the capacitors assuming an exponential model.

16.26 To investigate the performance of a logic circuit for a small electronic calculator, a laboratory puts 75 of the circuits on life test without replacement under specified environmental conditions, and the first 10 failures are observed after 28, 46, 50, 63, 81, 101, 116, 137, 159, and 175 hours. Using the Weibull model, estimate the mean life of such a circuit. How does this value compare with the mean life that would have been obtained under the exponential assumption?

16.27 Using the estimates of the parameters of the Weibull model obtained in Exercise 16.26, estimate the probability that this kind of circuit will perform satisfactorily for at least 100 hours.

16.28 With reference to Exercise 16.26, make

(a) a total time on test plot

(b) a Weibull plot

16.29 (*Stress-strength models for reliability*) An alternative model used in reliability treats the environmental stress as a random variable X, with probability density $f(x)$, and the strength of the component to withstand this stress as an independent random variable Y having probability density $g(y)$. Then, the reliability is defined as

$$R = P[Y > X] = \int_{-\infty}^{\infty} \int_{-\infty}^{y} f(x)\, g(y)\, dx\, dy$$

$$= \int_{-\infty}^{\infty} F(y)\, g(y)\, dy$$

where $F(x)$ is the distribution function of X. Evaluate this reliability when

(a) X has an exponential distribution with $\alpha = 0.01$ and Y has an exponential distribution with failure rate 0.005;

(b) X has an exponential distribution with $\alpha = 0.005$ and Y has an exponential distribution with failure rate 0.005;

(c) $\ln X$ has a normal distribution with $\mu = 60$ and $\sigma = 5$ and $\ln Y$ has a normal distribution with $\mu = 80$ and $\sigma = 5$.

Key Terms

BIBLIOGRAPHY

1. Theoretical Statistics

Miller, I., and M. Miller, *John E. Freund's Mathematical Statistics with Applications*, 8th ed. Pearson, 2012.

Hogg, R. V., A. T. Craig, and J. McKean, *Introduction to Mathematical Statistics*, 7th ed. Pearson, 2012.

2. Experimental Design and Analysis of Variance

Box, G. E., W. G. Hunter, and J. S. Hunter, *Statistics for Experimenters: Design, Innovation, and Discovery*, 2nd ed. John Wiley & Sons, Inc., 2005.

Wu, C. F., and M. S. Hamada, *Experiments: Planning, Analysis, and Optimization*, 2nd ed. John Wiley and Sons, Inc., 2009.

3. Quality Improvement and Assurance

Deming, W. E., *The Essential Deming: Leadership Principles from the Father of Quality*, McGraw-Hill, 2012.

Besterfield, D., *Quality Improvement*, 9th ed. Prentice Hall, 2012.

Taguchi, G., S. Chowdhury, and Y. Wu, *Taguchi's Quality Engineering Handbook*, John Wiley & Sons, Inc., 2004.

4. Special Topics

Box, G. E. P., and N. R. Draper, *Response Surfaces, Mixtures, and Ridge Analysis*. John Wiley & Sons, Inc., 2007.

Johnson, R. A., and G. K. Bhattacharyya, *Statistics: Principles and Methods*, 7th ed. John Wiley & Sons, Inc., 2014.

Johnson, R., and R. Li, Multivariate Statistical Process Control Schemes for Controlling a Mean, Chapter 18, 323–345. *Springer Handbook of Engineering Statistics*, Ed. H. Pham, Springer, 2006.

Johnson, R. A., and D. W. Wichern, *Applied Multivariate Statistical Analysis,* 6th ed. Prentice Hall, Inc., 2007.

Lawless, J. F., *Statistical Models and Methods for Lifetime Data*. John Wiley & Sons, Inc., 2002. Online 2012.

5. Software

MINITAB Inc., State College, PA.

R, downloadable at **http://cran.r-project.org**

SAS Institute Inc., Gary, NC.

APPENDIX B

STATISTICAL TABLES

Despite the increasing emphasis on using statistical software to obtain probabilities and percentiles, we still make several statistical tables available for download at the book's section of the website

http://www.perasonhighered.com/mathstatresources/

Appendix B lists these tables each of whose number ends in W. Then, for instance, in the text Table 5W refers to the website Table 5W which contains percentiles of the chi-square distribution. An expansion of Section 6.7 and Section 15.8 on Acceptance Sampling are also posted.

The binomial, normal and t tables are retained in the text.

Table I Binomial Distribution Function

$$B(x; n, p) = \sum_{k=0}^{x} \binom{n}{k} p^k (1-p)^{n-k}$$

n	x	0.05	0.10	0.15	0.20	0.25	0.30	0.35	0.40	0.45	0.50	0.55	0.60	0.65	0.70	0.75	0.80	0.85	0.90	0.95
2	0	0.9025	0.8100	0.7225	0.6400	0.5625	0.4900	0.4225	0.3600	0.3025	0.2500	0.2025	0.1600	0.1225	0.0900	0.0625	0.0400	0.0225	0.0100	0.0025
	1	0.9975	0.9900	0.9775	0.9600	0.9375	0.9100	0.8775	0.8400	0.7975	0.7500	0.6975	0.6400	0.5775	0.5100	0.4375	0.3600	0.2775	0.1900	0.0975
3	0	0.8574	0.7290	0.6141	0.5120	0.4219	0.3430	0.2746	0.2160	0.1664	0.1250	0.0911	0.0640	0.0429	0.0270	0.0156	0.0080	0.0034	0.0010	0.0001
	1	0.9927	0.9720	0.9393	0.8960	0.8438	0.7840	0.7183	0.6480	0.5748	0.5000	0.4252	0.3520	0.2818	0.2160	0.1563	0.1040	0.0607	0.0280	0.0073
	2	0.9999	0.9990	0.9966	0.9920	0.9844	0.9730	0.9571	0.9360	0.9089	0.8750	0.8336	0.7840	0.7254	0.6570	0.5781	0.4880	0.3859	0.2710	0.1426
4	0	0.8145	0.6561	0.5220	0.4096	0.3164	0.2401	0.1785	0.1296	0.0915	0.0625	0.0410	0.0256	0.0150	0.0081	0.0039	0.0016	0.0005	0.0001	0.0000
	1	0.9860	0.9477	0.8905	0.8192	0.7383	0.6517	0.5630	0.4752	0.3910	0.3125	0.2415	0.1792	0.1265	0.0837	0.0508	0.0272	0.0120	0.0037	0.0005
	2	0.9995	0.9963	0.9880	0.9728	0.9492	0.9163	0.8735	0.8208	0.7585	0.6875	0.6090	0.5248	0.4370	0.3483	0.2617	0.1808	0.1095	0.0523	0.0140
	3	1.0000	0.9999	0.9995	0.9984	0.9961	0.9919	0.9850	0.9744	0.9590	0.9375	0.9085	0.8704	0.8215	0.7599	0.6836	0.5904	0.4780	0.3439	0.1855
5	0	0.7738	0.5905	0.4437	0.3277	0.2373	0.1681	0.1160	0.0778	0.0503	0.0313	0.0185	0.0102	0.0053	0.0024	0.0010	0.0003	0.0001	0.0000	0.0000
	1	0.9774	0.9185	0.8352	0.7373	0.6328	0.5282	0.4284	0.3370	0.2562	0.1875	0.1312	0.0870	0.0540	0.0308	0.0156	0.0067	0.0022	0.0005	0.0000
	2	0.9988	0.9914	0.9734	0.9421	0.8965	0.8369	0.7648	0.6826	0.5931	0.5000	0.4069	0.3174	0.2352	0.1631	0.1035	0.0579	0.0266	0.0086	0.0012
	3	1.0000	0.9995	0.9978	0.9933	0.9844	0.9692	0.9460	0.9130	0.8688	0.8125	0.7438	0.6630	0.5716	0.4718	0.3672	0.2627	0.1648	0.0815	0.0226
	4	1.0000	0.9999	0.9999	0.9997	0.9990	0.9976	0.9947	0.9898	0.9815	0.9688	0.9497	0.9222	0.8840	0.8319	0.7627	0.6723	0.5563	0.4095	0.2262
6	0	0.7351	0.5314	0.3771	0.2621	0.1780	0.1176	0.0754	0.0467	0.0277	0.0156	0.0083	0.0041	0.0018	0.0007	0.0002	0.0001	0.0000	0.0000	0.0000
	1	0.9672	0.8857	0.7765	0.6554	0.5339	0.4202	0.3191	0.2333	0.1636	0.1094	0.0692	0.0410	0.0223	0.0109	0.0046	0.0016	0.0004	0.0001	0.0000
	2	0.9978	0.9841	0.9527	0.9011	0.8306	0.7443	0.6471	0.5443	0.4415	0.3438	0.2553	0.1792	0.1174	0.0705	0.0376	0.0170	0.0059	0.0013	0.0001
	3	0.9999	0.9987	0.9941	0.9830	0.9624	0.9295	0.8826	0.8208	0.7447	0.6563	0.5585	0.4557	0.3529	0.2557	0.1694	0.0989	0.0473	0.0158	0.0022
	4	1.0000	0.9999	0.9996	0.9984	0.9954	0.9891	0.9777	0.9590	0.9308	0.8906	0.8364	0.7667	0.6809	0.5798	0.4661	0.3446	0.2235	0.1143	0.0328
	5	1.0000	1.0000	1.0000	0.9999	0.9998	0.9993	0.9982	0.9959	0.9917	0.9844	0.9723	0.9533	0.9246	0.8824	0.8220	0.7379	0.6229	0.4686	0.2649
7	0	0.6983	0.4783	0.3206	0.2097	0.1335	0.0824	0.0490	0.0280	0.0152	0.0078	0.0037	0.0016	0.0006	0.0002	0.0001	0.0000	0.0000	0.0000	0.0000
	1	0.9556	0.8503	0.7166	0.5767	0.4449	0.3294	0.2338	0.1586	0.1024	0.0625	0.0357	0.0188	0.0090	0.0038	0.0013	0.0004	0.0001	0.0000	0.0000
	2	0.9962	0.9743	0.9262	0.8520	0.7564	0.6471	0.5323	0.4199	0.3164	0.2266	0.1529	0.0963	0.0556	0.0288	0.0129	0.0047	0.0012	0.0002	0.0000
	3	0.9998	0.9973	0.9879	0.9667	0.9294	0.8740	0.8002	0.7102	0.6083	0.5000	0.3917	0.2898	0.1998	0.1260	0.0706	0.0333	0.0121	0.0027	0.0002
	4	1.0000	0.9998	0.9988	0.9953	0.9871	0.9712	0.9444	0.9037	0.8471	0.7734	0.6836	0.5801	0.4677	0.3529	0.2436	0.1480	0.0738	0.0257	0.0038
	5	1.0000	1.0000	0.9999	0.9996	0.9987	0.9962	0.9910	0.9812	0.9643	0.9375	0.8976	0.8414	0.7662	0.6706	0.5551	0.4233	0.2834	0.1497	0.0344
	6	1.0000	1.0000	1.0000	1.0000	0.9999	0.9998	0.9994	0.9984	0.9963	0.9922	0.9848	0.9720	0.9510	0.9176	0.8665	0.7903	0.6794	0.5217	0.3017
8	0	0.6634	0.4305	0.2725	0.1678	0.1001	0.0576	0.0319	0.0168	0.0084	0.0039	0.0017	0.0007	0.0002	0.0001	0.0000	0.0000	0.0000	0.0000	0.0000
	1	0.9428	0.8131	0.6572	0.5033	0.3671	0.2553	0.1691	0.1064	0.0632	0.0352	0.0181	0.0085	0.0036	0.0013	0.0004	0.0001	0.0000	0.0000	0.0000
	2	0.9942	0.9619	0.8948	0.7969	0.6785	0.5518	0.4278	0.3154	0.2201	0.1445	0.0885	0.0498	0.0253	0.0113	0.0042	0.0012	0.0002	0.0000	0.0000
	3	0.9996	0.9950	0.9786	0.9437	0.8862	0.8059	0.7064	0.5941	0.4770	0.3633	0.2604	0.1737	0.1061	0.0580	0.0273	0.0104	0.0029	0.0004	0.0000
	4	1.0000	0.9996	0.9971	0.9896	0.9727	0.9420	0.8939	0.8263	0.7396	0.6367	0.5230	0.4059	0.2936	0.1941	0.1138	0.0563	0.0214	0.0050	0.0004
	5	1.0000	1.0000	0.9998	0.9988	0.9958	0.9887	0.9747	0.9502	0.9115	0.8555	0.7799	0.6846	0.5722	0.4482	0.3215	0.2031	0.1052	0.0381	0.0058
	6	1.0000	1.0000	1.0000	0.9999	0.9996	0.9987	0.9964	0.9915	0.9819	0.9648	0.9368	0.8936	0.8309	0.7447	0.6329	0.4967	0.3428	0.1869	0.0572
	7	1.0000	1.0000	1.0000	1.0000	1.0000	0.9999	0.9998	0.9993	0.9983	0.9961	0.9916	0.9832	0.9681	0.9424	0.8999	0.8322	0.7275	0.5695	0.3366

(continued on following page)

Table I (continued from page 523)

n	x	0.05	0.10	0.15	0.20	0.25	0.30	0.35	0.40	0.45	p 0.50	0.55	0.60	0.65	0.70	0.75	0.80	0.85	0.90	0.95
9	0	0.6302	0.3874	0.2316	0.1342	0.0751	0.0404	0.0207	0.0101	0.0046	0.0020	0.0008	0.0003	0.0001	0.0000	0.0000	0.0000	0.0000	0.0000	0.0000
	1	0.9288	0.7748	0.5995	0.4362	0.3003	0.1960	0.1211	0.0705	0.0385	0.0195	0.0091	0.0038	0.0014	0.0004	0.0001	0.0000	0.0000	0.0000	0.0000
	2	0.9916	0.9470	0.8591	0.7382	0.6007	0.4628	0.3373	0.2318	0.1495	0.0898	0.0498	0.0250	0.0112	0.0043	0.0013	0.0003	0.0000	0.0000	0.0000
	3	0.9994	0.9917	0.9661	0.9144	0.8343	0.7297	0.6089	0.4826	0.3614	0.2539	0.1658	0.0994	0.0536	0.0253	0.0100	0.0031	0.0006	0.0001	0.0000
	4	1.0000	0.9991	0.9944	0.9804	0.9511	0.9012	0.8283	0.7334	0.6214	0.5000	0.3786	0.2666	0.1717	0.0988	0.0489	0.0196	0.0056	0.0009	0.0000
	5	1.0000	0.9999	0.9994	0.9969	0.9900	0.9747	0.9464	0.9006	0.8342	0.7461	0.6386	0.5174	0.3911	0.2703	0.1657	0.0856	0.0339	0.0083	0.0006
	6	1.0000	1.0000	1.0000	0.9997	0.9987	0.9957	0.9888	0.9750	0.9502	0.9102	0.8505	0.7682	0.6627	0.5372	0.3993	0.2618	0.1409	0.0530	0.0084
	7	1.0000	1.0000	1.0000	1.0000	0.9999	0.9996	0.9986	0.9962	0.9909	0.9805	0.9615	0.9295	0.8789	0.8040	0.6997	0.5638	0.4005	0.2252	0.0712
	8	1.0000	1.0000	1.0000	1.0000	1.0000	1.0000	0.9999	0.9997	0.9992	0.9980	0.9954	0.9899	0.9793	0.9596	0.9249	0.8658	0.7684	0.6126	0.3698
10	0	0.5987	0.3487	0.1969	0.1074	0.0563	0.0282	0.0135	0.0060	0.0025	0.0010	0.0003	0.0001	0.0000	0.0000	0.0000	0.0000	0.0000	0.0000	0.0000
	1	0.9139	0.7361	0.5443	0.3758	0.2440	0.1493	0.0860	0.0464	0.0233	0.0107	0.0045	0.0017	0.0005	0.0001	0.0000	0.0000	0.0000	0.0000	0.0000
	2	0.9885	0.9298	0.8202	0.6778	0.5256	0.3828	0.2616	0.1673	0.0996	0.0547	0.0274	0.0123	0.0048	0.0016	0.0004	0.0001	0.0000	0.0000	0.0000
	3	0.9990	0.9872	0.9500	0.8791	0.7759	0.6496	0.5138	0.3823	0.2660	0.1719	0.1020	0.0548	0.0260	0.0106	0.0035	0.0009	0.0001	0.0000	0.0000
	4	0.9999	0.9984	0.9901	0.9672	0.9219	0.8497	0.7515	0.6331	0.5044	0.3770	0.2616	0.1662	0.0949	0.0473	0.0197	0.0064	0.0014	0.0001	0.0000
	5	1.0000	0.9999	0.9986	0.9936	0.9803	0.9527	0.9051	0.8338	0.7384	0.6230	0.4956	0.3669	0.2485	0.1503	0.0781	0.0328	0.0099	0.0016	0.0001
	6	1.0000	1.0000	0.9999	0.9991	0.9965	0.9894	0.9740	0.9452	0.8980	0.8281	0.7340	0.6177	0.4862	0.3504	0.2241	0.1209	0.0500	0.0128	0.0010
	7	1.0000	1.0000	1.0000	0.9999	0.9996	0.9984	0.9952	0.9877	0.9726	0.9453	0.9004	0.8327	0.7384	0.6172	0.4744	0.3222	0.1798	0.0702	0.0115
	8	1.0000	1.0000	1.0000	1.0000	1.0000	0.9999	0.9995	0.9983	0.9955	0.9893	0.9767	0.9536	0.9140	0.8507	0.7560	0.6242	0.4557	0.2639	0.0861
	9	1.0000	1.0000	1.0000	1.0000	1.0000	1.0000	1.0000	0.9999	0.9997	0.9990	0.9975	0.9940	0.9865	0.9718	0.9437	0.8926	0.8031	0.6513	0.4023
11	0	0.5688	0.3138	0.1673	0.0859	0.0422	0.0198	0.0088	0.0036	0.0014	0.0005	0.0002	0.0000	0.0000	0.0000	0.0000	0.0000	0.0000	0.0000	0.0000
	1	0.8981	0.6974	0.4922	0.3221	0.1971	0.1130	0.0606	0.0302	0.0139	0.0059	0.0022	0.0007	0.0002	0.0000	0.0000	0.0000	0.0000	0.0000	0.0000
	2	0.9848	0.9104	0.7788	0.6174	0.4552	0.3127	0.2001	0.1189	0.0652	0.0327	0.0148	0.0059	0.0020	0.0006	0.0001	0.0000	0.0000	0.0000	0.0000
	3	0.9984	0.9815	0.9306	0.8389	0.7133	0.5696	0.4256	0.2963	0.1911	0.1133	0.0610	0.0293	0.0122	0.0043	0.0012	0.0002	0.0000	0.0000	0.0000
	4	0.9999	0.9972	0.9841	0.9496	0.8854	0.7897	0.6683	0.5328	0.3971	0.2744	0.1738	0.0994	0.0501	0.0216	0.0076	0.0020	0.0003	0.0000	0.0000
	5	1.0000	0.9997	0.9973	0.9883	0.9657	0.9218	0.8513	0.7535	0.6331	0.5000	0.3669	0.2465	0.1487	0.0782	0.0343	0.0117	0.0027	0.0003	0.0000
	6	1.0000	1.0000	0.9997	0.9980	0.9924	0.9784	0.9499	0.9006	0.8262	0.7256	0.6029	0.4672	0.3317	0.2103	0.1146	0.0504	0.0159	0.0028	0.0001
	7	1.0000	1.0000	1.0000	0.9998	0.9988	0.9957	0.9878	0.9707	0.9390	0.8867	0.8089	0.7037	0.5744	0.4304	0.2867	0.1611	0.0694	0.0185	0.0016
	8	1.0000	1.0000	1.0000	1.0000	0.9999	0.9994	0.9980	0.9941	0.9852	0.9673	0.9348	0.8811	0.7999	0.6873	0.5448	0.3826	0.2212	0.0896	0.0152
	9	1.0000	1.0000	1.0000	1.0000	1.0000	1.0000	0.9998	0.9993	0.9978	0.9941	0.9861	0.9698	0.9394	0.8870	0.8029	0.6779	0.5078	0.3026	0.1019
	10	1.0000	1.0000	1.0000	1.0000	1.0000	1.0000	1.0000	1.0000	0.9998	0.9995	0.9986	0.9964	0.9912	0.9802	0.9578	0.9141	0.8327	0.6862	0.4312
12	0	0.5404	0.2824	0.1422	0.0687	0.0317	0.0138	0.0057	0.0022	0.0008	0.0002	0.0001	0.0000	0.0000	0.0000	0.0000	0.0000	0.0000	0.0000	0.0000
	1	0.8816	0.6590	0.4435	0.2749	0.1584	0.0850	0.0424	0.0196	0.0083	0.0032	0.0011	0.0003	0.0001	0.0000	0.0000	0.0000	0.0000	0.0000	0.0000
	2	0.9804	0.8891	0.7358	0.5583	0.3907	0.2528	0.1513	0.0834	0.0421	0.0193	0.0079	0.0028	0.0008	0.0002	0.0000	0.0000	0.0000	0.0000	0.0000
	3	0.9978	0.9744	0.9078	0.7946	0.6488	0.4925	0.3467	0.2253	0.1345	0.0730	0.0356	0.0153	0.0056	0.0017	0.0004	0.0001	0.0000	0.0000	0.0000
	4	0.9998	0.9957	0.9761	0.9274	0.8424	0.7237	0.5833	0.4382	0.3044	0.1938	0.1117	0.0573	0.0255	0.0095	0.0028	0.0006	0.0001	0.0000	0.0000
	5	1.0000	0.9995	0.9954	0.9806	0.9456	0.8822	0.7873	0.6652	0.5269	0.3872	0.2607	0.1582	0.0846	0.0386	0.0143	0.0039	0.0007	0.0001	0.0000
	6	1.0000	0.9999	0.9993	0.9961	0.9857	0.9614	0.9154	0.8418	0.7393	0.6128	0.4731	0.3348	0.2127	0.1178	0.0544	0.0194	0.0046	0.0005	0.0000
	7	1.0000	1.0000	0.9999	0.9994	0.9972	0.9905	0.9745	0.9427	0.8883	0.8062	0.6956	0.5618	0.4167	0.2763	0.1576	0.0726	0.0239	0.0043	0.0002
	8	1.0000	1.0000	1.0000	0.9999	0.9996	0.9983	0.9944	0.9847	0.9644	0.9270	0.8655	0.7747	0.6533	0.5075	0.3512	0.2054	0.0922	0.0256	0.0022
	9	1.0000	1.0000	1.0000	1.0000	1.0000	0.9998	0.9992	0.9972	0.9921	0.9807	0.9579	0.9166	0.8487	0.7472	0.6093	0.4417	0.2642	0.1109	0.0196
	10	1.0000	1.0000	1.0000	1.0000	1.0000	1.0000	0.9999	0.9997	0.9989	0.9968	0.9917	0.9804	0.9576	0.9150	0.8416	0.7251	0.5565	0.3410	0.1184
	11	1.0000	1.0000	1.0000	1.0000	1.0000	1.0000	1.0000	1.0000	0.9999	0.9998	0.9992	0.9978	0.9943	0.9862	0.9683	0.9313	0.8578	0.7176	0.4596

(continued on following page)

Table I (continued from page 524)

n	x	0.05	0.10	0.15	0.20	0.25	0.30	0.35	0.40	0.45	p 0.50	0.55	0.60	0.65	0.70	0.75	0.80	0.85	0.90	0.95
13	0	0.5133	0.2542	0.1209	0.0550	0.0238	0.0097	0.0037	0.0013	0.0004	0.0001	0.0000	0.0000	0.0000	0.0000	0.0000	0.0000	0.0000	0.0000	0.0000
	1	0.8646	0.6213	0.3983	0.2336	0.1267	0.0637	0.0296	0.0126	0.0049	0.0017	0.0005	0.0001	0.0000	0.0000	0.0000	0.0000	0.0000	0.0000	0.0000
	2	0.9755	0.8661	0.6920	0.5017	0.3326	0.2025	0.1132	0.0579	0.0269	0.0112	0.0041	0.0013	0.0003	0.0001	0.0000	0.0000	0.0000	0.0000	0.0000
	3	0.9969	0.9658	0.8820	0.7473	0.5843	0.4206	0.2783	0.1686	0.0929	0.0461	0.0203	0.0078	0.0025	0.0007	0.0001	0.0000	0.0000	0.0000	0.0000
	4	0.9997	0.9935	0.9658	0.9009	0.7940	0.6543	0.5005	0.3530	0.2279	0.1334	0.0698	0.0321	0.0126	0.0040	0.0010	0.0002	0.0000	0.0000	0.0000
	5	1.0000	0.9991	0.9925	0.9700	0.9198	0.8346	0.7159	0.5744	0.4268	0.2905	0.1788	0.0977	0.0462	0.0182	0.0056	0.0012	0.0002	0.0000	0.0000
	6	1.0000	0.9999	0.9987	0.9930	0.9757	0.9376	0.8705	0.7712	0.6437	0.5000	0.3563	0.2288	0.1295	0.0624	0.0243	0.0070	0.0013	0.0001	0.0000
	7	1.0000	1.0000	0.9998	0.9988	0.9944	0.9818	0.9538	0.9023	0.8212	0.7095	0.5732	0.4256	0.2841	0.1654	0.0802	0.0300	0.0075	0.0009	0.0000
	8	1.0000	1.0000	1.0000	0.9998	0.9990	0.9960	0.9874	0.9679	0.9302	0.8666	0.7721	0.6470	0.4995	0.3457	0.2060	0.0991	0.0342	0.0065	0.0003
	9	1.0000	1.0000	1.0000	1.0000	0.9999	0.9993	0.9975	0.9922	0.9797	0.9539	0.9071	0.8314	0.7217	0.5794	0.4157	0.2527	0.1180	0.0342	0.0031
	10	1.0000	1.0000	1.0000	1.0000	1.0000	0.9999	0.9997	0.9987	0.9959	0.9888	0.9731	0.9421	0.8868	0.7975	0.6674	0.4933	0.3080	0.1339	0.0245
	11	1.0000	1.0000	1.0000	1.0000	1.0000	1.0000	1.0000	0.9999	0.9995	0.9983	0.9951	0.9874	0.9704	0.9363	0.8733	0.7664	0.6017	0.3787	0.1354
	12	1.0000	1.0000	1.0000	1.0000	1.0000	1.0000	1.0000	1.0000	1.0000	0.9999	0.9996	0.9987	0.9963	0.9903	0.9762	0.9450	0.8791	0.7458	0.4867
14	0	0.4877	0.2288	0.1028	0.0440	0.0178	0.0068	0.0024	0.0008	0.0002	0.0001	0.0000	0.0000	0.0000	0.0000	0.0000	0.0000	0.0000	0.0000	0.0000
	1	0.8470	0.5846	0.3567	0.1979	0.1010	0.0475	0.0205	0.0081	0.0029	0.0009	0.0003	0.0001	0.0000	0.0000	0.0000	0.0000	0.0000	0.0000	0.0000
	2	0.9699	0.8416	0.6479	0.4481	0.2811	0.1608	0.0839	0.0398	0.0170	0.0065	0.0022	0.0006	0.0001	0.0000	0.0000	0.0000	0.0000	0.0000	0.0000
	3	0.9958	0.9559	0.8535	0.6982	0.5213	0.3552	0.2205	0.1243	0.0632	0.0287	0.0114	0.0039	0.0011	0.0002	0.0000	0.0000	0.0000	0.0000	0.0000
	4	0.9996	0.9908	0.9533	0.8702	0.7415	0.5842	0.4227	0.2793	0.1672	0.0898	0.0426	0.0175	0.0060	0.0017	0.0003	0.0000	0.0000	0.0000	0.0000
	5	1.0000	0.9985	0.9885	0.9561	0.8883	0.7805	0.6405	0.4859	0.3373	0.2120	0.1189	0.0583	0.0243	0.0083	0.0022	0.0004	0.0000	0.0000	0.0000
	6	1.0000	0.9998	0.9978	0.9884	0.9617	0.9067	0.8164	0.6925	0.5461	0.3953	0.2586	0.1501	0.0753	0.0315	0.0103	0.0024	0.0003	0.0000	0.0000
	7	1.0000	1.0000	0.9997	0.9976	0.9897	0.9685	0.9247	0.8499	0.7414	0.6047	0.4539	0.3075	0.1836	0.0933	0.0383	0.0116	0.0022	0.0002	0.0000
	8	1.0000	1.0000	1.0000	0.9996	0.9978	0.9917	0.9757	0.9417	0.8811	0.7880	0.6627	0.5141	0.3595	0.2195	0.1117	0.0439	0.0115	0.0015	0.0000
	9	1.0000	1.0000	1.0000	1.0000	0.9997	0.9983	0.9940	0.9825	0.9574	0.9102	0.8328	0.7207	0.5773	0.4158	0.2585	0.1298	0.0467	0.0092	0.0004
	10	1.0000	1.0000	1.0000	1.0000	1.0000	0.9998	0.9989	0.9961	0.9886	0.9713	0.9368	0.8757	0.7795	0.6448	0.4787	0.3018	0.1465	0.0441	0.0042
	11	1.0000	1.0000	1.0000	1.0000	1.0000	1.0000	0.9999	0.9994	0.9978	0.9935	0.9830	0.9602	0.9161	0.8392	0.7189	0.5519	0.3521	0.1584	0.0301
	12	1.0000	1.0000	1.0000	1.0000	1.0000	1.0000	1.0000	0.9999	0.9997	0.9991	0.9971	0.9919	0.9795	0.9525	0.8990	0.8021	0.6433	0.4154	0.1530
	13	1.0000	1.0000	1.0000	1.0000	1.0000	1.0000	1.0000	1.0000	1.0000	0.9999	0.9998	0.9992	0.9976	0.9932	0.9822	0.9560	0.8972	0.7712	0.5123
15	0	0.4633	0.2059	0.0874	0.0352	0.0134	0.0047	0.0016	0.0005	0.0001	0.0000	0.0000	0.0000	0.0000	0.0000	0.0000	0.0000	0.0000	0.0000	0.0000
	1	0.8290	0.5490	0.3186	0.1671	0.0802	0.0353	0.0142	0.0052	0.0017	0.0005	0.0001	0.0000	0.0000	0.0000	0.0000	0.0000	0.0000	0.0000	0.0000
	2	0.9638	0.8159	0.6042	0.3980	0.2361	0.1268	0.0617	0.0271	0.0107	0.0037	0.0011	0.0003	0.0001	0.0000	0.0000	0.0000	0.0000	0.0000	0.0000
	3	0.9945	0.9444	0.8227	0.6482	0.4613	0.2969	0.1727	0.0905	0.0424	0.0176	0.0063	0.0019	0.0005	0.0001	0.0000	0.0000	0.0000	0.0000	0.0000
	4	0.9994	0.9873	0.9383	0.8358	0.6865	0.5155	0.3519	0.2173	0.1204	0.0592	0.0255	0.0093	0.0028	0.0007	0.0001	0.0000	0.0000	0.0000	0.0000
	5	0.9999	0.9978	0.9832	0.9389	0.8516	0.7216	0.5643	0.4032	0.2608	0.1509	0.0769	0.0338	0.0124	0.0037	0.0008	0.0001	0.0000	0.0000	0.0000
	6	1.0000	0.9997	0.9964	0.9819	0.9434	0.8689	0.7548	0.6098	0.4522	0.3036	0.1818	0.0950	0.0422	0.0152	0.0042	0.0008	0.0001	0.0000	0.0000
	7	1.0000	1.0000	0.9994	0.9958	0.9827	0.9500	0.8868	0.7869	0.6535	0.5000	0.3465	0.2131	0.1132	0.0500	0.0173	0.0042	0.0006	0.0000	0.0000
	8	1.0000	1.0000	0.9999	0.9992	0.9958	0.9848	0.9578	0.9050	0.8182	0.6964	0.5478	0.3902	0.2452	0.1311	0.0566	0.0181	0.0036	0.0003	0.0000
	9	1.0000	1.0000	1.0000	0.9999	0.9992	0.9963	0.9876	0.9662	0.9231	0.8491	0.7392	0.5968	0.4357	0.2784	0.1484	0.0611	0.0168	0.0022	0.0001
	10	1.0000	1.0000	1.0000	1.0000	0.9999	0.9993	0.9972	0.9907	0.9745	0.9408	0.8796	0.7827	0.6481	0.4845	0.3135	0.1642	0.0617	0.0127	0.0006
	11	1.0000	1.0000	1.0000	1.0000	1.0000	0.9999	0.9995	0.9981	0.9937	0.9824	0.9576	0.9095	0.8273	0.7031	0.5387	0.3518	0.1773	0.0556	0.0055
	12	1.0000	1.0000	1.0000	1.0000	1.0000	1.0000	0.9999	0.9997	0.9989	0.9963	0.9893	0.9729	0.9383	0.8732	0.7639	0.6020	0.3958	0.1841	0.0362
	13	1.0000	1.0000	1.0000	1.0000	1.0000	1.0000	1.0000	1.0000	0.9999	0.9995	0.9983	0.9948	0.9858	0.9647	0.9198	0.8329	0.6814	0.4510	0.1710
	14	1.0000	1.0000	1.0000	1.0000	1.0000	1.0000	1.0000	1.0000	1.0000	1.0000	0.9999	0.9995	0.9984	0.9953	0.9866	0.9648	0.9126	0.7941	0.5367

(continued on following page)

Table I (continued from page 525)

n	x	0.05	0.10	0.15	0.20	0.25	0.30	0.35	0.40	0.45	p 0.50	0.55	0.60	0.65	0.70	0.75	0.80	0.85	0.90	0.95
16	0	0.4401	0.1853	0.0743	0.0281	0.0100	0.0033	0.0010	0.0003	0.0001	0.0000	0.0000	0.0000	0.0000	0.0000	0.0000	0.0000	0.0000	0.0000	0.0000
	1	0.8108	0.5147	0.2839	0.1407	0.0635	0.0261	0.0098	0.0033	0.0010	0.0003	0.0001	0.0000	0.0000	0.0000	0.0000	0.0000	0.0000	0.0000	0.0000
	2	0.9571	0.7892	0.5614	0.3518	0.1971	0.0994	0.0451	0.0183	0.0066	0.0021	0.0006	0.0001	0.0000	0.0000	0.0000	0.0000	0.0000	0.0000	0.0000
	3	0.9930	0.9316	0.7899	0.5981	0.4050	0.2459	0.1339	0.0651	0.0281	0.0106	0.0035	0.0009	0.0002	0.0000	0.0000	0.0000	0.0000	0.0000	0.0000
	4	0.9991	0.9830	0.9209	0.7982	0.6302	0.4499	0.2892	0.1666	0.0853	0.0384	0.0149	0.0049	0.0013	0.0003	0.0000	0.0000	0.0000	0.0000	0.0000
	5	0.9999	0.9967	0.9765	0.9183	0.8103	0.6598	0.4900	0.3288	0.1976	0.1051	0.0486	0.0191	0.0062	0.0016	0.0003	0.0000	0.0000	0.0000	0.0000
	6	1.0000	0.9995	0.9944	0.9733	0.9204	0.8247	0.6881	0.5272	0.3660	0.2272	0.1241	0.0583	0.0229	0.0071	0.0016	0.0002	0.0000	0.0000	0.0000
	7	1.0000	0.9999	0.9989	0.9930	0.9729	0.9256	0.8406	0.7161	0.5629	0.4018	0.2559	0.1423	0.0671	0.0257	0.0075	0.0015	0.0002	0.0000	0.0000
	8	1.0000	1.0000	0.9998	0.9985	0.9925	0.9743	0.9329	0.8577	0.7441	0.5982	0.4371	0.2839	0.1594	0.0744	0.0271	0.0070	0.0011	0.0001	0.0000
	9	1.0000	1.0000	1.0000	0.9998	0.9984	0.9929	0.9771	0.9417	0.8759	0.7728	0.6340	0.4728	0.3119	0.1753	0.0796	0.0267	0.0056	0.0005	0.0000
	10	1.0000	1.0000	1.0000	1.0000	0.9997	0.9984	0.9938	0.9809	0.9514	0.8949	0.8024	0.6712	0.5100	0.3402	0.1897	0.0817	0.0235	0.0033	0.0001
	11	1.0000	1.0000	1.0000	1.0000	1.0000	0.9997	0.9987	0.9951	0.9851	0.9616	0.9147	0.8334	0.7108	0.5501	0.3698	0.2018	0.0791	0.0170	0.0009
	12	1.0000	1.0000	1.0000	1.0000	1.0000	1.0000	0.9998	0.9991	0.9965	0.9894	0.9719	0.9349	0.8661	0.7541	0.5950	0.4019	0.2101	0.0684	0.0070
	13	1.0000	1.0000	1.0000	1.0000	1.0000	1.0000	1.0000	0.9999	0.9994	0.9979	0.9934	0.9817	0.9549	0.9006	0.8029	0.6482	0.4386	0.2108	0.0429
	14	1.0000	1.0000	1.0000	1.0000	1.0000	1.0000	1.0000	1.0000	0.9999	0.9997	0.9990	0.9967	0.9902	0.9739	0.9365	0.8593	0.7161	0.4853	0.1892
	15	1.0000	1.0000	1.0000	1.0000	1.0000	1.0000	1.0000	1.0000	1.0000	1.0000	0.9999	0.9997	0.9990	0.9967	0.9900	0.9719	0.9257	0.8147	0.5599
17	0	0.4181	0.1668	0.0631	0.0225	0.0075	0.0023	0.0007	0.0002	0.0000	0.0000	0.0000	0.0000	0.0000	0.0000	0.0000	0.0000	0.0000	0.0000	0.0000
	1	0.7922	0.4818	0.2525	0.1182	0.0501	0.0193	0.0067	0.0021	0.0006	0.0001	0.0000	0.0000	0.0000	0.0000	0.0000	0.0000	0.0000	0.0000	0.0000
	2	0.9497	0.7618	0.5198	0.3096	0.1637	0.0774	0.0327	0.0123	0.0041	0.0012	0.0003	0.0001	0.0000	0.0000	0.0000	0.0000	0.0000	0.0000	0.0000
	3	0.9912	0.9174	0.7556	0.5489	0.3530	0.2019	0.1028	0.0464	0.0184	0.0064	0.0019	0.0005	0.0001	0.0000	0.0000	0.0000	0.0000	0.0000	0.0000
	4	0.9988	0.9779	0.9013	0.7582	0.5739	0.3887	0.2348	0.1260	0.0596	0.0245	0.0086	0.0025	0.0006	0.0001	0.0000	0.0000	0.0000	0.0000	0.0000
	5	0.9999	0.9953	0.9681	0.8943	0.7653	0.5968	0.4197	0.2639	0.1471	0.0717	0.0301	0.0106	0.0030	0.0007	0.0001	0.0000	0.0000	0.0000	0.0000
	6	1.0000	0.9992	0.9917	0.9623	0.8929	0.7752	0.6188	0.4478	0.2902	0.1662	0.0826	0.0348	0.0120	0.0032	0.0006	0.0001	0.0000	0.0000	0.0000
	7	1.0000	0.9999	0.9983	0.9891	0.9598	0.8954	0.7872	0.6405	0.4743	0.3145	0.1834	0.0919	0.0383	0.0127	0.0031	0.0005	0.0000	0.0000	0.0000
	8	1.0000	1.0000	0.9997	0.9974	0.9876	0.9597	0.9006	0.8011	0.6626	0.5000	0.3374	0.1989	0.0994	0.0403	0.0124	0.0026	0.0003	0.0000	0.0000
	9	1.0000	1.0000	1.0000	0.9995	0.9969	0.9873	0.9617	0.9081	0.8166	0.6855	0.5257	0.3595	0.2128	0.1046	0.0402	0.0109	0.0017	0.0001	0.0000
	10	1.0000	1.0000	1.0000	0.9999	0.9994	0.9968	0.9880	0.9652	0.9174	0.8338	0.7098	0.5522	0.3812	0.2248	0.1071	0.0377	0.0083	0.0008	0.0000
	11	1.0000	1.0000	1.0000	1.0000	0.9999	0.9993	0.9970	0.9894	0.9699	0.9283	0.8529	0.7361	0.5803	0.4032	0.2347	0.1057	0.0319	0.0047	0.0001
	12	1.0000	1.0000	1.0000	1.0000	1.0000	0.9999	0.9994	0.9975	0.9914	0.9755	0.9404	0.8740	0.7652	0.6113	0.4261	0.2418	0.0987	0.0221	0.0012
	13	1.0000	1.0000	1.0000	1.0000	1.0000	1.0000	0.9999	0.9995	0.9981	0.9936	0.9816	0.9536	0.8972	0.7981	0.6470	0.4511	0.2444	0.0826	0.0088
	14	1.0000	1.0000	1.0000	1.0000	1.0000	1.0000	1.0000	0.9999	0.9997	0.9988	0.9959	0.9877	0.9673	0.9226	0.8363	0.6904	0.4802	0.2382	0.0503
	15	1.0000	1.0000	1.0000	1.0000	1.0000	1.0000	1.0000	1.0000	1.0000	0.9999	0.9994	0.9979	0.9933	0.9807	0.9499	0.8818	0.7475	0.5182	0.2078
	16	1.0000	1.0000	1.0000	1.0000	1.0000	1.0000	1.0000	1.0000	1.0000	1.0000	1.0000	0.9998	0.9993	0.9977	0.9925	0.9775	0.9369	0.8332	0.5819
18	0	0.3972	0.1501	0.0536	0.0180	0.0056	0.0016	0.0004	0.0001	0.0000	0.0000	0.0000	0.0000	0.0000	0.0000	0.0000	0.0000	0.0000	0.0000	0.0000
	1	0.7735	0.4503	0.2241	0.0991	0.0395	0.0142	0.0046	0.0013	0.0003	0.0001	0.0000	0.0000	0.0000	0.0000	0.0000	0.0000	0.0000	0.0000	0.0000
	2	0.9419	0.7338	0.4797	0.2713	0.1353	0.0600	0.0236	0.0082	0.0025	0.0007	0.0001	0.0000	0.0000	0.0000	0.0000	0.0000	0.0000	0.0000	0.0000
	3	0.9891	0.9018	0.7202	0.5010	0.3057	0.1646	0.0783	0.0328	0.0120	0.0038	0.0010	0.0002	0.0001	0.0000	0.0000	0.0000	0.0000	0.0000	0.0000
	4	0.9985	0.9718	0.8794	0.7164	0.5187	0.3327	0.1886	0.0942	0.0411	0.0154	0.0049	0.0013	0.0003	0.0000	0.0000	0.0000	0.0000	0.0000	0.0000
	5	0.9998	0.9936	0.9581	0.8671	0.7175	0.5344	0.3550	0.2088	0.1077	0.0481	0.0183	0.0058	0.0014	0.0003	0.0000	0.0000	0.0000	0.0000	0.0000
	6	1.0000	0.9988	0.9882	0.9487	0.8610	0.7217	0.5491	0.3743	0.2258	0.1189	0.0537	0.0203	0.0062	0.0014	0.0002	0.0000	0.0000	0.0000	0.0000
	7	1.0000	0.9998	0.9973	0.9837	0.9431	0.8593	0.7283	0.5634	0.3915	0.2403	0.1280	0.0576	0.0212	0.0061	0.0012	0.0002	0.0000	0.0000	0.0000
	8	1.0000	1.0000	0.9995	0.9957	0.9807	0.9404	0.8609	0.7368	0.5778	0.4073	0.2527	0.1347	0.0597	0.0210	0.0054	0.0009	0.0001	0.0000	0.0000
	9	1.0000	1.0000	0.9999	0.9991	0.9946	0.9790	0.9403	0.8653	0.7473	0.5927	0.4222	0.2632	0.1391	0.0596	0.0193	0.0043	0.0005	0.0000	0.0000
	10	1.0000	1.0000	1.0000	0.9998	0.9988	0.9939	0.9788	0.9424	0.8720	0.7597	0.6085	0.4366	0.2717	0.1407	0.0569	0.0163	0.0027	0.0002	0.0000

(continued on following page)

Table I (continued from page 526)

Cumulative binomial probabilities. Column headings are values of p (the "P" label is printed above the $p = 0.50$ column).

n	x	0.05	0.10	0.15	0.20	0.25	0.30	0.35	0.40	0.45	0.50	0.55	0.60	0.65	0.70	0.75	0.80	0.85	0.90	0.95
	11	1.0000	1.0000	1.0000	1.0000	0.9998	0.9986	0.9938	0.9797	0.9463	0.8811	0.7742	0.6257	0.4509	0.2783	0.1390	0.0513	0.0118	0.0012	0.0000
	12	1.0000	1.0000	1.0000	1.0000	1.0000	0.9997	0.9986	0.9942	0.9817	0.9519	0.8923	0.7912	0.6450	0.4656	0.2825	0.1329	0.0419	0.0064	0.0002
	13	1.0000	1.0000	1.0000	1.0000	1.0000	1.0000	0.9997	0.9987	0.9951	0.9846	0.9589	0.9058	0.8114	0.6673	0.4813	0.2836	0.1206	0.0282	0.0015
	14	1.0000	1.0000	1.0000	1.0000	1.0000	1.0000	1.0000	0.9998	0.9990	0.9962	0.9880	0.9672	0.9217	0.8354	0.6943	0.4990	0.2798	0.0982	0.0109
	15	1.0000	1.0000	1.0000	1.0000	1.0000	1.0000	1.0000	1.0000	0.9999	0.9993	0.9975	0.9918	0.9764	0.9400	0.8647	0.7287	0.5203	0.2662	0.0581
	16	1.0000	1.0000	1.0000	1.0000	1.0000	1.0000	1.0000	1.0000	1.0000	0.9999	0.9997	0.9987	0.9954	0.9858	0.9605	0.9009	0.7759	0.5497	0.2265
	17	1.0000	1.0000	1.0000	1.0000	1.0000	1.0000	1.0000	1.0000	1.0000	1.0000	1.0000	0.9999	0.9996	0.9984	0.9944	0.9820	0.9464	0.8499	0.6028
19	0	0.3774	0.1351	0.0456	0.0144	0.0042	0.0011	0.0003	0.0001	0.0000	0.0000	0.0000	0.0000	0.0000	0.0000	0.0000	0.0000	0.0000	0.0000	0.0000
	1	0.7547	0.4203	0.1985	0.0829	0.0310	0.0104	0.0031	0.0008	0.0002	0.0000	0.0000	0.0000	0.0000	0.0000	0.0000	0.0000	0.0000	0.0000	0.0000
	2	0.9335	0.7054	0.4413	0.2369	0.1113	0.0462	0.0170	0.0055	0.0015	0.0004	0.0001	0.0000	0.0000	0.0000	0.0000	0.0000	0.0000	0.0000	0.0000
	3	0.9868	0.8850	0.6841	0.4551	0.2631	0.1332	0.0591	0.0230	0.0077	0.0022	0.0005	0.0001	0.0000	0.0000	0.0000	0.0000	0.0000	0.0000	0.0000
	4	0.9980	0.9648	0.8556	0.6733	0.4654	0.2822	0.1500	0.0696	0.0280	0.0096	0.0028	0.0005	0.0001	0.0000	0.0000	0.0000	0.0000	0.0000	0.0000
	5	0.9998	0.9914	0.9463	0.8369	0.6678	0.4739	0.2968	0.1529	0.0777	0.0318	0.0109	0.0031	0.0007	0.0001	0.0000	0.0000	0.0000	0.0000	0.0000
	6	1.0000	0.9983	0.9837	0.9324	0.8251	0.6655	0.4812	0.3081	0.1727	0.0835	0.0342	0.0115	0.0031	0.0006	0.0001	0.0000	0.0000	0.0000	0.0000
	7	1.0000	0.9997	0.9959	0.9767	0.9225	0.8180	0.6656	0.4878	0.3169	0.1796	0.0871	0.0352	0.0114	0.0028	0.0005	0.0000	0.0000	0.0000	0.0000
	8	1.0000	1.0000	0.9992	0.9933	0.9713	0.9161	0.8145	0.6675	0.4940	0.3238	0.1841	0.0885	0.0347	0.0105	0.0023	0.0003	0.0000	0.0000	0.0000
	9	1.0000	1.0000	0.9999	0.9984	0.9911	0.9674	0.9125	0.8139	0.6710	0.5000	0.3290	0.1861	0.0875	0.0326	0.0089	0.0016	0.0001	0.0000	0.0000
	10	1.0000	1.0000	1.0000	0.9997	0.9977	0.9895	0.9653	0.9115	0.8159	0.6762	0.5060	0.3325	0.1855	0.0839	0.0287	0.0067	0.0008	0.0000	0.0000
	11	1.0000	1.0000	1.0000	1.0000	0.9995	0.9972	0.9886	0.9648	0.9129	0.8204	0.6831	0.5122	0.3344	0.1820	0.0775	0.0233	0.0041	0.0003	0.0000
	12	1.0000	1.0000	1.0000	1.0000	0.9999	0.9994	0.9969	0.9884	0.9658	0.9165	0.8273	0.6919	0.5188	0.3345	0.1749	0.0676	0.0163	0.0017	0.0000
	13	1.0000	1.0000	1.0000	1.0000	1.0000	0.9999	0.9993	0.9969	0.9891	0.9682	0.9223	0.8371	0.7032	0.5261	0.3322	0.1631	0.0537	0.0086	0.0002
	14	1.0000	1.0000	1.0000	1.0000	1.0000	1.0000	0.9999	0.9994	0.9972	0.9904	0.9720	0.9304	0.8500	0.7178	0.5346	0.3267	0.1444	0.0352	0.0020
	15	1.0000	1.0000	1.0000	1.0000	1.0000	1.0000	1.0000	0.9999	0.9995	0.9978	0.9923	0.9770	0.9409	0.8668	0.7369	0.5449	0.3159	0.1150	0.0132
	16	1.0000	1.0000	1.0000	1.0000	1.0000	1.0000	1.0000	1.0000	0.9999	0.9996	0.9985	0.9945	0.9830	0.9538	0.8887	0.7631	0.5587	0.2946	0.0665
	17	1.0000	1.0000	1.0000	1.0000	1.0000	1.0000	1.0000	1.0000	1.0000	1.0000	0.9998	0.9992	0.9969	0.9896	0.9690	0.9171	0.8015	0.5797	0.2453
	18	1.0000	1.0000	1.0000	1.0000	1.0000	1.0000	1.0000	1.0000	1.0000	1.0000	1.0000	0.9999	0.9997	0.9989	0.9958	0.9856	0.9544	0.8649	0.6226
20	0	0.3585	0.1216	0.0388	0.0115	0.0032	0.0008	0.0002	0.0000	0.0000	0.0000	0.0000	0.0000	0.0000	0.0000	0.0000	0.0000	0.0000	0.0000	0.0000
	1	0.7358	0.3917	0.1756	0.0692	0.0243	0.0076	0.0021	0.0005	0.0001	0.0000	0.0000	0.0000	0.0000	0.0000	0.0000	0.0000	0.0000	0.0000	0.0000
	2	0.9245	0.6769	0.4049	0.2061	0.0913	0.0355	0.0121	0.0036	0.0009	0.0002	0.0000	0.0000	0.0000	0.0000	0.0000	0.0000	0.0000	0.0000	0.0000
	3	0.9841	0.8670	0.6477	0.4114	0.2252	0.1071	0.0444	0.0160	0.0049	0.0013	0.0003	0.0000	0.0000	0.0000	0.0000	0.0000	0.0000	0.0000	0.0000
	4	0.9974	0.9568	0.8298	0.6296	0.4148	0.2375	0.1182	0.0510	0.0189	0.0059	0.0015	0.0003	0.0000	0.0000	0.0000	0.0000	0.0000	0.0000	0.0000
	5	0.9997	0.9887	0.9327	0.8042	0.6172	0.4164	0.2454	0.1256	0.0553	0.0207	0.0064	0.0016	0.0003	0.0000	0.0000	0.0000	0.0000	0.0000	0.0000
	6	1.0000	0.9976	0.9781	0.9133	0.7858	0.6080	0.4166	0.2500	0.1299	0.0577	0.0214	0.0065	0.0015	0.0003	0.0000	0.0000	0.0000	0.0000	0.0000
	7	1.0000	0.9996	0.9941	0.9679	0.8982	0.7723	0.6010	0.4159	0.2520	0.1316	0.0580	0.0210	0.0060	0.0013	0.0002	0.0000	0.0000	0.0000	0.0000
	8	1.0000	0.9999	0.9987	0.9900	0.9591	0.8867	0.7624	0.5956	0.4143	0.2517	0.1308	0.0565	0.0196	0.0051	0.0009	0.0001	0.0000	0.0000	0.0000
	9	1.0000	1.0000	0.9998	0.9974	0.9861	0.9520	0.8782	0.7553	0.5914	0.4119	0.2493	0.1275	0.0532	0.0171	0.0039	0.0006	0.0000	0.0000	0.0000
	10	1.0000	1.0000	1.0000	0.9994	0.9961	0.9829	0.9468	0.8725	0.7507	0.5881	0.4086	0.2447	0.1218	0.0480	0.0139	0.0026	0.0002	0.0000	0.0000
	11	1.0000	1.0000	1.0000	0.9999	0.9991	0.9949	0.9804	0.9435	0.8692	0.7483	0.5857	0.4044	0.2376	0.1133	0.0409	0.0100	0.0013	0.0000	0.0000
	12	1.0000	1.0000	1.0000	1.0000	0.9998	0.9987	0.9940	0.9790	0.9420	0.8684	0.7480	0.5841	0.3990	0.2277	0.1018	0.0321	0.0059	0.0004	0.0000
	13	1.0000	1.0000	1.0000	1.0000	1.0000	0.9997	0.9985	0.9935	0.9786	0.9423	0.8701	0.7500	0.5834	0.3920	0.2142	0.0867	0.0219	0.0024	0.0000
	14	1.0000	1.0000	1.0000	1.0000	1.0000	1.0000	0.9997	0.9984	0.9936	0.9793	0.9447	0.8744	0.7546	0.5836	0.3828	0.1958	0.0673	0.0113	0.0003
	15	1.0000	1.0000	1.0000	1.0000	1.0000	1.0000	1.0000	0.9997	0.9985	0.9941	0.9811	0.9490	0.8818	0.7625	0.5852	0.3704	0.1702	0.0432	0.0026
	16	1.0000	1.0000	1.0000	1.0000	1.0000	1.0000	1.0000	1.0000	0.9997	0.9987	0.9951	0.9840	0.9556	0.8929	0.7748	0.5886	0.3523	0.1330	0.0159
	17	1.0000	1.0000	1.0000	1.0000	1.0000	1.0000	1.0000	1.0000	1.0000	0.9998	0.9991	0.9964	0.9879	0.9645	0.9087	0.7939	0.5951	0.3231	0.0755
	18	1.0000	1.0000	1.0000	1.0000	1.0000	1.0000	1.0000	1.0000	1.0000	1.0000	0.9999	0.9995	0.9979	0.9924	0.9757	0.9308	0.8244	0.6083	0.2642
	19	1.0000	1.0000	1.0000	1.0000	1.0000	1.0000	1.0000	1.0000	1.0000	1.0000	1.0000	1.0000	0.9998	0.9992	0.9968	0.9885	0.9612	0.8784	0.6415

Table 4 Values of t_α

ν	$\alpha = 0.10$	$\alpha = 0.05$	$\alpha = 0.025$	$\alpha = 0.01$	$\alpha = 0.00833$	$\alpha = 0.00625$	$\alpha = 0.005$	ν
1	3.078	6.314	12.706	31.821	38.204	50.923	63.657	1
2	1.886	2.920	4.303	6.965	7.650	8.860	9.925	2
3	1.638	2.353	3.182	4.541	4.857	5.392	5.841	3
4	1.533	2.132	2.776	3.747	3.961	4.315	4.604	4
5	1.476	2.015	2.571	3.365	3.534	3.810	4.032	5
6	1.440	1.943	2.447	3.143	3.288	3.521	3.707	6
7	1.415	1.895	2.365	2.998	3.128	3.335	3.499	7
8	1.397	1.860	2.306	2.896	3.016	3.206	3.355	8
9	1.383	1.833	2.262	2.821	2.934	3.111	3.250	9
10	1.372	1.812	2.228	2.764	2.870	3.038	3.169	10
11	1.363	1.796	2.201	2.718	2.820	2.891	3.106	11
12	1.356	1.782	2.179	2.681	2.780	2.934	3.055	12
13	1.350	1.771	2.160	2.650	2.746	2.896	3.012	13
14	1.345	1.761	2.145	2.624	2.718	2.864	2.977	14
15	1.341	1.753	2.131	2.602	2.694	2.837	2.947	15
16	1.337	1.746	2.120	2.583	2.673	2.813	2.921	16
17	1.333	1.740	2.110	2.567	2.655	2.793	2.898	17
18	1.330	1.734	2.101	2.552	2.639	2.775	2.878	18
19	1.328	1.729	2.093	2.539	2.625	2.759	2.861	19
20	1.325	1.725	2.086	2.528	2.613	2.744	2.845	20
21	1.323	1.721	2.080	2.518	2.602	2.732	2.831	21
22	1.321	1.717	2.074	2.508	2.591	2.720	2.819	22
23	1.319	1.714	2.069	2.500	2.582	2.710	2.807	23
24	1.318	1.711	2.064	2.492	2.574	2.700	2.797	24
25	1.316	1.708	2.060	2.485	2.566	2.692	2.787	25
26	1.315	1.706	2.056	2.479	2.559	2.684	2.779	26
27	1.314	1.703	2.052	2.473	2.553	2.676	2.771	27
28	1.313	1.701	2.048	2.467	2.547	2.669	2.763	28
29	1.311	1.699	2.045	2.462	2.541	2.663	2.756	29
inf.	1.282	1.645	1.960	2.326	2.394	2.498	2.576	inf.

USING THE R SOFTWARE PROGRAM

Introduction to R

R is the name of widely used, powerful software free and available on the Web. If you wish, you can use it on your own computer. The program and help is maintained at *http://www.r-project.org*. Manuals can be found at *http://www.cran.r-project.org/doc/manuals*. Initially, you may find *R-intro.pdf* and *R-data.pdf* useful.

R is a program that is command-driven as opposed to menu-driven. You type in a command and R responds. R has many built-in functions that operate on objects generically called **x, y**, or a more descriptive name. These objects are strings of numbers, or vectors, where the order in which the numbers are entered is remembered. Learning a few functions will enhance your learning of introductory statistics. Typing a function name and hitting *Enter* will just display the function name. All functions are followed by ( ), and you must include an argument list between the parentheses. The list may be empty. To quit you need to type *q()*, not just *q*, and hit *Enter*.

Entering Data

When the data set is small, the easiest way to enter the data is to use the function **c** that *concatenates* numbers. For example, to create an object named **xmpg** for the mpg data in Example 18 of Chapter 2, after the prompt > enter the object name followed by **= c(** and then the data. If you do not finish on one line, R changes the prompt to a + until you finish entering the data and type).

> xmpg = c(19.7, 21.5, 22.5, 22.2, 22.6, 21.9, 20.5, 19.3, 19.9, 21.7,
+ 22.8, 23.2, 21.4, 20.8, 19.4, 22.0, 23.0, 21.1, 20.9, 21.3)

You can also download data from the Web site associated with this textbook. You may wish first to create a folder on your hard drive that will contain data files. The easiest files to read into R are the ASCII formatted files, or plain text files, that end with .dat or .txt. On your own computer, you need to select the File menu and make sure your working directory contains the data files.

For the data from Exercise 2.34, the command[1]

> Dat = read.table("2-34.txt",header=T)

will place the data in the object **Dat** and allow you to refer to the data by the name of variable(s) in the header line of the data file. In all the files for this text, the columns represent variables (often just one) and the rows represent observations. You can use the structure function **str()** to determine the name of each variable and their structure.

>str(Dat)

'data.frame': 29 obs. of 1 variable:
$ cost: num 1.41 1.7 1.03 0.99 1.68 1.09 1.68 1.94 1.53 2.25 ...

Here, cost is the name of the variable and there are 29 observations.

[1]On a windows machine, with the folder EngStatData on the desktop, the command is
read.table(file="C:/Users/Yourname/Desktop/EngStatDATA/2-34.txt",header=T)

Alternatively, if the data files are not in the same directory, the **file.choose** function will open a dialog box that will help you move through your folders to find the proper data file to read into R.

$$> \mathbf{Dat} = \mathbf{read.table(file.choose().header} = \mathbf{T)}$$

You can use the structure function **str()** to determine the name of each variable and their structure.

Arithmetic Operations

You can use R like a calculator where the asterisk (*) is the symbol for multiplication, and ^ is the symbol for exponentiation. Also, the colon (:) is a function, and **2:8** creates an array of numbers from 2 to 8 inclusive. For instance,

> **11*3-8/2+sqrt(25)**
[1]   34

where the answer follows the prompt [1].

> **3*(2:6)**
[1]   6   9   12   15   18

Arithmetic operations can also be applied to objects like **x** and **y**, which themselves are strings of numbers. For example, when $\mathbf{x} = c(1, 9, 4, 0)$ and $\mathbf{y} = c(-2, 5, -3, 7)$, we obtain the following results.

Arithmetic Function	Result
x + y	adds the corresponding entries in **x** and **y** −1, 14, 1, 7
x^2	squares each entry in **x** 1, 81, 16, 0
sqrt(x)	square root of each entry in **x** 1, 3, 2, 0

We now describe some functions useful for statistical analysis.

Descriptive Statistics

We illustrate the functions to obtain summary statistics using the data **xmpg** entered above.

Function	Result
mean(xmpg)	value of sample mean $\bar{x}$
var(xmpg)	calculates sample variance s^2
sd(xmpg)	calculates sample standard deviation s
median(xmpg)	value of sample median
summary(xmpg)	min Q_1 median Q_3 max

To summarize the data in Exercise 2.34, you would replace **xmpg** with **Dat$cost** where $ separates the name of the data and the variable name in the header.

Probability Distributions

The function for each probability distribution can have one of four prefixes:

Prefix	Meaning
p	cumulative probability distribution
q	quantile of distribution
r	randomly generated observations from distribution
d	probability assigned to a possible value, discrete case. height of the density function, continuous case.

The following functions can be used instead of tables of distributions:

Function	Result
dbinom(3,14,0.15)	$P[X = 3]$ binomial $n = 14$ and $p = 0.15$
pbinom(3,14,0.15)	$P[X \leq 3]$ binomial $n = 14$ and $p = 0.15$
ppois(3,2.1)	$P[X \leq 3]$ Poisson $\lambda = 2.1$

Normal Probability Calculations

Function	Result
pnorm(1.63)	$P[Z \leq 1.63]$ with Z standard normal
pnorm(9.23, 5.2, 1.7)	$P[X \leq 9.23]$ with X normal with mean 5.2 and standard deviation 1.7
qnorm(0.75, 5.2, 1.7)	$x_{0.75}$ where $P[X \leq x_{0.75}] = 0.75$ when X is normal with mean 5.2 and standard deviation 1.7
rnorm(5, 5.2, 1.7)	sample x_1, x_2, x_3, x_4, x_5 from a normal distribution with mean 5.2 and standard deviation 1.7
qqnorm(xmpg,main='MPG')	normal scores plot for mpg data

Sampling Distributions

Function	Result
pt (2.17, 8)	$P[t \leq 2.17]$ for students' t with 8 d.f.
qt (0.96, 8)	$t_{0.04}$ for students' t with 8 d.f.
pchisq (30.52, 27)	$P[\chi^2 \leq 30.52]$ for chi-square with 27 d.f.
qchisq (0.99, 27)	$\chi^2_{0.01}$ for chi-square with 27 d.f.
pf (3.68, 2, 19)	$P[F \leq 3.68]$ for F with (2, 19) d.f.
qf (0.90, 2, 19)	$F_{0.10}$ for F with (2, 19) d.f.

Confidence Intervals and Tests of Means

We illustrate the single sample calculations with the data in **x**. The other cases also use data in object **y**. These can be replaced by the variable names in the header, for instance, **Dat$variable** where variable is the name in the file header.

Function	
t.test(x)	A one sample two-sided test of $H_0: \mu = 0$. Also 95% confidence interval by default.
t.test(x,mu=20,alt="greater",conf.level =0.90)	One sample test of $H_0: \mu = 20$ versus $H_1: \mu > 20$. Also 90% confidence interval.
t.test(x,y,mu=15,alt="less",var.equal=T)	Two sample t test of $H_0: \mu_1 - \mu_2 = 15$ versus $H_1: \mu_1 - \mu_2 < 15$, pooled variance. Also 95% confidence interval for $\mu_1 - \mu_2$.
t.test(x,y,paired=T)	Matched paired t test of mean difference 0 versus two-sided alternative. Also 95% confidence interval for mean difference.

Inference about Proportions

Two commands produce confidence intervals for a proportion. The large sample version is **prop.test** while **binomial.test** guarantees at least the nominal coverage value.

We illustrate the chi square test for Example 8, Chapter 10. To input the table and variables names we encounter two new commands.

Function	Result
prop.test(42,100,conf.level=0.90)	90% confidence interval for proportion when 42 successes in 100 trials. The default test is a two-sided test of $H_0: p = 0.5$ and uses continuity correction.
binomial.test(42,100,conf.level=0.90)	A confidence interval with exact confidence at least 90%.
> **Xsq$expected**	gives the expected values as table.
> **Dat=as.table(rbind(c(41,27,22), c(79,53,78)))** > **dimnames(Dat) = list(Status = c("Crumbled", "Intact"),** **Material = c ("Material A","Material B", "Material C"))** > **(Xsq=chisq.test(Dat))** > **Xsq$residuals**	performs the χ^2 test. produces the Pearson residuals $(o_{ij} - e_{ij})/\sqrt{e_{ij}}$ whose square is the cell's contribution to the chi-square statistic.

Regression

We illustrate the regression commands with the data from Exercise 11.78 where y is damage, x_1 is weight(wt) and x_2 is distance.

> **Dat=read.table("11-78.txt",header=T)**

The linear models function **lm**, read 'el' 'em', is very versatile.

Function	Result
summary (lm(damage~wt,data=Dat)	Regresses y = damage on weight.
model = lm (damage ~ wt, data = Dat) **plot(fitted(model),residuals(model),** **xlab="Fitted",ylab="Residual")**	residuals versus fitted
qqnorm(residuals(model))	normal score plot of residuals
summary (lm(damage~wt+distance,data=Dat)	Multiple regression of y = damage on x_1 = weight and x_2 = distance.

One-Way Analysis of Variance (ANOVA)

We illustrate the commands using the data in Exercise 12.9, from a completely randomized design. After reading in the data having two columns headed times and arrange

> **Dat=read.table("12-9.txt", header=T)**

the command

> **anova(lm(times~arrange,data=Dat)**

produces the ANOVA table.

The analysis of variance for the 3×2 factorial design for recycled materials in Example 1 of Chapter 13 can be analyzed using the following steps.

```
Dat=read.table("C13Ex1.txt",header=T,
     colClasses=c("factor","factor","numeric","factor"))
model=lm(resilmod~A+B+A:B, data=Dat)
anova(model)
```

For the analysis of the 2^3 design on page 437, we illustrate creation of the data set and design as well.

```
> y=c(4.5,3.8,3.1,7.2,5.4,4.5,4.2,7.3,4.1,3.4,4.3,6.8,5.0,4.9,5.4,6.9)
> rate=rep(c(-1,1),8)
> additive=rep(c(-1,-1,1,1),4)
> nozzle=rep(c(rep(-1,4),rep(1,4)),2)
> cubeData=data.frame(rate,additive,nozzle,y)
> print(cubeData)
> cubeModel=lm(y~rate*additive*nozzle)
> summary(cubeModel)
> anova(cubeModel)
```

There are many other functions that apply to material covered in this book but we refer the interested reader to the Web sites mentioned at the start of this appendix. You may also type **help(stem)**, for stem-and-leaf diagram, where *stem* can be replaced with many other functions.

ANSWERS TO ODD-NUMBERED EXERCISES

CHAPTER 1

1.1 Statistical *population* could be air quality values for all U.S.-based flights during period of study. The *sample* is the measurements from the 158 flights.

1.3 (a) Laptop; (c) Collection of all laptop weights.

1.5 unit: hard drive, variable: distance, population: distances for drives made in hour, sample; distances for the 40 drives.

1.7 (a) $\bar{x} = 214.67$; (b) Below LCL.

CHAPTER 2

2.7 (b) 7.3 detached.

2.9 (a) yes; (b) no; (c) yes; (d) no; (e) no.

2.11 The cumulative "less than or equal to" frequencies are 0, 5, 16, 25, 43, 49 and 50.

2.13 The cumulative "less than" frequencies are 0, 10, 20, 29, 40, 52, 62, 68, 72, 76, 77, 79, 80.

2.15 The cumulative "less than" frequencies are 1, 12, 28, 43, 47, and 50.

2.19 No, because we tend to compare areas; the large sack should be modified so that its area is about double that of the small sack.

2.21 The large rectangle over (245, 325], the longest interval, is misleading. It makes it look like a larger proportion of the data are in that interval.

2.23

2	67 88 95
3	17 55 70 83 91
4	05 19 34 62
5	08 40
6	12

2.25

2	1 2
2	6 8
3	2 3 4 4
3	5 5 5 5 6 6 7 8 9
4	0 0 1 1 2 3 3 4
4	5 5 5 5 6 7 7 8 8 9
5	0 0 0 1 1 1 2 2 2 3 3 3 4 4
5	5 5 5 6 6 6 7 7 8 9 9
6	0 0 0 1 1 1 2 2 2 3 4
6	5 5 5 7 7 8 8 8 9
7	0 0 2 3 3 4 4 4
7	5 6 6 7 8 9
8	0 2 2 4
8	5 8

2.27 (b) Outlier high.   **2.29** Mean.

2.31 (a) $\bar{x} = -3$; (b) $s = 2.94$; (c) hole too small on average.

2.33 (a) $\bar{x} = 30.14$; (b) No, trend exists.

2.35 No, the total earnings are only $525,000.

2.37 (a) $\bar{x} = 0.3990$; (b) $s = 0.1817$.

2.39 (a) 3.35; (b) 3.25

2.41 $\bar{x} = 30.91$, $Q_1 = 30.4$, $Q_2 = 30.85$, $Q_3 = 31.2$.

2.43 min $= 1.11$, $Q_1 = 1.31$, Median $= 1.36$ $Q_3 - 1.44$, max $= 1.68$.

2.45 (a) $\bar{x} = 0.1673$; (b) $s = 0.0832$; (c) No major difference.

2.47 $\bar{x} = 87.40$, $s = 161.47$.

2.49 $v = 8.35\%$.   **2.53** (a) 90; (b) 5.00.

2.55 (a) $Q_1 = 279.55$, $Q_3 = 327.22$, and the interquartile range is 47.67; (b) $Q_1 = 3.00$, $Q_3 = 6.80$.

2.57 (a) 73.0; (b) 42.08%.

2.59 (a) $Q_1 = 1712$, $Q_2 = 1863$, $Q_3 = 2061$; (c) $Q_1 = 69.5$, $Q_2 = 70.55$, $Q_3 = 71.80$.

2.61 (a) The class frequencies are 1, 8, 19, 17, 9, 3, and 1.

2.63 (a) $\bar{x} = 5.4835$ and $s = 0.1904$; (b) median is 5.46, $Q_1 = 5.34$ and $Q_3 = 5.63$; (c) there is no apparent trend.

2.65 (a) median = 0.40, maximum = 0.57, minimum = 0.32 and the range = 0.25; (b) median = 0.51, maximum = 0.63, minimum = 0.47 and the range = 0.16.

2.67 (a) $Q_1 = 18.0$, $Q_2 = 27.0$, $Q_3 = 30.0$; (b) minimum = 12, maximum = 48, range = 36, and the interquartile range = 12.0.

2.69 (a) $Q_1 = 19$, $Q_2 = 28$, $Q_3 = 55$; (b) minimum = 12, maximum = 63, range = 51, and the interquartile range = 36.

2.73 (a) $\bar{x} = 0.2441$ (b) $s = 0.0036$; (c) $v = 1.5\%$; (d) the larger drive has $v = 17.9\%$ so is relatively more variable.

2.75 (a) $Q_1 = 497.2$, $Q_2 = 602.0$, $Q_3 = 743.3$; (b) 895.8.

2.77 No.   **2.79** (b) Median = 1 greater than $\bar{x} = 10/11$.

CHAPTER 3

3.1 (b) $R = \{(0, 0), (1, 1), (2, 2)\}$, $T = \{(0, 0), (1, 0), (2, 0), (3, 0)\}$, $U = \{(0, 1), (0, 2), (1, 2)\}$.

3.3 (a) $R \cup U = \{(0, 0), (1, 1), (2, 2), (0, 1), (0, 2), (1, 2)\}$ is the event that at least as many streams are contaminated as lakes.
(b) $R \cap T = \{(0, 0)\}$ is the event that none of the lakes or steams is contaminated;
(c) $\overline{T} = \{(0, 1), (1, 1), (2, 1), (3, 1), (0, 2), (1, 2), (2, 2), (3, 2)\}$ is the event that at least one of the streams is contaminated.

3.5 (a) $A \cup B = \{2, 3, 4\}$ is the event that work on the car is easy, average, or difficult; (b) $A \cap B = \{3\}$ is the event that work on the car is average; (c) $A \cup \overline{B} = \{1, 3, 4, 5\}$ is the event that work on the car is not easy; (d) $\overline{C} = \{1, 2, 3\}$ is the event that work on the car is very easy, easy, or average.

3.7 (b) B is the event that 3 graduate assistants are present, C is the event that as many professors as graduate assistants are present, D is the event that altogether 3 professors or graduate assistants are present; (c) $C \cup D = \{(1, 1), (1, 2), (2, 1), (2, 2)\}$, is the event that at most 2 graduate assistants are present;
(d) B and D are mutually exclusive.

3.9 Region 1 represents the event that the ore contains copper and uranium; region 2 represents the event that the ore contains copper but not uranium; region 3 represents the event that the ore contains uranium but not copper; region 4 represents the event that the ore contains neither uranium nor copper.

3.11 (a) Region 5 represents the event that the windings are improper, but the shaft size is not too large and the electrical connections are satisfactory; (b) regions 4 and 6 together represent the event that the electrical connections are unsatisfactory, but the windings are proper; (c) regions 7 and 8 together represent the event that the windings are proper and the electrical connections are satisfactory; (d) regions 1, 2, 3, and 5 together represent the event that the windings are improper.

3.17 (a) 7; (b) 12

3.19 (a) 90; (b) 75.

3.21 720.

3.23 105.

3.25 (a) 55; (b) 165.

3.27 280.

3.29 (a) 1/6; (b) 1/18; (c) 2/9; (d) 1/18; (e) 1/18; (f) 1/9.

3.31 0.359.

3.33 45.

3.35 (a) Yes; (b) no, sum exceeds 1; (c) no, $P(C)$ is negative; (d) no, sum is less than 1; (e) yes.

3.37 (b) 27/112, 45/112, and 5/14; (c) 25/56, 5/16, and 27/112.

3.41 (a) 0.55; (b) 0.75; (c) 0.45; (d) 0.25.

3.43 (a) 0.43; (b) 0.67; (c) 0.24; (d) 0.59.

3.45 (a) 15/32; (b) 13/32; (c) 5/32; (d) 23/32; (e) 8/32; (f) 9/32.

3.47 (a) 0.29; (b) 0.18.

3.51 (a) 4 to 3; (b) 19 to 1 against it; (c) 4 to 1.

3.53 (a) 0.60; (b) $0.75 \leq p < 0.80$.

3.55 $P(I \mid D) = 2/3$; $P(I \mid \overline{D}) = 4/97$.

3.57 (a) 62/85; (b) 74/84; (c) 29/51.

3.59 (a) 0.2133 different; (b) 0.36 different; (c) 0.2258 different.

3.65 (a) 33/59; (b) 7/118; (c) 45/118.

3.67 Yes.

3.69 (a) 1/256; (b) 1/648; (c) 1/243.

3.71 (a) 0.60; (b) 0.20; (c) 0.70.

3.73 (b) 0.498.

3.75 (a) 0.845; (b) 0.379.

3.77 (a) 0.686; (b) 0.171; (c) 0.0286.

3.79 (a) 0.28; (b) 0.04.

3.81 (a) $0.09 = 27/300$; (b) $0.074 = 28/380$.

3.83 (a) $\overline{A} = \{(0,0), (0,1), (0,2), (0,3), (1,0), (1,1), (1,2), (2,0), (2,1), (3,0)\}$ is the event that the salesman will not visit all 4 customers;
(b) $A \cup B = \{(4,0), (3,1), (2,2), (1,3), (0,4), (1,0), (2,0), (2,1), (3,0)\}$ is the event that the salesman will visit all 4 customers or more on the first day than on the second day;
(c) $A \cap C = \{(1,3), (0,4)\}$ is the event that he will visit all 4 customers but at most one on the first day;
(d) $\overline{A} \cap B = \{(1,0), (2,0), (2,1), (3,0)\}$ is the event that he will visit at most 3 of the customers and more on the first day than on the second day.

3.87 21.

3.89 (a) 0.50; (b) 0.20; (c) 0.10; (d) 0.80; (e) No.

3.91 There is a contradiction in his claim since total > 200.

3.95 (a) 20/58; (b) 8/58; (c) 12/58; (d) 27/58.

3.97 Purposeful action is most likely.

3.99 (a) 0.74; (b) 0.6923.

CHAPTER 4

4.1 1/12, 2/12, 3/12, 3/12, 2/12, and 1/12.

4.3 (a) Yes; (b) no, sum less than 1; (c) no, $f(4)$ is negative.

4.5 $k = 16/31$.

4.9 (a) Success: home has TV tuned to speech. Likely to hold.

4.11 (a) Trials not independent.

4.13 (a) 0.3915; (b) 0.1657; (c) 0.0152; (d) 0.0136; (e) 0.05774; (f) 0.7004.

4.15 27/128.

4.17 (a) 0.2969; (b) 0.0152; (c) 0.2061.

4.19 (a) 0.1501; (b) 0.7338; (c) 0.0982.

4.21 (a) 0.9571; (b) 0.7892; (c) 0.5614; (d) 0.3518.

4.23 (a) 0.5, 0.5; (b) 0, 0.5, 1.0.

4.25 (a) 0.4560; (b) 0.4291; (c) 0.1149.

4.27 (a) 0.15; (b) 0.0625.

4.29 (a) 0.4014; (b) 0.3917.

4.33 $\sigma^2 = 1.0$.

4.35 $\sigma^2 = 1.8$.

4.37 (a) $\mu = 2.8$ and $\sigma^2 = 0.84$.

4.39 (a) $\sigma^2 = 1.25$.

4.41 (a) $\mu = 338$ and $\sigma = 13$; (b) $\mu = 120$ and $\sigma = 10$; (c) $\mu = 24$ and $\sigma = 4.8$; (d) $\mu = 520$ and $\sigma = 13.49$.

4.45 The probability is greater than or equal to 8/9.

4.51 0.0498, 0.1494, 0.2240, 0.2240, 0.1680, 0.1008, 0.0504, 0.0216, 0.0081, 0.0027.

4.53 (a) 0.242; (b) 0.087; (c) 0.937.

4.55 (a) 0.182; (b) 0.857; (c) 0.648.

4.57 0.007.

4.59 (a) 0.981; (b) 0.577; (c) 0.978.

4.61 0.3010.

4.63 0.0478.

4.65 (a) 0.0384; (c) 0.0015.

4.67 (a) 0.217; (b) 0.108.

4.73 0.117.

4.75 (b) 0.195.

4.77 00–13, 14–41, 42–68, 69–86, 87–95, 96–99.

4.79 (a) 0.04; (b) 0.

4.81 (a) 2.12; (b) 0.9856; (c) 0.9928.

4.83 (a) Yes; (b) Yes; (c) No, sum exceeds 1.

4.85 (a) 0.1468; (b) 0.1468.

4.87 (a) 0.55; (b) 0.40; (c) 0.05.

4.89 (a) 0.72; (b) 0.72.

4.91 0.2707.

4.93 Probability at least 0.96.   **4.95** 0.5488.

4.97 (a) 0000–2465, 2466–5917, 5918–8334, 8335–9462, 9463–9857, 9858–9968, 9969–9994, 9995–9999.

CHAPTER 5

5.3 (a) 0.488; (b) 0.056.   **5.5** (a) 0.02; (b) 0.84.

5.7 (a) 0.5556; (b) 0.09.   **5.9** (a) 0.707; (b) 0.1339.

5.11 0.0916.

5.13 $\mu = 0.75$ and $\sigma^2 = 0.0375$.

5.15 $\mu = 4$ and σ^2 does not exist.

5.17 4.5 years.

5.19 (a) 0.9599; (b) 0.1056; (c) 0.0197; (d) 0.9656.

5.21 (a) 0.0653; (b) 6.5%; (c) 182.1.

5.25 $\sigma = 10.93 = 9.2/0.842$.

5.27 (a) 151.45; (b) 9.12.   **5.29** (a) 0.5391; (b) 0.0098.

5.31 83.15%.   **5.33** $\mu = 2.984$.

5.35 (a) 0.0199; (b) 0.0030.   **5.37** 0.1841.

5.39 (a) 4.5984; (b) 0.9911; (c) 0.0068.

5.45 $F(x) = 0$ for $x \le 0$, $F(x) = x$ for $0 < x < 1$, and $F(x) = 1$ for $x \ge 1$.

5.47 50%.   **5.49** 0.2646.

5.51 (a) 0.0807; (b) 0.0960.

5.53 0.496.

5.55 (a) 0.049; (b) 0.843.

5.57 No relative maximum when $0 < \alpha < 1$; maximum at $x = 0$ when $\alpha = 1$.

5.59 (a) 18.1%; (b) 36.8%.

5.61 $e^{-\alpha t}$.

5.65 (a) $\mu = 0.2$; (b) 0.3164.

5.67 0.6321.

5.69 0.2057.

5.71 (a) $\dfrac{\binom{2}{x_1}\binom{1}{x_2}\binom{2}{2 - x_1 - x_2}}{\binom{5}{2}}$ for $x_1 = 0, 1, 2$, $x_2 = 0, 1$,

and $0 \le x_1 + x_2 \le 2$; (b) 0.7; (c) $f_1(0) = 0.3$, $f_1(1) = 0.6$, $f_1(2) = 0.1$; (d) $f_1(0|0) = 1/6$, $f_1(1|0) = 4/6$, $f_1(2|0) = 1/6$.

5.73 (a) 1/4; (b) 1/24.

5.75 $F(x_1, x_2) = 0$ for $x_1 \le 0$ or $x_2 \le 0$,

$= \frac{1}{4} x_1^2 x_2^2$ for $0 < x_1 < 2$ and $0 < x_2 < 1$,

$= x_2^2$ for $0 < x_2 < 1$ and $x_1 \ge 2$,

$= \frac{1}{4} x_1^2$ for $0 < x_1 < 2$ and $x_2 > 1$, and $F(x_1, x_2) = 1$ for $x_1 \ge 2$ and $x_2 \ge 1$;

$F_1(x_1) = 0$ for $x_1 \le 0$, $F_1(x_1) = \frac{1}{4} x_1^2$ for $0 < x_1 < 2$, and $F_1(x_1) = 1$ for $x_1 \ge 2$;

$F_2(x_2) = 0$ for $x_2 \le 0$, $F_2(x_2) = x_2^2$ for $0 < x_2 < 1$, and $F_2(x_2) = 1$ for $x_2 \ge 1$; they are independent.

5.77 $F(x, y) = 0$ for $x \le 0$ or $y \le 0$,

$= \frac{3}{5} x^2 y + \frac{2}{5} x y^3$ for $0 < x < 1$ and $0 < y < 1$,

$= \frac{3}{5} x^2 + \frac{2}{5} x$ for $0 < x < 1$, and $y \ge 1$,

$= \frac{3}{5} y + \frac{2}{5} y^3$ for $x \ge 1$, and $0 < y < 1$,

and $F(x, y) = 1$ for $x \ge 1$ and $y \ge 1$.

5.79 (a) $f_1(x \,|\, y) = (x + y^2)/(\frac{1}{2} + y^2)$ for $0 < x < 1$ and $f_1(x \,|\, y) = 0$ elsewhere; (b) $f_1(x \,|\, \frac{1}{2}) = \frac{1}{3}(4x + 1)$ for $0 < x < 1$ and $f_1(x \,|\, y) = 0$ elsewhere; (c) 11/18.

5.81 (a) $1/3$; (b) $5/(6e) = 0.3066$.

5.83 (a) 0.3264; (b) 0.4712.

5.85 2.

5.87 $\mu = LW$ and $\sigma^2 = \frac{1}{12}\left(a^2 W^2 + b^2 L^2 + \frac{1}{12} a^2 b^2\right)$.

5.89 (a) 0; (b) 10.   **5.91** (a) -6; (b) 23.

5.93 (a) 200; (b) 60.

5.95 (a) $0.25\left(1 + 2e^t + e^{2t}\right)$; (b) $E(X) = 1$ and $E(X^2) = 1.5$.

5.97 (a) $(1 - t/2)^{-1}$; (b) $E(X) = 0.5$ and $E(X^2) = 0.5$.

5.99 (b) mean $= 6$ and variance $= 97$.

5.109 (a) 0.1465; (b) 0.3125; (c) $\mu = 3/8$ and $\sigma^2 = 0.0549$.

5.111 (a) 1; (b) 0.25.

5.113 0.000774.

5.115 (a) 1.28; (b) 3.09.

5.117 (a) 0.9997; (b) .0011.

5.121 $n = 25$

5.123 (a) $f(x) = 0.2\, e^{-0.2x}$ for $x > 0$ and mean $= 5$. $f(y) = 0.2\, e^{-0.2y}$ for $y > 0$ and mean $= 5$; (b) 10.

5.125 (a) -8; (b) 127.

5.127 (a) 0.0644; (b) 0.8613.

CHAPTER 6

6.3 (b) Grads with low incomes less likely to respond.

6.5 (a) 36; (b) 253.

6.7 (a) $\mu = 0$ and $\sigma^2 = 26/3$.

6.9 The samples are 1 and 1, 1, and 2, 1 and 3, 1 and 4, 2 and 1, 2 and 2 and 3, 2 and 4, 3 and 1, 3 and 2, 3 and 3, 3 and 4, 4 and 1, 4 and 2, 4 and 3, 4 and 4; the probabilities that $\bar{x}$ equals 1, 1.5, 2, 2.5, 3, 3.5, or 4 are 1/16, 1/8, 3/16, 1/4, 3/16, 1/8, and 1/16.

6.11 (a) It is divided by 2; (b) it is divided by 3/2; (c) it is multiplied by 3; (d) it is multiplied by 4.

6.15 approximately 0.7236.   **6.17** approximately 0.111.

6.21 $t = 1.15$; since $t_{0.10} = 1.476$ for 5 degrees of freedom, the data fail to reject the claim.

6.23 0.050.   **6.25** 0.02.   **6.27** 0.5249.   **6.29** 0.3125

6.31 t with 4 degrees of freedom.

6.35 $\prod_{i=1}^{5}(1 - 2t)^{-2i} = (1 - 2t)^{-30}$.

6.37 (b) mean $= -8$ and variance $= 89$.

6.39 (b) mean $= -27$ and variance $= 57$.

6.41 (b) Negative binomial $r = \sum_{i=1}^{n} r_i$ and success probability p.

6.43 $f(y) = (9\pi/2)^{-1/2} y^{-2/3} e^{-y^{2/3}/2}$ for $-\infty < y < \infty$.

6.45 $f(y) = e^y e^{-e^y}$ for $-\infty < y < \infty$.

6.47 $f(y) = e^{-y}$ for $y > 0$.

6.49 $f(y) = \dfrac{\Gamma(2\alpha)}{\Gamma(\alpha)\Gamma(\alpha)} \dfrac{y^{\alpha-1}}{(1+y)^{2\alpha}}$ for $0 < y < 1$.

6.51 (b) No, students from states with many participants, usually the larger states, have less chance.

6.53 (a) 1/66; (b) 1/190.

6.55 (a) not larger than 0.16; (b) 0.0062.

6.57 0.9876.   **6.59** 0.01.

6.61 The ratios of standard errors are (a) 0.707; (b) 0.816; (c) 2.0.

6.63 (b) Buses and trucks take longer to pass a fixed point.

CHAPTER 7

7.1 $E = 2.045(819.35)/\sqrt{30} = 305.9$.

7.3 $E - 1.96(1.250)/\sqrt{52} = 0.3398$.

7.5 $E = 2.326(3.057)/\sqrt{45} = 1.06$.

7.7 $E = 1.96(14,056)/\sqrt{50} = 3,896.1$.

7.9 84.7%.   **7.11** $n = 208$.

7.13 $0.816 < \mu < 1.852$.

7.15 $107.59 < \mu < 120.41$.

7.17 $1,791.7 < \mu < 2,025.8$.

7.19 $30.35 < \mu < 31.47$.

7.21 (a) $3.28 < \mu < 3.72$; (b) cannot tell μ unknown; (c) about 90%.

7.23 (a) $159.2 < \mu < 177.2$; (c) normal.

7.25 (a) $E = 22.14$; (b) $E = 4.46$.

7.31 (a) 0.8; (b) 0.64.

7.33 (a) $\widehat{\lambda} = 1.5$; (b) 0.0498.

7.35 (a) $\widehat{\mu} = 114$ and $\widehat{\sigma} = 7.860$; (b) 0.0689.

7.37 (a) $\widehat{\beta} = \overline{X}$; (b) $e^{-1/\overline{x}}$

7.39 (a) $H_0: \mu = 2$ and $H_1: \mu < 2$; (b) Type II; (c) Type I.

7.41 (a) $H_0: \mu = 56$ and $H_1: \mu \neq 56$; (b) Type II; (c) Type I.

7.43 (a) bridge unsafe; (b) 0.01 but prefer even smaller.

7.45 Type I; Type II.

7.47 (a) 0.1056; (b) 0.1056.

7.49 Reject when $\overline{x} < 95.58$.

7.51 (a) $\mu \neq 1,250$; (b) $\mu < 1,250$; (c) $\mu > 1,250$.

7.53 (a) $Z = -1.35$; cannot reject H_0; (b) Type II.

7.55 (a) $Z = -2.49$; reject H_0; (b) Type I.

7.57 (a) $T = 2.52$; reject H_0; (b) Type I.

7.59 $Z = 1.76$; reject H_0.

7.61 $T = 3.652$; reject H_0.

7.63 $T = 5.66$; reject H_0.

7.65 (a) $Z = 2.02$; reject H_0; (b) $T = 3.82$; reject H_0.

7.67 (a) Reject H_0; (b) Fail to reject H_0; (c) Fail to reject.

7.69 (a) Fail to reject H_0; (b) reject H_0; (c) Fail to reject.

7.71 (a) $\gamma(77) = 0.523$; (b) 0.491.

7.77 $70.23 < \mu < 71.16$.

7.79 $24.92 < \mu < 27.88$.

7.81 $0.992 < \mu < 1.048$.   **7.83** $n = 11$.

7.85 (a) $c = 1651, 1$; (b) For 1620, …, .87, .66, .37, .14, .036, .006, .0005, .00003.

7.87 (a) $1.530 < \mu < 1.770$

CHAPTER 8

8.1 $z = -2.15$; reject H_0.

8.3 $z = 4.69$; reject H_0.

8.5 (a) $z = -2.28$; reject H_0. (b) 0.05.

8.7 (b) 7.333.

8.9 $t = 0.96$; cannot reject H_0

8.11 $t = 1.03$; cannot reject H_0.

8.13 (a) $t' = -1.30$ with 13 degrees of freedom; cannot reject H_0; (b) $t' = -0.145$ with 8 degrees of freedom; cannot reject H_0.

8.15 $t = 1.58$ with 4 degrees of freedom; cannot reject H_0.

8.17 (a) $0.048 < \mu_D < 1.848$; (b) $t = 2.35$; reject H_0.

8.19 $0.12 < \mu_D < 1.28$.

8.21 $t = 3.94$ with 15 degrees of freedom; reject H_0.

8.23 (a) Select 3 elevators by random drawing; (b) flip coin for each of the 6 elevators. If heads, elevator gets the modified circuit board first. After some time, it is replaced by the original board. If tails, elevator gets modified circuit board second.

8.25 Randomly select 25 cars, and install the modified air-pollution device. The other 25 cars use the current device.

8.27 $-0.183 < \mu_1 - \mu_2 < -0.037$.

8.29 $t = 2.082$ with 8 degrees of freedom; fail to reject H_0.

8.31 n should be 24.

8.33 (a) Randomly select 10 cars to use the modified spark plugs. The other 10 cars use the regular spark plugs; (b) select 7 specimens by random drawing to try in the old oven.

8.37 $2.8 < \mu_D < 7.6$

CHAPTER 9

9.1 (a) $s = 8.34$; (b) 8.75.

9.3 (a) 1.787; (b) 2.144.

9.5 $0.0067 < \sigma^2 < 0.4831$.

9.7 $\chi^2 = 5.832$; cannot reject H_0.

9.9 $\chi^2 = 122.5$; reject H_0.

9.11 (a) $\chi^2 = 10.89$; cannot reject H_0. (b) distribution invalid

9.13 $F = 1.496$; cannot reject H_0.

9.15 $F = 2.42$; cannot reject H_0.

9.17 $0.22 < \sigma < 0.53$.

9.19 $\chi^2 = 72.22$; reject H_0.

9.21 $F = 1.81$; cannot reject H_0.

CHAPTER 10

10.1 $0.352 < p < 0.488$.

10.3 $0.514 < p < 0.642$.

10.5 $0.090 < p < 0.244$.

10.7 90.9%.   **10.9** $n = 267$.

10.11 $n = 1,354$.

10.13 $0.563 < p < 0.943$.

10.17 $0.12 < p < 0.14$.

10.19 $z = 2.19$; reject H_0.

10.21 $z = -1.83$; cannot reject H_0.

10.23 $z = 1.489$; cannot reject H_0.

10.25 $z = -1.886$; reject H_0.

10.27 $\chi^2 = 2.37$; reject H_0.

10.29 $\chi^2 = 9.39$; cannot reject H_0.

10.33 $-0.005 < p_1 - p_2 < 0.145$.

10.35 $0.170 < p_1 - p_2 < 0.374$.

10.39 $\chi^2 = 15.168$; reject H_0.

10.41 $\chi^2 = 54.328$; reject H_0.

10.43 $\chi^2 = 0.657$; cannot reject H_0.

10.45 $\chi^2 = 7.91$; cannot reject H_0.

10.49 $z = -0.50$; cannot reject H_0.

10.51 $z = -0.99$; cannot reject H_0.

10.53 $z = 2.00$; cannot reject H_0 at $\alpha = 0.01$.

10.55 $z = 3.71$; reject H_0 in favor of $p_1 > p_2$.

10.57 $z = -1.746$; reject H_0.

10.59 (a) $\chi^2 = 8.190$; reject H_0;
(b) $0.256 < p_1 < 0.611$; $0.108 < p_2 < 0.425$; $0.461 < p_2 < 0.806$.

10.61 $\chi^2 = 10.481$; cannot reject H_0.

10.63 $\chi^2 = 47.862$; reject H_0.

CHAPTER 11

11.1 (b) Extrapolation beyond x values used.

11.3 (b) $\widehat{y} = 39.05 + 0.764x$; $\widehat{y} = 65.8$.

11.5 (a) $11.86 < \beta < 17.11$; (b) 40.0 to 63.71.

11.7 $t = -1.212$; cannot reject H_0.

11.9 (a) $\widehat{y} = 3.452 + 0.4868x$; (b) $\widehat{y} = 3.695$.

11.11 $t = 3.30$; reject H_0.

11.13 7.23 to 12.39.

11.15 (a) $\Sigma\, xy / \Sigma\, x^2$;   (b) 14.75.

11.17 $4.6 < \alpha < 52.2$.

11.19 (a) $\widehat{y} = 3.214 - .446x$; 2.10. (b) $\widehat{y} = 2.95 - .2369x$; 2.00.

11.23 (a) $\widehat{y} = 87.9 + 2.46x$; (b) $t = 9.58$; reject H_0: $\beta = 0$;
(c) (325.65, 342.98).

11.25 (b) $\log_{10}\widehat{y} = 4.842 + 0.0604x$ or $\widehat{y} = 69{,}502.4(1.149)^x$;
(c) 1,122,018.

11.27 44.86.

11.29 $\widehat{y} = \exp[\exp(0.000041x + 1.21)]$.

11.31 $\widehat{\alpha} = 0.240$.

11.33 (a) $t = -2.28$; cannot reject $\beta_1 = 0$; (b) $F = 38.59$; reject $\beta_2 = 0$.

11.37 $\widehat{y} = 64.5$.

11.39 $\widehat{y} = 2.266 + 0.225x_1 + 0.0623x_2$; $\widehat{y} = 8.37$.

11.43 Normal scores plot nearly straight.

11.45 A serious violation, time trend.

11.47 No; population size.

11.49 (b) $\mathcal{Z} = 4.89$; reject H_0: $\rho = 0$.

11.51 $\mathcal{Z} = 5.15$; reject H_0: $\rho = 0$.

11.53 $0.717 < \rho < 0.976$.   **11.55** $r = 0.810$.

11.57 $\mathcal{Z} = -1.120$; cannot reject H_0: $\rho = -0.4$.

11.61 (a) 2,812.4; (b) 233.9; (c) 0.958.

11.63 $r = 0.738$.

11.65 (a) $\widehat{y} = -10 + 20x$; (b) 80; (c) model may not hold outside experimental range.

11.67 $t = 3.35$, reject H_0: $\beta = 5$.

11.69 (a) $\widehat{y} = 20.4 - 1.80x$; (b) $t = -7.79$; reject H_0: $\beta = 0$;
(c) $(-4.12, 12.52)$ (d) outside range.

11.71 $r^2 = 0.953$.

11.73 0.619 ± 0.427 or $0.192 < \alpha < 1.046$.

11.75 $\widehat{\gamma} = 1.499$.   **11.77** $\widehat{\tau} = 0.9284$.

11.79 (a) 3.67 to 3.72; (b) 3.63 to 3.76.

11.81 The first linear relationship is roughly twice as strong.

11.83 (a) $0.446 < \rho < 0.923$; (b) $-0.797 < \rho < -0.346$;
(c) $-0.173 < \rho < 0.476$.

11.85 (a) $\widehat{y} = -0.075 + 0.480x$; (b) $0.44 < \beta < 0.52$;
(c) $t = -1.08$, cannot reject H_0; (d) The variance appears to increase somewhat with x.

CHAPTER 12

12.1 (b) Use equal sample sizes.

12.3 $SS(Tr) = 416$.

12.5 $F = 5.75$, significant at the 0.01 level.

12.7 (a) $SS(Tr) = 204$, with 3 degrees of freedom; $SSE = 34$, with 11 degrees of freedom; $SST = 238$, with 14 degrees of freedom. (b) $F = 22.0$, significant at the 0.05 level.

12.9 $F = 11.3$, significant at the 0.01 level.

12.11 $F = 15.7$, significant at the 0.05 level.

12.17 (a) For brands, $F = 1.047$. (b) $t = -1.023$.

12.19 (a) $b = 4$, $SS(Tr) = 56$. (b) For treatments, $F = 5.25$, significant at the 0.05 level.

12.21 For technicians, $F = 5.91$, not significant at the 0.01 level; for days, $F = 1.11$, not significant at the 0.01 level.

12.23 (b) $SS(Tr) = 70.173$ with 4 degrees of freedom; $SS(Bl) = 0.330$ with 3 degrees of freedom; $SSE = 23.315$ with 12 degrees of freedom; $SST = 95.818$, with 19 degrees of freedom. For treatments, $F = 8.314$, significant at the 0.01 level. Blocks are not significant.

12.25 $F = 1.95$, not significant at the 0.05 level.

12.27 For machines, $F = 0.05$, not significant at the 0.05 level; for workers, $F = 1.346$, not significant at the 0.05 level.

12.31 $Tr_1 - Tr_2$: $(-9.40, -0.60)$ $Tr_1 - Tr_3$: $(-6.40, 2.40)$
$Tr_1 - Tr_4$: $(-5.40, 3.40)$ $Tr_2 - Tr_3$: $(-1.40, 7.40)$
$Tr_2 - Tr_4$: $(-.40, 8.40)$ $Tr_3 - Tr_4$: $(-3.40, 5.40)$

12.33 $\mu_1 - \mu_2: -14 \pm 8.7$; $\mu_1 - \mu_2: -4 \pm 8.7$; $\mu_2 - \mu_3: 10 \pm 8.7$.

12.35 (a) $T_1 - T_2$: $(-7.82, -2.13)$ $T_1 - T_3$: $(-5.70, -0.002)$
$T_1 - T_4$: $(-5.87, -0.18)$ $T_1 - T_5$: $(-3.07, 2.63)$
$T_2 - T_3$: $(-0.72, 4.97)$ $T_2 - T_4$: $(-.90, 4.80)$
$T_2 - T_5$: $(1.90, 7.60)$ $T_3 - T_4$: $(-3.02, 2.67)$
$T_3 - T_5$: $(-0.22, 5.47)$ $T_4 - T_5$: $(-0.05, 5.65)$

12.37 For treatments, $F = 19.21$ with 2 and 5 degrees of freedom. Reject the null hypothesis of equal treatment means at $\alpha = 0.05$ For the covariate, $F = 22.28$ with 1 and 5 degrees of freedom. Reject the null hypothesis $\beta = 0$ at $\alpha = 0.05$

12.39 For track designs, $F = 6.44$, significant at the 0.01 level. The estimated effect of usage on breakage resistance is 0.43.

12.43 $F = 2.80$, not significant at the 0.05 level.

12.45 (b) $SS(Tr) = 56$, with 2 degrees of freedom; $SS(Bl) = 138$ with 3 degrees of freedom; $SSE = 32$, with 6 degrees of freedom; $SST = 226$, with 11 degrees of freedom. (c) For treatments, $F = 5.25$, significant at the 0.05 level; for blocks, $F = 8.63$, significant at the 0.05 level.

12.47 (a) For agencies, $F = 4.84$ significant at the 0.05 level. (b) For sites, $F = 101.75$ significant at the 0.05 level.

12.49 For treatments, $F = 12.09$ with 3 and 23 degrees of freedom. Reject the null hypothesis of equal treatment means at $\alpha = 0.05$. For the covariate, $F = 35.73$ with 1 and 23 degrees of freedom. Reject the null hypothesis $\beta = 0$ at $\alpha = 0.05$

12.51 (a) For surface treatments, $F = 10.65$, significant at the 0.05 level, (b) Both show treatments are significant. But, the coefficient of traffic volume is significant and the P-value is about half the value for the analysis of variance.

CHAPTER 13

13.1 Interaction ($F = 24.3$) is significant at $\alpha = 0.050$ but $A: (F = 0.58)$ and $B: (F = 0.11)$ are not.

13.3 Concentration ($F = 44.04$), temperature ($F = 10.56$) and their interaction ($F = 3.50$) all significant. $(37.32, 46.02)$ at 10 grams and 75 degrees.

13.5 Detergents ($F = 0.05$) and interaction ($F = 0.86$) are not significant at the 0.05 level. Engines ($F = 7.33$) is significant.

13.7 Defoliation ($F = 32.44$) and Treatment ($F = 6.14$) are significant at $\alpha = 0.05$. Surface ($F = 1.58$) and the two and three

factor interactions ($F = 1.72$, $F = 2.94$, $F = 0.33$, $F = 0.18$) are not.

13.11 (b) Factor A: 3.4 ± 0.80; Factor B: 4.3 ± 0.80; AB interaction: 1.6 ± 0.80.

13.13 (b) Factor A: -2.635 ± 0.677; Factor B: -0.665 ± 0.677; AB interaction: 0.175 ± 0.677.

13.15 (b) A: -0.803 ± 0.211; B: -0.360 ± 0.211; C: 0.178 ± 0.211; AB: 0.010 ± 0.211; AC: -0.058 ± 0.211; BC: -0.115 ± 0.211; ABC: 0.030 ± 0.211.

13.17 (b) A: -2.024 ± 0.131; B: 0.114 ± 0.131; C: 0.052 ± 0.131; AB: -0.133 ± 0.131; AC: -0.057 ± 0.131; BC: 0.051 ± 0.131; ABC: 0.034 ± 0.131.

13.19 Minimum at $(x_1, x_2) = (31.5, 113.9)$

13.21 $\widehat{y} = -40.8750 + 1.5036x_1 + 0.5604x_2 - 0.0037x_1^2 - 0.0006x_2^2 - 0.0070x_1x_2$ The constant, x_2 and x_2^2 terms are not significant.

13.23 Fuel ($F = 13.34$) and interaction ($F = 27.94$) are significant at the 0.01 level. Launcher ($F = 0.39$) is not siginificant for any reasonable level. Because the interaction is significant, summarize by a two-way table of means.

13.25 Nickel ($F = 6.05$), carbon ($F = 15.37$), and their interaction ($F = 22.44$) are significant at $\alpha = 0.01$ and so is manganese ($F = 34, 22$). The other interactions are not significant even at level 0.05. Summarize with a two-way nickel-carbon table of means and the two means for manganese.

13.27 (b) Only the main effect of A is non-zero. Factor A: 5 ± 4.40, Factor B: 1 ± 4.40 AB interaction: -1 ± 4.40

13.29 Pressure: 9.75 ± 7.38 Temperature: -1.25 ± 7.38 Interaction: 0.75 ± 7.38

13.31 Factor A: -2.325 ± 1.107 Factor B: -2.225 ± 1.107 Factor C: -1.475 ± 1.107 AB interaction: $3.275 + 1.107$ AC interaction: $-.575 \pm 1.107$ BC interaction: $-.275 \pm 1.107$ ABC interaction: $.425 \pm 1.107$

13.33

$$\text{Total } SS = SS_A + SS_B + SS_C + SS_{AB} + SS_{AC} + SS_{BC} + SS_{ABC} + SSE$$
$$27.32 = 4.84 + 5.76 + 2.56 + 11.56 + .16 + .16 + .36 + 1.92$$

13.35 $\widehat{y} = 99.95 + 29.34x_1 + 27.10x_2 + 25.571x_1^2 + 34.48x_1x_2 - 3.429x_2^2$ but x_2^2 term is not significant. No maximum in region.

CHAPTER 14

14.1 $P(10 \text{ or more}) = 0.2272$; cannot reject H_0.

14.3 $z = -3.91$; reject H_0.

14.5 $z = 2.92$; difference is significant.

14.7 $z = 1.62$; cannot reject H_0.

14.9 $H = 26.0$; the populations are not identical.

14.11 $z = -0.244$; cannot reject H_0.

14.13 $z = -4.00$; reject H_0.

14.15 Maximum difference is about 0.27; cannot reject H_0.

14.17 $W_1 = 58$ and $z = -0.447$, cannot reject H_0.

14.19 $H = 0.904$; cannot reject H_0.

14.21 $z = -1.797$; reject H_0.

14.23 $W_1 = 25$ so $U_1 = 19$; reject H_0.

14.25 $z = -2.24$; reject H_0.

CHAPTER 15

15.1 (a) Central line $= 0.150$, $UCL = 0.153$, $LCL = 0.147$; (b) central line $= 0.005$, $UCL = 0.010$, $LCL = 0$; (c) $\bar{x}$: eighth, sixteenth, and seventeenth sample values outside limits; R: all sample values within limits.

15.3 (a) Central line $= 48.1$, $UCL = 50.3$, $LCL = 46.0$; (b) central line $= 2.95$, $UCL = 6.7$, $LCL = 0$; (c) process mean out of control, process variability in control; (d) $z = -2.24$, there is a trend; (e) no, process is not in control.

15.5 (a) $\bar{x}$: central line $= 21.7$, $UCL = 25.2$, $LCL = 18.2$; σ: central line $= 1.05$, $UCL = 3.74$, $LCL = 0$; (b) yes, process is in control. For σ chart, 17-th point above limit.

15.7 Central line $= 0.0017$, $UCL = 0.0042$, $LCL = 0$.

15.9 (a) Central line $= 0.037$, $UCL = 0.094$, $LCL = 0$.

15.11 Yes, central line for c chart is 4.9, $UCL = 11.6$ and $LCL = 0$.

15.13 We can assert with 95% confidence that 99% of the pieces will have yield strength between 36,843 and 68,757 psi.

15.15 (a) 0.1063 ± 0.0008; (b) 0.1063 ± 0.0001.

15.17 (a) Central line $= 4.1$, $UCL = 4.17$, $LCL = 4.01$; (b) central line $= 0.116$, $UCL = 0.25$, $LCL = 0$; (c) $\bar{x}$: many sample values are outside limits; R: sixth sample value outside limits.

15.19 Central line $= 0.02$, $UCL = 0.050$, $LCL = 0$. The standard is not being met.

15.21 (a) $UCL = 3.48$, $LCL = 0$; (b) all the 10-foot sections are within the control limits except the 19th section, which is out of the limits.

15.23 We can assert with 95% confidence that 90% of the inter-request times will be between 887 and 54,377 microseconds.

15.25 (b) $L = 186.86$; (c) the cardboard strength data seems to be sampled from a normal distribution.

15.27 (a) 1.515; (b) 1.45.

CHAPTER 16

16.1 $R = 0.9936$.

16.3 $R = 0.9983$.

16.5 (a)
$$f(t) = \begin{cases} \beta(1 - t/\alpha)\exp[-\beta(t - t^2/(2\alpha))] & \text{for } 0 < t < \alpha \\ 0 & \text{otherwise} \end{cases}$$

$$F(t) = \begin{cases} 1 - \exp[-\beta(t - t^2/(2\alpha))] & \text{for } 0 < t < \alpha \\ 1 & \text{for } t > \alpha \end{cases}$$

16.7 (a) 0.6703; (b) 0.6703.

16.9 0.0106.

16.11 (a) $3{,}253.1 < \mu < 38{,}005.6$; (b) $T_r = 40{,}970 < 45{,}767.5$; cannot reject H_0.

16.13 (a) $254.9 < \mu < 1{,}182.8$; (b) $T_r = 3{,}329$ so we cannot reject H_0 at level 0.01.

16.17 287,156.8.

16.19 0.8758.

16.23 (a) 0.6753; (b) 0.4066.

16.25 627,197.5.

16.27 0.9520.

16.29 (a) 0.6667; (b) 0.50; (c) 0.9977.

Table 3 Standard Normal Distribution Function

$$F(z) = \frac{1}{\sqrt{2\pi}} \int_{-\infty}^{z} e^{-t^2/2}\, dt$$

z	0.00	0.01	0.02	0.03	0.04	0.05	0.06	0.07	0.08	0.09
−5.0	0.0000003									
−4.0	0.00003									
−3.5	0.0002									
−3.4	0.0003	0.0003	0.0003	0.0003	0.0003	0.0003	0.0003	0.0003	0.0003	0.0002
−3.3	0.0005	0.0005	0.0005	0.0004	0.0004	0.0004	0.0004	0.0004	0.0006	0.0003
−3.2	0.0007	0.0007	0.0006	0.0006	0.0006	0.0006	0.0006	0.0005	0.0005	0.0005
−3.1	0.0010	0.0009	0.0009	0.0009	0.0008	0.0008	0.0008	0.0008	0.0007	0.0007
−3.0	0.0013	0.0013	0.0013	0.0012	0.0012	0.0011	0.0011	0.0011	0.0010	0.0010
−2.9	0.0019	0.0018	0.0018	0.0017	0.0016	0.0016	0.0015	0.0015	0.0014	0.0014
−2.8	0.0026	0.0025	0.0024	0.0023	0.0023	0.0022	0.0021	0.0021	0.0020	0.0019
−2.7	0.0035	0.0034	0.0033	0.0032	0.0031	0.0030	0.0029	0.0028	0.0027	0.0026
−2.6	0.0047	0.0045	0.0044	0.0043	0.0041	0.0040	0.0039	0.0038	0.0037	0.0036
−2.5	0.0062	0.0060	0.0059	0.0057	0.0055	0.0054	0.0052	0.0051	0.0049	0.0048
−2.4	0.0082	0.0080	0.0078	0.0075	0.0073	0.0071	0.0069	0.0068	0.0066	0.0064
−2.3	0.0107	0.0104	0.0102	0.0099	0.0096	0.0094	0.0091	0.0089	0.0087	0.0084
−2.2	0.0139	0.0136	0.0132	0.0129	0.0125	0.0122	0.0119	0.0116	0.0113	0.0110
−2.1	0.0179	0.0174	0.0170	0.0166	0.0162	0.0158	0.0154	0.0150	0.0146	0.0143
−2.0	0.0228	0.0222	0.0217	0.0212	0.0207	0.0202	0.0197	0.0192	0.0188	0.0183
−1.9	0.0287	0.0281	0.0274	0.0268	0.0262	0.0256	0.0250	0.0244	0.0239	0.0233
−1.8	0.0359	0.0351	0.0344	0.0336	0.0329	0.0322	0.0314	0.0307	0.0301	0.0294
−1.7	0.0446	0.0436	0.0427	0.0418	0.0409	0.0401	0.0392	0.0384	0.0375	0.0367
−1.6	0.0548	0.0537	0.0526	0.0516	0.0505	0.0495	0.0485	0.0475	0.0465	0.0455
−1.5	0.0668	0.0655	0.0643	0.0630	0.0618	0.0606	0.0594	0.0582	0.0571	0.0559
−1.4	0.0808	0.0793	0.0778	0.0764	0.0749	0.0735	0.0721	0.0708	0.0694	0.0681
−1.3	0.0968	0.0951	0.0934	0.0918	0.0901	0.0885	0.0869	0.0853	0.0838	0.0823
−1.2	0.1151	0.1131	0.1112	0.1093	0.1075	0.1056	0.1038	0.1020	0.1003	0.0985
−1.1	0.1357	0.1335	0.1314	0.1292	0.1271	0.1251	0.1230	0.1210	0.1190	0.1170
−1.0	0.1587	0.1562	0.1539	0.1515	0.1492	0.1469	0.1446	0.1423	0.1401	0.1379
−0.9	0.1841	0.1814	0.1788	0.1762	0.1736	0.1711	0.1685	0.1660	0.1635	0.1611
−0.8	0.2119	0.2090	0.2061	0.2033	0.2005	0.1977	0.1949	0.1922	0.1894	0.1867
−0.7	0.2420	0.2389	0.2358	0.2327	0.2296	0.2266	0.2236	0.2206	0.2177	0.2148
−0.6	0.2743	0.2709	0.2676	0.2643	0.2611	0.2578	0.2546	0.2514	0.2483	0.2451
−0.5	0.3085	0.3050	0.3015	0.2981	0.2946	0.2912	0.2877	0.2843	0.2810	0.2776
−0.4	0.3446	0.3409	0.3372	0.3336	0.3300	0.3264	0.3228	0.3192	0.3156	0.3121
−0.3	0.3821	0.3783	0.3745	0.3707	0.3669	0.3632	0.3594	0.3557	0.3520	0.3483
−0.2	0.4207	0.4168	0.4129	0.4090	0.4052	0.4013	0.3974	0.3936	0.3897	0.3859
−0.1	0.4602	0.4562	0.4522	0.4483	0.4443	0.4404	0.4364	0.4325	0.4286	0.4247
−0.0	0.5000	0.4960	0.4920	0.4880	0.4840	0.4801	0.4761	0.4721	0.4681	0.4641

(continued on following page)